AF556669

ELECTRON OPTICAL SYSTEMS
for
MICROSCOPY, MICROANALYSIS & MICROLITHOGRAPHY

Proceedings of the
3rd Pfefferkorn Conference, held
April 9 to 14, 1984, at the
Carousel Hotel, Ocean City, MD

Edited By

John J. Hren **Friedrich A. Lenz** **Eric Munro** **Peter B. Sewell**

Managing Editor

Sudha A. Bhatt

Published By
Scanning Electron Microscopy, Inc.
P.O. Box 66507
AMF O'Hare, IL 60666 U.S.A.

In quoting the papers in this book, it is strongly recommended that the following format be used: "Author(s). (1984). Title of Article, in: Electron Optical Systems, J.J. Hren, F.A. Lenz, E. Munro, P.B. Sewell (eds.), SEM Inc., AMF O'Hare, IL 60666, page numbers."

SEM, Inc. is a not-for-profit organization with the following goals:

a. Promotion of advancement of science of SEM and related characterization techniques;
b. Promotion of application of these techniques in existing and new areas of applications;
c. Promotion of these techniques so that their users obtain the best information of the highest quality from their instruments.

SEM, Inc. sponsors annual meetings on Scanning Electron Microscopy and Pfefferkorn Conferences on basic subjects related to SEM. SEM, Inc. publishes a quarterly journal, "Scanning Electron Microscopy," and a semi-annual journal, "Food Microstructure."

The 4th Pfefferkorn Conference, The Science of Biological Specimen Preparation, will take place from March 25 to 30, 1985, in Grand Canyon, Arizona.

For more information or other inquiries contact:

Dr. Om Johari,
SEM, Inc.,
P.O. Box 66507,
AMF O'Hare, IL
60666 U.S.A.
(312) 529-6677

ISBN: 0-931288-34-7

Printed in the United States of America

3rd Pfefferkorn Conference

ELECTRON OPTICAL SYSTEMS FOR MICROSCOPY, MICROANALYSIS AND MICROLITHOGRAPHY

Sponsored By:	Scanning Electron Microscopy, Inc.
Conference Director:	Om Johari
Editorial Assistants:	Joseph Staschke, Frank Rose

Scanning Electron Microscopy, Inc.

President	John D. Fairing
Vice President	Robert P. Becker
Secretary-Treasurer	Om Johari
Advisors	Godfried M. Roomans, Takashi Makita

FOREWORD

This third conference in the series dedicated to Professor Gerhard E. Pfefferkorn continues the tone set by the first two: in depth analyses of important areas of electron microscopy. Since the community of electron microscopists has expanded and fragmented as it has grown, it is only through topical meetings and publications of this kind that we take the time to come back together. The cycle indeed seems to be completing itself, if the present meeting on "Electron Optical Systems" (at Ocean City, Maryland, from April 4 to April 14, 1984) is a representative indicator.

The present cycle had its beginnings in the 1930's and postwar 1940's, when research on electron optics was not yet neatly categorized. Different groups (in Germany, Belgium, Japan, Canada, and the U.S.A.) had different interests, it is true, but they still communicated extensively. Thus, related work on cathode ray tubes, television, scanning electron microscopes, transmission electron microscopes, etc. was reported and discussed in common forums and there were only a few journals to follow. All of the areas named and much larger ones, e.g. in microelectronics, have now been subdivided, hold their own topical meetings, publish separately in so many journals we hardly recognize them by title, etc. Of course, specialization has its rewards--we need only look at the products of it. However, the participants have lost track of one another's thinking and achievements as a result.

The rediscovery of the SEM and AEM in recent years has gotten several disparate groups once again in communication. The surface science community, in turn, has become interested in spatial resolution, cementing another link. In turn, the need for inspection of microelectronics and microlithography has benefited from these developments and given back information in kind. Microanalysis, microfabrication and microinspection, after all, do have in common the prefix micro- and much more. Perhaps the last major link to be made is with the display and viewing industry. The commonality of interests with the rest has been somewhat loose, but that is indeed changing too. Lithography and displays on a CRT do have a certain commonality, sources are surely of common interest, rapid pulsing and blanking operations are of widespread interest to all of us, computer codes designed to track electron beams in a TV tube can sometimes be used to help design a lens, etc. Finally, many of us are becoming interested in ions for microanalysis and microfabrication. Even here we have an interconnection or two, e.g. electrostatic lenses are of interest again.

Accordingly, the aim of this conference was to bring together active researchers in various areas related to electron optical systems. Topics discussed in depth included electron optics, electron lenses, deflection systems, electron and ion sources, electron detectors, display and recording systems, image processing, and a few areas of systems and applications. This was a most stimulating meeting for all of us. We participated not as authors, but concentrated on bringing together select groupings, encouraging participation, and asking questions, and thus we learned a great deal from one another. We sincerely hope you enjoy reading these papers as much as the participants enjoyed presenting and hearing them.

Our thanks for the fine cooperation and essential prodding of Om Johari. This meeting was a truly cooperative effort by the participants, the organizers, and SEM Inc.--we were delighted to have been a part of it. Indeed, we all made new acquaintances and friends (particularly from other countries), which is bound to benefit us all.

The Organizing Committee

Prof. John J. Hren, University of Florida, Gainesville
Dr. Eric Munro, Imperial College, London, U.K.
Dr. Peter B. Sewell, National Research Council, Ottawa, Canada
Prof. Friedrich A. Lenz, University of Tübingen, W. Germany

TABLE OF CONTENTS

SCANNING ELECTRON MICROSCOPY, Inc.

This not-for-profit organization was established with the following goals:

a. *Promotion of advancement of science of SEM and related material characterization techniques;*

b. *Promotion of applications of these techniques in existing and new areas of applications;*

c. *Promotion of these techniques so that their users obtain the best information of the highest quality from their instruments.*

In an effort to fulfill these goals SEM, Inc. sponsors the annual SEM meetings, and the Pfefferkorn Conferences. It publishes the journals **"Scanning Electron Microscopy"** and **"Food Microstructure"** and other related publications. Suggestions on activities which SEM may sponsor can be communicated to any one of the following persons:

John D. Fairing *President, SEM, Inc.* 314-694-5007
809 Westwood Dr., Ballwin, MO 63011

Robert P. Becker *Vice President, SEM, Inc.* 312-996-7215
1S 640 Brook Court, Glen Ellyn, IL 60137

Om Johari *Secretary-Treasurer & Director of SEM Meetings*
SEM, Inc., P.O. Box 66507 312-529-6677
AMF O'Hare (Chicago), IL 60666

Takashi Makita *Advisor* 0839-22-6111 x438
3-4-203 Kumano-cho, 753 Yamaguchi, Japan

Godfried M. Roomans *Advisor* 46-8-340860 x251
Wenner-Gren Institute, University of Stockholm
Norrtullsgatan 16, S-11345 Stockholm, Sweden

1985 MEETINGS SPONSORED BY SEM, INC.

(i) The annual SEM meeting (called **Scanning Electron Microscopy/1985)** will take place during **March 31 to April 5, 1985 at Riviera Hotel in Las Vegas, Nevada.**

SEM 1985 meetings highlights will include a **comprehensive equipment exhibition** with almost all suppliers of SEM, related analytical equipment, and other suppliers/services in attendance.

In addition to the **general session, programs of common interest** will include *Analytical Electron Microscopy (including STEM, Energy Loss Spectroscopy, High Resolution Imaging, Diffraction, Microanalysis, etc.), Secondary Ion Microscopy and Secondary Ion Mass Spectroscopy, Electron Sources, Microprobe Surface Analytical Techniques & Their Applications, Image Analysis, etc.* Several programs devoted to SEM applications in physical and biological sciences, as well as food microstructure are being planned.

Specific biological programs being planned for SEM/1985 include biological microanalysis, cell culture, chromosomes, developmental biology, skin biology, collagen gels, inner ear, ocular tissue, neurobiology, blood vascular systems, clinical applications, stones & crystals in disease, ultrastructural effects of radiation, etc.

Additional details (Letter of Intent form, Instructions for Authors, Registration forms, etc.) are available from SEM office.

(ii) 4th Pfefferkorn Conference on **"The Science of Biological Specimen Preparation for Microscopy and Microanalysis."**

Immediately preceding the SEM/1985 meetings, a special program on biological specimen preparation has been planned during **March 25 to 30, 1985**. The emphasis at this program will be on **The Science of Specimen Preparation** (and not on recipes and tips). Preparation of specimens for **all types of microscopy (light optical, SEM, TEM, STEM)** and **microanalysis** (x-ray, LAMMA, etc.) will be covered. The location of the conference will be Grand Canyon Squire Inn, at Grand Canyon, AZ.

Prof. Alan Boyde (Dept. Anatomy, Univ. College, Gower St., London WC1E 6BT, U.K., phone 01-3877050 ext. 635) and **Prof. Robert P. Becker** (Dept. Anatomy, Univ. Illinois Medical School, Chicago, IL 60612, phone 312-996-7215) as co-chairmen; and **Dr. Martin Mueller**, ETH-Zurich, Switzerland (phone 41-1-256 3937), and **Prof. John Wolosewick**, Univ. Illinois Medical School, Chicago (phone 312-996-6022) are the organizers of this program. Additional details are available from SEM office.

International Journal —"Scanning Electron Microscopy"

SEM publishes the quarterly International Journal "Scanning Electron Microscopy." The Journal publishes papers submitted for publication only, as well as papers presented at SEM meetings. From 1980, four parts are issued for each volume. In **1980 and 1981** the first 3 parts contained papers organized by specific subject areas: Part I (physical sciences), Parts II and III (biological sciences). Late papers on topics of first 3 parts (both physical and biological) were included in Part IV. **From 1982,** the policy of organizing papers by specific subject areas in different parts has been discontinued.

Papers for publication are invited. Of maximum interest are papers emphasizing topics of general interest (techniques, theory, instrumentation, interpretation, novel or unusual applications, etc. of SEM and related techniques). Sample Tables of Contents are available on request.

Prices for SEM/1980, 1981, 1982, 1983, 1984 or 1985 each:	**U.S. Delivery**	**Elsewhere**
☐ Complete Set (Parts 1-4)	$109.00	$119.00
☐ any 2 different parts	$ 84.00	$ 87.50
☐ any 1 part, also for 1979	$ 52.00	$ 55.00
Prices for SEM/1979 & 1978:	(SEM/1979/III is out of print)	
☐ SEM/1979/I+II	$ 65.00	$ 71.50
☐ SEM/1978/I (physical)	$ 37.00	$ 41.00
☐ SEM/1978/II (biological)	$ 40.50	$ 44.50
☐ SEM/1978/2 part set	$ 67.50	$ 74.00
Pfefferkorn Conferences:		
☐ 1st, Electron Beam Interactions with Solids	$ 51.00	$ 54.00
☐ 2nd, Science of Biological Specimen Preparation	$ 40.00	$ 43.00
☐ 3rd, Electron Optical Systems	$ 44.00	$ 47.00
☐ Preparation of Biological Specimens	$ 32.00	$ 35.00
☐ SEM of Cells in Culture	$ 29.00	$ 32.00
☐ Biological X-ray Microanalysis	$ 22.00	$ 25.00
☐ Ultrastructural Effects of Radiation	$ 18.00	$ 20.50
☐ Cell Surface Labeling	$ 10.00	$ 12.00
☐ Brain Ventricular Surfaces	$ 15.00	$ 17.00
☐ Studies of Food Microstructure	$ 49.00	$ 52.00

SEM/1986 **May 4 to 9 at Clarion Hotel, New Orleans, LA.**

Electron Optical Systems (pp. 1-14)
SEM Inc., AMF O'Hare (Chicago), IL 60666-0507, U.S.A.

0-931288-34-7/84$1.00+.05

MAGNETIC ELECTRON LENSES

Wolfgang Dieter Riecke

Fraunhofer-Institute for Information
and Data Processing
Sebastian-Kneipp-Strasse 12/14
7500 Karlsruhe, W. Germany
Phone No.: 0721-6091294

Abstract

The electron optical properties of magnetic lenses are determined by the ratio S/D of gap width S to bore diameter D, by the actual size of the gap and by the lens strength $NJ/\sqrt{U^*}$. Objective lenses of particularly small aberrations are obtained if the specimen is positioned close to the center of the gap and if the width of the gap is made as small as possible. For projector lenses the best performance results if the lens is employed at its minimum focal length. Another aspect important for high resolution lenses is that a stigmator is available for the correction of axial astigmatism. With ferromagnetic superconducting lenses, both permendure and rare earth metals (these cooled to liquid helium temperatures) are employed for the pole pieces in combination with an iron casing which enshrouds the superconducting coil. With the superconducting shielding lens, the superconducting coil is encapsulated by a superconducting shield which has about the same general shape as the iron circuit casing of a normal lens, whereby the narrow gap in the shielding tube around the axis allows the magnetic field there to penetrate into the space close to the axis and form the electron lens.

Key Words: Magnetic electron lens, objective lens, projector lens, spherical aberration, chromatic aberration, optimum objective lens, condenser-objective lens, second-zone lens, chromatic field aberration, chromatic aberration of magnification, chromatic aberration of image rotation, image distortion, radial distortion, spiral distortion, stigmator, super-conducting lens, rare-earth metal pole pieces, persistent current mode, superconducting shielding lens

Introduction

The user of an electron optical instrument, be it a scanning electron microscope, a transmission electron microscope or an electron beam lithographic machine, usually is not so much interested in acquiring a thorough knowledge of the art of high level instrument design. But it should be realized that for actually pushing the instrument's performance to the limit of its capabilities, some insight into e.g. the construction and the optical behaviour of the single lenses and the complete optic system is indispensable.

The present contribution is a user's guide to magnetic electron lenses, their electron optical and physical properties, their design and construction, and also provides a few glimpses into the history of their development trends. Due to the limited space, the treatment had to be rather sketchy on most subjects, is primarily intended to give some insight into the general aspects of the magnetic lens and gives hints to the points the user must particularly observe.

First predictions on how to design an electron lens and some later consequences

The art of constructing magnetic electron lenses goes back to the late 1920s and early 1930s, when a fundamental theoretical investigation by BUSCH /4/, /5/ had led to the conclusion that a strong and concentrated rotationally symmetric magnetic field would act as a focussing means for an electron beam, and this in a manner similar to the action of glass lenses on a beam of visible light. It also became clear that the basic notions of light optics: focal lengths, focal points and principal planes, could likewise be applied successfully to the magnetic electron lens. It was now understood that the concentration coils employed already for several years with cathode ray oscillographs were indeed electron lenses, although, if taken as a component of the optical system of the cathode ray oscillograph, the concentration coils were certainly little more ambitious (optically speaking) than a burning glass for focussing light rays.

The essential result of BUSCH's work was that he developed an equation

$$\frac{1}{f} = \frac{e}{8m\,U}\int_{-\infty}^{+\infty} B(z)^2 dz = \frac{e}{8m}\, C\frac{(NJ)^2}{U}\int_{-\infty}^{+\infty}\left(\frac{B(z)}{B_o}\right)^2 dz \qquad (1)$$

for the refractive power 1/f of a magnetic lens. This equation allows the focal length f to be predicted for a lens which is employed to focus an electron beam accelerated by the voltage U to a velocity $v = \sqrt{2eU/m}$ and consists of a rotationally symmetric lens field specified by the characteristic value B_o of the magnetic flux density and the relative distribution $B(z)/B_o$ of the field along the beam axis z. (e and m denote the charge and mass of the electron, $\sqrt{C}=B_o/NJ$ is a factor of proportionality between the ampere turns NJ of the field generating coil and a characteristic value B_o of the magnetic flux density on the field axis z, e.g. its maximum value; the specific value of $\sqrt{C}$ is determined by the actual shape of the coil and its associated iron circuit.)

According to equation (1) the focal length f decreases in inverse proportion to the square of the magnetic field strength. This expectation has since been the starting point for many projects which aimed at improving the performance of objective and projector lenses by driving the magnetic lens field to higher and higher strength and thus hopefully to shorter and shorter focal length.

Actually, for short focal length lenses used in electron microscopes, the above conclusions based on equation (1) must not be taken too literally although they show the proper trend. The reason for this is that equation (1) has been derived under the assumption that the distance from the lens axis of the individual electron ray does not change much throughout the region of high field strength. As a consequence, equation (1) applies to electron optical situations where the focal points, the object and the image all remain well outside of the lens field and the focal length is long in comparison to the field extension.

When the field strength increases more and more the focal points gradually move into the lens field proper, so that part of the field becomes ineffective for lens action, viz. that part of the field which - if seen in the direction of the beam - lies before the front focal plane. It turns out that when increasing the field strength the front focal plane moves in the direction of the beam and into the lens field in just such a manner that the focal length cannot be made much smaller than the half width of the field distribution. Thus, although lens fields with extremely high field strength have been realized, especially by employing liquid helium cooled rare earth pole pieces and superconducting lens coils /2/, /3/, little benefit has resulted from it for making the focal length shorter (as discussed later).

Nevertheless, still today BUSCH's equation (1) may be successfully applied for designing condenser lenses and similar long focal length devices as long as $(NJ)^2/U$ remains well below about 50 $amps^2/volt$.

Two scaling laws for magnetic lenses

A most useful concept for predicting the performance of magnetic electron lenses is available as the so-called scaling laws. By employing them a new lens of specified electron optical properties can often be derived from an already existing or theoretically established lens of known behaviour.

So, as an example it can be seen from equation (1) that for one and the same 'normalized' distribution $B(z)/B_o$ of the magnetic field its performance as an electron lens is actually determined by the quotient $(NJ)^2/U$, the square root $NJ/\sqrt{U}$ of which is called the 'lens strength' of the magnetic lens. This implies an <u>Electromagnetic Scaling Law for magnetic lenses</u>:

- The path of the electron beam remains the same if at a variation of the acceleration voltage U the current in the lens coil is correspondingly followed-up in such a manner that $NJ/\sqrt{U}$ remains constant. (If the acceleration voltage of the beam becomes larger than about U = 50 kVolts it is imperative to employ in the place of U the so-called relativistically corrected voltage $U^* = U\,(1 + 10^{-6}\,U/\text{ Volts})$.)

As a likewise important relation the Geometric Scaling Law applies if the whole magnetic lens device : coils and iron circuit and with it the field distribution, is blown-up or compressed with a scaling factor s. If at this, also the sampling point for the determination of B_o is moved correspondingly, then the 'normalized' flux density distribution B/B_o retains the same values, but these values are 'stretched' over an s-times longer piece of the beam axis z. Thus, the integral $\int (B/B_o)^2 dz$ in equation (1) assumes the s-fold value in comparison to the 'unscaled' situation.

On the other hand, if during the scaling operation NJ remains the same the flux density values all throughout the field and at corresponding pairs of points of the unscaled and the scaled distributions will vary from the unscaled value B to B/s, and so likewise from B_o to B_o/s. This is a consequence of Ampere's Law:

$$\mu_o NJ = \int_{-\infty}^{+\infty} B dz \qquad (2)$$

Because of $\sqrt{C} = B_o/NJ$, the unscaled C must therefore be substituted by C/s^2 at scaling. At inspection of equation (1) it now becomes immediately clear that by scaling the lens design up s-times (s >1 means blowing it up, s <1 making it smaller) the focal length will be increased by the same factor from f to sf, and this remains true even if a simultaneous change in ampere turns NJ is performed under the condition that a corresponding change of the acceleration voltage U keeps the lens strength $NJ/\sqrt{U^*}$ at the previous value.

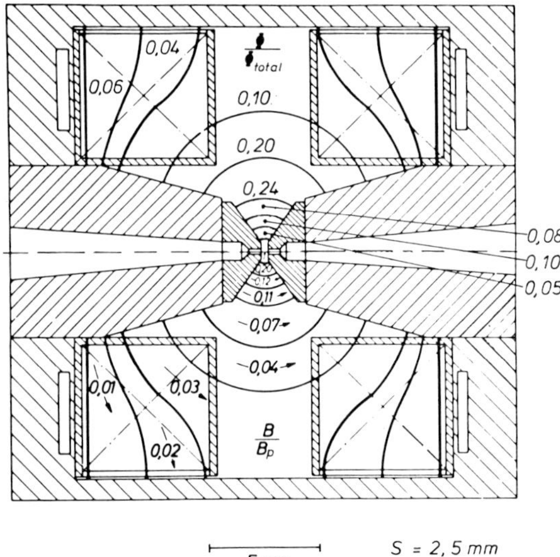

Fig. 1: Simplified cross-section of a symmetric electron lens. The continuous curves trace the path of the magnetic field lines. In the upper half of the drawing the numbers indicate the fraction of the total magnetic flux passing through the respective partial volumes of the space enclosed by the casing. In the lower half, the magnetic flux density B is indicated in units of the field B_p between the parallel faces of the pole pieces. The symmetrical design shown here is often employed with objective lenses, especially with condenser-objective lenses, because the two-coil design accommodates a high number of ampere turns excitation and the application of a side entry stage. (But not every symmetrical objective lens is necessarily a condenser-objective lens!)

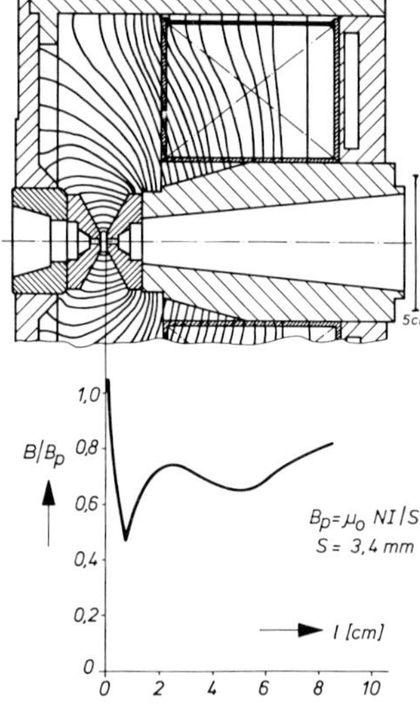

Fig. 2: Simplified cross-section of an asymmetrical electron lens. The curves in the space enclosed by the casing indicate the path of the magnetic field lines. Through each of the partial volumes corresponding to two consecutive field lines, about 3 $^o/o$ of the total magnetic flux of the lens passes between the core and the casing or between the two pole pieces. Lenses of the asymmetric type are widely used in electron optical systems if only a moderate number of ampere turns is required, so that using a single coil will not lead to thermal problems. Usually, they are employed as condenser and projector lenses and as objective lenses which are operated in the 'normal-objective mode' (see further below).

In a more general way, the Geometric Scaling Law can be expressed as follows:

- If the design of a magnetic electron lens is enlarged or reduced in size by the magnification factor s, the path of the electron rays is likewise expanded or contracted, whereat the scaled ray path passes through a correspondingly scaled set of points within the lens volume. At scaling, the numerical values of the paraxial electron optical parameters, such as: focal length, distances of the focal points and the principal planes from the lens field center also become s times larger than the original unscaled quantities. The same is also true for the coefficient C_s of spherical aberration but neither for the coma coefficient nor for the coefficients of radial and spiral distortion. The above statements stay valid if the ampere turns NJ generating the lens field and the (relativistically corrected) acceleration voltage U* are varied in such a manner that the lens strength $NJ/\sqrt{U^*}$ remains the same.

The use of the above scaling laws is not restricted to the comparatively weak ('thin') lenses of the BUSCH type. The laws also apply to stronger lenses where both focal points and usually either the object or the image are immersed in the lens field and lie within the region of high magnetic flux density. They are also the basis for the 'universal' representation of theoretically calculated electron optical properties of series of magnetic electron lenses the data of which are often given as function of the Electromagnetic Scaling Law parameter lens strength $NJ/\sqrt{U^*}$ (or other parameters derived from it in a straightforward fashion), and are normalized with respect to a characteristic dimension within the field forming lens design whereat the Geometric Scaling Law comes into play (cf. e.g. /10/, /13/, /14/, /19/, /21/, /22/).

Determination and representation of lens properties

The distribution of the magnetic flux density within the lens field is determined both by the shape of the lens coil and by the contour of the ferromagnetic casing which enshrouds the coil. The casing is composed of an outer can and of central lens cores which surround the lens axis and usually terminate in a pair of truncated pole pieces (Figs. 1 and 2). The principal purpose of the casing is to short circuit the magnetic stray field around the coil and thereby concentrate the magnetomotive force NJ (generated by a current J flowing through a coil of N turns) into the space between the pole faces.

Now, as long as the gap width S between the plane pole piece faces remains smaller at least by an order of magnitude in comparison to the coil's cross section and to its inner diameter, then the distribution of the magnetic flux composing the lens field depends only on the gap width S and on the diameter D of the bore, which has to be provided in the pole pieces to allow the electron beam to enter and to leave the lens field (cf. Fig. 3). Given these qualifications it also does not matter if the pole piece system is of a symmetric or asymmetric design (Fig. 3), and if the whole lens: pole piece system, coils and casing has a symmetric or asymmetric configuration (cf. Fig. 4).

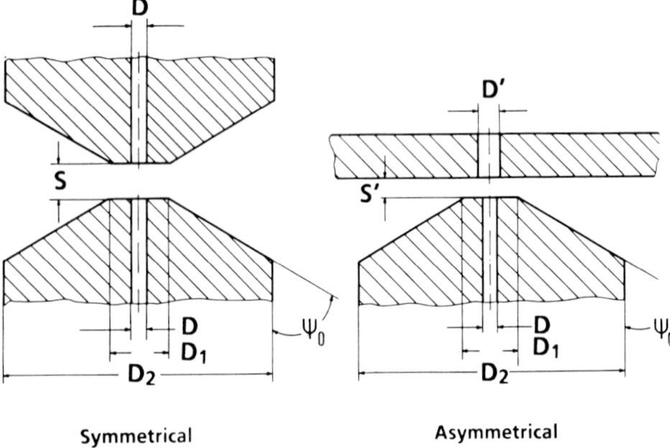

Fig. 3: Typical configurations of pole piece systems.

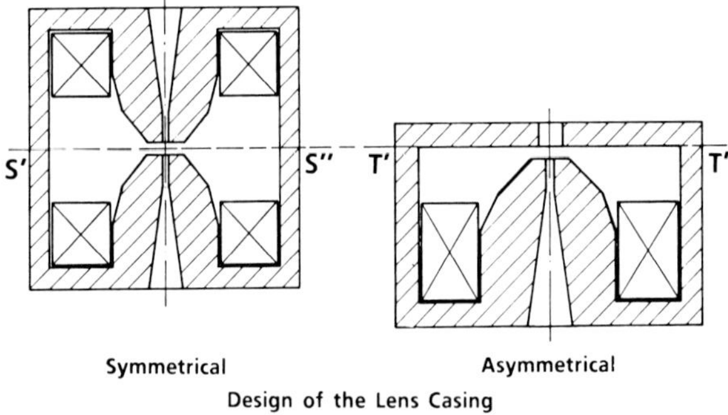

Fig. 4: Typical configurations of the lens casing design and the coil positions.

Under these assumptions, the actual field distribution within that narrow cylindrical space close to the lens axis through which the electron rays pass and which so represents the lens proper, can be calculated in good approximation from a model. For this, the following assumptions are made:

- The plane pole faces are postulated to extend very far out from the lens axis and up to a distance orders of magnitude larger than the diameter of the bore.
- The magnetization of the pole pieces is kept to remain well below magnetic saturation, so that each of the pole pieces attains a constant magnetic potential.
- The lens casing is understood to have been made so thick that its magnetic resistance is small by orders of magnitude in comparison to the magnetic resistance of the gap, and that therefore the magnetomotive force at the gap, i.e. the magnetic potential difference between the pole pieces is equal to the ampere turns NJ generated in the lens coil.

With these assumptions, the distribution in space of the magnetic field of the lens can be determined as a potential problem either numerically by employing the relaxation method /20/ or by means of a resistance network analogue /35/.

Using these results the path of the electron rays can be determined by numerically solving the paraxial beam equation /16/. Another possibility for obtaining the ray paths is to approximate the numerically determined or measured (cf. e.g. /11/) field distribution by a suitable mathematical field model which allows a solution of the paraxial ray equation as an analytical closed expression /14/, /19/. From a combination of two linearly independent paraxial ray paths, the cardinal elements of the lens: focal length, position of focal points and principal planes, can now be calculated straight away and also the coefficient C_c of chromatic aberration.

The same two lines of approach: the numerical method on the one hand and the predominantly analytical method on the other have also been adopted for the determination of the third order (or 'Seidel') aberrations of the lens (for details cf. eg. /14/, /19/, /21/). The most important of these are spherical aberration and coma for objective lenses and radial and spiral distortion for projectors.

Values of the coefficients which characterize the magnitude of these aberrations have been published in graphical (cf. e.g. /10/, /14/, /21/) or tabular (cf. e.g. /13/, /15/) form as functions of the lens strength in its differently used shapes ($NJ/\sqrt{U^*}$, k^2 etc.; cf. e.g. /19/), with the aspect ratio S/D being employed as a parameter.

The amount of data published on the electron optical performance of magnetic lenses is rather extensive and cannot be discussed here, so that the reader must be referred to the original publications quoted above. But as an example, a graphical representation of lens data can be seen further below in Fig. 9, where the variation of the focal length of a projector is shown as a function of the lens strength.

Although quite general practice, describing the lens data as function of the lens strength is evidently not the only possibility. Other independent variables may be employed if they are unambiguously related to the lens strength. So, with objective lenses it is more instructive to represent the lens data as function of the position z_o of the front focal plane which is nearly identical with the position of the specimen at high magnification imaging. (As examples of this kind of representation, cf. Figs. 5 and 6).

Optimum objective lenses

All through electron microscope development, and nearly from the early days on, a continuous struggle was waged to overcome spherical aberration /32/. Rather soon, in a famous paper of 1936, SCHERZER /31/ had demonstrated that in principle spherical aberration cannot be reduced to zero in an electron lens composed of a rotationally symmetric magnetic field which is free of electrical space charge and constant in time. Nevertheless, there remained some hope to noticeably reduce the absolute value of the coefficient C_s of spherical aberration by inventing a particularly clever design for the pole piece system. But, calling back to mind the well-known formula:

$\delta \approx 0.4 \cdot \sqrt[4]{C_s \lambda^3}$ (with λ = electron wave length)

for the limit of the resolving power of an electron objective lens it becomes immediately clear that nothing less than decreasing C_s by about an order of magnitude would produce a tangible improvement of the resolution.

That there is little hope in this respect can be understood from Fig. 5 where (with still other data) the smallest attainable coefficient C_{smin} of spherical aberration is shown as a function of the relative front focal point or specimen position z_o/S. (The relative position z_o/S means the distance z_o of the focal point from the lens center, expressed in units of gap width S.) So, the independent variable in Fig. 5 has a form which reflects the Geometric Scaling Law of the above, and any relative focal point position z_o/S whatever corresponds to one and only one particular value of the lens strength $NJ/\sqrt{U^*}$, and this correspondence is independent of the actual value S of the gap width. An analogous correspondence exists between the values of C_s/S on the one hand and z_o/S or $NJ/\sqrt{U^*}$ on the other. So, for every relative focal point position z_o/S and lens strength $NJ/\sqrt{U^*}$ we have, in principle, a whole range of values available for C_s according to the individual gap width S chosen, and whereby evidently C_s is scaled in direct proportion to S.

In actual practice, in order to avoid magnetic oversaturation of the pole piece material the gap width cannot be made smaller than a critical value S_{min}. The corresponding value $C_{smin} = (C_s/S)S_{min}$ is the smallest coefficient of spherical aberration that can be attained with an object placed at $z_o = (z_o/S)S_{min}$.

As in Fig. 5, the critical data are best represented in a universal form such as e.g. $S_{min} / (\sqrt{U^*}/B_s)$. Here, B_s is the magnetic flux density of the pole piece material at magnetic

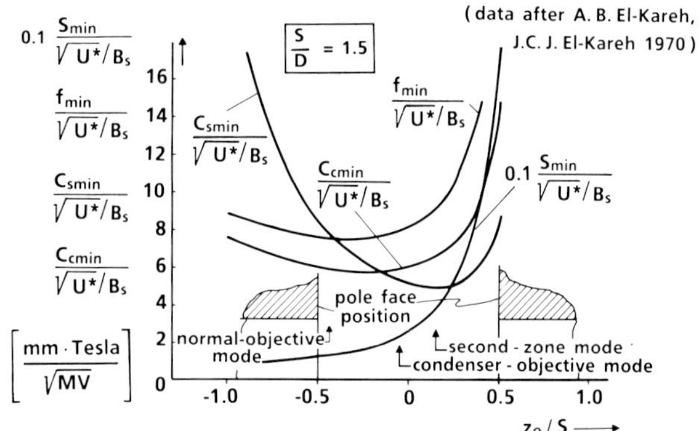

Fig. 5: Minimum attainable values of the objective focal length f_{min} and of the coefficients C_{smin} of spherical and C_{cmin} of chromatic aberration, represented as functions of the distance z_o of the front focal plane respectively the specimen from the lens center. Also shown is the smallest gap width S_{min} that can be assumed without overrunning the magnetic saturation B_s.

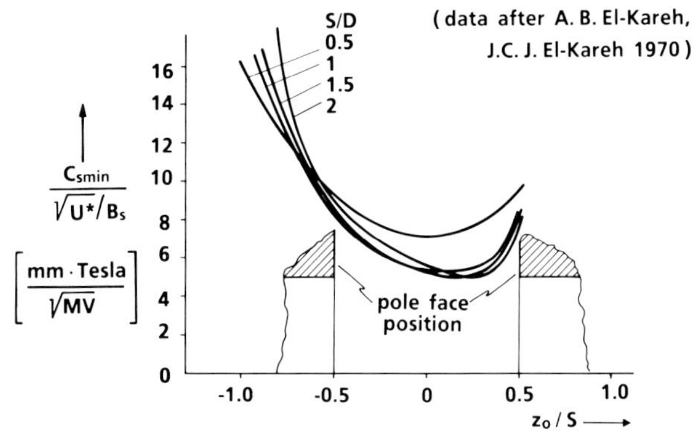

Fig. 6: Minimum value of the coefficient C_{smin} of spherical aberration that can be obtained with objective lenses of different aspect ratios S/D. C_{smin} is represented as function of the distance z_o of the front focal plane respectively the specimen from the lens center.

saturation, so that $S_{min} = \mu_o NJ/B_s$ with $\mu_o = 4\pi \cdot 10^{-7}$ Vsec/Am and as a consequence $S_{min} / (\sqrt{U^*}/B_s) = \mu_o NJ/\sqrt{U^*}$. Thus, $S_{min}/(\sqrt{U^*}/B_s)$ is a universal expression for the critical gap width S_{min} and directly proportional to the lens strength $NJ/\sqrt{U^*}$. Moreover, it does not depend on the aspect ratio S/D. On the other hand, $f_{min}/(\sqrt{U^*}/B_s)$, $C_{smin}/(\sqrt{U^*}/B_s)$ and $C_{cmin}/(\sqrt{U^*}/B_s)$ depend on S/D, although comparatively little in the range of operating modes suitable to be used with objective lenses. This is shown as example for $C_{smin}/(\sqrt{U^*}/B_s)$ in Fig.6.

Now, graphical representations as shown in Fig.5 for the coefficient C_{smin} of minimum spherical aberration yield nearly at a glance the answer to the old question of what is the optimum objective lens: There is not one particular design and lens strength that really can be called an 'optimum'. Rather, if we confine ourselves to the

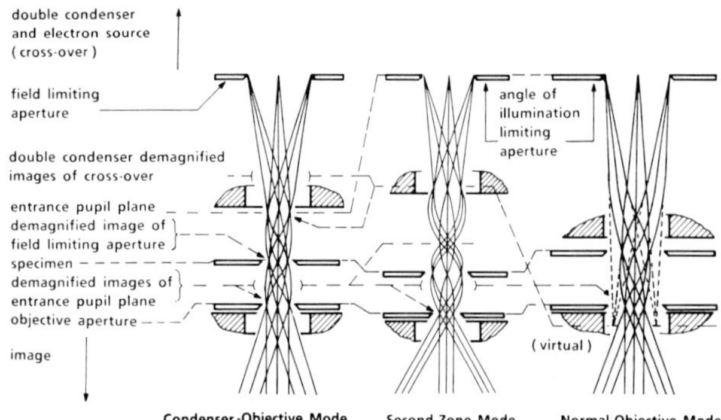

Fig. 7: Electron beam paths of basic operation modes of magnetic electron objective lenses.

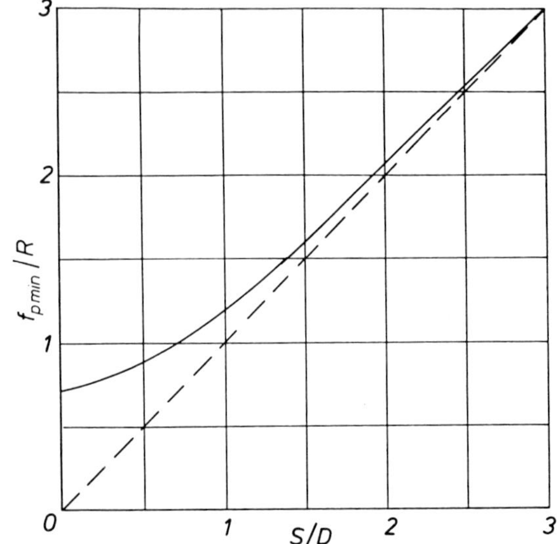

Fig. 8: Smallest attainable projector focal length f_{pmin}, shown in units of bore radius R, for a pole piece system of fixed gap width S and bore diameter D = 2R /22/.

range of aspect ratios 1/2 < S/D < 2 and specimen positions $- S/2 < z_o < S/4$, then a sufficient recipe for designing an 'optimum' objective lens is to make the gap width as small as possible down to S_{min}. Discussing the design with respect to minimizing the coefficient C_c of chromatic aberration clearly gives a similar result, as can also be deduced from Fig. 5.

Different positions z_o/S of the specimen within the favourable gap space correspond to different operation modes of the objective lens. But not every specimen position is compatible with an ob-objective lens that can be operated easily. Difficulties in handling the lens usually increase with the amount of magnetic field extending before the specimen plane which generates the so-called pre-field lens. In Fig. 7, the ray paths for the three preferred operating modes for objective lenses have been reproduced (cf. /29/). With reference to Fig. 5, the particular advantages of the three modes can be characterized as follows:

- With the normal objective mode, chromatic aberration C_c is particularly small and there is little pre-field and little complications due to it for illuminating the specimen with a parallel beam.

- With the condenser objective mode /25/ extremely small electron probes can be projected into the specimen plane /28/ and spherical aberration C_s is particularly small. Parallel-beam specimen illumination must be carefully controlled because of the complications due to the strong pre-field lens action /27/.

- With the second-zone mode /36/,/37/, C_s is still small but C_c tends to increase a little. This design offers a particularly large gap and much space for specimen handling equipment. The lens action of the pre-field is nearly telecentric so that illumination alignment and beam spot size are similar to the conditions of the normal-objective mode.

Optimum projector lenses

With projector lenses, the imaging aperture is so small that aperture dependent aberrations do not matter at all. The aberrations that really count here are the chromatic field aberrations that blur the image through a small oscillatory movement of the image points and the image distortion. In contrast to the objective lens which is operated practically at almost constant focal length, the focal length f_p of the projector in principle can be varied over a more or less extended range, although usually at the expense of having to put up with more or less pronounced aberrations. However, it turns out that these aberrations become particularly small if the projector is operated at about its maximum refractive power, i.e. its minimum focal length f_{pmin}.

It should be clearly understood that f_{pmin} is not the absolute minimum projector focal length that can be attained. Rather, it is the smallest focal length that is assumed by a projector having a fixed diameter D of the bore and width S of the gap when the lens strength $NJ/\sqrt{U^*}$ is varied over a sufficiently extended range.

The actual value of f_{pmin} is a function of the aspect ratio S/D and of a characteristic length within the pole piece system, such as e.g. the bore radius R. As a consequence of the Geometric Scaling Law, a single curve for the minimum focal length results if f_{pmin} is normalized with respect to R and f_{pmin}/R is represented as a function of S/D /22/ (Fig.8). Moreover, it is found /22/ that another single universal curve results for the projector focal length f_p as a function of the ampere turns NJ, if these quantities are normalized with respect to the pair of values f_{pmin}, NJ_{min} corresponding to the minimum focal length (cf. Fig. 9). It should be understood, however, that the f_p/f_{pmin}-curve of Fig.9 is not 'universal' in the rigorous sense of the word, but throughout the aspect ratio range 0.5 < S/D < 2 the deviations of the actual values from the curve remain within a few percent.

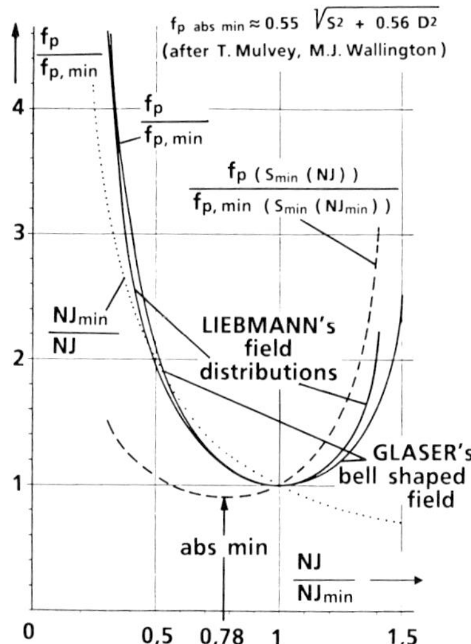

Fig. 9: Dependence of the projector focal length f_p (shown in units of the minimum focal length f_{pmin}, cf. Fig. 8) on the relative lens strength NJ/NJ_{min}. At this, NJ_{min} are the ampere turns required to attain f_{pmin}. (Values after LIEBMANN /22/; a theoretical curve employing GLASER's bell-shaped field /14/ is also shown). If the pole piece system is not kept at fixed dimensions but is scaled down so as to be always just saturated the interrupted curve results which yields a still smaller absolute minimum of the focal length /23/.

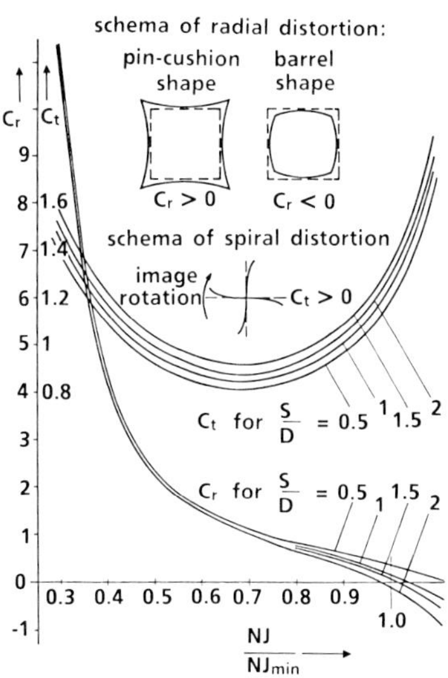

Fig. 11: Dependence on the relative lens strength NJ/NJ_{min} of the coefficients C_r of radial and C_t of spiral (tangential) distortion (curves calculated from values given in /22/).

We can now come back to the question of how to find the 'optimum' projector lens which should have particularly small aberrations. For this, let us first discuss the chromatic field aberrations which are schematically represented in Figs. 10 and 11. The chromatic aberrations of magnification

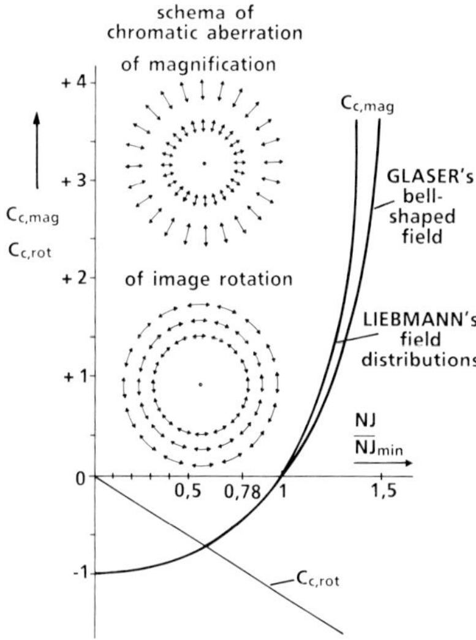

Fig. 10: Dependence on the relative lens strength NJ/NJ_{min} of the coefficients $C_{c,mag}$ of the chromatic aberration of magnification and $C_{c,rot}$ of the chromatic aberration of image rotation.

and of image rotation come about if the (relativistically corrected) beam voltage U* changes by ΔU* and/or the lens coil current varies by ΔJ.

The chromatic aberration of magnification causes a varying radial displacement δ_r as indicated in the upper schematic drawing in Fig. 10. This is equivalent to a continuous small variation of the projector magnification M_p:

$$\frac{\delta_r}{r_i} = \frac{\Delta M_p}{M_p} = C_{c,mag}\left(\frac{\Delta U^*}{U^*} - 2\,\frac{\Delta J}{J}\right).$$

(r_i is the off-axis distance of the image point just under consideration.)

Correspondingly, the chromatic aberration of image rotation can be described as a small variation of the angle of image rotation whereby the individual image points are displaced by δ_{rot} in tangential direction. The chromatic aberration of image rotation is represented schematically in the lower drawing of Fig. 10 and can be described by:

$$\frac{\delta_{rot}}{r_i} = C_{c,rot}\left(\frac{\Delta U^*}{U^*} - 2\,\frac{\Delta J}{J}\right).$$

The coefficients $C_{c,mag}$ and $C_{c,rot}$ are represented in Fig. 10. $C_{c,rot}$ is genuinely universal: it does not depend on S/D and also is not varied by scaling. Its absolute value is just half the angle of image rotation (in rad). $C_{c,mag}$ depends a little on S/D, but is universal for all practical purposes in the same sense as the projector focal length f_p curve of Fig.9.

From the shape of the above equations it is easily seen that the chromatic field aberration and the corresponding blurring of the image points increases in direct proportion to the off-axis

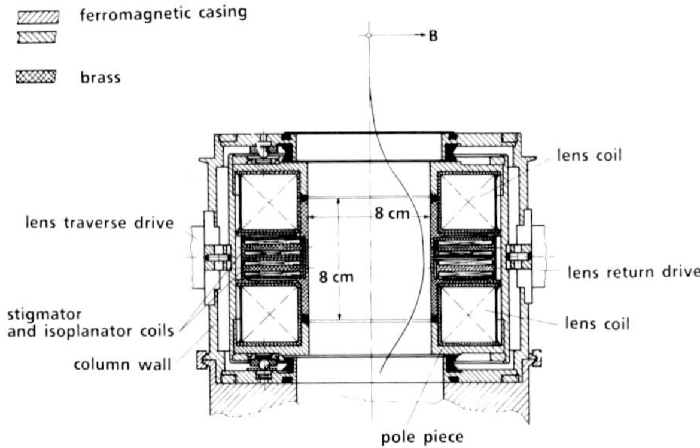

Fig. 12: Long-focal-length condenser lens having its focal length f and coefficient C_s of spherical aberration in the same order of magnitude. For the usually employed range of focal lengths between about 60 and 100 mm, C_s was found to vary from 20 to 160 mm. The construction of the stigmator and of the isoplanator is schematically represented in Fig. 13 /24/ /26/.

distance r_i of the image points. It is a first order aberration. It should be noted that the amount of blurring is completely independent both of the image distance L of the projector and of the projector magnification M_p. The distortions, in contrast to the chromatic aberrations, are of the third order, which means that here the image points are displaced proportional to the cube of their off-axis distance r_i. The total third order distortion is composed of radial and spiral distortion.

The radial distortion is represented schematically in the top two drawings in Fig. 11 and can be described quantitatively by:

$$\frac{\Delta r_i}{r_i} = C_r\left(\frac{r_i}{L}\right)^2 \quad .$$

If $C_r > 0$, the distortion moves the image points further away from the lens axis so that an axis-centered square is deformed into the shape of a pin-cushion. If $C_r < 0$, the image points move towards the axis and a distortion of the square into the shape of a barrel results.

The spiral (tangential) distortion is represented schematically in the lower drawing of Fig. 11. Here, straight lines through the axis point are deformed into the shape of short pieces of cubic parabolae for which the direction of the tangential image point displacement is always in the same sense as an increase of image rotation. The amount Δt_i of tangential displacement can be described as:

$$\frac{\Delta t_i}{r_i} = C_t\left(\frac{r_i}{L}\right)^2 \quad ,$$

whereby always $C_t > 0$. As has been already seen to be the case with the field aberrations we also see here that the distortion does not depend on the projector magnification M_p, but now for the same off-axis distance r_i of the image point the distortion can be made smaller rather effectively if the projector image distance L is made larger. This aspect can become rather important if electron diffraction patterns are to be projected with high precision.

From Figs. 10 and 11 it can now be concluded that radial distortion C_r and chromatic aberration of magnification C_{cmag} become particularly small if the reduced lens strength NJ/NJ_{min} is about 1. They both vanish completely at $NJ/NJ_{min} = 1$ if the pole piece system has the aspect ratio $S/D = 1.6$, and the focal length assumes its minimum value f_{pmin} at the same time. This evidently is the optimum projector lens and it is operated at a lens strength $NJ/\sqrt{U^*} = 13.5\ A/\sqrt{V}$. It is interesting to note that for the whole range of aspect ratios $0.5 < S/D < 2$ the lens strength required to obtain the minimum focal length deviates from that value only by a few percent /12/, /23/.

Condenser lenses

Condenser lenses are employed for concentrating the illuminating electron beam into that partial area of the specimen which is investigated. If this area is comparatively large, say of the order of a few 0.01 mm in diameter, neither spherical aberration nor astigmatism plays a role and the design of the long focal length condenser lens is quite uncritical: gap width S and bore diameter D can be chosen to be of the order of a few mm, and the fact that the coefficient of spherical aberration C_s easily amounts to a few 10 m just does not matter.

Things are different, if at about unit magnification a small electron probe must be projected into the specimen plane with a probe diameter of about 1 µm or even below. Then, spherical aberration becomes the limiting factor for the current density that can be attained in the probe. It turns out that if no loss of brightness of the electron beam shall be incurred, the coefficient C_s of spherical aberration has to be of the same order of magnitude as the focal length, i.e., both have to be of the order of a few cm /26/. This can only be achieved if the gap width S and the bore diameter D are also of this order of magnitude, and about the same number of ampere turns is required as for an objective lens working in the normal-objective mode. In fact, a low aberration long focal length condenser lens can be regarded as being a drastically scaled-up objective lens. A condenser-lens having these properties is reproduced in Fig. 12, and it has been shown that it can be employed to generate high current density electron probes with diameters well below 0.1 µm /26/.

Another important point with high quality condensers is that they must be equipped with a stigmator device capable of correcting the axial astigmatism, an aberration particularly serious with high quality long focal length lenses (see below). In conclusion some remarks shall be added about spherical aberration for lenses working approximately at unit magnification. The coefficients C_s of spherical aberration discussed in the literature usually concern high magnification objective lenses and they are "referred

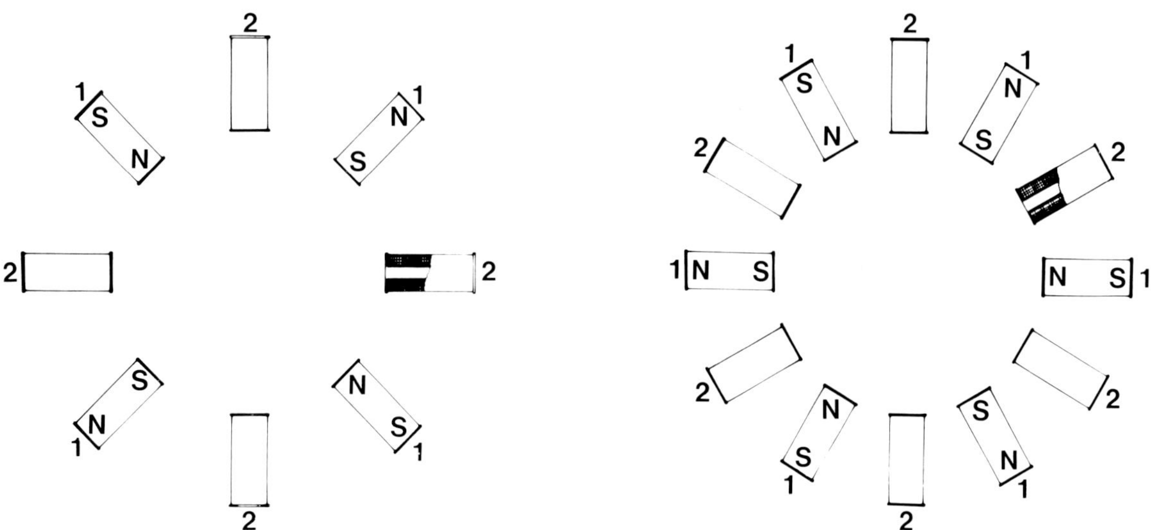

Fig. 13: Schematic representation of the construction of a stigmator (left) and of an isoplanator (right), capable of correcting a field disturbance of two-fold (stigmator) and of three-fold (isoplanator) symmetry. The axis of the lens to be corrected would pass through the centers of the coil sets and be at right angles to the plane of the drawing /24/, /26/, /29/.

back" to the object plane. So, with an imaging aperture α and for spherical aberration alone the blurring in the image would correspond to a circle of least confusion of a diameter of $0.5\ C_s\ \alpha^3$ in the object plane.

With condenser lenses where the size of the probe in their image plane counts and which operate at magnifications M_c of the order of 1, C_s has to be employed with a correcting factor $(1+M_c)^4$. So, for a probe projected with an illuminating aperture β into a specimen plane, the circle of least confusion generated there by spherical aberration has a diameter of $0.5\ C_s\ (1+M_c)^4\ \beta^3$. Usually, this limits the smallness of the probe that can be reached.

Stigmators

It is well known that no high quality electron lens can be operated successfully without a stigmator capable of compensating its axial astigmatism. The origin of the astigmatism may be quite diverse. It might be e.g. an 'elliptical' distortion of the magnetic field of the lens which can be caused by a corresponding ellipticity of the lens bores or by a waviness of the pole faces. Another origin could be a charging-up of apertures or the instrument's walls which face the beam so that an unround electrical lens is generated.

During the development of high resolution objective lenses it was first tried to 'distort the magnetic field back into rotationally symmetric shape' by introducing minute morsels of ferromagnetic material into the pole piece bore or into the gap and provide some rather artful mechanical devices to move them into the proper positions /18/, /30/.

Nowadays, these 'magnetomechanical' stigmators are rarely used any more. The presence of axial astigmatism means that the focal length f of the lens differs a little by Δf_a in different azimuthal orientations of the lens, and the idea behind the modern stigmator is to employ a compensating lens which counteracts the disturbance Δf_a. Because a lens with axial astigmatism can be modelled as a combination of a round lens with an additional weak cylindrical lens, such a stigmator should essentially be a cylindrical electron lens which can be adjusted in its strength and azimuthal orientation.

The desired electron optical action can be generated by employing electromagnetic quadrupoles as depicted schematically on the left of Fig. 13. A quadrupole consists of four small coils which have their axes at right angles both with respect to each other and to the electron beam. The direction of the current in the coils is such that as seen from the beam they present to it one pair of opposite magnetic north poles and at right angles to it another pair of magnetic south poles. The strength of the quadrupole can of course be adjusted by appropriately controlling the current in the coils. Two methods have been employed for adjusting the azimuthal orientation: either using one quadrupole which can be rotated, or providing two fixed quadrupoles which include an angle of 45° between them so that a rotation of the cylinder lens can be effected by properly adjusting the current in the two sets.

Numerous designs have been published which embody these basic ideas (cf. e.g. Figs. 14 and 15). It does not matter much if the stigmator coils are positioned within the lens field proper as in Fig. 14, or if the coils follow well after

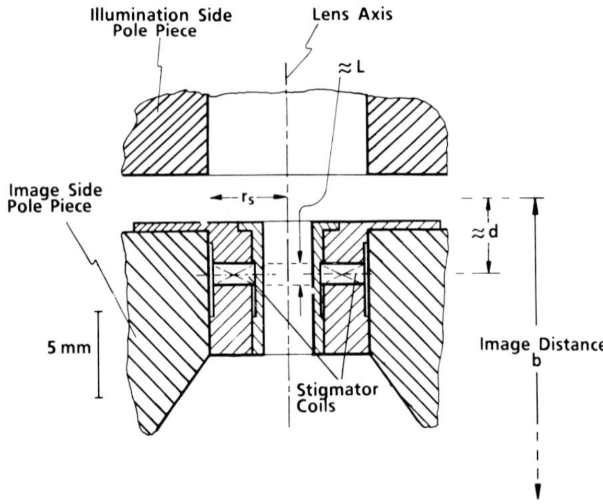

Fig. 14: Small stigmator positioned within the image side bore of an objective lens /38/.

the lens field. The stigmator placed within the lens field is sometimes believed to be superior because then 'the correcting action takes place where the defect arises'. Actually, the difference between the two positions is that a stigmator placed at a distance d after the center of the lens field will cause an elliptical distortion within the image. This means that the magnification M is a little different in two orthogonal directions. The amount of the distortion can be calculated from

$$\frac{\Delta M}{M} = \frac{d}{f} \cdot \frac{\Delta f_a}{f}$$

and usually remains at the order of 1 °/o at most.

Finally, a simple formula shall be reported which will enable the user of the instrument to calculate the ampere turns required for the coils of the stigmator if he decides to construct one himself.

If the stigmator shall be placed at a distance d after a lens which has a focal length f, an image distance b and an astigmatic focal length difference Δf_a to be corrected, then the required lens strength for the quadrupole is:

$$q^2 = \frac{1}{2} \frac{\Delta f_a}{f} \frac{1}{Lf\ (1-d/b)^2} ,$$

and the corresponding ampere turns for each coil

$$N_s J_s = 1{,}34\ \sqrt{U^*/V} \quad r_s^2\ q^2\ \text{Amp.}$$

Here, L is the diameter of the coils, and r_S the radius of the ferromagnetically short circuiting wall which surrounds the stigmator coils.

Superconducting electron lenses

Superconducting electron lenses are a rather recent development. Originally the extremely

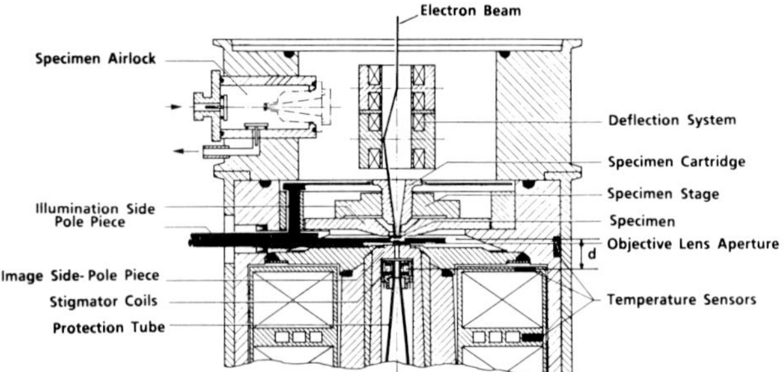

Fig. 15: Cross section of the objective lens of the SIEMENS ELMISKOP 101 electron microscope, which is equipped with an 8-coil stigmator placed d=20 mm after the lens gap /1/.

high stability of the lens current that can be attained in the persistent current mode by placing a superconducting short circuit across the leads of a superconducting coil at first seemed to be a feature of great value. But this did not suffice to noticeably reduce chromatic aberration, because beam voltage fluctuation and energy loss experienced by the beam electrons at interaction with the specimen just remained and turned out to be the dominating effect in any case. Also, the hope did not materialize that an extremely strong and simultaneously short magnetic lens field could be generated.

At present, the main advantages derived from superconducting lenses seem to come from fringe benefits. For example, the vacuum around the specimen is dramatically improved by the cryogenic pumping action of the helium cooled walls, thereby protecting the specimen against hydrocarbon contamination, and the resistance of the specimen against radiation damage appears to be markedly increased at liquid helium temperatures in comparison to ambient temperature microscopy /8/. Thermal drift of specimen stages is strongly reduced at low temperatures since then the thermal conductivity of suitable stage materials is higher by orders of magnitude /9/.

The superconducting lenses can be classified into two groups which are based on totally different physical principles:

- the superconducting lenses with ferromagnetic pole pieces,

and

- the superconducting shielding lens.

With the first group, a superconducting coil is enshrouded by an essentially classical lens casing and employs a conventional pole piece system. Examples of such lenses are shown in the Figs. 16 and 17.

With the lens of Fig. 16, only the coils are cooled by liquid helium. They are freely suspended within the iron circuit /17/, /33/. This design permits ambient temperature specimen stages, airlocks and other devices to be used on which a lot of experience is available from conventional electron optical systems. The system shown in Fig. 17 has an objective lens with its specimen stage and two projector lenses combined into a solid block which is cooled down as a whole. Here,

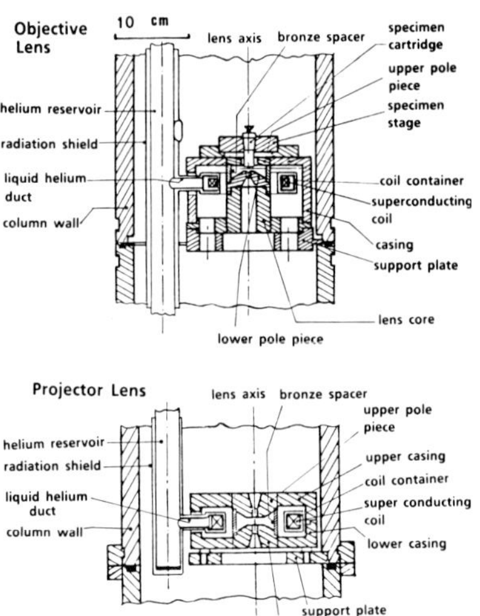

Fig. 16: Simplified cross-section of two superconducting electron lenses which employ liquid helium cooled coils but pole pieces and lens casings at room temperature. The enshrouding casing is divided into two halves which can be separated to allow access to the toroidal coil containers /17/, /33/.

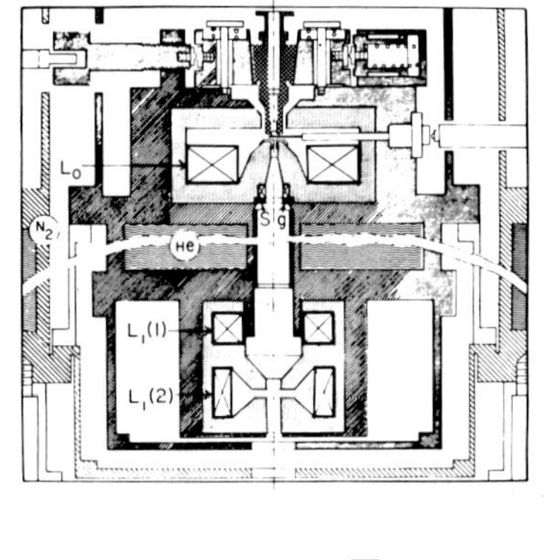

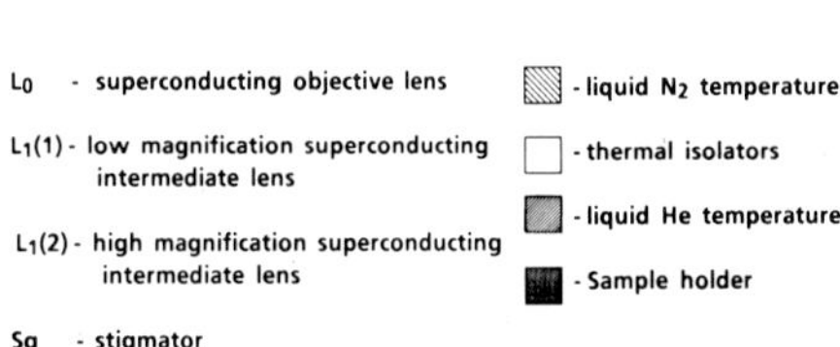

Fig. 17: Simplified cross-section of a superconducting electron lens system composed of an objective lens and two intermediate projectors. Here, the lenses are not actually immersed in liquid helium. They are clamped in a copper block containing the liquid-helium bath, and both the lenses and the specimen stage are cooled by thermal conduction. Thus, mechanical and thermal integrity of the design is achieved. The liquid-helium cooled block is surrounded by a liquid-nitrogen cooled radiation shield. The magnetic construction of all three lenses is essentially classical. (after /34/).

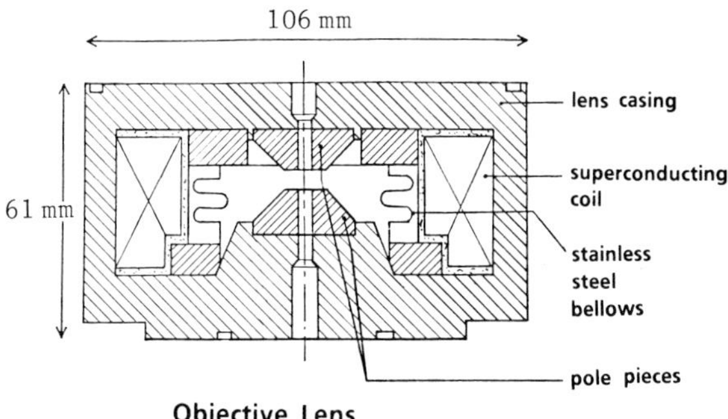

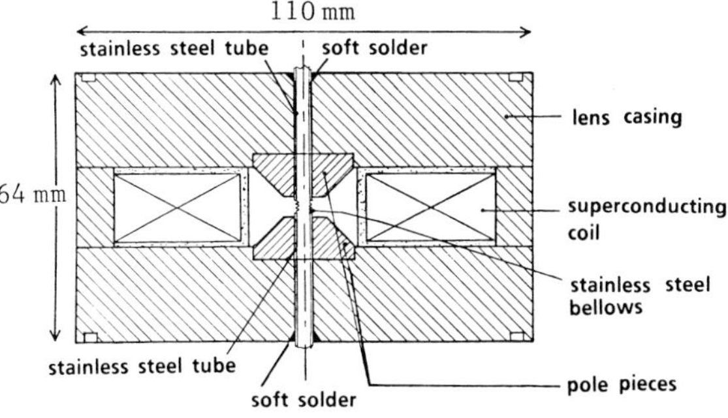

Fig. 18: Simplified cross-sections of two superconducting electron lenses equipped with pole-pieces fabricated from rare-earth metals /3/.

the fringe benefits are gained of having the specimen at low temperature by making use of the cryogenic pumping action /34/.

A still more unconventional design is represented in Fig. 18. Here again we have lenses which are intended to be cooled down as a whole, but the interesting point is that the pole pieces are not fabricated from conventional high saturation cobalt iron such as permendure. They are made from dysprosium or holmium which are metals of the rare-earth group /2/, /3/. At low temperatures they display a saturation magnetization well above that one available with permendure type materials (cf. Fig. 19). Unfortunately, as is seen from Fig. 19, the permeability of the rare-earth metals remains rather low and this leads to a quite undesirable spreading out of the "foot" of the lens field distribution so that the advantages in comparison to soft iron-pole pieces appear to remain rather limited (cf. Fig. 20).

The superconducting shielding lens works on a totally different physical principle in comparison to the superconducting lenses with ferromagnetic circuit. Now, the superconducting coils are encapsulated by a superconducting shield made from niobium-tin (cf. Fig. 21) which at liquid-helium temperatures repels the magnetic field generated by the coil and confines it to the interior volume of the shroud. The coil field

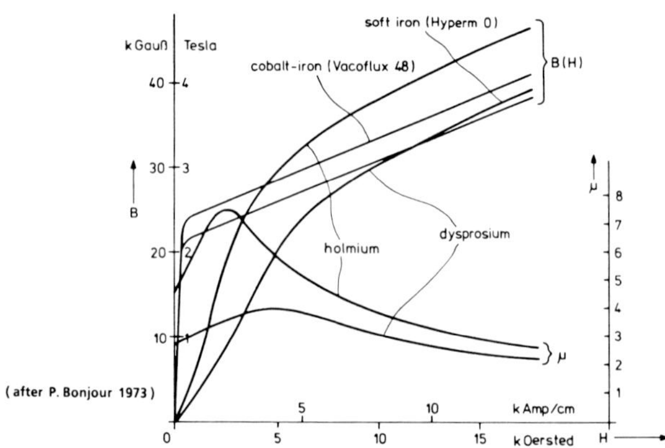

Fig. 19: Comparison of the magnetic flux density B carried by polycrystalline holmium and dysprosium with the flux density that can be obtained using conventional pole piece materials. Both are shown as a function of the magnetic field strength H applied.

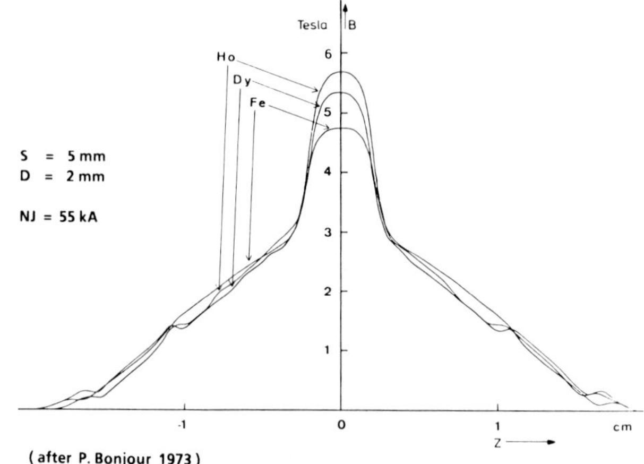

Fig. 20: Magnitude and distribution of the magnetic flux density on the axis of a lens employing pole pieces fabricated from polycrystalline dysprosium and holmium and from soft-iron. Magnetic saturation of the soft-iron pole pieces begins at about NJ=10 kA /3/.

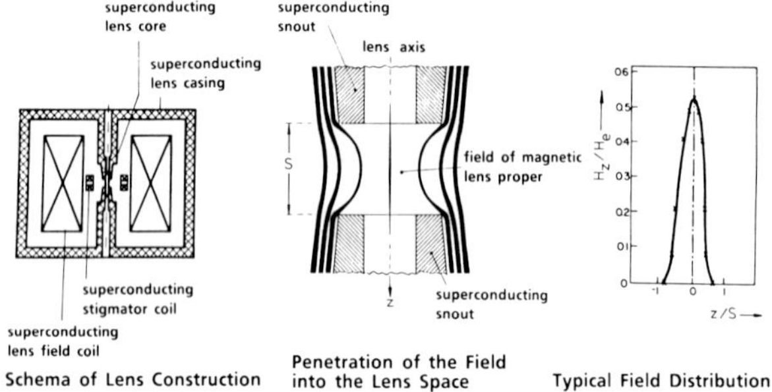

Fig. 21: Basic construction and main physical properties of the superconducting shielding lens /6/.

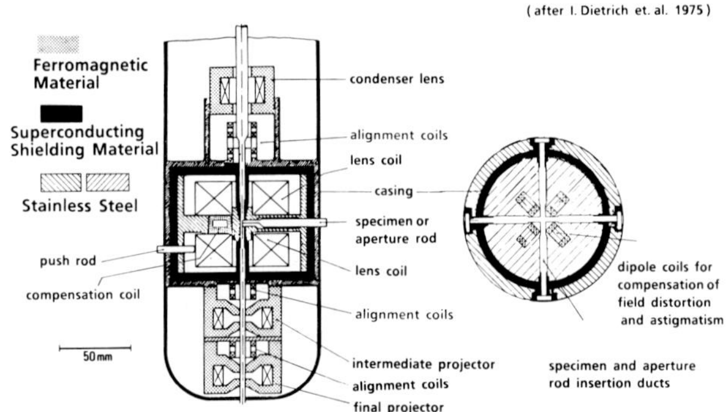

Fig. 22: Simplified cross-section of a superconducting lens system consisting of a superconducting shielding objective lens and superconducting iron shrouded condenser and projector lenses /7/.

is also shielded from the lens axis by means of two superconducting cores which terminate into a superconducting 'pole piece system'. Thus, the gap between the two superconducting pole pieces is the only place where the magnetic field can penetrate to the axial region and form there the magnetic lens field proper /6/. The device is completed by providing superconducting dipole coils which are employed for beam alignment and as a stigmator. An actual superconducting lens system which consists of a superconducting shielding objective lens and conventional iron circuit superconducting condenser and projector lenses is shown for illustration in Fig. 22 /7/.

References

/1/ Asmus A, Herrmann KH, Wolff O. (1969). Elmiskop 101 - a new high-power electron microscope. Siemens Rev. 36, 57-67.

/2/ Bonjour P, Septier A. (1967). Lentilles magnêtiques supraconductrices à pièces polaires en métaux des terres rares. C.R. Acad. Sci. Paris B 264, 747-750.

/3/ Bonjour P. (1973). Un nouveau type de lentille magnétique supraconductrice à pôles d'holmium pour très haute tension. Thèse, Orsay.

/4/ Busch H. (1926). Berechnung der Bahn von Kathodenstrahlen im axialsymmetrischen elektromagnetischen Felde. Ann. Physik (Leipzig) 4, 81, 974-993.

/5/ Busch H. (1927). Über die Wirkungsweise der Konzentrierungsspule bei der Braunschen Röhre. Arch. Elektrotech. 18, 583-594.

/6/ Dietrich I, Knapek E, Weyl R, Zerbst H. (1975). Superconducting lenses in electron microscopy. Cryogenics 15, 691-699.

/7/ Dietrich I, Fox F, Knapek E, Lefranc G, Nachtrieb N, Weyl R, Zerbst H. (1975). Supraleitende Linsen für Höchstspannungsmikroskope. Forschungsber. T75-44. Bundesministerium für Forschung und Technologie, Bonn, 35-37, 75-77.

/8/ Dietrich I. (1976). Superconducting Electron Optic Devices. Plenum Press, New York, 23, 26.

/9/ Dietrich I, Fox F, Knapek E, Lefranc G, Nachtrieb K, Weyl R, Zerbst H. (1977). Improvements in electron microscopy by application of superconductivity. Ultramicroscopy 2, 241-249.

/10/ Dugas J, Durandeau P, Fert C. (1961). Lentilles électroniques magnétiques symétriques et dyssymétriques. Rev. Optique 40, 277-305.

/11/ Durandeau P. (1957). Étude sur les lentilles électroniques magnétiques. Thèse, Toulouse.

/12/ Durandeau P, Fert C. (1957). Lentilles électroniques magnétiques. Rev. Opt. 36, 205-234.

/13/ El-Kareh AB, El-Kareh JCJ. (1970). Electron Beams, Lenses and Optics. Vol. 1, 2. Academic Press, New York. Vol. 1: 255-275, 293-405; Vol. 2: 72-80, 89-127, 281-286, 295-313.

/14/ Glaser W. (1952). Grundlagen der Elektronenoptik. Springer-Verlag, Wien. 184-207, 262-267, 301-314, 373-384, 413-428.

/15/ Jandeleit O, Lenz F. (1959). Berechnung der Bildfehlerkoeffizienten magnetischer Elektronenlinsen in Abhängigkeit von der Polschuhgeometrie und den Betriebsdaten. Optik 16, 87-107.

/16/ Kasper E. (1982). Magnetic Field Calculation and the Determination of Electron Trajectories. In: Magnetic Electron Lenses, P.W. Hawkes (ed). Springer-Verlag, Heidelberg, New York. 57-118.

/17/ Laberrigue A, Levinson P, Homo JC. (1971). Microscope électronique 400 kV à lentilles supraconductrices. I. Description génerale du cryostat et du montage. Rev. Physique Appl. 6, 453-458.

/18/ Leisegang S. (1953). Zum Astigmatismus von Elektronenlinsen. Optik 10, 5-14.

/19/ Lenz F. (1982). Properties of Electron Lenses. In: Magnetic Electron Lenses, P.W. Hawkes (ed). Springer-Verlag, Heidelberg New York. 119-161.

/20/ Liebmann G. (1950). Solution of partial differential equations with a resistance network analogue. Brit.Jour.Appl.Physics 1, 92-103.

/21/ Liebmann G, Grad EM. (1951). Imaging properties of a series of magnetic electron lenses. Proc. Physic. Soc. London B 64, 956-971.

/22/ Liebmann G. (1952). Magnetic electron microscope projector lenses. Proc. Physic. Soc. London B 65, 94-108.

/23/ Mulvey T, Wallington MJ. (1969). The focal properties and aberrations of magnetic electron lenses. Jour. Physics E 2, 466-472.

/24/ Riecke WD. (1960). Über eine neue Einrichtung zur Feinstrahlbeugung. In: Vierter Int. Cong. für Elektronenmikroskopie, Berlin 1958. W. Bargmann, G. Möllenstedt, H. Niehrs, D. Peters, E. Ruska, C. Wolpers (eds). Springer-Verlag, Berlin Göttingen Heidelberg. Vol. 1, 89-194.

/25/ Riecke WD. (1962). Ein Kondensorsystem für eine starke Objektivlinse. In: Proc. Fifth Int. Cong. on Electron Microscopy, Philadelphia 1962, S.S. Breese (ed). Academic Press, New York London, Vol. 1. KK-5.

/26/ Riecke WD. (1962). Feinstrahl-Elektronenbeugung mit dreistufigem Kondensor und langbrennweitiger letzter Kondensorstufe. Optik 19, 81-116.

/27/ Riecke WD. (1968). On the alignment of an electron microscope with condenser-objective lens. In: Electron Microscopy 1968, Fourth Reg. Conf., Rome, D.S. Bocciarelli (ed). Tipografia Poliglotta Vaticana, Rome. Vol. 1, 207-208.

/28/ Riecke WD. (1969). Beugungsexperimente mit sehr feinen Elektronenstrahlen. Z. angew. Physik 27, 155-165.

/29/ Riecke WD. (1982). Practical Lens Design. In: Magnetic Electron Lenses, P.W. Hawkes (ed). Springer-Verlag, Heidelberg New York. 163-357.

/30/ Ruska E. Wolff O. (1956). Ein hochauflösendes 100-kV-Elektronenmikroskop mit Kleinfelddurchstrahlung. Z. wiss. Mikroskopie 62, 465-509.

/31/ Scherzer O. (1936). Über einige Fehler von Elektronenlinsen. Z. Physik 101, 593-603.

/32/ Septier A. (1966). The Struggle to Overcome Spherical Aberration in Electron Optics. In: Advances in Optical and Electron Microscopy. Vol. 1, R. Barer, V.E. Cosslett, (eds). Academic Press, London New York. 204-274.

/33/ Severin C, Génotel D, Girard A, Laberrigue A. (1971). Microscope électronique 400 kV à lentilles supraconductrices. II. Caractéristiques électrooptiques et functionnement. Rev. Physique Appl. 6, 459-465.

/34/ Siegel BM. (1976). The transmission electron microscope system: Characteristics for high resolution and optimum signal to noise. In: Proc. 6th Eur. Cong. on Electron Microscopy, Jerusalem 1976. D.G.Brandon (ed). Tal International, Jerusalem. Vol I, 105-108.

/35/ Southwell RV. (1946). Relaxation Methods in Theoretical Physics. Clarendon Press, Oxford.

/36/ Suzuki S, Akashi K, Tochigi H. (1968). Objective lens properties of very high excitation. Proc. 26th Annual EMSA Meeting, New Orleans, La., C.J. Arceneaux (ed). Claitor, Baton Rouge, La., 320-321.

/37/ Suzuki S, Ishikawa A. (1978). On the magnetic electron lens of minimum spherical aberration. In: Electron Microscopy 1978, Ninth Int. Cong. on Electron Microscopy, Toronto, J.M. Sturgess (ed). Microscopical Soc. of Canada, Toronto, Vol. 1, 24-25.

/38/ Watanabe M, Someya T. (1963) Electromagnetic Stigmator for magnetic lens. Optik 20, 99-108.

Electron Optical Systems (pp. 15-27)
SEM Inc., AMF O'Hare (Chicago), IL 60666-0507, U.S.A.

0-931288-34-7/84$1.00+.05

MAGNETIC ELECTRON LENSES II

T. Mulvey

Department of Mathematics and Physics
The University of Aston in Birmingham
Birmingham B4 7ET (UK)
Phone No. 021-359 3611

Abstract

Conventional magnetic electron lenses have evolved to their present highly developed state under the pressure of meeting the exacting requirements of high resolution electron microscopy. More recently, however, the desire to extract quantitative analytical information from the specimen has led to significant changes in the design of electron optical systems. The introduction of efficient lanthanum hexaboride cathodes and high beam current field-emission sources has strengthened this tendency. In addition, more complex lens systems than previously envisaged are now possible since microprocessors can be employed to assist in the rapid and reliable readjustment of the lens system, including the extensive alignment procedures. The use of high current density, e.g. superconducting coils, is also paving the way for new lens configurations. Furthermore, the increasing demands placed on the lens systems in electron beam lithography are bound to bring benefits to electron optical systems in general.

KEYWORDS: Magnetic Lens, Finite Element Program, Spherical Aberration, Image Distortion, Mini-lenses, Projector System, De-scan Coils, Rotation-free Lens.

Introduction

Last year we celebrated the fiftieth anniversary of high resolution electron microscopy. In November 1933 Ernst Ruska achieved a resolution better than that of the optical microscope with the aid of a new kind of magnetic electron lens - the iron shrouded polepiece lens - which he and von Borries[22] had patented in 1932 and which was to be decisive for the future progress of all forms of electron optical equipment. There is no consensus as to who is the inventor of the magnetic electron lens or for that matter the electron microscope, but there is, however, general agreement that iron-shrouded magnetic electron lenses sprang almost accidentally from technology rather than from science. The researches of Gabor[8] into the measurement of fast surges on high voltage transmission lines at the Institut für Hochstspannungstechnik in 1924-6 gave birth to a crude form of what was later recognised as an iron-shrouded magnetic electron lens, the forerunner of our modern high resolution magnetic lens, the essential element in a high resolution electron microscope as well as in many other forms of electron optical instruments. Gabor's chief inspiration was to dispense with the conventional concentrating coil of the high voltage oscillograph and to replace it with a short coil encased in an iron shroud except over the axial region. His chief reason for doing so was to contain the magnetic field as far as possible within the confines of the lens itself. The main idea was to prevent any stray magnetic field from adversely affecting the operation of the cold cathode source and that of the deflecting plates used to scan the electron beam. Unwittingly he had stumbled across a way of making an efficient focussing element that worked in an entirely different way from that of the long solenoid. However, since he could not give an adequate explanation of the focussing action he is not generally considered to be the inventor of the magnetic electron lens.

The correct explanation of how such a short coil focusses the electron beam was provided later in that year by Busch[5] who thus became the founding father of electron optics. However, Busch found that he could not get satisfactory agreement between his theory and experimental results that he had obtained previously on a short coil. The problem of resolving the discrepancies between experimentally measured focal properties of magnetic electron lenses and the theory calculated by Busch was resolved by Knoll and Ruska[10], who in 1931 succeeded in constructing a crude electron microscope with a magnification of some 12 times using two iron free solenoids. The invention of the polepiece lens by Ruska and von Borries[22] was based on the idea of using iron polepieces to confine the field in a narrow gap thereby creating very high axial flux densities. Taken to its logical conclusion this led Riecke and Ruska[17] in 1966 to the idea of the high resolution condenser-objective lens in which the specimen is placed at the centre of the magnetic field distribution whose half-width is as small as possible and whose axial field strength is as high as possible. This lens is now widely used both in high resolution TEMs and STEMs. This lens has excellent performance but is not easy to manufacture, align, or to operate. There are also difficulties, because of the narrow objective polepiece bores and gap, in extracting x-rays, Auger electrons and other emissions from the sample.

It seems therefore that classical magnetic electron lenses and the associated electron optical systems have, after a period of fifty years of development, reached the peak of their performance. However, the demands on electron optical systems and on the lenses themselves, far from being satisfied, are becoming more pressing. This is largely due to the widespread use of electron optical instruments for analytical purposes where a great deal of information has to be extracted from the sample and the different devices such as Auger spectrometers, energy loss and x-ray spectrometers, have to be interfaced to the electron optical column. Furthermore, it is desirable that the electron optical system can be housed in a normal laboratory. Conventional lenses and columns have the grave disadvantage of occupying an enormous volume of space. High voltage microscopes are even more demanding on space and weight. This is not the only disadvantage. The large size of each lens unit restricts severely the number of lenses in the column and also their optimum placing. In the early days this was not a serious disadvantage because for manual operation it is more convenient to have as few lenses as possible. However, this does mean many electron optical compromises when the mode of operation of the instrument is changed and the same lens has to perform an incompatible number of roles. The operational difficulties of aligning and setting a multi-lens system can be largely overcome by microprocessor or computer-controlled procedures where each lens is interfaced to a central computer which stores the necessary alignment data. This paper considers some of the steps that have already been taken at the research and development level to implement the changeover to multi-lens columns of modest size and of enhanced electron optical performance.

Electron sources

Thermionic cathodes using tungsten filaments have held sway for approximately fifty years although in most commercial instruments they are still in a comparatively crude state of development. The chief advantage is that the crude hairpin cathode is comparatively cheap, non-critical in alignment and tolerant of poor vacuum. Much better performance could in fact be obtained from carefully aligned pointed cathodes of oriented crystal material. For a STEM instrument a field emission cathode is essential for high resolution work but unfortunately such cathodes cannot usually produce sufficient current for analytical work with probes in the range of hundreds of nanometres. In any case a field emission system demands a superb vacuum (10^{-11} mbar) and this reduces the speed of changing specimens. Lanthanum hexaboride cathodes represent a good compromise for TEM and STEM systems. Ideally the vacuum should be just as good as for field emission systems and the temperature of the emitting crystal should be controlled by specially designed electronic circuitry. With a suitably designed LaB_6 cathode, it should be possible to obtain an order of magnitude improvement in source brightness compared with that of a tungsten filament. The improvement of brightness is an overriding consideration in electron optical systems since probe currents and/or exposure times will increase by the same proportion. Alternatively, for a given probe current or exposure time, the design limitations of the lenses can be correspondingly relaxed making it easier to carry out analysis more conveniently on a specimen. From an electron optical point of view, field emission guns can be improved by placing a magnetic lens in the vicinity of the emitting tip. In a field emitting gun the effective size of the cathode is of the order of nanometres. The electron beam is therefore sensitive to the effect of spherical aberration. This can be minimised by placing a suitable magnetic electron lens in the vicinity of the tip. This is easier said than done since this is a critical region for the high vacuum and also the tip is usually at a high negative potential. Nevertheless Troyon and Laberrigue[20] succeeded in placing a miniature magnetic electron lens just below the tip, as shown in Fig. 1. The technical difficulties associated with this construction have now been overcome and the source is now in commercial production. Another method proposed by Smith and Swann[18] is to immerse the emitting tip in the field of a single polepiece electron lens placed in such a way that the field strength is a maximum in the vicinity of the emitting tip, falling off gradually in the direction of the emitted electron beam. Such a lens has low spherical aberration in this configuration. There are of course practical difficulties with this arrangement. Although very promising from a theoretical point of view, it has not yet found its way into production. An interesting compromise however, between the approaches of Troyon and Laberrigue and of Smith and Swann, is shown in Fig. 2. The design is due to Venables and Archer[21] in which a single-polepiece lens of fairly large bore size is placed in the vicinity of the extraction electrode of the field emitting source; the shape of the axial magnetic field is not ideal, but the design is compatible with high vacuum operation and is comparatively simple to

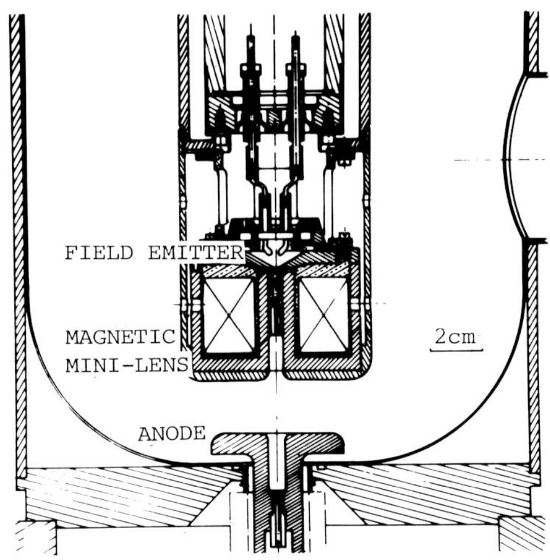

Fig. 1. Field emission electron gun (Troyon and Laberrigue 1977) with emitting tip immersed in the magnetic field of a double-polepiece mini-lens. Courtesy of the authors.

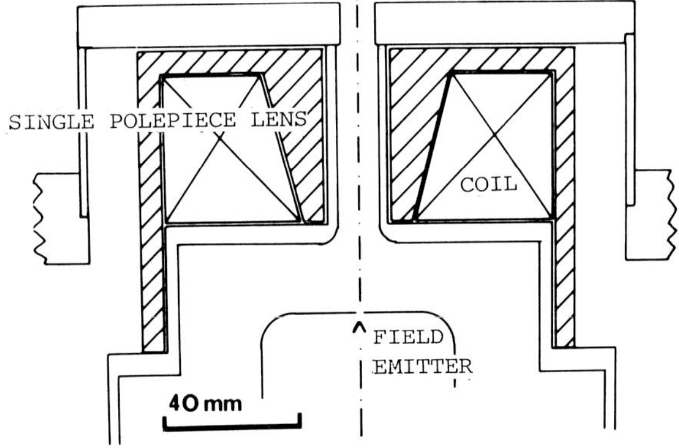

Fig. 2. Field emission electron gun (Venables and Archer 1980) with emitting tip immersed in the field of a single-polepiece lens to reduce aberrations and increase total beam current. Courtesy of J A Venables.

implement on existing field emission guns. Such a gun is capable of producing an appreciably greater current than is possible in the absence of the magnetic field. Furthermore, by concentrating most of the electron beam into the axial region, secondary benefits arise due to the minimisation of collisions of beam electrons with exposed metal surfaces. There is of course a limit to the improvement of the performance that can be expected merely by the use of magnetic field concentration of this kind, since with large currents, coulomb and other interactions may lead to a larger chromatic spread in the beam. Nevertheless it seems that if the vacuum is sufficiently good and the tip radius can be carefully controlled, a field emission gun with a suitably designed magnetic electron lens can give an appreciably greater beam current than is now possible with conventional field emission guns.

Calculation of magnetic electron lenses

The exacting specifications of magnetic electron lenses preclude the possibility of determining the final design purely by previous experience or even by a purely experimental investigation in which modifications are carried out to a well-known design until the required performance is obtained. Such procedures are time consuming and do not necessarily converge on the required solution. Fortunately great progress has been made in the last few years in numerical methods of determining lens properties and in the general area of computer-aided design. The starting point of such an investigation is the determination of the magnetic flux density distribution in the lens and especially on the lens axis. The designer must of course supply details of the structure he wishes to analyse as a basis for further refinement. Two main methods are available for determining the magnetic field distribution of a given structure. These are the finite difference method and the finite element method. The latter method is the most popular and suitable programs are generally available.

The finite element method is also preferred because its principle of operation, namely to minimise the energy in the magnetic structure, has perhaps an appeal on physical grounds especially where boundary problems arise, e.g. between iron and vacuum, between iron and copper, etc. Two forms of the finite element method are available: the differential form largely introduced into electron optics by Munro[15] in which the whole of the magnetic field distribution is divided up into finite elements and the vector potential associated with each element is determined by solving a large matrix. Boundary conditions are automatically taken into consideration and need no special attention from the program user. This means that "open" magnetic structures, whose fields extend in principle to infinity, call for very large matrices and hence very large computers if errors are to be avoided. On the other hand, the integral form of finite element method, associated in the UK with Trowbridge[19] and colleagues at the Rutherford-Appleton Laboratory, Harwell, follows a different approach. Here, only the coil and the magnetic circuit itself are divided into finite elements. This completely avoids the difficulty of having to divide the whole of space into finite elements. Instead, the field at any point in the magnetic circuit can be thought of as consisting of two components, one due to the coil (which can be readily calculated by the Biot-Savart law) and the other component due to the magnetisation of the iron. This magnetisation of the iron arises from the field in the iron due to the coil. The total field therefore is the sum of these two components. The iron does not itself contribute any ampere turns to the circuit, but simply modifies the flux density distribution produced by the coil. This is a valuable concept and can often be applied to check results obtained by the differential form of the finite element method. However, the price to be paid by the apparent simplicity of the integral form of the method, is that the essential information is concentrated into the small volume of the iron circuit rather

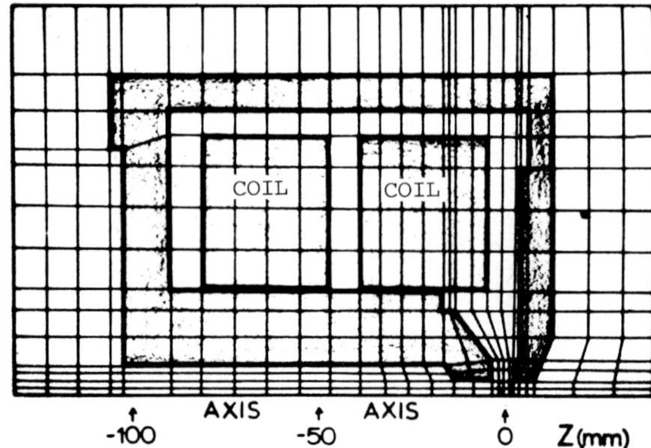

Fig. 3. Example of an early mesh layout for determining the magnetic flux distribution of a conventional magnetic objective lens. Courtesy of E Munro.

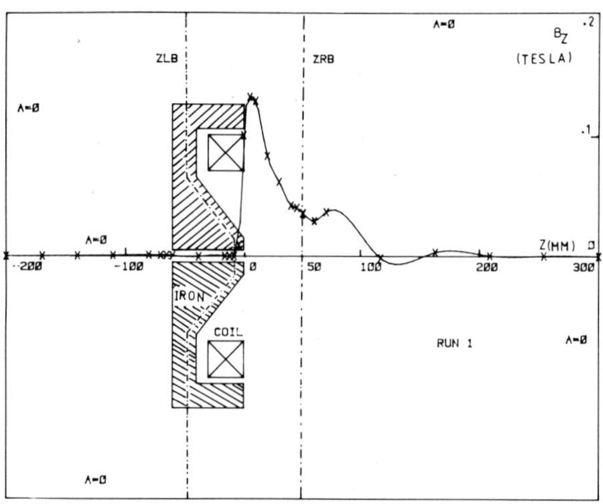

Fig. 4. Axial field distribution B_z of a single-polepiece lens by the standard finite element method with 19 (radial) × 29 (axial) element network. Spline-fitted curve through calculated points. ZLB and ZRB (dotted lines) are chosen as intermediate boundaries for subsequent refinement of the calculation[13].

than being spread through space. This results in a very dense matrix to be inverted, with the possibility of strongly localised errors. There is therefore no saving in computer store required and so far a critical comparison has not been made of the two methods which, in the opinion of the author, should be regarded as complementary rather than competitive. Fig. 3 shows the application of Munro's programme to the determination of the field distribution in a typical conventional lens. The outer shell of the lens is unbroken except for a small air gap. This means that the external field is very small and the condition that the vector potential A = 0 immediately outside the lens is satisfied. Note that whereas the lens action takes place in a volume of only a few cubic millimetres, the coil itself occupies the bulk of the space around the lens thereby restricting the possibility of placing other lenses near to the first one. The reason for this is that in the past electronic circuits were not capable of supplying large currents or large amounts of power so that the current density in the coil was low, cooling was inefficient and therefore the coil was bulky. In lenses in which one polepiece has a wide bore or in the limiting case of a single polepiece lens, as illustrated in Fig. 4, serious difficulties arise in the differential version of the finite element program. Fig. 4 shows a typical single polepiece lens with a small localised coil. Since the field from this lens extends a considerable distance away from the polepiece it is necessary to place the boundary of the area to be discretized as far away as possible. Otherwise, the boundary will appear to absorb a considerable fraction of the lens excitation. The physical explanation for this is that a surface at which the vector potential A = 0 has a vanishingly small permeability, and thus acts as a super-conducting screen. If this screen is placed too near the magnetic structure it will not only remove ampere-turns from the system but will considerably distort the field distribution. This may not seriously affect the calculated focal lengths and chromatic aberration, but will almost certainly introduce serious errors into the calculation of spherical aberration. In a computer with a limited core store and hence a limited number of elements available in the axial and radial directions, further troubles will arise if the limited number of mesh points are spaced too widely. Although it is true that such effects as loss of ampere-turns and irregularities in the calculated field distribution can be minimised as the number of mesh points is increased, the errors cannot in fact be reduced to negligible proportions; furthermore each problem requires separate consideration. For computer-aided design, especially in the initial stages, great accuracy is not required provided that the resulting field distributions are smooth. What is needed most is speed of operation and the ability to interact directly with the computer. The finalised design can of course be computed in greater detail offline. By attention to detail and the introduction of some diagnostic checks, the differential finite element method can be made vastly superior to any other method for calculating electron lenses. It is also possible to carry out quite complicated calculations on a quite modest computer. Thus the field distribution in Fig. 4 was carried out on a Commodore PET Microcomputer making full use of the disk store. The axial field distribution shown in Fig. 4 is fairly smooth near the polepiece where the mesh points are fairly closely spaced and exhibits large discrete errors in the far field where fewer mesh points are available. A simple method of overcoming these defects is shown in Fig. 5. Here the previous boundary on which A=0 (now shown by the dotted boundary) is replaced by another boundary (shown by the solid line) placed much closer to the lens. The vector potentials along this boundary are known from the first calculation and these are inserted on the left-hand and right-hand side respectively in Fig. 5. The calculation is re-run using the whole of the computing power within this much smaller boundary.

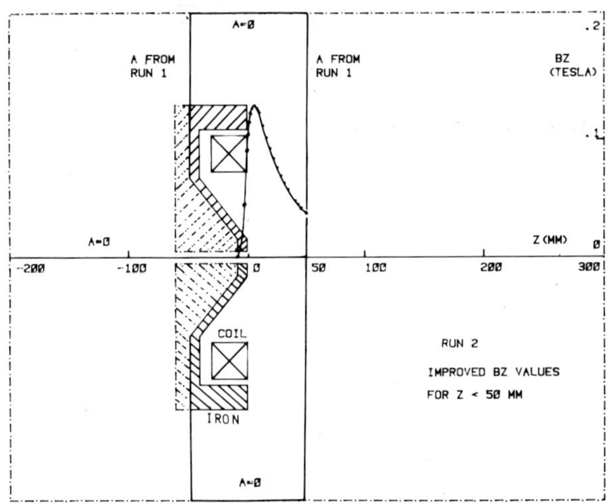

Fig. 5. Improved B_Z values using 19 × 29 network inside the intermediate boundary of Fig. 4.

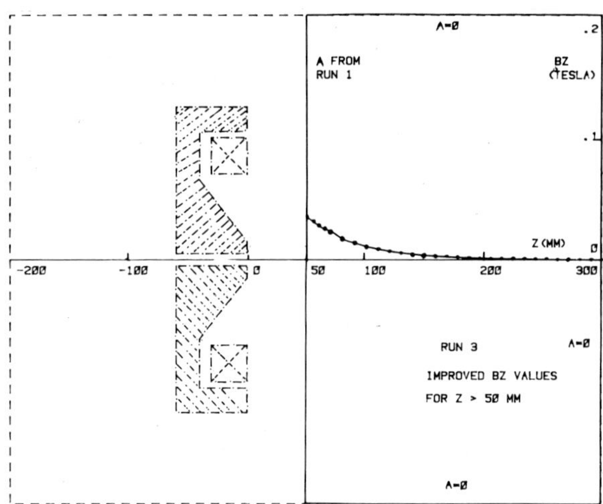

Fig. 6. Improved B_Z values using 19 × 29 network outside the chosen boundary of Fig. 4.

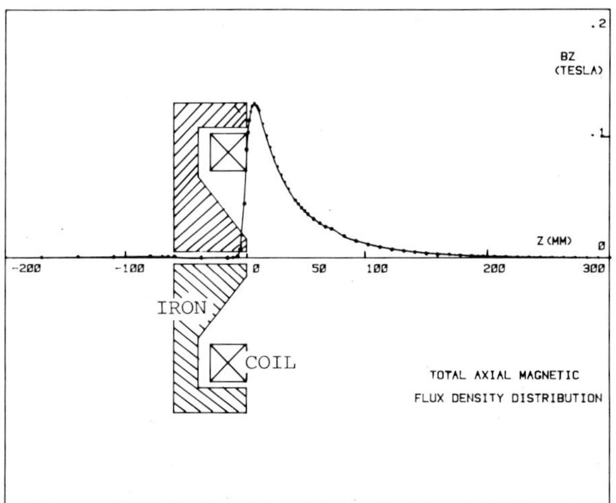

Fig. 7. Finally improved total axial field distribution of single-polepiece lens by the selected intermediate boundary method. Effective network 19 × 58 (Mulvey and Nasr 1981).

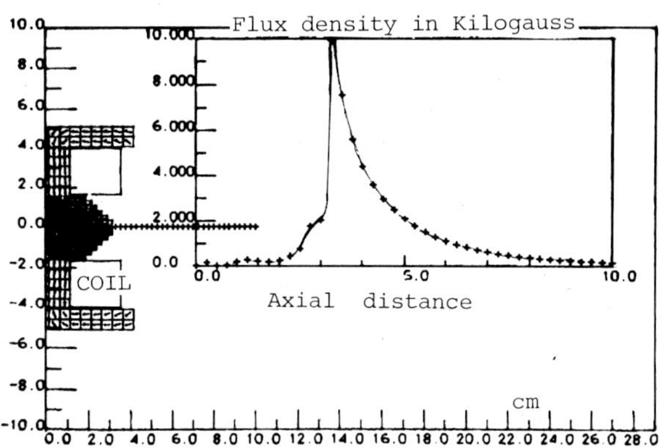

Fig. 8. Axial field distribution of single-polepiece lens calculated by the GFUN magnet design program of the Rutherford-Appleton Laboratory (courtesy of CW Trowbridge). Direction of magnetisation in iron circuit indicated by arrows. Boundary of coil also shown.

The resulting field distribution is clearly much smoother and the peak has slightly increased. The whole field can now be improved by transferring all the computer power to the right-hand side of the field as shown in Fig. 6. Here the left-hand boundary of the field, shown as a solid line, is set to the correct vector potential as found from Fig. 5 and the calculation repeated. The result is a smooth curve, as shown in Fig. 6.

Finally the total axial field distribution is shown in Figure 7. The effective network has therefore been increased to 19 × 58 without increase in core store.

In this method the effective core store of the computer is increased at the expense of time. Its chief advantage is that it produces a reliably smooth field with minimal computer resources. The method is in fact analogous to that used in electrolytic tank solutions of Laplace's equation in which the potential field is first obtained over a big area from a small model in order to determine the potentials on a much more localised boundary. The calculation is then repeated using a larger model surrounded by a more local boundary at which the potentials have been determined. It can be seen from Fig. 6 that the problem has been divided into two parts. In the first region the coil windings and associated magnetic circuit are completely contained. In the second calculation the field is determined in a region where there are no iron elements or exciting windings. A more recent and elegant method is that of Lencová and Lenc[12] who use a mathematical approach to determine the vector potential on the intermediate boundary between the two regions.

The Integral Method

Fig. 8 shows by contrast the calculation of the axial field distribution of a single pole lens by the G-FUN magnet design program[19] of the Rutherford-Appleton Laboratory. The iron circuit is divided into finite elements; arrows indicate the direction of magnetisation in each element.

Since the current density in the coil is assumed constant, it is only necessary to include the outline of the coil. No artificial boundary is imposed and the field at any point in space may be calculated directly. In particular, the field distribution outside the lens is perfectly smooth as would be expected since this space is not discretized. Within the polepiece region itself care has to be taken with the arrangement of the finite elements especially where the field is changing very rapidly as, for example, the sharp rise at the pole face. Fig. 9 shows a refinement of this area. These results emphasise the complementarity of the two methods. The differential method is at its weakest near the artificially imposed boundary; in addition, the smoothness of the field is liable to exhibit kinks and discontinuities even in regions remote from the exciting coil and the iron circuit because of the discretization of the whole of space. The method also tends to create errors concerning lens excitation since the area under the axial field distribution curve invariably differs from that calculated from the known lens excitation. This error usually manifests itself in an apparent loss in ampere-turns but sometimes the error can be positive indicating an apparent gain in ampere-turns. This cannot happen in the integral method but some discontinuities in the field distribution may be expected in the region occupied by the iron circuit.

The Differential-Integral Method[13]

The foregoing discussion suggests that the differential method can be considerably improved at the expense of only a trivial increase in computing time, as illustrated in Fig. 10. Here the axial flux density distribution as calculated by the differential method using 29 meshes in the axial direction and 19 in the radial direction is indicated by the crosses. In addition the axial field B_{coil} due to the coil has been calculated by the Biot-Savart law. If this is subtracted from the total field distribution the result will be the magnetic field B_{Fe} produced along the axis by the iron circuit. The calculated field from the coil is exact and not affected by the position of the artificial boundary A=0. If now the field due to the iron, smoothed if necessary, is added to that of the field from the coil, an improved total field will result with only a trivial additional computing effort. The differential-integral method thus overcomes many of the weaknesses of the pure differential method. It is especially useful at the initial design stage where rapid interactive computing is essential.

Rotation-Free Projector Lenses

Compact windings with efficient water cooling[14] not only reduce considerably the size and weight of the electron optical column but it enables lenses to be grouped in pairs with the exciting coils wound in opposite directions, exactly compensating image rotation at all currents[9]. Fig. 11 shows the focal properties of a pair of projector lenses of conventional design but miniaturised in construction. The two lens gaps are separated by a distance of 50 mm. The focal properties of such a pair can be readily simulated by the square top magnetic field model. The images at the top of the figure show typical images formed by this doublet. At low magnification (A) a distortion free picture of the grid is easily obtained. At very high magnification (D) essentially distortion free magnification is obtained. The lens system in this region has the same aberrations as that of the final projector lens acting on its own; a range of magnification of roughly three times can therefore be obtained with adequately low distortion. At the lower end of this range of magnification (Fig. B) characteristic pin-cushion distortion makes its appearance.

Single-Polepiece Projector Lenses

Single-polepiece lenses can have very favourable electron-optical and constructional properties. A very simple construction for a rotation-free single polepiece projector doublet is shown in Fig. 12. Here the lens body is machined from a single piece of iron. Coils are inserted in each of the lens units and the end faces sealed off with a non-ferromagnetic lid. The bore can be made quite large so that a vacuum liner tube can be used as shown in Fig. 13 which shows two such units installed in an experimental electron microscope. The upper lens unit serves as a rotation-free diffraction lens while the lower one serves as the rotation-free main projector. Fig. 14 shows a typical selected area diffraction pattern taken by a double exposure in which the diffraction lens operates as a weak lens to acquire the diffraction pattern and as a strong lens to acquire the resulting image of the molybdenum trioxide crystal. In conventional lens systems selected area diffraction patterns are subject to severe disorientation between the image and the corresponding diffraction pattern. A rotation-free projector system automatically preserves the correct orientation and incidentally eliminates chromatic aberration of rotation from the image.

Distortion-Free Wide-Angle Projector Systems

For the past fifty years, conventional projector lenses have been restricted by spiral distortion to a semi-angle α_p of projection of about some 5°. This leads to excessively long viewing chambers (500-1000 mm) in TEM and difficulties of interfacing energy loss spectrometers in STEM. A wide-angle (α_p=30°) system would go a long way to solving these problems, and single polepiece lenses are uniquely suited to this purpose. The simple type of rotation-free doublet described above is not, however, optimised for this purpose. The reason for this is that the aberrations of a single-polepiece lens are lowest when the polepiece faces the incoming beam and largest when it faces away from the direction of the incident beam. In a correcting system the polepiece of the final projector lens must therefore face the incoming electron beam in order to produce minimum aberration at the fluorescent screen. The corrector lens on the other hand must face away from the direction of the electron beam in order to produce as much radial and spiral distortion as possible so that even after a magnification of 3x by this lens sufficient distortion will still be available to cancel that of the final projector lens. An early experimental scheme for producing a wide-angle

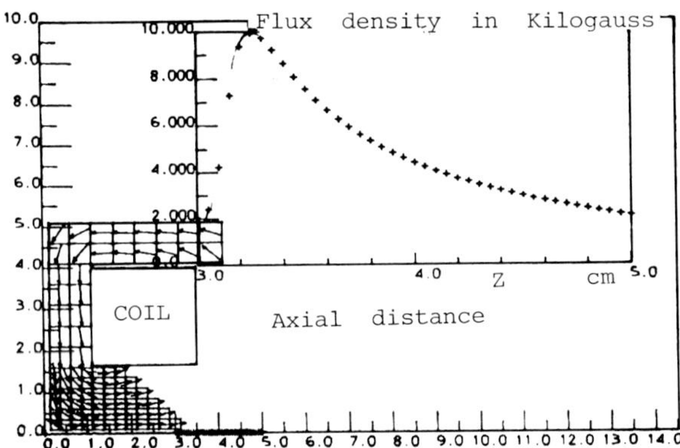

Fig. 9. Expanded detail from Fig. 8 with extra calculated points in the polepiece region giving improved accuracy.

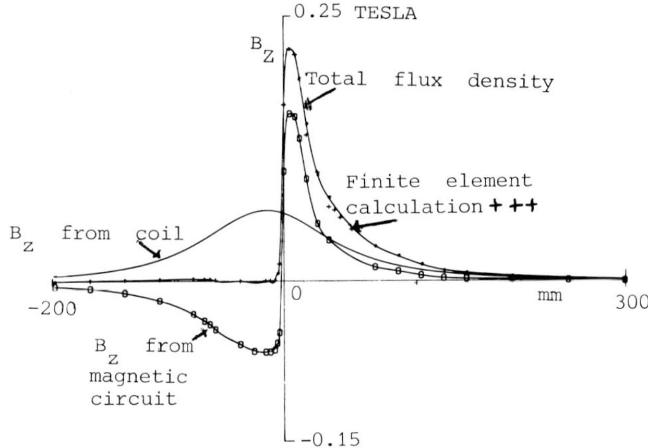

Fig. 10. Differential-Integral method (Mulvey and Nasr 1981) for improving the accuracy of the differential finite element method (Munro 1971). ++++ calculated values from Munro program, —— B_Z due to coil, θθθθθθ B_Z due to iron, ♦♦♦♦ total B_Z.

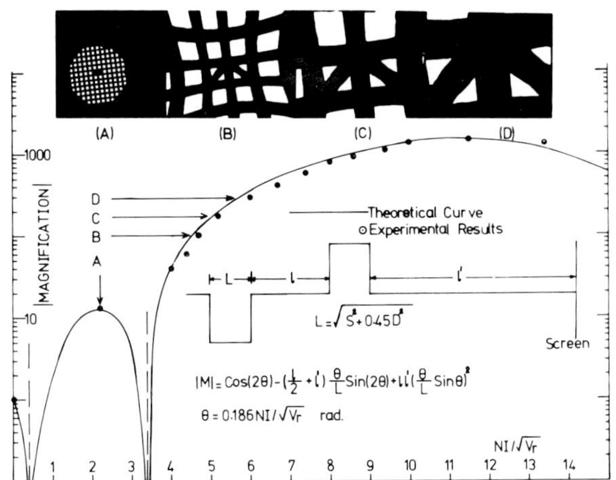

Fig. 11. Electron-optical characteristics of rotation-free projector doublet comprising two mini-lenses with conventional polepieces. Magnification can be calculated from square top model as indicated. Top magnification 1200 × for ℓ' = 450 mm. Micrographs A, B, C, D show distortion characteristics at different magnifications. Juma and Mulvey (1978) [9].

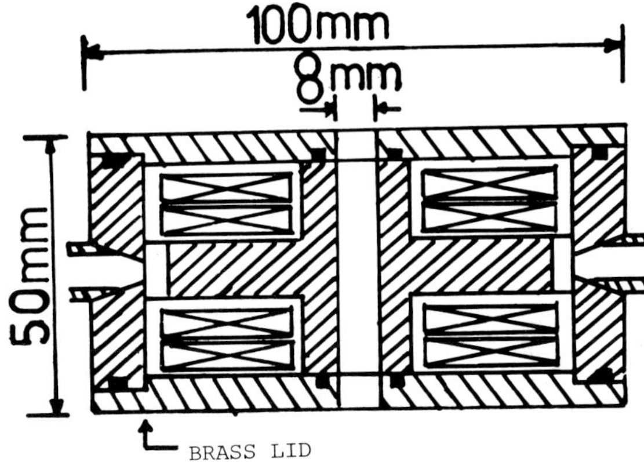

Fig. 12. Rotation-free miniature single-pole projector lens doublet for 100 kV electron microscope. Note the wide bore (8 mm) for vacuum liner. Juma and Mulvey (1978) [9].

single polepiece lens doublet[11] is shown in Fig. 15. Here the electron beam passes through a corrector lens in the form of the lens of Fig. 12 but in which only the lower coil is energised. The beam then passes through a specially designed single-polepiece lens of low aberration provided with a conical exit in the lower polepiece to allow the passage of the beam of some 30° semi-angle. This experiment demonstrated the feasibility of making a wide-angle projector lens. It also confirmed calculations that the corrector lens needs about twice the excitation required by the projector lens resulting in considerable field cancellation by the two single polepieces of opposite polarities. This problem was overcome[7]

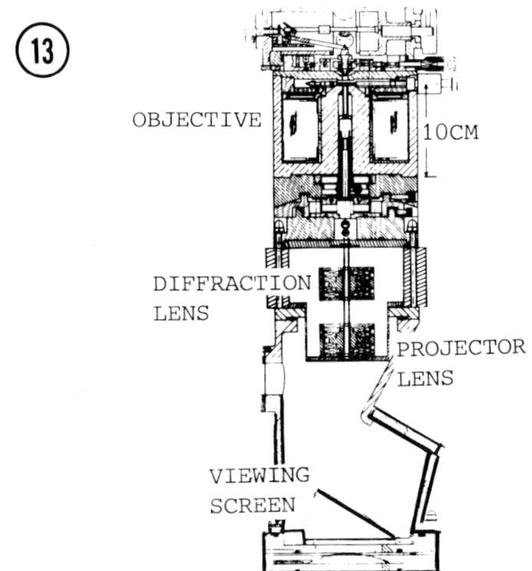

Fig. 13. Miniature rotation-free single-polepiece doublets as diffraction lens and final projector lens in a 100 kV electron microscope with vacuum liner tube fitted. Mulvey and Juma (1978) [9].

in the design shown in Fig. 16 in which the corrector lens is physically larger than the final projector lens; a magnetic screen was also introduced between the two polepieces. The magnetic screen must be kept as far as possible from the projector lens polepiece in order to maintain the favourable field distribution for minimum spiral and radial distortion. The corrector lens must produce approximately ten times more distortion than that of the final projector, but of opposite sign, assuming that a corrector lens magnification of approximately 3.3 is required. The remarkable improvement in distortion-free operation is shown in Fig. 17. On the right is shown the calculated distortion pattern of the final projector alone with a semi-angle of 30°. Considerable spiral distortion is noticeable. The inner circle shows the virtually distortion-free pattern that is obtained in such a lens at a semi-angle of 5° as with a conventional projector lens. The left hand image is an image of a rectangular grid taken in an experimental electron microscope fitted with a wide-angle projector lens operating with a semi-angle of 30°. The final adjustment of this lens had to be carried out by trial and error methods since the marginal rays differed significantly from those calculated from the paraxial ray equation. Similarly, the presence of higher order aberrations made the image differ markedly from the predictions of third order aberration theory. It was therefore decided in a subsequent investigation to use the methods of computer-aided design assisted by the general ray equation[1] so that the real electron trajectories could be plotted directly without the need for third or higher order aberration theory.

Guided by the experience gained with the correcting system shown in Fig. 16, the projector lens doublet shown in Fig. 18 was designed and constructed. It is of integral construction and is shown mounted, for testing, inside the viewing chamber of a JEOL electron microscope type JEM50 between the existing final projector and the fluorescent screen giving the possibility of forming a wide-angle image of 30° semi-angle on a transmission fluorescent screen. It is an integral construction machined from a solid block of soft iron. Each end face carries a single polepiece and is also machined from a solid piece of iron. The single polepiece of the corrector lens is separated from the intermediate magnetic screen by a non-ferromagnetic spacer, and is essentially a very asymmetric double-pole lens designed to produce some 100% of spiral distortion, permitting a magnification of some 3.3 times while still being able to correct 10% of spiral distortion in the final image. The right-hand end plate contains a carefully designed single polepiece of exceptionally low spiral distortion; the polepiece is shaped to permit a wide-angle beam to traverse the lens freely. The field distribution of this lens is essentially that of the spherical field[4] model which has the lowest known spiral and radial distortion coefficient of any lens. The calculated field distribution through this lens and the corresponding paraxial trajectory for a ray of height 1 mm are shown in Fig. 19. Such a ray would leave the projector at a semi-angle of 28° as shown. These

Fig. 14. Electron micrograph of molybdenum crystal with selected area diffraction pattern in the correct orientation by the use of rotation-free single-polepiece diffraction lens system.

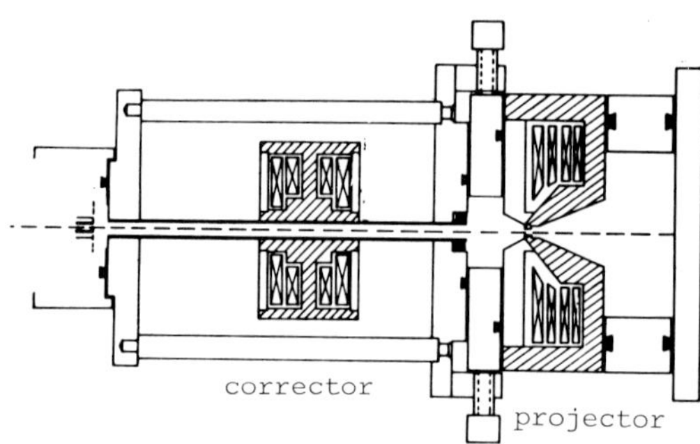

Fig. 15. Early experimental arrangement with two single-polepiece lenses for correcting spiral distortion in a wide-angle projector lens. Transmission fluorescent screen. Intermediate lens, mounted outside the vacuum, slides on vacuum liner for the cancellation of spiral distortion. Lambrakis et al (1977)[11].

trajectories show that, at least to a first approximation, the shaping of the final polepiece was just sufficient to allow passage of the electron beam. This is an important point because the presence of too large a bore in a single polepiece lens degrades the desired field distribution and increases the spiral distortion coefficient. Calculation of the spiral and radial distortion coefficients for this lens on the basis of third order aberrations are shown in Fig. 20. Here a normalised distortion coefficient of radial distortion (solid line) and the corresponding quantity for spiral distortion (dotted line) are plotted against the excitation parameter $NI/V_r^{\frac{1}{2}}$ of the corrector lens. This indicates that at an excitation parameter $NI/V_r^{\frac{1}{2}}$ of 18 the radial and spiral distortion vanish simultaneously.

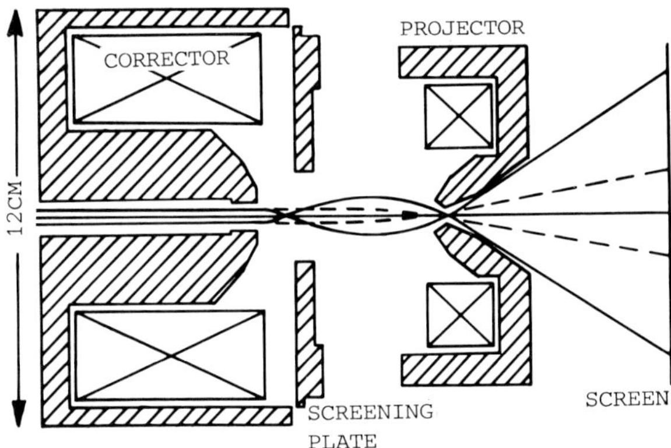

Fig. 16. Improved design of wide-angle projector lens. Semi-angle α_p = 30°. Full lines: paraxial ray calculation for parallel rays entering the corrector lens. Dashed lines: trajectories of same paraxial rays entering the system with the corrector lens de-energized. Note adjustable magnetic screen plate for avoiding field cancellation effects[7].

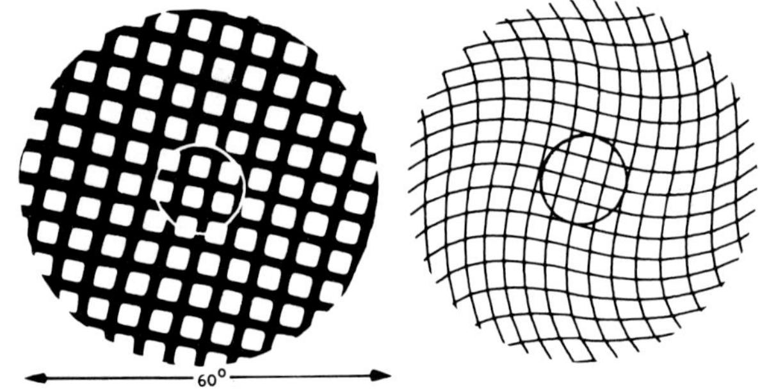

Fig. 17. Left. Experimentally obtained distortion-free image of a rectangular grid in a 100 kV electron microscope. Total angular field 60°. Inner circle indicates distortion-free field of view of a conventional projector lens. Right. Calculated distortion pattern of the projector lens acting alone over a total field of 60°. Micrograph by H El-Kamali.

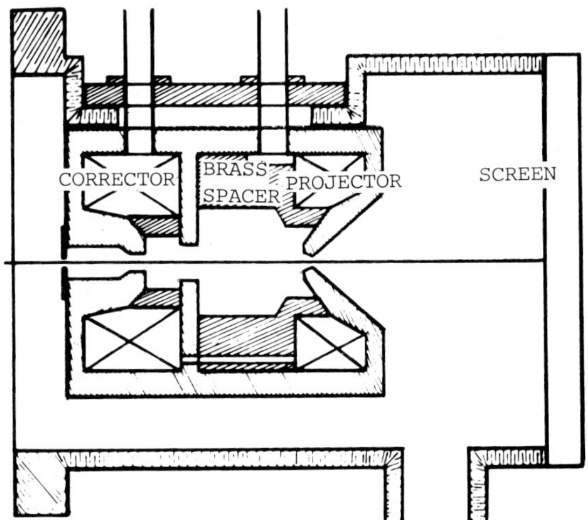

Fig. 18. Wide-angle projector lens unit[1] of integral construction mounted inside the viewing chamber of a JEOL JEM50 electron microscope. Semi-projection angle α_p = 30°.

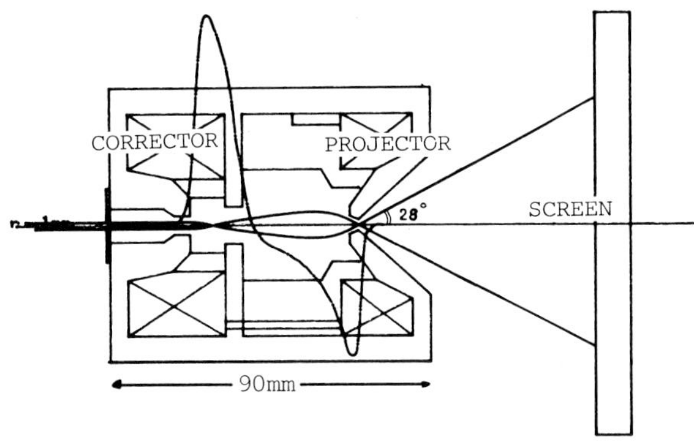

Fig. 19. Calculated axial field distribution and trajectories, calculated by the paraxial ray equation for an incoming ray of height 1 mm. Exit angle α_p = 28°.

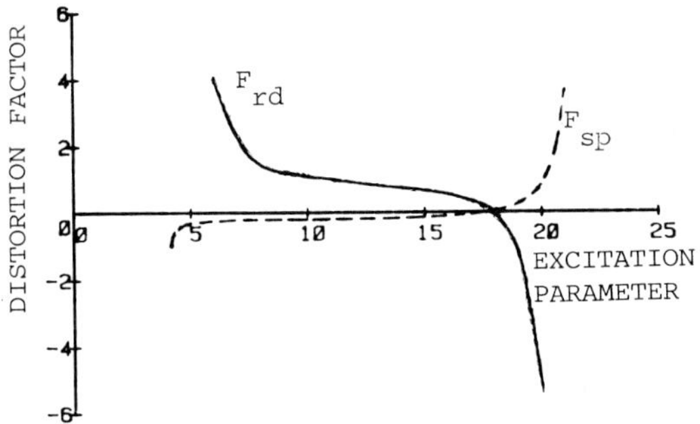

Fig. 20. Calculated radial distortion factor (solid line) and spiral distortion factor (dashed line) for the integral wide-angle projector unit as a function of the excitation parameter $NI/V_r^{\frac{1}{2}}$ of the corrector lens.

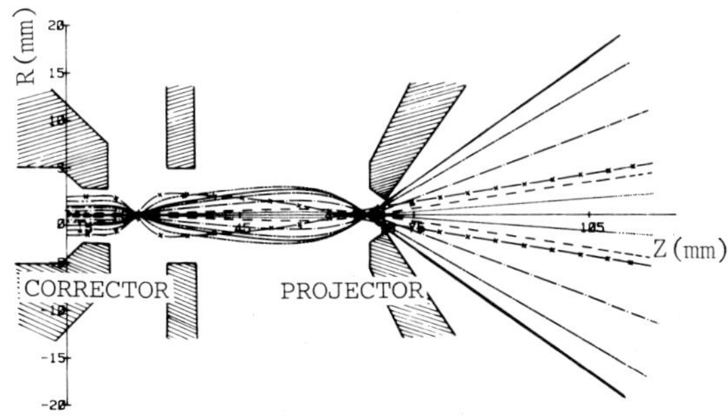

Fig. 21. Actual electron trajectories through the wide-angle projector system as calculated from the general ray equation. Trajectories indicate distortion-free operation up to a semi-angle α_p = 30°.

Moreover, the correction of the spiral distortion is not critical and remains at a fairly low value right up to the point of correction. This is a very useful property of this design since the correction point can be readily found experimentally by concentrating on the radial distortion in the image. However, this is a case in electron optical instrumentation in which the third order aberration theory can only be regarded as a rough guide. This was borne out by the experimental behaviour of the lens, which was broadly in line with the calculated values but there were important differences especially concerning the behaviour of the marginal rays. These took the form of an unwanted and highly distorted image inconveniently superimposed on an otherwise perfectly corrected image on the final screen. An explanation for this phenomenon was found when the real trajectories were plotted through the system from the general ray equation as shown in Fig. 21. Parallel rays entering the corrector lens are brought to a focus at the centre of the corrector lens and enter the field of the projector lens as a nearly parallel bundle of rays, as indicated by the (dotted) paraxial rays, forming an image at the centre of the fluorescent screen. This indicates that the corrector lens in this mode is forming a virtual image located to the left of the corrector lens. This means that the projector lens is effectively accepting a beam of approximately parallel incident electrons thereby reducing its own coefficients to a minimum. The exit angle of the ray is proportional to the radial height in the corrector lens up to the maximum semi-angle of the exit cone, as indicated by the solid line which just touches the inner edge of the polepiece of the projector lens. However, the bore of the corrector lens as designed will admit rays of even larger radius. For such rays, however, such as the one marked with a cross the aberrations of the corrector lens suddenly become excessively large and deliver a converging beam which strikes the principal plane of the projector lens and so is hardly refracted. This is the cause of the unwanted image originally seen at the centre of the fluorescent screen. The cure is simply to restrict the extreme marginal rays by an aperture of some 2 mm in diameter placed in the bore of the corrector lens. It can also be seen from Fig. 21 that the shaping of the projector polepiece in terms of paraxial rays has not been fully optimised for the real rays and minor changes in its shape could produce some further small improvements. The effectiveness of this corrector unit was in every way comparable with that of the previous experimental corrector unit shown in Figure 16, and images of the same quality as that of Figure 17 were obtained but without the need for any mechanical adjustment of the lens system. It also confirmed the view that an exit semi-angle $\alpha_p = 30°$ is probably the upper limit for a corrector device of this type. If such a lens were used in a conventional electron microscope with the normal viewing distance of some 450 mm, distortion-free operation of this type would be possible on a screen roughly half a metre in diameter. This investigation has shown that the use of the general ray equation can be very useful in the design of real electron optical systems since it can often explain the apparently unusual behaviour of the electron optical system compared with the design expectations based on paraxial ray theory.

Scanning Transmission Electron Microscopes with Advanced Electron-Optical Systems

The STEM was invented in 1938 by von Ardenne but lay in abeyance until the late 1960s when Crewe and his colleagues introduced an experimental STEM with a field emission electron gun. In its original form, Crewe's system was very simple consisting of a field emission gun, a condenser lens and a final probe forming lens. Interestingly Crewe chose the Riecke/Ruska condenser-objective lens as a final probe forming lens. The first part of this lens (the condenser part) was used in conjunction with the preceding condenser lens to focus the incoming beam on the specimen; the second (objective) part was used to converge the scattered beam from the specimen conveniently into the electron detector. Such an instrument is particularly well suited to high resolution dark field microscopy and is capable of the same resolution as a TEM with an objective lens of the same spherical aberration coefficient at the same accelerating voltage. The output from a STEM is automatically in a form that is suitable for direct interfacing to a computer for subsequent image processing. It is also possible to allow the inelastically scattered electrons to pass into an electron velocity spectrometer whose output can also be displayed as an image on the display tube. So far, the accelerating voltage of STEM instruments has been restricted to 50-100 kV and so it has not yet been possible to compare STEM and TEM at very high resolution.

Analytical STEMS

In the meantime attention has been turning more to improving the STEM as a micro-analytical tool for the quantitative examination of micro-regions in thin specimens. Operators of analytical TEMs are accustomed to being able to obtain, in addition to the image, both conventional and convergent beam diffraction patterns from selected micro-regions. One might also wish to obtain characteristic x-ray spectra by means of an energy dispersive detector or an electron energy loss spectrometer. In the latter case it is desirable to match the angular spread of the electrons leaving the specimen to that of the electron beam that can enter the spectrometer. This can only be done by adding a number of post-specimen projector lenses. At the same time it is advantageous to have a means of converting the scanned electron beam leaving the specimen into a static beam falling on to the various fixed detectors which can then include a fluorescent screen for recording diffraction patterns. The latter is almost essential since the normal serial method of acquiring a diffraction pattern in a STEM is extremely time-consuming. Fig. 22 shows an experimental analytical STEM of this type[6] designed by Professor Ferrier and his team at Glasgow University. A standard Vacuum Generator's field emission gun STEM forms the basis of the instrument. The lower, probe-forming, part of the column consists of the field emission gun, two condenser lenses and a Riecke/Ruska lens as a final probe-forming lens. Two condenser lenses are used to enable greater freedom in operating

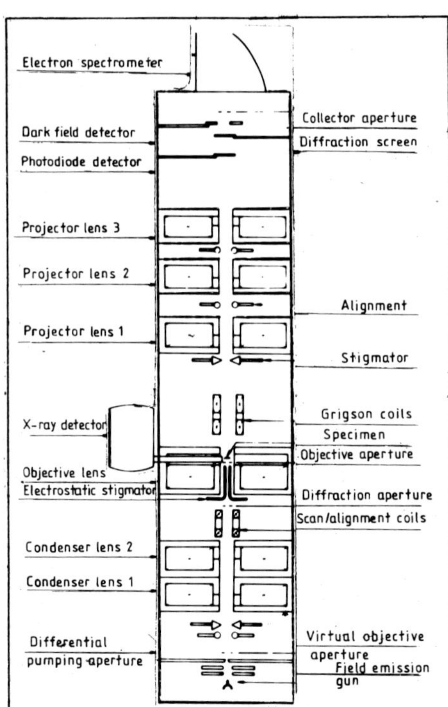

Fig. 22. Schematic arrangement of an analytical STEM[6] at Glasgow University. Note the projector lens system for providing a static diffraction pattern and an interface between the specimen and the energy loss spectrometer.

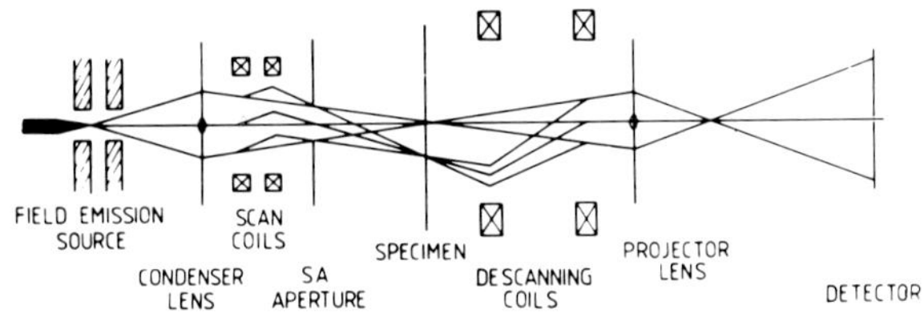

Fig. 23. Schematic arrangement of the complete ray path in the analytical STEM[6] showing the action of the scanning coils and the "de-scanning" coils for producing a static image.

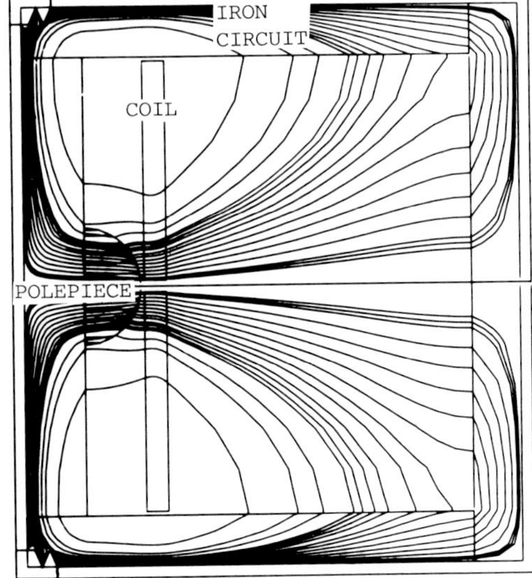

Fig. 24. Flux distribution in a single-polepiece lens with a spherical pole-tip energized by an optimised coil design[3]. Peak flux density on the axis 3.4 Tesla. Outside diameter 72 mm. Excitation 30 k A-t.

the latter lens. The x-ray detector is an energy dispersive (EDX) system that allows a characteristic x-ray spectrum to be obtained from a chosen point on the specimen. Above the specimen are the normal diffraction scan coils which can also be used to "de-scan" the electron beam leaving the specimen as indicated in Fig. 23 which shows schematically the complete ray path of the beam from source to detector. The "de-scan" removes the scanning motion of the electron beam leaving the sample so that a steady image of a diffraction pattern may be viewed on a fluorescent screen or recorded on a photo-diode detector. The three post-specimen projector lenses between the specimen and the fluorescent screen enable the magnification of the various images to be adjusted. Similarly the angular aperture of the beam entering the spectrometer can be optimised to that leaving the specimen. All the problems previously mentioned in connection with wide-angle projectors are relevant here. In addition there is the extra requirement that the lens units must be fully compatible with high vacuum procedures such as "bakeout". To control an instrument of this type manually would be extremely time-consuming and so computer control has become essential both in adjusting the instrument and in data handling.

Future Developments in Magnetic Electron Lenses and Lens Systems

The future development of high resolution magnetic electron lenses must lie in the greater attention to detail in the design of the exciting coil in order to achieve higher flux densities. In a conventional lens the exciting coil makes a negligible contribution to the axial field distribution, nearly all of which is produced by the magnetisation of the iron polepieces. As the lens excitation is increased these polepieces and often other parts of the magnetic circuit begin to saturate. Further increase of lens excitation leads to a broadening of the field distribution, and an effective limitation to the maximum flux density that can be achieved. Many of these effects can be reduced by the optimum placing of the coil[2]. Figure 24, for example, shows a single polepiece lens with a spherical tip in which a thin coil of high ratio of outer to inner diameter is placed in close proximity to the tip[3]. The resulting field distributions are shown in Figure 25. It can be seen that even at high peak axial flux densities approaching 4 Tesla the field broadening is remarkably small. The reason for this is that in this particular design the saturation magnetization of the iron is strongly localised at the tip. Hence in the vicinity of the polepiece the saturation flux density is simply added to the field produced by the coil. In this type of lens, therefore, there is no limit to the maximum flux density that can be produced except that set by the maximum permissible current density in the exciting coil. With super-conducting windings, for example, this permissible current density is of the order of 10^{10} A/m^2. These lenses, therefore, are not limited so much by the properties of the iron but largely by the technology of super-conducting windings. Similar principles can be applied to the double polepiece lens of the condenser-objective type as shown in Fig. 26, which shows the flux distribution in a twin-polepiece lens with a central coil of high ratio of outside to inside diameter. Here again high fluxes can be produced at the specimen position in the centre of the lens as shown in Fig. 27 which also shows the magnetization component of the axial flux density distribution created by the iron. Fig. 28 shows the axial flux density distribution in this lens for a vanishingly small polepiece bore. These results suggest that an increase in maximum flux density up to 4 Tesla is feasible for high resolution objective lenses. However, it should be mentioned that, for a given accelerating voltage, the excitation of such a lens is a fixed quantity. Thus the only way to achieve a higher flux density in an objective lens of optimised shape is to reduce its size. This is largely a question of superconductor technology. For complete electron-optical columns, intermediate lenses can conveniently be rotation-free lenses of miniature construction and modest flux density. These can often be conveniently accommodated within the internal bores of conventional lenses as described for example by Podbrdsky[16]. The alignment of such lenses and the setting of the excitation can readily be controlled by a mini-computer. Such systems will provide and record a vast amount of quantitative data from the specimen and will be physically more compact than present designs. In addition they will lend themselves to automatic or semi-automatic operation under computer-control.

Acknowledgements

The author would like to place on record the considerable contribution of a long line of postgraduate researchers who have contributed ideas, experiments and theory to the investigations reported in this paper and whose original work is cited in the text and in the references. These include Drs Christopher Newman, Richard Bassett, Fathi Marai, Emmanuel Lambrakis, Sabah Juma, Peter Harris, Stelios Christofides, Adil Al Shwaikh, Hisham El-Kamali, Hamid Nasr, Shatha Al-Hilly, Muna Al Khashab and In'am Al-Nakeshli. He would also like to thank Professor Armin Delong, Dr Josef Podbrdský, Dr Jiři Komrska, Dr Bohumila Lencová and Dr Michael Lenc of the Institute of Scientific Instruments at Brno, Czechoslovakia for stimulating discussions on the practical realisation of completely new forms of electron optical columns. He is also indebted to Dr Eric Munro, of Imperial College, London, and Mr C W Trowbridge of the Rutherford-Appleton Laboratory, Didcot, UK, for valuable discussions on the mathematical basis of the finite element method. He would also like to thank Mr Wenxiong Cheng of Beijing University for his contribution to computing methods in electron optics during his sabbatical year at Aston University. This list would not be complete without mentioning the skilful workmanship and technical know-how of Messrs Howard Arrowsmith, Ken Bates, Roland Keen and Frank Lane of the Physics Workshop in the practical realisation of new forms of magnetic electron lenses.

References

1. Al-Hilly SM, Mulvey T. (1981). Wide-angle projector systems for the TEM. Inst. Phys. Conf. Series No. 61, M J Goringe (ed), 103-106.

2. Al Khashab M. (1983). The electron optical limits of performance of single polepiece magnetic electron lenses. PhD dissertation, University of Aston in Birmingham.

3. Al-Nakeshli IS, Juma SM, Mulvey T. (1983). High flux density single polepiece electron lenses. Inst. Phys. Conf. Series No. 68, 475-478.

4. Alshwaikh A, Mulvey T. (1977). The magnetised iron sphere, a realistic theoretical model for single-polepice lenses. Inst. Phys. Conf. Series No. 36, 25-28.

5. Busch H. (1927). On the mode of action of the concentrating coil in the Braun tube. Arch. Elektrotech. 18, 583-594.

6. Chapman JN, Morrison GR, Ferrier RP. (1980). Proc. 7th Eur. Conf. Elec. Micros. The Hague 1, 90-91. 7th Eur. Conf. Foundation. Leiden, 1980.

7. El Kamali H, Mulvey T. (1980). A wide-angle TEM projection system. Proc. 7th Eur. Conf. Elec. Micros. The Hague 1, 74-75. 7th Eur. Conf. Foundation. Leiden, 1980.

8. Gabor D. (1926). Doctoral Dissertation, Technological University of Berlin. Oscillographic recording of travelling waves with the cathode ray oscillograph. In German.

9. Juma SM, Mulvey T. (1978). Rotation-free magnetic electron lenses. J. Phys. E. 11, 759-764.

10. Knoll M, Ruska E. (1931). Contribution to geometrical electron optics. Ann. Phys. 12, 607-640.

11. Lambrakis E, Marai FZ, Mulvey T. (1977). Correction of spiral distortion in the TEM. Inst. Phys. Conf. Series No. 36, 35-38.

12. Lencová B, Lenc M. (1984). The computation of open electron lenses by coupled finite element and boundary integral methods. Optik (in press).

13. Mulvey T, Nasr H. (1981). An improved finite element method for calculating the magnetic field distribution in magnetic electron lenses and electromagnets. Nucl. Instr. Meth. 187, 201-208.

14. Mulvey T. (1982). Chapter 5 in Magnetic Electron Lenses, ed. PW Hawkes, Springer-Heidelberg, 351-412.

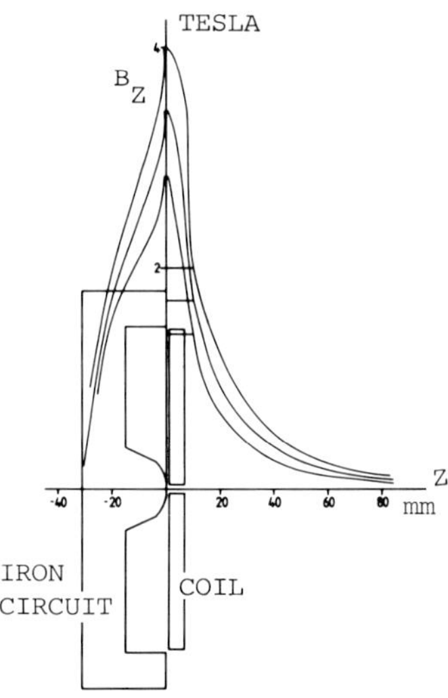

Fig. 25. Axial flux density distributions as a function of excitation for optimised single-polepiece lens[3], for excitations of 40, 60, and 80 kA-t. Peak axial flux 4 Tesla with relatively small broadening.

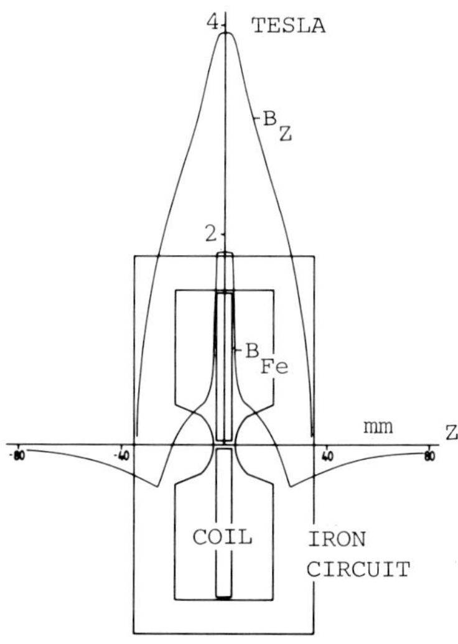

Fig. 27. Axial flux density distribution B_Z in the air gap and the iron circuit of twin polepieces without a bore[3]. B_{Fe} is the axial flux density distribution contributed by the iron for a polepiece having a vanishingly small bore[3].

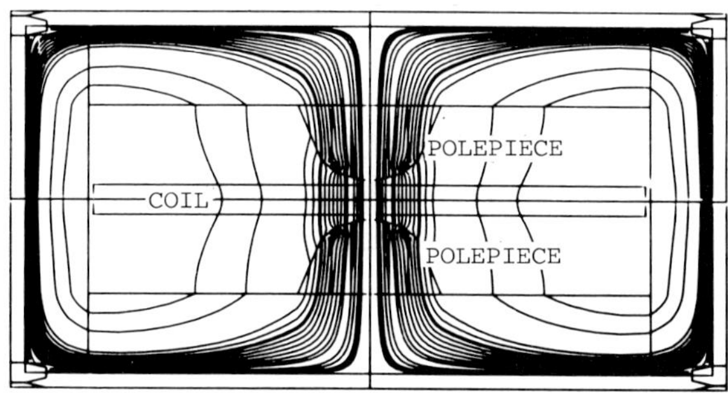

Fig. 26. Flux distribution in twin-polepiece lens with optimised exciting coil[3]. Peak axial flux density 4 Tesla.

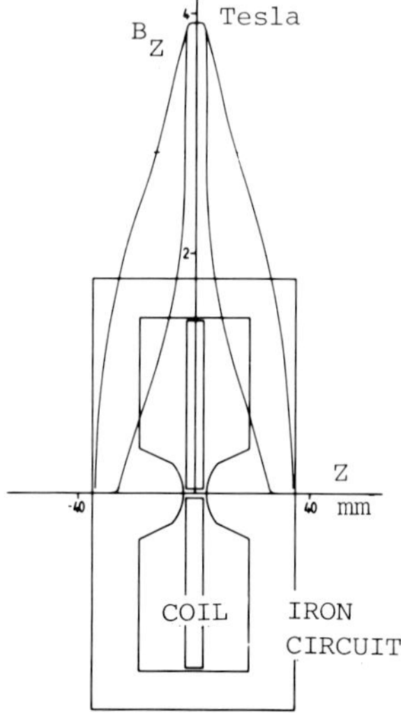

Fig. 28. Axial flux density distribution in a double polepiece lens[3] of vanishingly small bore. Lens excitation 60 kA-t. Soft iron polepieces.

15. Munro E. (1981). PhD Dissertation. Computer design methods in electron optics, University of Cambridge, UK.

16. Podbrdský J. (1980). The optical system of the analytical electron microscope with mini-lenses. Proc. 7th Eur. Conf. Elec. Micros. The Hague 1, 66-67. 7th Eur. Conf. Foundation. Leiden, 1980.

17. Riecke WD, Ruska E. (1966). A 100 kV TEM with single field condenser-objective lens. Proc. Int. Conf. Kyoto 1, 19-20. Maruzen. Tokyo, 1966.

18. Smith KCA, Swann DJ. (1969). UK Patent 1291221.

19. Trowbridge CW. (1976). Applications of integral equation methods for the numerical solution of magnetostatic and eddy current problems. Int. Conf. Elec. and Mag. field problems Santa Margherita, Italy, 1-22. (Available from Rutherford Laboratory, Didcot, OX11 0QX (UK).)

20. Troyon M, Laberrigue A. (1977). A field emission gun with high beam current stability. J. Micros. Spectrosc. Electron. 2, 7-11.

21. Venables JA, Archer GD. (1980). On the design of FEG-SEM columns. Proc. 7th Eur. Conf. Elec. Micros. The Hague 1, 54-55. 7th Eur. Conf. Foundation. Leiden, 1980.

22. Von Borries B, Ruska E. (1932). German Patent No. 680284. Magnetic converging lens of short focal length.

Electron Optical Systems (pp. 29-35)
SEM Inc., AMF O'Hare (Chicago), IL 60666-0507, U.S.A.

0-931288-34-7/84$1.00+.05

VARIOUS ATTEMPTS TO REDUCE THE DIMENSIONS OF MAGNETIC ELECTRON LENSES USED FOR HIGH VOLTAGE MICROSCOPES

J.L. Balladore*, R. Murillo, J. Trinquier

Laboratoire d'Optique Electronique du C.N.R.S.
B.P. 4347 - 31055 Toulouse cedex, France
Laboratoire conventionné à l'Université Paul Sabatier

Abstract

Magnetic fields are used in electron microscopy for a variety of purposes : image formation, energy analysis and correction of astigmatism, for example.

We are interested in the problems of designing short focal length lenses to be used for focusing high energy electrons in the energy range 1-3 MeV.

These lenses are generally very large and we have tried to reduce their dimensions to simplify their construction and use. From this point of view, it is necessary to diminish the magnetic coil and the magnetic circuit iron.

Several solutions have been proposed in the case of the coil. We have obtained good results with superconducting coils.

The reduction of the magnetic circuit is more difficult when we try to use a smaller volume of iron ; we find that for high values of magnetizing current iron saturation appears. And we can observe a deterioration of the electron optical characteristics.

These problems can be solved by using a special magnetic circuit composed of elements of anisotropic magnetic material. These new types of lenses will be invaluable if we need to focus electrons with energies greater than 3 MeV and could lead to considerable simplification of the mechanical design of all high voltage instruments.

KEY WORDS : Magnetic lens, Electron optical characteristics, Superconducting coil, Anisotropic magnetic circuit.

*Address for correspondence:
For reprints and other information please contact J.L. Balladore at the above address. Phone no.: (61) 52 65 96.

Introduction

In very high tension electron microscopy, lenses of short focal length are needed to focus electrons with energies between 1 and 3 MeV. The chromatic and spherical aberration coefficients of these lenses should be small in order to obtain good images. Generally such lenses are very large and create difficulties in construction and use. We present here various techniques developed in our laboratory for reducing the dimensions of these lenses.

Conventional lenses

In order to focus electrons with energies in the MeV range, the maximum field on the lens axis must lie between 2 and 3 T. Such lenses have been realized in our laboratory and give excellent results (Dupouy et al. 1970, Jouffrey et al. 1979). Nevertheless their size increases rapidly as the accelerating voltage is raised. As an example, Fig. 1 shows a meridional section through the objective lenses of the 1 MeV and 3 MeV Toulouse microscopes.

In order to decrease the total bulk we might reduce the dimensions of the gap and consequently the dimensions of the magnetizing coil. If the magnetic field is concentrated in a more reduced space, for a given number of Ampere turns NI, the maximum value B_M of B becomes higher as the dimensions of the gap are made smaller. The increase of B_M obtained by reducing the gap is limited for high values of NI by iron saturation.

In Fig. 2 we present curves giving B_M at the center of the gap as a function of NI for various values of the bore diameter D of the pole pieces and of the gap width S.

We observe,for a given NI, that as S and D decrease B_M increases. This result allows us to obtain smaller focal lengths f and chromatic C_c and spherical C_s aberration coefficients, as shown in Figs. 3, 4 and 5. These values are obtained for an accelerating voltage of 1 MV.

For example, for a focal length of 10 mm, 9600 At are necessary if S = D = 8 mm and 20,000 At if S = D = 20 mm. Consequently if S and D are small we can diminish the size of the magnetizing coil.

However, modern electron microscopes are expected to provide information of many kinds about

List of symbols

B	magnetic field on the lens axis
B_M	maximum value of B
NI	number of Ampere turns
D	bore diameter of pole pieces
s	gap width
f	focal length
C_c	chromatic aberration coefficient
C_s	spherical aberration coefficient
μ_t	magnetic permeability tangent to a given curve
At	Ampere turns

the object. It is therefore necessary to accommodate specimen stages, allowing us to tilt the specimen, to modify its temperature, or to exert a mechanical constraint on it. Such devices are relatively cumbersome and since they are placed at the center of the gap the volume available between the pole pieces cannot be much reduced. In very high tension electron microscopes, the problem is complicated by the large diameter of the lenses : the specimen lies far from the outside of the microscope. The specimen stages must be relatively large in order to avoid mechanical vibrations, which occur when the object is placed at the tip of a long specimen holder.

The dimensions of the objective lens are thus severely conditioned by the nature of the experiments that we want to carry out. Under these conditions, in order to reduce the lens size it is necessary to make the magnetizing coil smaller while maintaining the number of Ampère turns needed to obtain short focal lengths with different gap volumes.

Reduction of the size of the coil

Several solutions have been proposed whereby the intensity in the current coil is increased in order to reduce the number of windings.

We mention the works of Mulvey and colleagues (Mulvey and Newman 1972) who use very efficient water cooling and have achieved an appreciable reduction of the lens dimensions.

However, the most spectacular results have been obtained with superconducting coils : Fernandez-Moran 1965, Laberrigue et al. 1976, Dietrich 1978 and Balladore 1972 have shown the advantages of using superconducting coils. For example in order to obtain 20,000 At the superconducting coil cross-section is equal to 2 cm^2. The coil cross-section on Fig. 1 a is approximately 10 times greater. Although this coil is much smaller than a conventional one the superconducting wires must of course be held at liquid helium temperature : a cryostat is therefore required.

Such methods enable us to reduce the size of the magnetizing coil, and hence the dimensions of the magnetic circuit surrounding the coil.

It is however possible to go much further in the reduction of the iron circuit.

Reduction of the magnetic circuit

The reduction of the magnetic circuit is more difficult : when we try to use a smaller volume of iron we find that for high values of the magnetizing current, iron saturation appears : the curve representing the magnetic field variations spreads along the lens axis z'z as shown in Fig.6.

These curves correspond to the lens presented in Fig. 1b. This broadening leads to a deterioration of the electron optical characteristics. On Fig. 7 can be seen the variations of the focal length as a function of NI.

We observe that as NI increases, f begins to decrease, passes through a minimum and then increases. This phenomenom is mainly due to the iron saturation. The other optical characteristics, the chromatic and spherical aberration coefficients, vary similarly. When NI is increased, therefore, magnetic saturation of the circuit appears which in turn leads to deterioration of the optical characteristics of the lens (Liebmann and Grad 1951, Murillo and Balladore 1974). In order to avoid such a deterioration, it is very interesting to know the state of saturation of the different parts of the circuit. In order to establish this, we utilize a computer method of Munro (1973) which yields the value of the magnetic field at any point of a lens with symmetry of revolution. We present in Fig. 8a the results obtained for the magnetic circuit of the superconducting lens. This figure represents one quarter of the circuit. In Fig. 8b we have drawn the flux lines.

The magnetization is equal to 31,500 A turns. For such a value the iron begins to be saturated. We have plotted the values of the magnetic field in Tesla. As we can see, the metal is not uniformly permeated by the magnetic field. Although the field distribution is practically uniform near the polepieces, it is by no means uniform farther inside the circuit. The magnetic field is concentrated in some zones along the line D, for example, the field drops from 1.63 to 0.83 T. The flux lines take the shortest path and almost all avoid part of the magnetic circuit. This basic result has been confirmed with other lenses.

In order to avoid this it is necessary to constrain the magnetic field to occupy all the iron cross-section uniformly. The aim is not to minimize saturation effects as such but to minimize the size and weight of the lens under the constraint that the magnetic flux density nowhere exceeds some given tolerable limit.

For this, we have proposed channelling the field lines by using the phenomenon of magnetic anisotropy (Balladore and Murillo 1977 ; Balladore et al. 1981). For example, let us consider a magnetic material having a magnetic permeability μ_t in the direction tangent at all points to the curve Γ, that is very much greater than the permeability in the direction δ_n perpendicular to Γ(Fig.9).

If the flux lines enter the material along μ_t their direction cannot in practice alter : they follow the preferential direction Γ. If the field is uniform along Δ_o, the same will be true along Δ_i by virtue of the anisotropy.

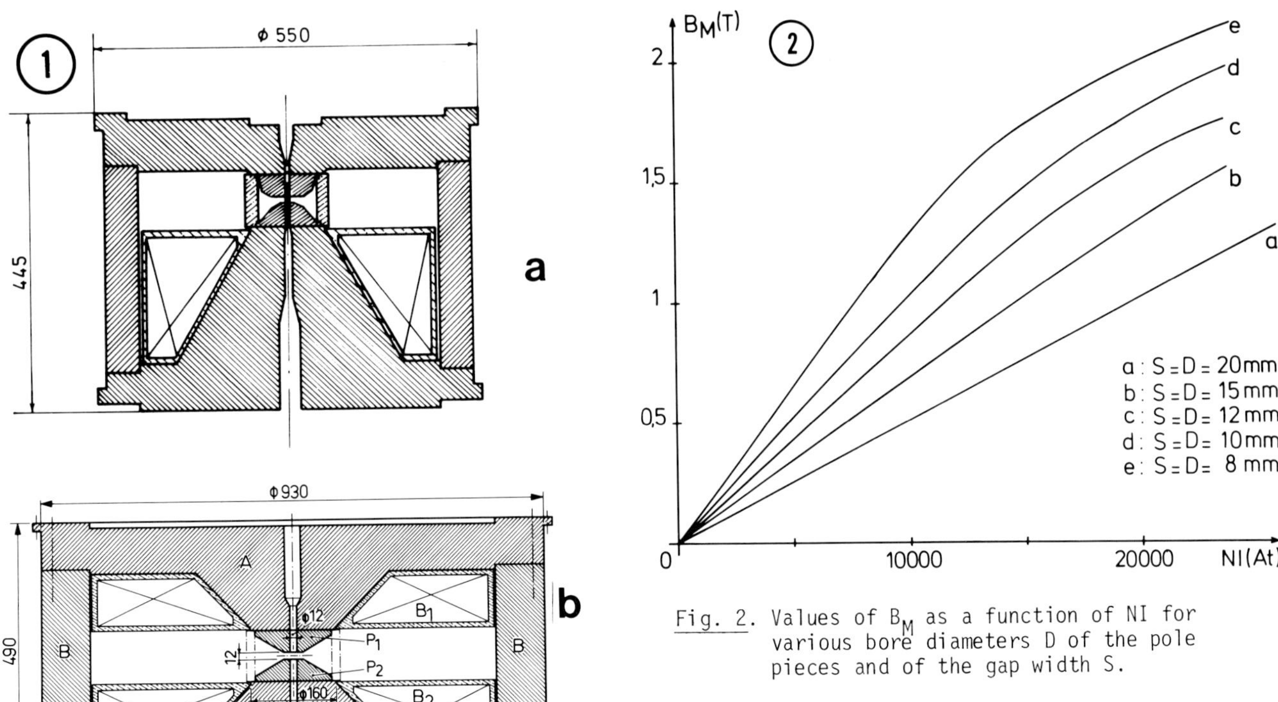

Fig. 2. Values of B_M as a function of NI for various bore diameters D of the pole pieces and of the gap width S.

Fig. 1. a) 1 MeV objective lens
b) 3 MeV objective lens.

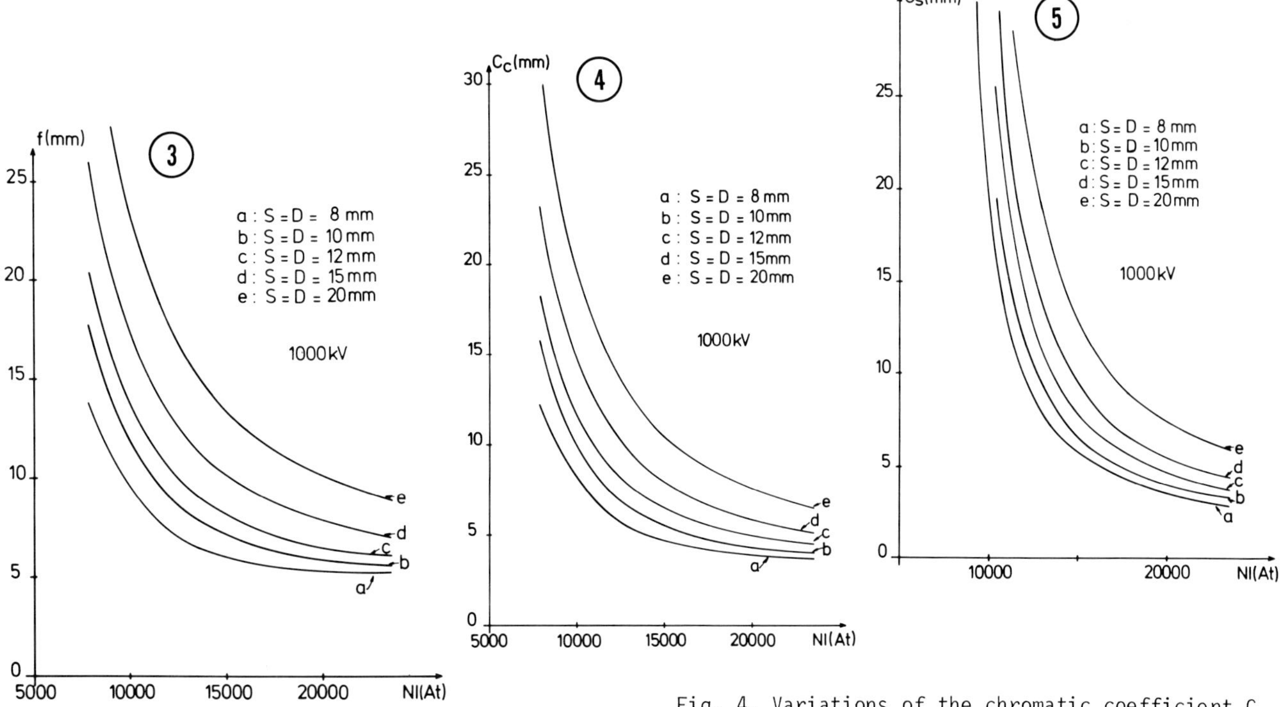

Fig. 4. Variations of the chromatic coefficient C_c.

Fig. 3. Variations of the focal length f.

Fig. 5. Variations of the spherical coefficient C_s.

NI (A.t)	A B_M(T)	f(mm)	C_c(mm)	B B_M(T)	f(mm)	C_c(mm)
7518	1.109	11.9	10.6	1.120	12	11
9941	1.451	7.6	6.4	1.458	7.5	6.3
12516	1.751	5.8	4.6	1.825	5.5	4.4
15035	1.907	5.3	4.1	2.080	4.8	3.7
17946	2.014	5.1	3.8	2.260	4.4	3.3
20034	2.099	5.2	3.6	2.390	4.3	3.2

Table 1. Optical characteristics of an experimental lens (A) and a conventional lens (B).

In order to realize such a circuit, we have simulated a magnetic anisotropic domain by making two rotationally symmetric cuts in the circuit of Fig. 8. The extremities of these cuts are in regions where the field is as uniform as possible (Fig. 10).

We have in fact two air gaps. The magnetic resistance due to these air gaps in a plane perpendicular to Γ created anisotropy : the flux lines confined to one of the metallic sections are prevented by the air gap from shifting to neighbouring sections. We note that along the line D, the field gradient is practically equal to zero. The magnetization is equal to 31,500 A turns. In the upper region of the circuit of Fig. 8, the field varies from 0.82 to 0.20 T when we go from point A to B. We see on Fig. 10 that this variation is now from 1.69 to 0.79 T between the same points : the upper part of the circuit now contributes to the flux conduction as well as the lower region. Although the volume has fallen by 30%, the computation of the maximum field on the lens axis shows that it is practically unchanged (1.947 T instead of 1.968 T).

New type of magnetic lens

A particularly small lens can therefore be constructed if discrete elements of anisotropic material having a constant cross sectional area are placed around the axis z'z. Under these conditions we have shown (Balladore et al. 1981) that the entire circuit outside the polepieces will be in the same magnetic state and there will be no region that is incompletely used by the flux lines. All the regions which do not contribute to flux conduction are excluded. Consequently this new circuit uses the minimum metal for a given maximum induction.

In Fig. 11, we present a lens designed according to these principles. The circuit is composed of two parts : two polepieces in "Hyperm 0" iron identical with those used in high voltage microscopy ; around these polepieces we have arranged eight anisotropic multilayered elements ; the number and geometry of these can be varied according to the polepiece shape and maximum field. We utilize sheets of an iron-silicon alloy with oriented grains. The magnetic permeability of this material is very much higher in the direction parallel to the sheet than in the perpendicular directions. We attempt to set these circuit end faces in a region where the polepieces are not saturated and where the field is as homogeneous as possible. The field lines enter along the direction of highest permeability and cannot shift to the next foil.

In this lens the magnetic coil consists of 1900 turns of copper wire. A very efficient water cooling is used. We can obtain 23,000 At. at a water temperature equal to 70°C.

We have determined the optical characteristics of this lens for an electron accelerating voltage of 1 MV. The values obtained are plotted in table 1. Part A represents the experimental lens (Fig. 11) and part B the normal objective lens (Fig. 1a). The values are of the same order of magnitude.

In conclusion, we can say that this magnetic circuit is very interesting for electron lenses because it enables us to reduce the dimensions of the magnetic circuit without saturation setting in. At the present time the problem that remains unsolved is the determination of the magnetic field at any point of an anisotropic circuit. The computer method used for isotropic material must be modified to take into account the anistropy. This is a far from easy problem on which we are currently working.

Conclusion

In high voltage electron microscopes, the conventional lenses that are employed are very large. The problem of reducing their volume can be solved by using superconducting or water-cooled coils and an anisotropic magnetic circuit. These new types of lenses will be invaluable if we need to focus electrons with energies greater than 3 MeV and could lead to considerable simplification of the mechanical design of all high-voltage instruments, at the cost of some loss in radiation shielding. A reduction in the total weight of the lens by a factor of the order of three should be attainable.

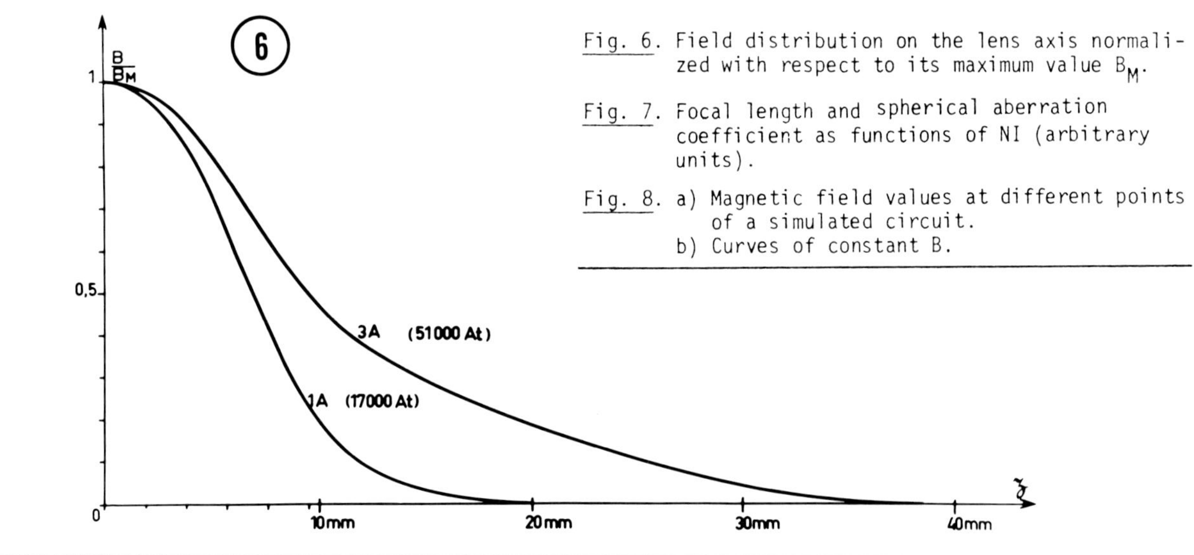

Fig. 6. Field distribution on the lens axis normalized with respect to its maximum value B_M.

Fig. 7. Focal length and spherical aberration coefficient as functions of NI (arbitrary units).

Fig. 8. a) Magnetic field values at different points of a simulated circuit.
b) Curves of constant B.

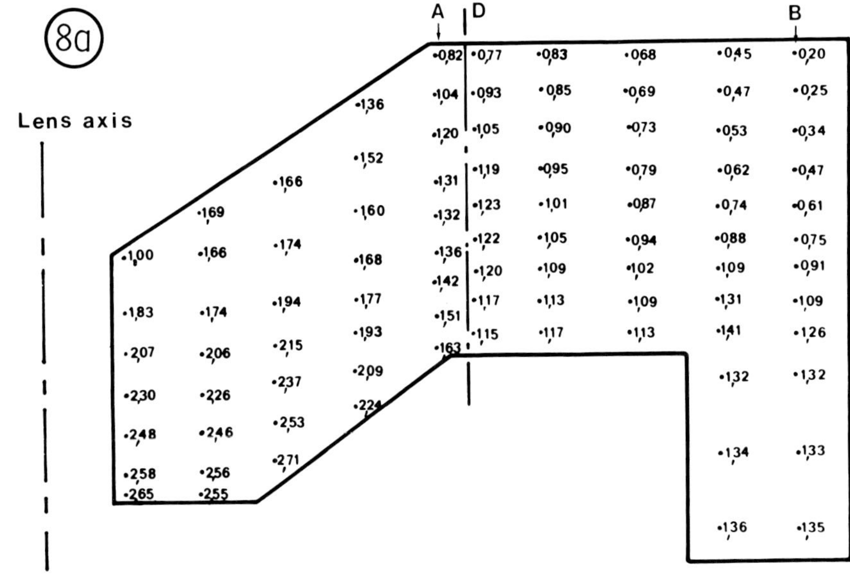

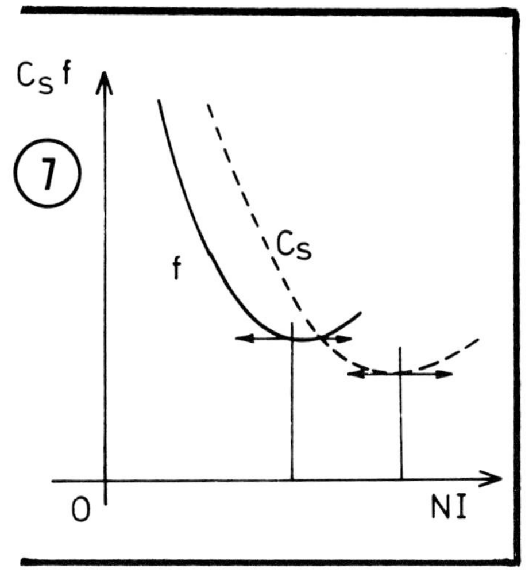

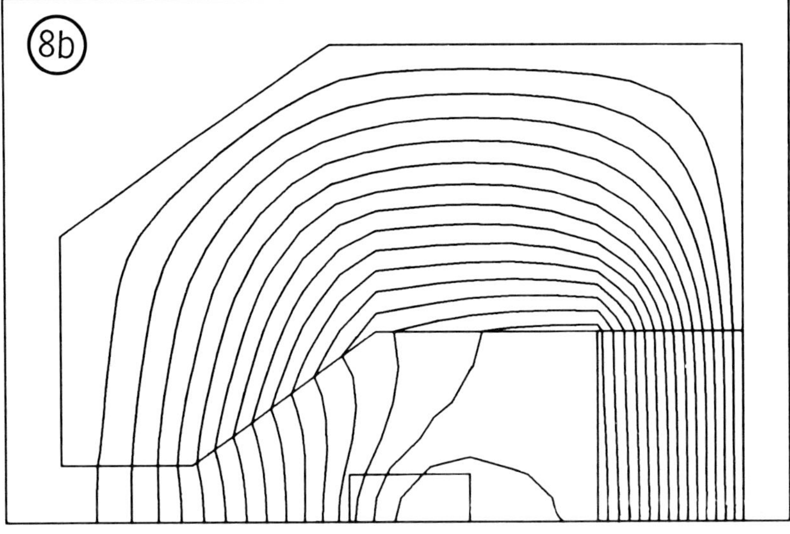

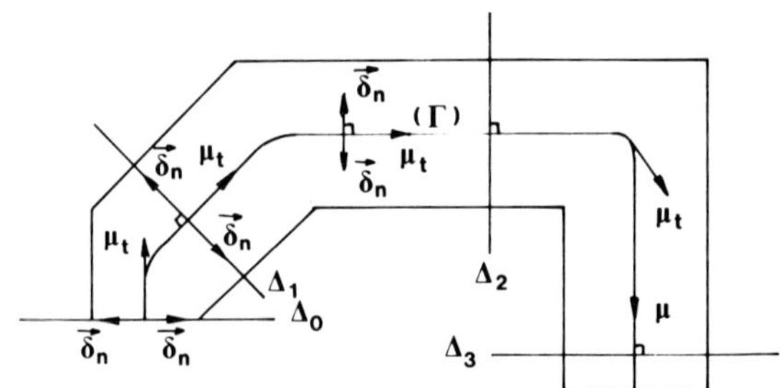

Fig. 9. Anisotropic circuit.

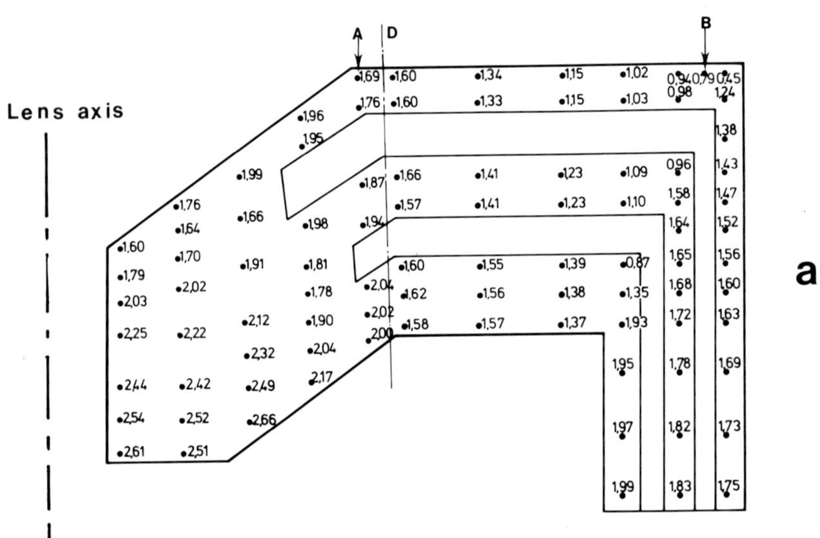

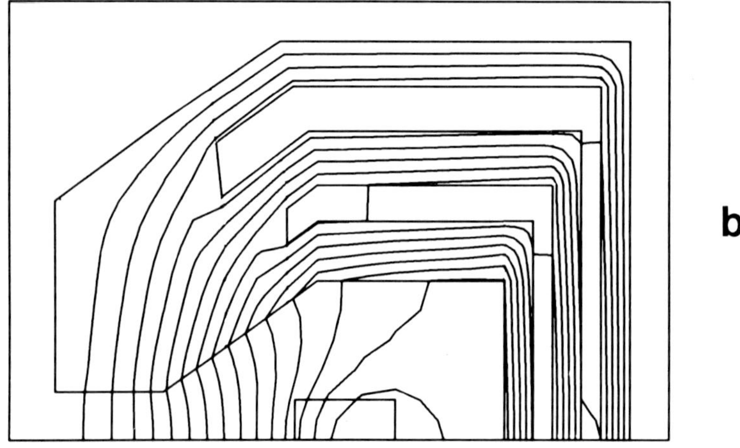

Fig. 10. Simulation of an anisotropic circuit

a) Magnetic field values at different points of a simulated circuit

b) Direction of the flux density B.

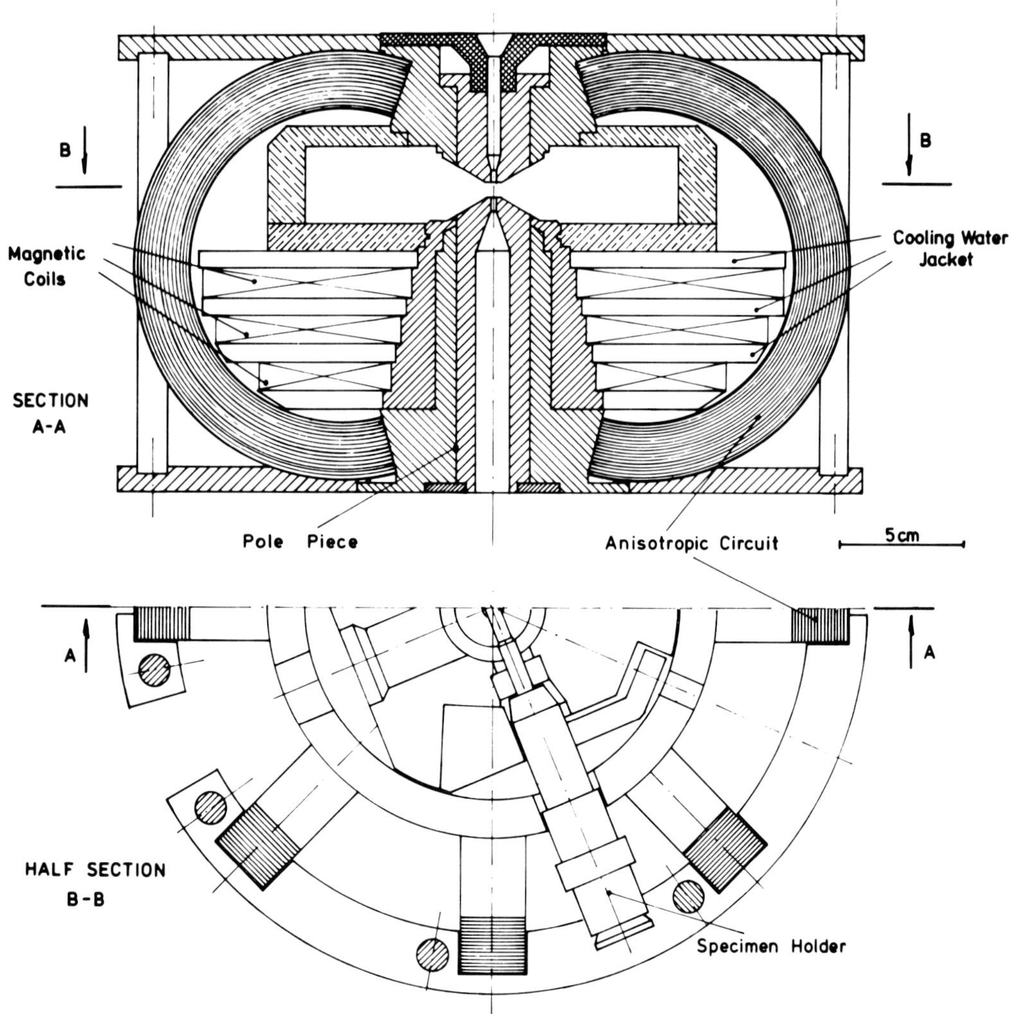

Fig. 11. New type of magnetic lens.

References

Balladore JL. (1972). Propriétés et applications des lentilles électroniques à enroulement supraconducteur, thèse Toulouse France.

Balladore JL, Murillo R. (1977). Nouveau type de lentille magnétique pour microscope électronique, J. Micro. Spectr. électr. 2, 211-222.

Balladore JL, Murillo R, Hawkes PW, Trinquier J, Jouffrey B. (1981). Magnetic electron lenses for high voltage microscopes, Nucl. Inst. and Methods 187,209-215.

Dietrich I. (1978). Superconducting lenses, 9th Intl. Congr. on Electr. Micr., Canadian Microscopical Society, Toronto 2, 173-184.

Dupouy G, Perrier F, Durrieu L. (1970). Microscope électronique 3 millions de Volts, J. Micro. 9, 575-592.

Fernandez-Moran M. (1965). Electron microscopy with high field superconducting solenoid lenses, Proc. Nat. Acad. Sci. USA 53, 2, 445-454.

Jouffrey B, Trinquier J, Balladore JL. (1979). Double focus electron probe forming lens for STEM. J. Micr. Spect. electr. 4, 89-93.

Laberrigue A, Balossier G, Genotel D, Homo JC. (1976). Développement et amélioration d'un microscope 400 kV à lentilles supraconductrices, 6th Europ. Congr. on Electr. Microsc. Tal Publishing Co., Jerusalem 1, 338-339.

Liebmann G, Grad EM. (1951). Imaging properties of a series of magnetic electron lenses, Proc. Phys. Soc. B, 54, 956-971.

Mulvey T, Newman CD. (1972). Versatile miniature electron lenses, 5th Europ.Congr. on Elect. Micr. Manchester, The Inst. of Physics, London, 116-117.

Munro E. (1973). Computer-aided-design methods in electron optics, Thesis Cambridge, England.

Murillo R, Balladore JL. (1974). Etude des éléments cardinaux d'un objectif à pièces polaires saturées, J. Micr. 20, 1-14.

Electron Optical Systems (pp. 37-43)
SEM Inc., AMF O'Hare (Chicago), IL 60666-0507, U.S.A.

0-931288-34-7/84$1.00+.05

ELECTROSTATIC AND MAGNETIC IMAGING WITHOUT IMAGE ROTATION

Zhou Li-wei*, Ni Guo-qiang, Fang Er-lun[1]
Optical Engineering Department, Beijing Institute of Technology
P.O.Box 327 Beijing, China
Tel. 890321 Ext.413

ABSTRACT

A method is discussed for designing electrostatic and magnetic imaging systems without image rotation. Tsukkerman's condition for an imaging system free of rotation does not appear to be applicable to practical designs. We have modified his theory to obtain two other conditions for the normalized axial electric potential V(z), the normalized axial magnetic induction G(z), and a lens strength parameter λ which can be interpreted as an eigen-value. Some analytical examples and numerical results are given.

Keywords: Electron optics, magnetic electron lenses, cathode lenses, imaging systems, imaging sections

* Address for correspondence
1 Xian Institute of Chemical Industries No.3, Xian, China

INTRODUCTION

Tsukkerman's theory [2-6] can be outlined as follows:

1. The vector of position $\mathbf{r}(z)=\mathbf{r}_*(z)+\mathbf{p}(z)$ of an electron trajectory is written as the sum of that of a principal trajectory $\mathbf{r}_*(z)$ and an increment $\mathbf{p}(z)$ for which simpler equations are obtained.

2. The axial potential V(z) and the magnetic field distribution G(z) are subject to conditions simplifying the "paraxial" ray equation. For a given object plane $z=z_0$ and a given image plane $z=z_i$ the simplified ray equation can be treated as an eigen-value problem.

Tsukkerman's theory does not appear to be applicable to the case where $\phi(z_0)=0$, i.e. where the object plane is the cathode surface. Zhou[7] has modified Tsukkerman's theory to make it applicable to that case. In the present paper we have applied this modified theory to establish two other conditions for V(z), G(z) and λ.

FUNDAMENTAL THEORY OF THE IMAGING SYSTEM

For an axisymmetric cathode lens we express the complex coordinates $\mathbf{r}_*(z)$ of the principal ray and $\mathbf{p}(z)$ of the increment by their moduli and arguments:

$$\mathbf{r}_*(z)=r_*(z)\,e^{i\theta_*(z)},$$

$$\mathbf{p}(z)=p(z)e^{i\beta(z)}. \quad (1)$$

Their sum

$$\mathbf{r}(z)=\mathbf{r}_*(z)+\mathbf{p}(z) \quad (2)$$

describes a ray neighboring to the principal ray (cf. Fig.1).

LIST OF SYMBOLS

$B(z)$ = Axial magnetic induction

B_0 = Magnetic induction at the cathode surface

e = Electron charge

$e\varepsilon_0$ = Initial energy of emission electrons, $e\varepsilon_0 = (m_0/2)v_0^2$

$e\varepsilon_r^*$ = Relative transversal initial energy of emission electrons, $e\varepsilon_r^* = e\varepsilon_0 \sin^2\alpha_0/\phi_i$

$e\varepsilon_z^*$ = Relative axial initial energy of emission electrons, $e\varepsilon_z^* = e\varepsilon_0 \cos^2\alpha_0/\phi_i$

$e\varepsilon_{z1}^*$ = Relative axial initial energy of emission electrons, which can be focused ideally at the image plane

$G(z)$ = Relative axial magnetic induction, $G(z) = B(z)/B_0$

l = Axial distance from cathode surface to image plane

$\mathbf{M}$ = Magnification vector

M = Magnification

m_0 = Electron mass

m = Integer number, $0 < m \leqslant n$

n = Total of loops

$\mathbf{p}(z)$ = Vector of increment of electron ray

$p(z)$ = Module of $\mathbf{p}(z)$

$\mathbf{q}(z)$ = Vector of increment of electron ray in a twisted coordinate system (x,y)

$\mathbf{r}(z)$ = Vector of electron ray

$r(z)$ = Module of $\mathbf{r}(z)$

r_0 = Off-axis height from axis z at the cathode surface

$\mathbf{r}_*(z)$ = Vector of principal electron ray

$r_*(z)$ = Module of $\mathbf{r}_*(z)$

$\mathbf{u}(z)$ = Vector of electron ray in a twisted coordinate system (x,y)

$\mathbf{u}_*(z)$ = Vector of principal electron ray in a twisted coordinate system (x,y)

u = Variable in eq.(24)

$V(z)$ = Relative axial electric potential, $V(z) = \phi(z)/\phi_i$

$v(z), w(z)$ = Two independent special solutions of the "paraxial" ray equation

v_0 = Initial velocity of electrons

(X,Y,z) = Fixed Cartesian coordinate system

(x,y,z) = Twisted Cartesian coordinate system

z = Axial coordinate measured in unit of the distance l between cathode surface and image plane

z_0, z_i = Coordinates of cathode surface and image plane respectively

α_0 = Emission angle of electrons

$\beta(z)$ = Angle of $\mathbf{p}(z)$ turned around to the axis X

β_0 = Initial value of $\beta(z)$

γ = Parameter in eq.(46)

$\theta(z)$ = Angle of $\mathbf{r}(z)$ turned around to the axis X

$\theta_*(z)$ = Angle of $\mathbf{r}_*(z)$ turned around to the axis X

θ_0 = Initial value of $\theta_*(z)$

λ = Dimensionless lens strength parameter

$\phi(z)$ = Axial electric potential

ϕ_i = Electric potential at the image plane

$\chi(z)$ = Angle of rotation of the twisted coordinate system (x,y) with respect to the fixed coordinate system (X,Y)

χ_0 = Initial value of $\chi(z)$

ψ = Variable in eq.(25)

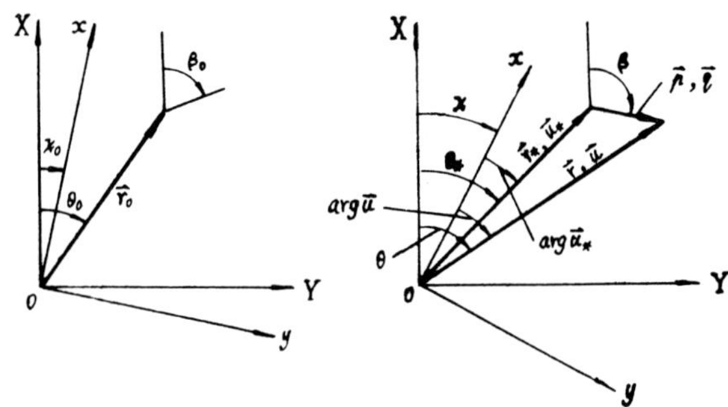

Fig.1 The principal trajectory and its increment in fixed Cartesian coordinate system (X,Y) and twisted Cartesian coordinate system (x,y)

From the "paraxial" ray equation for r(z) two equations for the principal ray

$$\sqrt{V(z)+\varepsilon_z^*}\frac{d}{dz}\left(\sqrt{V(z)+\varepsilon_z^*}\frac{dr_*}{dz}\right)$$

$$+\left\{\frac{1}{4}V''(z)+\lambda\left[G^2(z)-\frac{r_0^4}{r_*^4}\right]\right\}r_*=0, \tag{3}$$

$$\sqrt{V(z)+\varepsilon_z^*}\frac{d\theta_*}{dz}=\sqrt{\lambda}\left(G(z)-\frac{r_0^2}{r_*^2}\right); \tag{4}$$

and two others for the increment

$$\sqrt{V(z)+\varepsilon_z^*}\frac{d}{dz}\left(\sqrt{V(z)+\varepsilon_z^*}\frac{dp}{dz}\right)$$

$$+\left[\frac{1}{4}V''(z)+\lambda G^2(z)\right]p=0, \tag{5}$$

$$\frac{d\beta}{dz}=\sqrt{\lambda}\frac{G(z)}{\sqrt{V(z)+\varepsilon_z^*}} \tag{6}$$

are derived [7]. z is the axial coordinate measured in unit of the distance l between cathode and image plane, i.e. $z_0=0$, $z_i=1$. Primes denote differentiation with respect to z, $\phi(z)$ is the axial electric potential, $\phi(0)=0$, $\phi_i=\phi(1)$, $B_0=B(0)$, $V(z)=\phi(z)/\phi_i$, $G(z)=B(z)/B_0$ $r_0=r(0)$, $e\varepsilon_0=(m_0/2)v_0^2$ is the initial energy of the electron, α_0 the angle between its initial tangent and the z-axis, $\varepsilon_z^*=\varepsilon_0\cos^2\alpha_0/\phi_i$, $\varepsilon_r^*=\varepsilon_0\sin^2\alpha_0/\phi_i$ and

$$\lambda=\frac{e}{8m_0}\frac{B_0^2 l^2}{\phi_i} \tag{7}$$

is a dimensionless lens strength parameter.

In order to solve the equations (3) to (6) we introduce a twisted coordinate system by using the complex coordinate $\mathbf{u}=\mathbf{r}\exp(-i\chi(z))$ where

$$\chi(z)=\sqrt{\lambda}\int_0^z\frac{G(z)}{\sqrt{V(z)+\varepsilon_z^*}}dz+\chi_0. \tag{8}$$

In the twisted coordinate system the differential equations for the principal ray and the increment assume the same form [7]:

$$[V(z)+\varepsilon_z^*]\mathbf{u}''+\frac{1}{2}V'(z)\mathbf{u}'+[\frac{1}{4}V''(z)$$
$$+\lambda G^2(z)]\mathbf{u}=0. \tag{9}$$

Their solutions can be expressed using two linearly independent solutions v(z) and w(z) of this ray equation satisfying the initial conditions:

$$\begin{aligned} &v(z_0=0)=0, && v'(z_0=0)\sqrt{\varepsilon_z^*}=1;\\ &w(z_0=0)=1, && w'(z_0=0)\sqrt{\varepsilon_z^*}=0. \end{aligned} \tag{10}$$

For electrons emitted from the cathode surface with the initial conditions:

$$\mathbf{r}_0=r_0e^{i\theta_0}, \qquad \mathbf{r}_0'=l\cdot\tan\alpha_0 e^{i\beta_0}; \tag{11}$$

the solution of the principal trajectory equation can be written as

$$r_*(z)=r_0\sqrt{w^2+\lambda v^2}, \tag{12}$$

$$\theta_*(z)=\theta_0+\sqrt{\lambda}\int_0^z\frac{G(z)}{\sqrt{V(z)+\varepsilon_z^*}}dz$$
$$-\text{Arctan}\frac{\sqrt{\lambda}v}{w}, \tag{13}$$

whereas the solution for the increment becomes

$$p(z)=\sqrt{\varepsilon_r^*}\,l\,|v|, \tag{14}$$

$$\beta(z)=\sqrt{\lambda}\int_0^z\frac{G(z)}{\sqrt{V(z)+\varepsilon_z^*}}dz+\beta_0$$
$$+(m-1)\pi \tag{15}$$

with $\theta_0=\theta_*(z_0=0)$, $\beta_0=\beta(z_0=0)$. The integer number m which depends on z is defined by the inequality:

$$(m-1)\pi<\sqrt{\lambda}\int_0^z\frac{G(z)}{\sqrt{V(z)+\varepsilon_z^*}}dz\leqslant m\pi,$$

its value for z=1 is m=n. If

$$v(z_i,\varepsilon_{z1}^*)=0, \tag{16}$$

then all electrons having the same initial axial velocity component, i.e. the same value of $\varepsilon_z^*=\varepsilon_{z1}^*$, are ideally focused at the image plane z=1. The complex magnification becomes

$$\mathbf{M}=\frac{\mathbf{r}_*(z_i,\varepsilon_{z1}^*)}{\mathbf{r}_0}=w(z_i,\varepsilon_{z1}^*)e^{i(\chi_i-\chi_0)}$$
$$=\frac{1}{v'(z_i,\varepsilon_{z1}^*)\sqrt{1+\varepsilon_{z1}^*}}e^{i(\chi_i-\chi_0)} \tag{17}$$

and its module

$$M=|w(z_i,\varepsilon_{z1}^*)|.$$

For the angle of image rotation we obtain

$$\theta_*(z_i)-\theta_0=\chi_i-\chi_0-n\pi \tag{18}$$

where $\chi_i=\chi(1)$ follows from eq.(8) for z=1. The system is free of image rotation if $\theta_*(z_i)=\theta_0$, i.e.

$$\chi_i-\chi_0=n\pi, \tag{19}$$

The module of the magnification becomes

$$M=(-1)^n w(z_i,\varepsilon_{z1}^*)$$
$$=(-1)^n\frac{1}{v'(z_i,\varepsilon_{z1}^*)\sqrt{1+\varepsilon_{z1}^*}} \tag{20}$$

Eqs.(17) and (20) can also be applied to the case V(0)=0 and $\varepsilon_{z1}^*=0$.

A NEW APPROACH TO REALIZE AN IMAGING SYSTEM WITHOUT IMAGE ROTATION

Tsukkerman has pointed out that a system without image rotation can be realized if the condition

$$G(z)=\frac{r_0^2}{r_*^2(z)} \tag{21}$$

is satisfied. Then eq.(3) assumes the form of the "paraxial" ray equation for a purely electrostatic field. It can be solved with the initial conditions $r_*(0)=r_0$, $r'_*(0)=$

0, and the magnification becomes $M = r_*(1)/r_0$. The principal trajectory is then a meridional ray following a magnetic field line.

Since we think that the condition (21) can hardly be realized, we suggest another solution of the problem. In order to solve the "paraxial" ray equation in the twisted coordinate system

$$\frac{d}{dz}\left(\sqrt{V(z)+\varepsilon_{z1}^*}\,\frac{dv}{dz}\right)+\left(\frac{V''(z)}{4\sqrt{V(z)+\varepsilon_{z1}^*}}\right.$$
$$\left.+\lambda\frac{G^2(z)}{\sqrt{V(z)+\varepsilon_{z1}^*}}\right)v=0, \tag{22}$$

for $v(z)$ with the boundary condition

$$v(z_0=0)=0,\quad v(z_i=1,\ \varepsilon_{z1}^*)=0; \tag{23}$$

we introduce the new coordinates

$$u=\sqrt{G}\,v \tag{24}$$

and

$$\psi=\int_0^z\frac{G(z)}{\sqrt{V(z)+\varepsilon_{z1}^*}}dz. \tag{25}$$

The ray equation in terms of the new coordinates can be written as

$$\frac{d^2u}{d\psi^2}+\lambda u+\frac{u}{4G^2}\left\{\left(\frac{dG}{d\psi}\right)^2-2G\frac{d^2G}{d\psi^2}\right.$$
$$+\frac{G}{V+\varepsilon_{z1}^*}\left[G\frac{d^2V}{d\psi^2}+\frac{dG}{d\psi}\frac{dV}{d\psi}\right.$$
$$\left.\left.-\frac{G}{2[V+\varepsilon_{z1}^*]}\left(\frac{dV}{d\psi}\right)^2\right]\right\}=0 \tag{26}$$

and

$$u(0)=0,\quad u(\psi_i,\ \varepsilon_{z1}^*)=0; \tag{27}$$

where

$$\psi_i=\psi(z_i=1).$$

If the contents of the curved brackets in eq.(26) vanish, eq.(26) is reduced to the Sturm-Liouville equation:

$$\frac{d^2u}{d\psi^2}+\lambda u=0. \tag{28}$$

The solution which satisfies the condition (27) is

$$u=\sin(\sqrt{\lambda}\,\psi), \tag{29}$$

where

$$\sin(\sqrt{\lambda}\,\psi_i)=0. \tag{30}$$

Equation (30) is equivalent to the condition (19).

It allows to define the n-th eigen-value λ_n for which $\sqrt{\lambda_n}\,\psi_i=n\pi$. For $\lambda=\lambda_n$, there are $(n-1)$ intermediate foci between the cathode surface $z_0=0$ and the image plane $z_i=1$.

If the contents of the curved brackets in eq. (26) are expressed as a function of z and set equal to zero, we obtain:

$$\frac{3[V+\varepsilon_{z1}^*]}{G^2}\left(\frac{dG}{dz}\right)^2-\frac{2[V+\varepsilon_{z1}^*]}{G}\frac{d^2G}{dz^2}$$
$$-\frac{1}{G}\frac{dv}{dz}\frac{dG}{dz}+\frac{d^2V}{dz^2}=0. \tag{31}$$

The n-th eigen-value must satisfy the condition:

$$\lambda_n=\frac{e}{8m_0}\frac{B_0^2l^2}{\phi_i}$$
$$=\frac{n^2\pi^2}{\left[\int_0^1\frac{G(z)}{\sqrt{V(z)+\varepsilon_{z1}^*}}dz\right]^2}. \tag{32}$$

If these two conditions are satisfied, the cathode surface $z_0=0$ is imaged onto the image plane $z_i=1$ without image rotation. Our conditions are more general than Tsukkerman's condition (21) since they do not imply that $\theta_*(z)=\theta_0$ for the principal ray but only that $\theta_*(z_i)=\theta_0$ in the image plane.

From eqs.(24) and (28) together with the initial conditious (10) it follows that

$$v(z,\varepsilon_{z1}^*)=\frac{1}{\sqrt{\lambda G(z)}}\sin\left(\sqrt{\lambda}\int_0^z\frac{G(z)}{\sqrt{V(z)+\varepsilon_{z1}^*}}dz\right), \tag{33}$$

$$w(z,\varepsilon_{z1}^*)=-\sqrt{\varepsilon_{z1}^*}\,v(z,\varepsilon_{z1}^*)\left[\frac{d}{dz}\frac{1}{\sqrt{G(z)}}\right]_{z=0}$$
$$+\frac{1}{\sqrt{G(z)}}\cos\left(\sqrt{\lambda}\int_0^z\frac{G(z)}{\sqrt{V(z)+\varepsilon_{z1}^*}}dz\right). \tag{34}$$

For $z=z_i$ we find the magnification

$$M=\frac{1}{\sqrt{G(z_i)}}=\sqrt{B_0/B(z_i)}. \tag{35}$$

The same result has been obtained by Tsukkerman [3,4,6] but we think that our treatment is more general and more realistic.

DESIGN OF AN IMAGING SYSTEM WITHOUT IMAGE ROTATION

The design procedure is as follows:

1. $V(z)$ or $G(z)$ is given;
2. Use condition (31) to determine $G(z)$ from given $V(z)$ or $V(z)$ from given $G(z)$. V and G must satisfy the conditions:

$$V(z_0=0)=0, \quad V(z_i=1)=1,$$

$$G(z_0=0)=1, \quad G(z_i=1)=\frac{1}{M^2};$$

3. Determine M;
4. Choose suitable values of ϕ_i, B_0, l and n which must be in agreement with eq.(32), and evaluate $\lambda=\lambda_n$;
5. The solutions v and w of eq.(22) are then given by eqs.(33) and (34);
6. Determine $r(z)=r_*(z)+p(z)$ using the solutions $v(z)$ and $w(z)$;
7. Design electrodes and magnetic circuit in a way that they generate the fields $V(z)$ and $G(z)$ with the constants V_i and B_0.

We think that our method is more general and flexible than Tsukkerman's since his method is restricted to the case that $V(z)$ is given and that $G'(0)=0$, and it yields only one solution $G(z)$ and one value for the magnification for a given $V(z)$. On the other hand, we can, for a given $V(z)$, choose an arbitrary value for the magnification M, and then solve eq.(31) for $G(z)$ with the boundary conditions $G(0)=1$; $G(1)=1/M^2$.

AN ANALYTIC EXAMPLE FOR THE DESIGN OF AN IMAGING SYSTEM WITHOUT IMAGE ROTATION

If we choose

$$G(z)=\frac{1}{[1-(1-M)z]^2}, \tag{36}$$

and

$$V(z)=\frac{Mz}{1-(1-M)z}, \tag{37}$$

then we have

$$G\frac{d^2G}{dz^2}-\frac{3}{2}\left(\frac{dG}{dz}\right)^2=0,$$

$$G\frac{d^2V}{dz^2}-\frac{dV}{dz}\frac{dG}{dz}=0, \tag{38}$$

i.e. eq.(31) is satisfied [1,7].

From eqs.(33) and (34) we obtain

$$v(z)=\frac{1}{\sqrt{\lambda}}[1-(1-M)z]\sin(\chi-\chi_0), \tag{39}$$

$$w(z)=[1-(1-M)z]\{\cos(\chi-\chi_0) + \sqrt{\frac{\varepsilon_z^*}{\lambda}}(1-M)\sin(\chi-\chi_0)\}; \tag{40}$$

where

$$\chi=\frac{2\sqrt{\lambda}}{M}\sqrt{V(z)+\varepsilon_z^*}, \quad \chi_0=\frac{2\sqrt{\lambda}}{M}\sqrt{\varepsilon_z^*}; \tag{41}$$

with v and w from eqs. (39) and (40) we obtain from eqs.(12) to (15)

$$r_*(z)=r_0[1-(1-M)z]\left\{1+(1-M)\sqrt{\frac{\varepsilon_z^*}{\lambda}}\sin2(\chi-\chi_0) + (1-M)^2\frac{\varepsilon_z^*}{\lambda}\sin^2(\chi-\chi_0)\right\}^{1/2}, \tag{42}$$

$$\theta_*(z)=\theta_0+\chi-\chi_0 -\operatorname{Arctan}\frac{\sin(\chi-\chi_0)}{\cos(\chi-\chi_0)+(1-M)\sqrt{\frac{\varepsilon_z^*}{\lambda}}\sin(\chi-\chi_0)} \tag{43}$$

$$p(z)=\sqrt{\frac{\varepsilon_r^*}{\lambda}}\, l[1-(1-M)]\,|\sin(\chi-\chi_0)|, \tag{44}$$

$$\beta(z)=\chi-\chi_0+\beta_0+(m-1)\pi. \tag{45}$$

For M=1, the example describes the well-known system using parallel homogeneous electric and magnetic fields. Fig.2 shows graphs for the case M=1.4, n=1. In Fig.3, $V(z)$ and $G(z)$ are plotted for a number of different magnifications in the range $0.6\leqslant M\leqslant 2.5$.

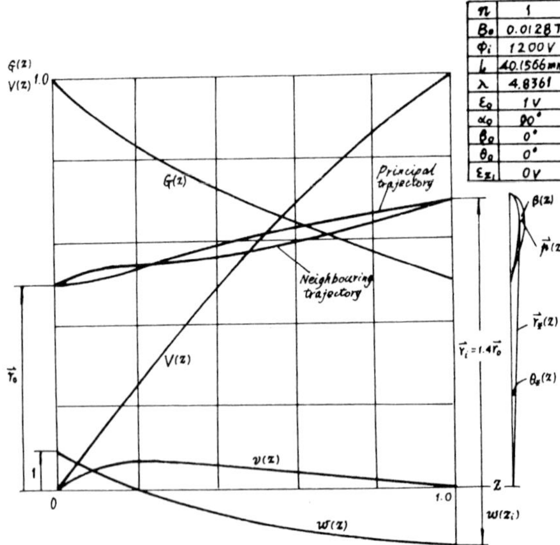

Fig.2 An analytic example of imaging system with M=1.4, n=1 (graphs of G(z), V(z), v(z), w(z), $r_*(z)$, $\theta_*(z)$, p(z), β(z))

A COMPUTATIONAL EXAMPLE FOR THE DESIGN OF AN IMAGING SYSTEM WITH VARIABLE MAGNIFICATION

For a given V(z), an arbitrary value for the magnification M may be chosen, and then eq.(31) may be solved numerically for G(z) with the boundary conditions G(0)=1, G(1)=1/M². We have worked out a computer program to do this. As an example, Fig.4 shows for the given function

$$V(z)=\frac{\gamma z}{1-(1-\gamma)z} \tag{46}$$

with γ=0.6 a number of solutions G(z) for different values of the given magnification in the range 0.6≤M≤2.5.

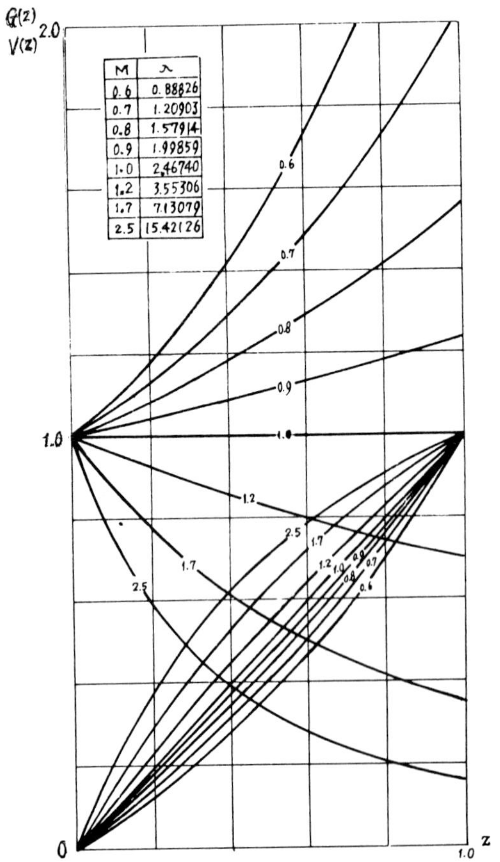

Fig.3 Some analytic curves V(z) and G(z) for different magnification M

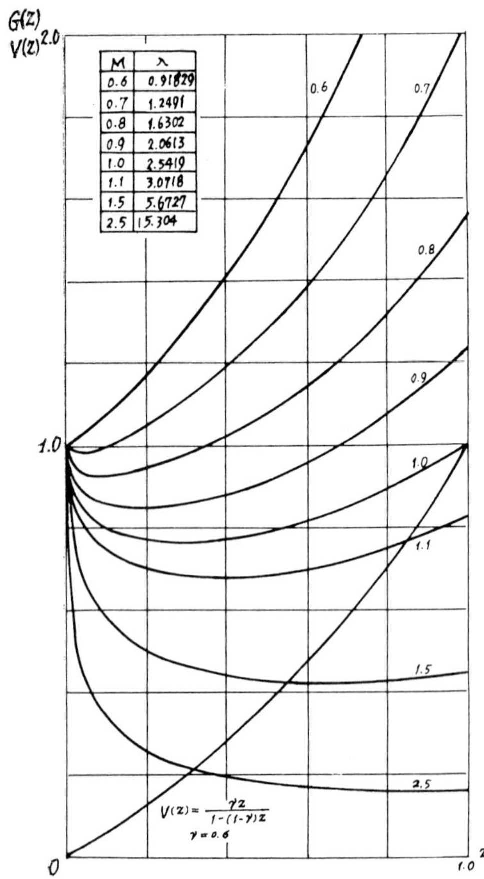

Fig.4 A computational example for a system with variable magnification

SUMMARY

Tsukkerman's method of finding combined electrostatic and magnetic imaging systems without image rotation is generalized in order to make it applicable to cathode lenses where the electric potential in the cathode surface vanishes and to magnetic fields which are not necessarily perpendicular to the cathode surface. A few analytical and numerical examples are given.

ACKNOWLEDGEMENT

We should like to thank Prof. F. Lenz from Tubingen University for a critical review of the manuscript.

REFERENCES

1. Chou LW(Zhou Li-wei).(1979).Electron Optics of Concentric Electromagnetic Focusing Systems, in: Advances in Electronics and Electron Physics, L.Marton and C. Marton(ed): Academic Press, NY, 52, 119-132.

2. Tsukkerman II. (1952). On the "non-optical" theory of focusing in magnetic field of rotation. Zh. Tekh. Fiz. 22, 1843-1847. (in Russian)

3. Tsukkerman II. (1953). On the magnetic focusing in Camera tubes with image section. Zh. Tekh. Fiz. 23, 1228-1238. (in Russian)

4. Tsukkerman II. (1955). Magnetic electron optical systems with variable magnification but without image rotation. Zh. Tekh. Fiz. 25, 950-952. (in Russian)

5. Tsukkerman II. (1957). Electron-optical method for scaling television picture. Radiotekh. 12, 3-9. (in Russian)

6. Tsukkerman II. (1961). Electron Optics in Television. Pergamon Press, NY, Chapter 2.6.

7. Zhou Li-wei. (1983). On the theory of combined electrostatic and magnetic focusing imaging system. Journal of Beijing Institute of Technology, No.3, 12-24. (in Chinese)

Electron Optical Systems (pp. 45-62)
SEM Inc., AMF O'Hare (Chicago), IL 60666-0507, U.S.A.

0-931288-34-7/84$1.00+.05

OPTICS OF WIDE ELECTRON BEAM FOCUSING

Zhou Li-wei
Optical Engineering Department, Beijing Institute of Technology
P.O.Box 327 Beijing, China
Tel. 890321 Ext. 413

ABSTRACT

In the present paper, we have further studied an approach to the theoretical treatment of wide electron beam focusing, based on the application of the theory of focusing fields with arbitrary curved optical axis, considering the initial condition emitted from the cathode surface. It is assumed that the magnetic field vector **B** will be used instead of the magnetic scalar potential Ω and the electrostatic potential φ and the magnetic field vector **B** are known functions either by computation or experimentally in the laboratory system of coordinates. The new approach to the mathematical treatment appears to be a useful method for solving problems of wide electron beam focusing with arbitrary electrostatic and magnetic fields.

The theoretical treatment presented here can be extended to solve problems of deflection defocusing in cathode-ray tubes at large deflection angles.

Key words: Electron optics, electron beams, electron lenses, cathode lenses, imaging systems, deflection systems

INTRODUCTION

In electron optical systems involving cathode lenses, where the photoemitter is used as the object surface, the cathode is situated in both magnetic and electrostatic fields and electrons leave the cathode with small velocity and large inclination. The theory of conventional narrow electron beam focusing in a curvilinear coordinate system cannot be applied to deal with such cases involving wide electron beam focusing with large cathode surface.

An approach to the theoretical treatment of wide electron beam focusing has been derived [9], based on the application of the theory of focusing fields with arbitrary curved optical axes, considering the initial conditions emitted from the cathode surface. In that paper[9], equations of the principal trajectory and "paraxial" trajectories are deduced. The orthogonal condition of a curvilinear "paraxial" system and the electron optical properties of an orthogonal system have been studied, some practical examples of wide electron beam focusing satisfying the orthogonal condition have been given. The results of computation showed that the theoretical treatment is efficient and suitable for the electrostatic cathode lenses, particularly for the determination of field curvature and astigmatism.

It has been found that in the above mentioned article the magnetic field is expressed by magnetic scalar potential Ω, but it is not suitable for the practical computation. In the present paper, the magnetic field vector **B** and the electrostatic potential φ are known functions either experimentally or by computation in the laboratory system of coordinates. The generalized theory of wide electron beam focusing is further studied. The new approach to the mathematical treatment appears to be a useful method for solving problems of wide electron beam focusing with arbitrary electrostatic and magnetic fields.

The theoretical treatment presented here can be exten-

LIST OF SYMBOLS

A^l, C^l	=Parameters in eqs.(16)
$a_{11}, a_{12}, a_{21}, a_{22}$	=Parameters in eqs.(72)
$\mathbf{B}$	=Vector of magnetic induction
B_1, B_2, B_3	=Components of the vector **B** in the Frenet local coordinate system(x^1, x^2, x^3)
B_X, B_Y, B_Z	=Components of the vector **B** in the laboratory Cartesian coordinate system(x, y, z)
B_n, B_{n2}, B_{n3}; B_b, B_{b2}, B_{b3}; B_t, B_{t2}, B_{t3}	=Parameters in eqs.(32) and (35)
B_{X2}, B_{X3}; B_{Y2}, B_{Y3}; B_{Z2}, B_{Z3}	=Parameters in eqs.(33)
C	=Constant in eqs.(95a,b)
ds	=Element of arc length of the principal trajectory
ds^*	=Element of arc length of the neighboring trajectory
$\mathbf{E}$	=Vector of electric field intensity
e	=Electron charge
$e\varepsilon_0$	=Initial energy of electrons emitted from the cathode surface
$e\varepsilon_s$	=Initial energy of principal electrons emitted from the cathode surface
$e\varepsilon_{s1}$	=Initial energy of principal electron which can be focused ideally at the image foci
$e\varepsilon_z$	=Axial initial energy of emission electrons emitted from the cathode surface
$e\varepsilon_{z1}$	=Axial initial energy of emission electron which can be focused ideally at the image plane
F_1, G_1, F_2, G_2	=Parameters in eqs.(92)
f_1, f_2; g_1, g_2; h_1, h_2	=Parameters in eqs.(66)
g	=Determinant of the components g_{ij}
g_{ij}	=Components of the covariant metric tensor
g^{ij}	=Components of the contravariant metric tensor
G(i,j)	=Sub-determinant of the components g_{ij}
i, **j**, **k**	=Unit vectors of the laboratory Cartesian coordinate
M	=Module of magnification
M_u, M_v	=Module of magnification in two directions
m_0	=Electron mass
n, K, F, R, N, Q	=Parameters in eqs.(47)
n_X, n_Y, n_Z; b_X, b_Y, b_Z; t_X, t_Y, t_Z	=Parameters in eqs.(40)
p(s)	=Vector of increment of electron trajectory from point N to point N*
$\mathbf{r}_N$	=Vector of space curved axis at point N
$\mathbf{r}^*$	=Vector of the neighboring trajectory at point N*
t, **n**, **b**	=Unit vectors in the tangent, principal normal and binormal directions respectively
u, v	=Variables in a twisted coordinate system
u	=u+iv
$u_\alpha, u_\beta; v_\alpha, v_\beta$	=Two sets of special solutions of eqs.(74a,b)
U, V	=Parameters in eqs.(75)
v	=Vector of velocity of principal electron
v	=Module of **v**
v^*	=Velocity of electron at the neighboring trajectory
V_0	=Constant electrostatic potential
$w_1^{(i)}, w_2^{(i)}$, $w_3^{(i)}, w_4^{(i)}$	=A sets of solutions of eqs.(65)
x_N, y_N, z_N	=Components of the space curve $\mathbf{r}_N$ in the laboratory coordinate system
(x,y,z)	=Laboratory Cartesian coordinate system
(x^1, x^2, x^3)	=Frenet local coordinate system, $(x^1, x^2, x^3)=(s, p_2, p_3)$
x^1	=Arc length along the curved axis from the initial point A, $x^1=s$
z	=Axial coordinate
Γ_{ij}^l	=Christoffel symbols of the second kind
$1/\rho$	=Curvature of the principal trajectory
$1/\tau$	=Torsion of the principal trajectory
φ	=Space electrostatic potential
φ_*	=Normalized electrostatic potential
φ_i, φ_{ij}	=Parameters in eqs.(25)
$\phi(s)$	=Electrostatic potential along the curved axis
ϕ_X, ϕ_Y, ϕ_Z	$=\phi_X=\partial\phi/\partial x, \phi_Y=\partial\phi/\partial y, \phi_Z=\partial\phi/\partial z$
χ	=Angle of rotation of the twisted coordinate system (u,v) with respect to the fixed coordinate system (p_2, p_3)
Ω	=Scalar magnetic potential

ded to solve problems of deflection defocusing in cathode-ray tubes at large deflection angles.

ELECTRON MOTION EQUATION AND ELECTRON TRAJECTORY EQUATION IN CURVILINEAR COORDINATE SYSTEM

Let $\mathbf{r}_N=\mathbf{r}_N(x^1)$ be a natural representation of a space curve, consisting of the curved axis of an electron beam (principal trajectory), emitted from a point A at the object (cathode) surface, where x^1 is the arc length along the curved axis from the initial point A. The normal plane through an arbitrary point N on the curved axis intersects the neighboring curvilinear trajectory at point N^*. It follows that the position of N^* can be determined by a vector $\mathbf{p}$ from N to N^*. If the expressions of $\mathbf{p}=\mathbf{p}(x^1)$ are known, then the curvilinear neighboring trajectory will be completely defined.

Introduce Frenet local coordinate system (x^1, x^2, x^3) (indices 1,2,3 are here used as superscripts, not exponents) and denote the unit vectors at the point N by $\mathbf{t}$, $\mathbf{n}$, $\mathbf{b}$ for the tangent, principal normal and binormal directions respectively. The curvature radius and torsion radius of a space curve (principal trajectory) $\mathbf{r}_N$ are expressed by $\rho=\rho(x^1)$ and $\tau=\tau(x^1)$. As mentioned above, the vector $\mathbf{p}$ is placed in the normal plane orthogonal to the tangent of the principal trajectory, therefore its components at the principal normal and binormal directions are x^2 and x^3 respectively, as shown in Fig.1.

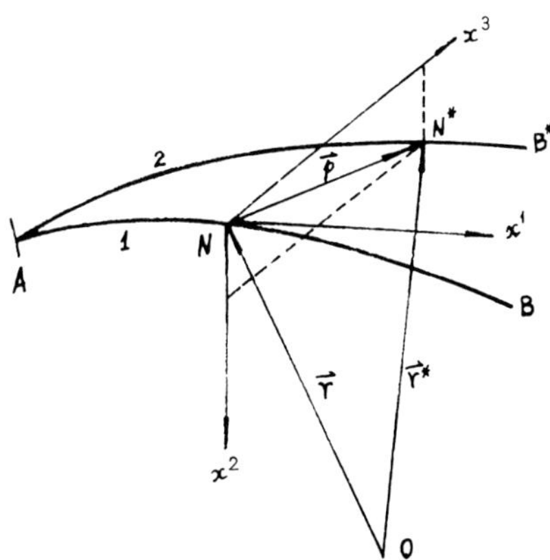

Fig.1 The Frenet local curvilinear coordinate system, principal trajectory and its neighboring trajectories

1——principal trajectory

2——neighboring trajectory

Suppose the principal trajectory (central trajectory) in a laboratory Cartesian coordinate system can be expressed as the following vector form:

$$\mathbf{r}_N=\mathbf{r}(x^1)=x_N(x^1)\mathbf{i}+y_N(x^1)\mathbf{j}+z_N(x^1)\mathbf{k}, \tag{1}$$

where $\mathbf{i},\mathbf{j},\mathbf{k}$ are the unit vectors in the laboratory Cartesian coordinate system.

Curvature and torsion of the space curve of principal trajectory (1) are given as:

$$\frac{1}{\rho}=\frac{|\mathbf{r}_N'\times\mathbf{r}_N''|}{|\mathbf{r}_N'|^3}, \tag{2}$$

$$\frac{1}{\tau}=\frac{|\mathbf{r}_N'\times\mathbf{r}_N''|\cdot\mathbf{r}_N'''}{|\mathbf{r}_N'\times\mathbf{r}_N''|^2}. \tag{3}$$

It can be written in a scalar form:

$$\frac{1}{\rho}=[(y_N'z_N''-z_N'y_N'')^2+(z_N'x_N''-x_N'z_N'')^2 + (x_N'y_N''-y_N'x_N'')^2]^{1/2}, \tag{4}$$

$$\frac{1}{\tau}=\left\{(y_N'z_N''-z_N'y_N'')x_N'''+(z_N'x_N''-x_N'z_N'')y_N''' + (x_N'y_N''-y_N'x_N'')z_N'''\right\}\Big/\left\{(y_N'z_N''-z_N'y_N'')^2 + (z_N'x_N''-x_N'z_N'')^2+(x_N'y_N''-y_N'x_N'')^2\right\}. \tag{5}$$

here and below the prime indicates derivatives with respect to arc length x^1.

From analytical geometry, the system of unit vectors in the Frenet triad is defined by

$$\begin{aligned} \mathbf{t}&=\mathbf{r}_N', &&\text{(tangent)}\\ \mathbf{n}&=\rho\mathbf{r}_N'', &&\text{(normal)}\\ \mathbf{b}&=\rho(\mathbf{r}_N'\times\mathbf{r}_N''). &&\text{(binormal)} \end{aligned} \tag{6}$$

Frenet-Serret's formulae are

$$\begin{aligned} \frac{d\mathbf{t}}{dx^1}&=\frac{1}{\rho}\mathbf{n},\\ \frac{d\mathbf{n}}{dx^1}&=-\frac{1}{\rho}\mathbf{t}+\frac{1}{\tau}\mathbf{b},\\ \frac{d\mathbf{b}}{dx^1}&=-\frac{1}{\tau}\mathbf{n}. \end{aligned} \tag{7}$$

The vector $\mathbf{r}^*$ at point N^* on the neighboring trajectory can be expressed by

$$\mathbf{r}^*=\mathbf{r}_N+\mathbf{p}=\mathbf{r}_N+x^2\mathbf{n}+x^3\mathbf{b}. \tag{8}$$

Now let us calculate the element of arc length ds^* in the selected curvilinear coordinate system. Differentiating eq.(8), using Frenet-Serret's formulae (7), considering that $(ds^*)^2=(dr^*)^2$ and expressing it in tensor form, we have[5]

$$(ds^*)^2=g_{ij}dx^idx^j, \tag{9}$$

where g_{ij} are the components of the covariant metric tensor, which can be written as follows:

$$(g_{ij})=\begin{pmatrix} (1-\frac{1}{\rho}x^2)^2+(\frac{1}{\tau}x^2)^2+(\frac{1}{\tau}x^3)^2 & -\frac{1}{\tau}x^3 & \frac{1}{\tau}x^2 \\ -\frac{1}{\tau}x^3 & 1 & 0 \\ \frac{1}{\tau}x^2 & 0 & 1 \end{pmatrix} \tag{10}$$

In the non-relativistic case, the contravariant components of Lorentz's equation have the form [4]:

$$m_0(\ddot{x}^l+\Gamma_{ij}{}^l\dot{x}^i\dot{x}^j)=e(A^l-C^l),\quad (l=1,2,3) \tag{11}$$

where e is the magnitude of the electron charge, m_0, the electron mass, $\Gamma_{ij}{}^l$, the Christoffel symbols of the second kind, can be written as

$$\Gamma_{ij}{}^l=\frac{1}{2}g^{kl}\left(\frac{\partial g_{kj}}{\partial x^i}+\frac{\partial g_{ik}}{\partial x^j}-\frac{\partial g_{ij}}{\partial x^k}\right), \tag{12}$$

where g^{ij} are the components of the contravariant metric tensor, i.e.

$$g^{ij}=\frac{G(i,j)}{g}, \tag{13}$$

In formula (13), g and G(i,j) denote the determinant and sub-determinant of the components g_{ij}. From (10), we have

$$g=\det|g_{ij}|=(1-\frac{1}{\rho}x^2)^2, \tag{14}$$

$$(g^{ij})=\begin{pmatrix} \frac{1}{g} & \frac{\frac{1}{\tau}x^3}{g} & \frac{-\frac{1}{\tau}x^2}{g} \\ \frac{\frac{1}{\tau}x^3}{g} & \frac{(1-\frac{1}{\rho}x^2)^2+(\frac{1}{\tau}x^3)^2}{g} & \frac{-(\frac{1}{\tau})^2x^2x^3}{g} \\ \frac{-\frac{1}{\tau}x^2}{g} & \frac{-(\frac{1}{\tau})^2x^2x^3}{g} & \frac{(1-\frac{1}{\rho}x^2)^2+(\frac{1}{\tau}x^2)^2}{g} \end{pmatrix} \tag{15}$$

Using formulae (10) (14) and (15), considering that $\Gamma_{ij}{}^l=\Gamma_{ji}{}^l$, we can get each value of $\Gamma_{ij}{}^l$. A^l, C^l $(l=1,2,3,)$ in (11) have the form:

$$A^l=g^{li}\frac{\partial\varphi}{\partial x^i},$$

$$C^1=\frac{1}{\sqrt{g}}(\dot{x}_2B_3-\dot{x}_3B_2),$$

$$C^2=\frac{1}{\sqrt{g}}(\dot{x}_3B_1-\dot{x}_1B_3),\quad C^3=\frac{1}{\sqrt{g}}(\dot{x}_1B_2-\dot{x}_2B_1), \tag{16}$$

where

$$\dot{x}_i=g_{ij}\dot{x}^j, \tag{17}$$

and φ is the potential of the electrostatic field. The magnetic field vector **B** can be defined by three components B_1, B_2, B_3 at the Frenet local coordinate system.

For convenience, we can use symbols (s, p_2, p_3) instead of (x^1, x^2, x^3). From (11), we have the electron motion equations in curvilinear coordinate system (s, p_2, p_3):

$$\frac{m_0}{e}\left\{\ddot{s}+\frac{\frac{1}{\rho\tau}p_3+\frac{1}{\rho^2}\rho'p_2}{1-\frac{1}{\rho}p_2}\dot{s}^2-\frac{\frac{2}{\rho}}{1-\frac{1}{\rho}p_2}\dot{s}\dot{p}_2\right\}$$

$$=\frac{1}{(1-\frac{1}{\rho}p_2)^2}\left\{\frac{\partial\varphi}{\partial s}+\frac{1}{\tau}p_3\frac{\partial\varphi}{\partial p_2}-\frac{1}{\tau}p_2\frac{\partial\varphi}{\partial p_3}\right\}$$

$$-\frac{1}{1-\frac{1}{\rho}p_2}\left\{(\dot{p}_2-\frac{1}{\tau}p_3\dot{s})B_3-(\dot{p}_3+\frac{1}{\tau}p_2\dot{s})B_2\right\},$$

$$\frac{m_0}{e}\left\{\ddot{p}_2+(\frac{1}{\rho}-\frac{1}{\rho^2}p_2-\frac{1}{\tau^2}p_2+\frac{1}{\tau^2}\tau'p_3\right.$$

$$+\frac{\frac{1}{\rho\tau^2}p_3^2+\frac{1}{\rho^2\tau}\rho'p_2p_3}{1-\frac{1}{\rho}p_2})\dot{s}^2-\frac{2\frac{1}{\rho\tau}p_3}{1-\frac{1}{\rho}p_2}\dot{s}\dot{p}_2$$

$$\left.-\frac{2}{\tau}\dot{s}\dot{p}_3\right\}=\frac{1}{(1-\frac{1}{\rho}p_2)^2}\left\{\frac{1}{\tau}p_3\frac{\partial\varphi}{\partial s}\right.$$

$$\left.+\left[(1-\frac{1}{\rho}p_2)^2+\frac{1}{\tau^2}p_3^2\right]\frac{\partial\varphi}{\partial p_2}-\frac{1}{\tau^2}p_2p_3\frac{\partial\varphi}{\partial p_3}\right\}$$

$$-\frac{1}{1-\frac{1}{\rho}p_2}\left\{(\dot{p}_3+\frac{1}{\tau}p_2\dot{s})B_1-\left\{\left[(1-\frac{1}{\rho}p_2)^2\right.\right.\right.$$

$$\left.\left.\left.+\frac{1}{\tau^2}p_2^2+\frac{1}{\tau^2}p_3^2\right]\dot{s}-\frac{1}{\tau}p_3\dot{p}_2+\frac{1}{\tau}p_2\dot{p}_3\right\}B_3\right\},$$

$$\frac{m_0}{e}\left\{\ddot{p}_3+\left[-\frac{1}{\tau^2}\tau' p_2-\frac{1}{\tau^2}p_3-\frac{1}{1-\frac{1}{\rho}p_2}\left(\frac{1}{\tau\rho^2}\rho' p_2^2+\frac{1}{\tau^2\rho}p_2p_3\right)\right]\dot{s}^2+\frac{2\frac{1}{\tau}\dot{s}\dot{p}_2}{1-\frac{1}{\rho}p_2}\right\}=\frac{1}{(1-\frac{1}{\rho}p_2)^2}\left\{-\frac{1}{\tau}p_2\frac{\partial\varphi}{\partial s}-\frac{1}{\tau^2}p_2p_3\frac{\partial\varphi}{\partial p_2}+\left[(1-\frac{1}{\rho}p_2)^2+\frac{1}{\tau^2}p_2^2\right]\frac{\partial\varphi}{\partial p_3}\right\}$$

$$-\frac{1}{1-\frac{1}{\rho}p_2}\left\{\left\{\left[(1-\frac{1}{\rho}p_2)^2+\frac{1}{\tau^2}p_2^2+\frac{1}{\tau^2}p_3^2\right]\dot{s}-\frac{1}{\tau}p_3\dot{p}_2+\frac{1}{\tau}p_2\dot{p}_3\right\}B_2-(\dot{p}_2-\frac{1}{\tau}\dot{s}\,p_3)B_1\right\}.$$

(18a,b,c)

(Note that, here the upper indices denote exponents.)

To transform the general equations (18a, b, c) of electron motion trajectory equations, let $v^*=ds^*/dt$, the velocity of electrons at the neighboring trajectory. From the standpoint of energy, we have

$$ds^*/dt=\sqrt{(2e/m_0)\varphi_*}, \qquad (19)$$

where $\varphi_*=\varphi+\varepsilon_0$, $e\varepsilon_0$ is the initial energy of electrons emitted from the cathode surface, φ_* is called the normalized potential. The element of arc length ds of the principal trajectory and ds^* of its neighboring trajectory are so related such that:

$$\left(\frac{ds^*}{ds}\right)^2=(1-\frac{1}{\rho}p_2)^2+\frac{1}{\tau^2}(p_2^2+p_3^2)-2\frac{1}{\tau}p_3p_2'+2\frac{1}{\tau}p_2p_3'+p_2'^2+p_3'^2. \qquad (20)$$

Transforming each of the parameters $\dot{s}$, $\dot{p}_2$, $\ddot{p}_2$, we have

$$\dot{s}=\frac{ds}{ds^*}\frac{ds^*}{dt}=\sqrt{\frac{2e}{m_0}\varphi_*}\,\frac{ds}{ds^*},$$

$$\dot{p}_2=\sqrt{\frac{2e}{m_0}\varphi_*}\,\frac{ds}{ds^*}p_2', \qquad (21)$$

$$\ddot{p}_2=\frac{e}{m_0}\left\{2\varphi_*\left(\frac{ds}{ds^*}\right)^2p_2''+p_2'\frac{d}{ds}\left[\varphi_*\left(\frac{ds}{ds^*}\right)^2\right]\right\}$$

Similar expressions can be obtained for $\dot{p}_3$ and $\ddot{p}_3$. Substituting these into eqs.(18b,c), we can derive the trajectory equations in curvilinear coordinate system, convenient for the study of wide electron beam focusing and its aberrations.

$$\frac{2\varphi_*}{(1-\frac{1}{\rho}p^2)^2+(\frac{1}{\tau}p_2)^2+(\frac{1}{\tau}p_3)^2-2\frac{1}{\tau}p_3p_2'+2\frac{1}{\tau}p_2p_3'+p_2'^2+p_3'^2}p_2''$$

$$+p_2'\frac{d}{ds}\left[\frac{\varphi_*}{(1-\frac{1}{\rho}p_2)^2+(\frac{1}{\tau}p_2)^2+(\frac{1}{\tau}p_3)^2-2\frac{1}{\tau}p_3p_2'+2\frac{1}{\tau}p_2p_3'+p_2'^2+p_3'^2}\right]$$

$$+\frac{2\varphi_*}{(1-\frac{1}{\rho}p_2)^2+(\frac{1}{\tau}p_2)^2+(\frac{1}{\tau}p_3)^2-2\frac{1}{\tau}p_3p_2'+2\frac{1}{\tau}p_2p_3'+p_2'^2+p_3'^2}\left(\frac{1}{\rho}\right.$$

$$\left.-\frac{1}{\rho^2}p_2-\frac{1}{\tau^2}p_2+\frac{\tau'}{\tau^2}p_3+\frac{\frac{1}{\rho\tau^2}p_3^2+\frac{1}{\rho^2\tau}\rho' p_2p_3-2\frac{1}{\rho\tau}p_3p_2'}{1-\frac{1}{\rho}p_2}-2\frac{1}{\tau}p_3'\right)$$

$$=\frac{1}{(1-\frac{1}{\rho}p_2)^2}\left\{\frac{1}{\tau}p_3\frac{\partial\varphi}{\partial s}+\left[(\frac{1}{\tau}p_3)^2+(1-\frac{1}{\rho}p_2)^2\right]\frac{\partial\varphi}{\partial p_2}-\frac{1}{\tau^2}p_2p_3\frac{\partial\varphi}{\partial p_3}\right\}$$

$$-\frac{\sqrt{\frac{2e}{m_0}}}{(1-\frac{1}{\rho}p_2)}\left(\frac{\varphi_*}{(1-\frac{1}{\rho}p_2)^2+(\frac{1}{\tau}p_2)^2+(\frac{1}{\tau}p_3)^2-2\frac{1}{\tau}p_3p_2'+2\frac{1}{\tau}p_2p_3'+p_2'^2+p_3'^2}\right)^{\frac{1}{2}}$$

$$\times\left\{(p_3'+\frac{1}{\tau}p_2)B_1-\left\{\left[(1-\frac{1}{\rho}p_2)^2+(\frac{1}{\tau}p_2)^2+(\frac{1}{\tau}p_3)^2\right]-\frac{1}{\tau}p_3p_2'+\frac{1}{\tau}p_2p_3'\right\}B_3\right\}.$$

$$\frac{2\varphi_*}{(1-\frac{1}{\rho}p_2)^2+(\frac{1}{\tau}p_2)^2+(\frac{1}{\tau}p_3)^2-2\frac{1}{\tau}p_3p_2'+2\frac{1}{\tau}p_2p_3'+p_2'^2+p_3'^2}p_3''$$

$$+p_3'\frac{d}{ds}\left[\frac{\varphi_*}{(1-\frac{1}{\rho}p_2)^2+(\frac{1}{\tau}p_2)^2+(\frac{1}{\tau}p_3)^2-2\frac{1}{\tau}p_3p_2'+2\frac{1}{\tau}p_2p_3'+p_2'^2+p_3'^2}\right]$$

$$+\frac{2\varphi_*}{(1-\frac{1}{\rho}p_2)^2+(\frac{1}{\tau}p_2)^2+(\frac{1}{\tau}p_3)^2-2\frac{1}{\tau}p_3p_2'+2\frac{1}{\tau}p_2p_3'+p_2'^2+p_3'^2}\left(-\frac{\tau'}{\tau^2}p_2\right.$$

$$\left.-\frac{1}{\tau^2}p_3-\frac{\frac{1}{\tau\rho^2}\rho'p_2^2+\frac{1}{\tau^2\rho}p_2p_3}{1-\frac{1}{\rho}p_2}+\frac{2\frac{1}{\tau}p_2'}{1-\frac{1}{\rho}p_2}\right)$$

$$=\frac{1}{(1-\frac{1}{\rho}p_2)^2}\left\{-\frac{1}{\tau}p_2\frac{\partial\varphi}{\partial s}-\frac{1}{\tau^2}p_2p_3\frac{\partial\varphi}{\partial p_2}+\left[(1-\frac{1}{\rho}p_2)^2+(\frac{1}{\tau}p_2)^2\right]\frac{\partial\varphi}{\partial p_3}\right\}$$

$$-\frac{1}{1-\frac{1}{\rho}p_2}\sqrt{\frac{2e}{m_0}}\left(\frac{\varphi_*}{(1-\frac{1}{\rho}p_2)^2+(\frac{1}{\tau}p_2)^2+(\frac{1}{\tau}p_3)^2-2\frac{1}{\tau}p_3p_2'+2\frac{1}{\tau}p_2p_3'+p_2'^2+p_3'^2}\right)^{\frac{1}{2}}$$

$$\times\left\{\left\{\left[(1-\frac{1}{\rho}p_2)^2+(\frac{1}{\tau}p_2)^2+(\frac{1}{\tau}p_3)^2\right]-\frac{1}{\tau}p_3p_2'+\frac{1}{\tau}p_2p_3'\right\}B_2-(p_2'-\frac{1}{\tau}p_3)B_1\right\}. \quad (22a,b)$$

If the magnetic field is defined by scalar magnetic potential Ω, then $-\frac{\partial\Omega}{\partial x^i}(x^1=s, x^2=p_2, x^3=p_3)$ can be used instead of $B_i (i=1,2,3)$ in the above equations, we may obtain those equations derived in ref.[9].Eqs.(22a,b) are convenient for studying wide electron beam focusing and narrow electron beam focusing and their aberrations in the curvilinear coordinate system.

ELECTROSTATIC AND MAGNETIC FIELD EXPANSIONS ALONG THE CURVED OPTICAL AXIS

In what follows mathematical expressions are going to be derived in terms of the electrostatic potential and magnetic induction along curved optical axis.

As is known, the electrostatic potential must satisfy Laplace equation:

$$\nabla^2\varphi=0. \quad (23)$$

we assume that there is no singular point near to the optical axis, the electrostatic potential $\varphi(s,p_2,p_3)$ is an analytical function.

Expanding electrostatic potential φ in a power series of p_2 and p_3 along the curved optical axis with coefficients that are functions of s, we have

$$\varphi(s,p_2,p_3)=\phi(s)+p_2\varphi_2(s)+\frac{1}{2}p_2^2\varphi_{22}(s)+$$

$$+p_2p_3\varphi_{23}(s)+\frac{1}{2}p_3{}^2\varphi_{33}(s)+\cdots\cdots \qquad (24)$$

where the symbols introduced are:

$$\varphi\Big|_{\substack{p_2=0\\p_3=0}}=\varphi(s,0,0)=\phi(s),$$

$$\frac{\partial\varphi}{\partial p_i}\Big|_{\substack{p_2=0\\p_3=0}}=\varphi_i,\quad \frac{\partial^2\varphi}{\partial p_i\partial p_j}\Big|_{\substack{p_2=0\\p_3=0}}=\varphi_{ij}; \qquad (25)$$

here

$$\varphi_{\bullet}=\varphi(s,p_2,p_3)+\varepsilon_0,\quad \varphi(0,0,0)=\phi(0)=0,\varphi_1=\frac{d\phi}{ds}.$$

The quantities of φ_i, φ_{ij} in (24) can be expressed by the known partial derivatives in the laboratory Cartesian coordinate system:

$$\varphi_2=\left[n_x\frac{\partial\varphi}{\partial x}+n_Y\frac{\partial\varphi}{\partial y}+n_z\frac{\partial\varphi}{\partial z}\right]_{x=x_N,\ y=y_N,\ z=z_N}$$

$$\varphi_3=\left[b_x\frac{\partial\varphi}{\partial x}+b_Y\frac{\partial\varphi}{\partial y}+b_Z\frac{\partial\varphi}{\partial z}\right]_{x=x_N,\ y=y_N,\ z=z_N}$$

$$\varphi_{22}=\left[n_x{}^2\frac{\partial^2\varphi}{\partial y^2}+n_Y{}^2\frac{\partial^2\varphi}{\partial y^2}+n_Z{}^2\frac{\partial^2\phi}{\partial z^2}+2n_xn_Y\frac{\partial^2\varphi}{\partial x\partial y}\right.$$

$$\left.+2n_xn_Z\frac{\partial^2\varphi}{\partial x\partial z}+2n_Yn_Z\frac{\partial^2\varphi}{\partial y\partial z}\right]_{x=x_N,\ y=y_N,\ z=z_N}$$

$$\varphi_{23}=\left[n_x\left(\frac{\partial^2\varphi}{\partial x^2}b_x+\frac{\partial^2\varphi}{\partial x\partial y}b_Y+\frac{\partial^2\varphi}{\partial x\partial z}b_Z\right)\right.$$

$$+n_Y\left(\frac{\partial^2\varphi}{\partial x\partial y}b_x+\frac{\partial^2\varphi}{\partial y^2}b_Y+\frac{\partial^2\varphi}{\partial y\partial z}b_Z\right)$$

$$\left.+n_Z\left(\frac{\partial^2\varphi}{\partial x\partial z}b_x+\frac{\partial^2\varphi}{\partial y\partial z}by+\frac{\partial^2\varphi}{\partial z^2}b_Z\right)\right]_{\substack{x=x_N,\\y=y_N,\\z=z_N}}$$

$$\varphi_{33}=\left[b_x{}^2\frac{\partial^2\varphi}{\partial x^2}+b_Y{}^2\frac{\partial^2\varphi}{\partial y^2}+b_Z{}^2\frac{\partial^2\varphi}{\partial z^2}+2b_xb_Y\frac{\partial^2\varphi}{\partial x\partial y}\right.$$

$$\left.+2b_xb_Z\frac{\partial^2\varphi}{\partial x\partial z}+2b_Yb_Z\frac{\partial^2\varphi}{\partial y\partial z}\right]_{x=x_N,y=y_N,z=z_N}. \qquad (26)$$

Substituting (24) into (23) yields the relations between some of the coefficients

$$-\left[\varphi_{22}+\varphi_{33}\right]+\frac{1}{\rho}\varphi_2=\frac{d^2\phi}{ds^2}. \qquad (27)$$

Because the direction of curved optical axis at point N coincides with unit vector **t**, we have

$$\varphi_1=\phi'=\left(\frac{d\varphi}{ds}\right)_N=\text{grad}\varphi\cdot\mathbf{t}.$$

Differentiating it,

$$\frac{d^2\phi}{ds^2}=\frac{d}{ds}(\text{grad}\varphi)\cdot\mathbf{t}+\text{grad}\varphi\cdot\frac{d\mathbf{t}}{ds},$$

and using Frenet-Serret's formulae (7), we obtain

$$\frac{d^2\phi}{ds^2}=\varphi_{11}+\frac{1}{\rho}\varphi_2. \qquad (28)$$

From (27) and (28), we have

$$\varphi_{11}+\varphi_{22}+\varphi_{33}=0. \qquad (29)$$

For the magnetic field, similar to the treatment in ref.[3], we assume that the components of vector **B** are known in the laboratory Cartesian coordinate system:

$$\mathbf{B}(x,y,z)=B_x(x,y,z)\mathbf{i}+B_Y(x,y,z)\mathbf{j}+B_Z(x,y,z)\mathbf{k}. \qquad (30)$$

It means that at the principal trajectory, $\mathbf{B}(s)=\mathbf{B}(0,0,s)=\mathbf{B}(x_N,y_N,z_N)$ is known.

For the study of the motion of electrons near the curved optical axis, we must expand the field vector $\mathbf{B}(p_2,p_3,s)$ in a power series of p_2 and p_3 with coefficients that are functions of s. Since we investigate the focusing properties of a system with wide electron beam focusing, so we limit our analysis to first order terms, then we have

$$\mathbf{B}(p_2,p_3,s)=\mathbf{B}(s)+p_2(\mathbf{n}\cdot\nabla)\mathbf{B}+p_3(\mathbf{b}\cdot\nabla)\mathbf{B}+\cdots\cdots \qquad (31)$$

where $\nabla=\frac{\partial}{\partial x}\mathbf{i}+\frac{\partial}{\partial y}\mathbf{j}+\frac{\partial}{\partial z}\mathbf{k}$,

and all the derivatives are calculated at the point (x_N, y_N, z_N) of the principal trajectory.

From (31) we can get the component of the field vector **B** in the direction of unit vector **n**:

$$B_2=\mathbf{B}(s)\mathbf{n}+p_2\mathbf{n}(\mathbf{n}\cdot\nabla)\mathbf{B}+p_3\mathbf{n}(\mathbf{b}\cdot\nabla)\mathbf{B}$$

$$=B_n(s)+B_{n2}p_2+B_{n3}p_3; \qquad (32)$$

where

$$B_n(s)=n_xB_x+n_YB_Y+n_ZB_Z,$$

$$B_{n2}=n_xB_{x2}+n_YB_{Y2}+n_ZB_{Z2},$$

$$B_{n3}=n_xB_{x3}+n_YB_{Y3}+n_ZB_{Z3}; \qquad (33)$$

and

$$B_x=B_x(x,y,z)\Big|_{x=x_N,\ y=y_N,\ z=z_N},$$

$$B_{x2}=\left[n_x\frac{\partial B_x}{\partial x}+n_Y\frac{\partial B_Y}{\partial y}+n_Z\frac{\partial B_Z}{\partial z}\right]_{x=x_N,y=y_N,z=z_N},$$

$$B_{x3}=\left[b_x\frac{\partial B_x}{\partial x}+b_Y\frac{\partial B_Y}{\partial y}+b_Z\frac{\partial B_Z}{\partial z}\right]_{x=x_N,y=y_N,z=z_N}. \qquad (34)$$

Similarly, the other coefficients B_Y, B_{Y2}, B_{Y3} ; B_Z B_{Z2} , B_{Z3} will be obtained if in (34) B_x is replaced by B_Y, B_Z.

Similar to (32), we can get the components of the field vector **B** in the direction of unit vectors **t** and **b**.

$$B_1 = B_t(s) + B_{t2}p_2 + B_{t3}p_3,$$

$$B_3 = B_b(s) + B_{b2}p_2 + B_{b3}p_3; \tag{35}$$

where $B_t(s)$, B_{t2}, B_{t3} and $B_b(s)$, B_{b2}, B_{b3} can be obtained if we use t_x, t_y, t_z or b_x, b_Y, b_z instead of n_x, n_Y, n_Z in (33).

Since the magnetic field vector B must satisfy the relations:

$$\nabla \cdot \mathbf{B} = 0, \tag{36}$$

$$\nabla \times \mathbf{B} = 0; \tag{37}$$

substituting (31) into (36) and (37) yields the relations between some of the coefficients:

$$B_t'(s) + B_{n2} + B_{b3} - \frac{1}{\rho} B_n(s) = 0; \tag{38}$$

$$B_{n3} - B_{b2} = 0,$$

$$B_{t3} - B_b'(s) = 0,$$

$$B_{t2} - B_n'(s) = 0. \tag{39}$$

In the above mentioned expressions n_x, n_Y, n_Z; b_x, b_Y, b_Z and t_x, t_Y, t_Z can be obtained by the following formulae:

$$t_x = x_N', \quad t_Y = y_N', \quad t_Z = z_N';$$

$$n_x = \rho x_N'', \quad n_Y = \rho y_N'', \quad n_Z = \rho z_N'';$$

$$b_x = \rho(y_N' z_N'' - z_N' y_N''),$$

$$b_Y = \rho(z_N' x_N'' - x_N' z_N''),$$

$$b_Z = o(x' y_N'' - y_N' x_N''); \tag{40}$$

where the prime denotes derivatives with respect to the arc length s.

EQUATIONS OF PRINCIPAL TRAJECTORY AND ITS NEIGHBORING "PARAXIAL" TRAJECTORIES

In an electron-optical system with wide electron beam focusing, we consider the cathode with null electrostatic potential to be where the electron has small initial velocity, as mentioned above.

Similar to the cathode lenses with axial symmetry[8, 10], we assume that in the system with wide electron beam focusing, the curvilinear trajectories, being adjacent to the principal trajectory, satisfy the following "curved paraxial conditions" everywhere:

$$p^2(s) \approx 0 \;,\; p(s) \ll \rho(s) \;,\; p(s) \ll \tau(s);$$

$$\dot{p}^2(s) \ll 1 \tag{41}$$

and $\varphi(0) = 0$. It is not necessary to assume that the condition $p'^2(s) \ll 1$ is satisfied everywhere.

We introduce the "curved paraxial conditions" (41) for wide electron beam focusing into (22a,b), regarding p_2, p_3 and $\dot{p}_2$, $\dot{p}_3$ as first order terms and using the method given in refs.[8,10], investigate each of the coefficients in (22a,b), we can thus arrive at equations for principal trajectory and its neighboring "paraxial" trajectories.

Now study the common coefficient of first terms on the right hand of equations (22a,b), which is related to the tangent velocity $\dot{s}$. It may be seen that this coefficient at the initial point can be written as

$$\left\{ \varphi. \Big/ \left[(1 - \frac{1}{\rho} p_2)^2 + (\frac{1}{\tau} p_2)^2 + (\frac{1}{\tau} p_3)^2 - 2\frac{1}{\tau} p_3 p_2' + 2\frac{1}{\tau} p_2 p_3' + p_2'^2 + p_3'^2 \right] \right\}_{s=0, p_2=0, p_3=0}$$

$$= \left(\frac{\varepsilon_0}{1 + p_2'^2 + p_3'^2} \right) = \varepsilon_s. \tag{42}$$

Therefore, the $0+1$ order approximation can be expressed by

$$\left\{ \varphi. \Big/ \left[(1 - \frac{1}{\rho} p_2)^2 + (\frac{1}{\tau} p_2)^2 + (\frac{1}{\tau} p_3)^2 - 2\frac{1}{\tau} p_3 p_2' + 2\frac{1}{\tau} p_2 p_3' + p_2'^2 + p_3'^2 \right] \right\}_{0+1}$$

$$= \varepsilon_s + \phi + p_2\varphi_2 + p_3\varphi_3 + (\varepsilon_s + \phi)\frac{2}{\rho} p_2. \tag{43}$$

From the $0+1+2$ order approximation of this coefficient, we may obtain the condition, by which the formula was set up.

$$p^2 \approx 0, \qquad \frac{\frac{m_0}{2e}(\dot{p}^2 - \dot{p}_0^2)}{\phi + \varepsilon_s} \ll 1 \tag{44}$$

This is called the "paraxial" condition. Therefore, when investigating the system with wide electron beam focusing, we regard p and $\sqrt{\frac{m_0}{2e}}\,\dot{p} \Big/ \sqrt{\phi + \varepsilon_s}$ as first order terms and the products as second order terms, which can be neglected.

Similar to those discussed above and for other coefficients in (22a,b), after a series of rather complicated manipulations, we have the trajectory equation of zero order approximation:

$$\frac{2}{\rho}(\phi+\varepsilon_s)=\varphi_2+\sqrt{\frac{2e}{m_0}}\sqrt{\phi+\varepsilon_s}\,B_b(s),$$

$$\varphi_3=\sqrt{\frac{2e}{m_0}}\sqrt{\phi+\varepsilon_s}\,B_n(s); \tag{45a,b}$$

and the trajectory equations of first-order approximation:

$$\frac{d}{ds}\left(n\frac{dp_2}{ds}\right)=Fp_2+Np_3+2K\frac{dp_3}{ds},$$

$$\frac{d}{ds}\left(n\frac{dp_3}{ds}\right)=Rp_3-Qp_2-2K\frac{dp_2}{ds}; \tag{46a,b}$$

where n, K, F, R, N and Q express the following functions at the curve of principal trajectory:

$$n=\sqrt{\phi+\varepsilon_s},\qquad K=\frac{\sqrt{\phi+\varepsilon_s}}{\tau}-\sqrt{\frac{e}{8m_0}}B_t(s),$$

$$F=\sqrt{\phi+\varepsilon_s}\left(\frac{1}{\tau^2}-\frac{1}{\rho^2}\right)+\frac{\varphi_{22}}{2\sqrt{\phi+\varepsilon_s}}-\frac{\varphi_2}{\rho\sqrt{\phi+\varepsilon_s}}$$

$$-\sqrt{\frac{e}{2m_0}}\left(\frac{1}{\tau}B_t(s)-B_{b2}-\frac{\varphi_2 B_b(s)}{2(\phi+\varepsilon_s)}\right),$$

$$R=\frac{1}{\tau^2}\sqrt{\phi+\varepsilon_s}+\frac{\varphi_{33}}{2\sqrt{\phi+\varepsilon_s}}-\sqrt{\frac{e}{2m_0}}\left(\frac{1}{\tau}B_t(s)\right.$$

$$\left.+B_{b2}+\frac{\varphi_3 B_n(s)}{2(\phi+\varepsilon_s)}\right),$$

$$N=\frac{\varphi_1}{2\tau\sqrt{\phi+\varepsilon_s}}-\frac{\sqrt{\phi+\varepsilon_s}}{\tau^2}\frac{d\tau}{ds}-\frac{\varphi_3}{\rho\sqrt{\phi+\varepsilon_s}}$$

$$+\frac{\varphi_{23}}{2\sqrt{\phi+\varepsilon_s}}+\sqrt{\frac{e}{2m_0}}\left(B_{b3}+\frac{\varphi_3 B_b(s)}{2(\phi+\varepsilon_s)}\right),$$

$$Q=\frac{\varphi_1}{2\tau\sqrt{\phi+\varepsilon_s}}-\frac{\sqrt{\phi+\varepsilon_s}}{\tau^2}\frac{d\tau}{ds}-\frac{\varphi_{23}}{2\sqrt{\phi+\varepsilon_s}}$$

$$+\sqrt{\frac{e}{2m_0}}\left(B_{n2}+\frac{\varphi_2 B_n(s)}{2(\phi+\varepsilon_s)}\right). \tag{47}$$

It must be pointed out that the following relationship among coefficients N, Q and K in (47) holds[7]:

$$\frac{N+Q}{2}=\frac{dK}{ds}. \tag{48}$$

(45a,b) are called equations for the principal trajectory. The coordinate curve s derived from eqs. (45a,b) when $p_2=p_3=0$ thus represents one of the electron trajectories——principal trajectory (it might be the axis of the system).

Eqs.(46a,b) can be used to describe the neighboring trajectories with rather large initial slope, emitting from the cathode surface with potential $\phi(0)=0$. we will call them curvilinear "paraxial" trajectory equations.

It is to be noted that the value of coefficients n, F, N, R, Q and K must be taken from the principal trajectory when solving eqs.(46a,b). Thus, we usually solve eqs.(45a,b) and (46a,b) simultaneously. Besides, either in the principal trajectory equations (45a,b) or in the curvilinear "paraxial" trajectory equations (46a,b), the same value for ε_s must be taken.

In what follows we will further discuss the physical meaning of the principal trajectory equations expressed in the form of (45a,b). From the fundamental equation of electron motion

$$\frac{d}{dt}(m_0\mathbf{v})=-e\mathbf{E}-e(\mathbf{v}\times\mathbf{B}), \tag{49}$$

we investigate the projection of motion equation of principal trajectory on the Frenet local coordinate system. For the motion of principal electron, $\mathbf{v}=v\mathbf{t}$, $v=ds/dt$, we have

$$\frac{d}{dt}(m_0\mathbf{v})=m_0v\left(\frac{dv}{ds}\mathbf{t}+v\frac{d\mathbf{t}}{ds}\right). \tag{50}$$

Using Frenet-Serret's formulae (7), equation (49) can be transformed to

$$m_0v\frac{dv}{ds}+m_0\frac{v^2}{\rho}\mathbf{n}=e\,\mathrm{grad}\varphi-e(v\mathbf{t}\times\mathbf{B}). \tag{51}$$

Multiply eq.(51) by **t**, **n**, **b** as scalar multiplication, we obtain

$$m_0v\frac{dv}{ds}=e\frac{d\varphi}{ds}, \tag{52}$$

$$\frac{m_0v^2}{\rho}=e(\mathrm{grad}\varphi\cdot\mathbf{n})+ev(\mathbf{B}\cdot\mathbf{b}), \tag{53}$$

$$\mathrm{grad}\varphi\cdot\mathbf{b}=v(\mathbf{B}\cdot\mathbf{n}). \tag{54}$$

Integrate (52) and consider the initial energy $e\varepsilon_s$ of principal electron emitted from the cathode surface and $\varphi(0,0,s)=\phi(s)$, $\phi(s=0)=0$, we have the velocity of principal electron

$$v=\sqrt{\frac{2e}{m_0}(\phi+\varepsilon_s)}. \tag{55}$$

Differentiate eqs.(53) and (54) with respect to s, we have

$$\frac{d}{ds}\left(\frac{m_0v^2}{\rho}\right)=e\left\{\frac{d}{ds}\mathrm{grad}\varphi\cdot\mathbf{n}-\frac{1}{\rho}\mathrm{grad}\varphi\cdot\mathbf{t}+\frac{1}{\tau}\mathrm{grad}\varphi\cdot\mathbf{b}\right.$$

$$\left.+v\left(\frac{d}{ds}\mathbf{B}\cdot\mathbf{b}-\frac{1}{\tau}\mathbf{B}\cdot\mathbf{n}\right)+\frac{e}{m_0v}\frac{d\varphi}{ds}\mathbf{B}\cdot\mathbf{b}\right\} \tag{56}$$

$$\frac{d}{ds}\mathrm{grad}\varphi\cdot\mathbf{b}-\frac{1}{\tau}\mathrm{grad}\varphi\cdot\mathbf{n}=v\left(\frac{d}{ds}\mathbf{B}\cdot\mathbf{n}-\frac{1}{\rho}\mathbf{B}\cdot\mathbf{t}+\frac{1}{\tau}\mathbf{B}\cdot\mathbf{b}\right)$$

$$+\frac{e}{m_0v}\frac{d\varphi}{ds}\mathbf{B}\cdot\mathbf{n}. \tag{57}$$

Using (55), eqs.(53) (54) (56) (57) can be written as

$$\frac{2(\phi+\varepsilon_s)}{\rho}=\varphi_2+\sqrt{\frac{2e}{m_0}}\sqrt{\phi+\varepsilon_s}\,B_b(s), \tag{58}$$

$$\varphi_3=\sqrt{\frac{2e}{m_0}}\sqrt{\phi+\varepsilon_s}\,B_n(s), \tag{59}$$

$$\frac{d}{ds}\left(\frac{2(\phi+\varepsilon_s)}{\rho}\right)=\varphi_{12}-\frac{1}{\rho}\varphi_1+\frac{1}{\tau}\varphi_3+\sqrt{\frac{2e}{m_0}}\sqrt{\phi+\varepsilon_s}\,(B_{t3}$$

$$-\frac{1}{\tau}B_n(s))+\sqrt{\frac{e}{2m_0}}\frac{1}{\sqrt{\phi+\varepsilon_s}}\varphi_1B_b(s), \tag{60}$$

$$\varphi_{13}-\frac{1}{\tau}\varphi_2=\sqrt{\frac{2e}{m_0}}\sqrt{\phi+\varepsilon_s}\Big(B_{t2}-\frac{1}{\rho}B_t(s)$$

$$-\frac{1}{\tau}B_b(s)\Big)+\sqrt{\frac{e}{2m_0}}\frac{1}{\sqrt{\phi+\varepsilon_s}}\varphi_1B_n(s). \tag{61}$$

It will be seen that the four equations (58)~(61) are derived from the motion equation, and equations (58) and (59) are just the same equations (45a,b) derived above.

Problems of wide electron beam focusing can be put forward in two ways. Usually, given the electrostatic and magnetic fields, the electron trajectories can be calculated through their differential equations, thus determining the focusing properties. At that time, eq.(45a) or (58) indicates the equation of principal trajectory, and the other three equations (59)(60)(61) will be automatically satisfied. But, on the contrary, the field distributions may also be defined from the given beams or trajectories. At this moment, (58) (59) (60) (61) will be the four complemental equations to be satisfied, which show the relations between the coefficients of electrostatic and magnetic fields, that are not independent of each other.

In general, the principal trajectory equation (45a), that are functions of arc length s, is not convenient to solve here, we transform it into a form using the laboratory Cartesian coordinate system. Because of

$$\varphi_2=n_x\phi_x+n_y\phi_y+n_z\phi_z,$$

$$B_b(s)=b_xB_x+b_yB_y+b_zB_z;$$

where $\phi_x=\dfrac{\partial\phi}{\partial x}$, $\phi_y=\dfrac{\partial\phi}{\partial y}$, $\phi_z=\dfrac{\partial\phi}{\partial z}$;

let b_x, n_x be expressed by (40), ρ by (4) and transform each derivative with respect to s into the derivative with respect to z, using

$$ds=\sqrt{1+[x'(z)]^2+[y'(z)]^2}\,dz, \tag{62}$$

then the principal trajectory equation (45a) can be written as

$$\frac{d}{dz}\left(\frac{\sqrt{\phi+\varepsilon_s}}{\sqrt{1+x'^2+y'^2}}x'\right)=\frac{\sqrt{1+x'^2+y'^2}}{2\sqrt{\phi+\varepsilon_s}}\frac{\partial\phi}{\partial x}$$

$$+\sqrt{\frac{e}{2m_0}}(B_y-y'B_z),$$

$$\frac{d}{dz}\left(\frac{\sqrt{\phi+\varepsilon_s}}{\sqrt{1+x'^2+y'^2}}y'\right)=\frac{\sqrt{1+x'^2+y'^2}}{2\sqrt{\phi+\varepsilon_s}}\frac{\partial\phi}{\partial y}$$

$$+\sqrt{\frac{e}{2m_0}}(x'B_z-B_x)\,. \tag{63a,b}$$

where the prime indicates derivatives with respect to z.

If eqs. (63a,b) are set up, suppose

$$\varphi_3=b_x\phi_x+b_y\phi_y+b_z\phi_z,$$

$$B_n(s)=n_xB_x+n_yB_y+n_zB_z,$$

substitute them into (45b), and through a series of transformations, we can prove that the equation (45b) is automatically satisfied.

For equations (46a,b) of neighboring "paraxial" trajectories, it is convenient to transform them into equations that are functions of z. By using (62), (46a,b) will have the form:

$$\frac{d}{dz}\left(\sqrt{\phi+\varepsilon_s}\,\frac{dp_2}{dz}\right)=(1+x'^2+y'^2)(Fp_2+Np_3)$$

$$+2K\sqrt{1+x'^2+y'^2}\,\frac{dp_3}{dz}+\sqrt{\phi+\varepsilon_s}\,\frac{x'x''+y'y''}{1+x'^2+y'^2}\,\frac{dp_2}{dz},$$

$$\frac{d}{dz}\left(\sqrt{\phi+\varepsilon_s}\,\frac{dp_3}{dz}\right)=(1+x'^2+y'^2)(Rp_3-Qp_2)$$

$$-2K\sqrt{1+x'^2+y'^2}\,\frac{dp_2}{dz}+\sqrt{\phi+\varepsilon_s}\,\frac{x'x''+y'y''}{1+x'^2+y'^2}\,\frac{dp_3}{dz}\,. \qquad (64a,b)$$

It is to be noted that the value of coefficients n, K, F, R, N and Q and x', y', x'', y'' must be taken from the principal trajectory when solving eqs.(64a,b).

Eqs.(64a,b) are two coupled, linear, second-order differential equations which may be transformed to a system of four first order equations. Let

$$w_1=p_2,\quad w_2=p_2',\quad w_3=p_3,\quad w_4=p_3';$$

then eqs.(64a,b) become

$$w_1'=w_2,$$
$$w_2'=g_1w_1+g_2w_3+f_1w_2+f_2w_4,$$
$$w_3'=w_4,$$
$$w_4'=h_1w_1+h_2w_3+f_1w_4-f_2w_2; \qquad (65)$$

where

$$g_1=\left(\frac{1+x'^2+y'^2}{\sqrt{\phi+\varepsilon_s}}F\right)_{x=x_N,\ y=y_N,\ z=z_N}$$

$$g_2=\left(\frac{1+x'^2+y'^2}{\sqrt{\phi+\varepsilon_s}}N\right)_{x=x_N,\ y=y_N,\ z=z_N}$$

$$h_1=\left(\frac{1+x'^2+y'^2}{\sqrt{\phi+\varepsilon_s}}Q\right)_{x=x_N,\ y=y_N,\ z=z_N}$$

$$h_2=-\left(\frac{1+x'^2+y'^2}{\sqrt{\phi+\varepsilon_s}}R\right)_{x=x_N,\ y=y_N,\ z=z_N}$$

$$f_1=\left(\frac{x'x''+y'y''}{1+x'^2+y'^2}-\frac{1}{2(\phi+\varepsilon_s)}\frac{d\phi}{dz}\right)_{x=x_N,\ y=y_N,\ z=z_N}$$

$$f_2=\left(2\frac{\sqrt{1+x'^2+y'^2}}{\sqrt{\phi+\varepsilon_s}}K\right)_{x=x_N,\ y=y_N,\ z=z_N} \qquad (66)$$

By numerical method, one may determine a set of solutions of (65):

$$w_1^{(i)},\ w_2^{(i)},\ w_3^{(i)},\ w_4^{(i)};\ (i=1,2,3,4)$$

which satisfy the following initial conditions:

i	1	2	3	4
$w_1^{(i)}=p_2^{(i)}$	1	0	0	0
$w_2^{(i)}=(p_2^{(i)})'$	0	$1/\sqrt{\varepsilon_s}$	0	0
$w_3^{(i)}=p_3^{(i)}$	0	0	1	0
$w_4^{(i)}=(p_3^{(i)})'$	0	0	0	$1/\sqrt{\varepsilon_s}$

(67)

The solutions of (64a,b) may then be written as

$$p_2(z)=p_{20}w_1^{(1)}+\sqrt{\varepsilon_s}p_{20}'w_1^{(2)}+p_{30}w_1^{(3)}+\sqrt{\varepsilon_s}\,p_{30}'\,w_1^{(4)},$$

$$p_3(z)=p_{20}\,w_3^{(1)}+\sqrt{\varepsilon_s}\,p_{20}'\,w_3^{(2)}+p_{30}w_3^{(3)}+\sqrt{\varepsilon_s}\,p_{30}'\,w_3^{(4)}; \qquad (68a,b)$$

and these satisfy the initial conditions:

$$p_2(z_0)=p_{20},\qquad p_2'(z_0)=p_{20}';$$
$$p_3(z_0)=p_{30},\qquad p_3'(z_0)=p_{30}'.$$

THE ORTHOGONAL CONDITION OF SYSTEMS WITH WIDE ELECTRON BEAM FOCUSING

The orthogonal condition of systems with narrow electron beam focusing was first given by Sturrock [6]. We will try here to derive the orthogonal condition of system with wide electron beam focusing. It is necessary first of all to simplify eqs.(46a,b), in order that the term p_3' does not appear in eq.(46a) and p_2' in eq.(46b).

Introducing coordinate transformation

$$p_2+ip_3=(u+iv)e^{i\chi} \qquad (69)$$

to eqs.(46a,b), and supposing that

$$\chi'=\frac{d\chi}{ds}=-\frac{K}{n}, \qquad (70)$$

we can get the curvilinear "paraxial" trajectory equations in a rotating coordinate system (u, v):

$$\frac{d}{ds}\left(n\frac{du}{ds}\right)=a_{11}u+a_{12}v,$$

$$\frac{d}{ds}\left(n\frac{dv}{ds}\right)=a_{21}u+a_{22}v; \qquad (71a,b)$$

where

$$a_{11}=F\cos^2\chi+R\sin^2\chi+(N-Q)\sin\chi\cos\chi-\frac{K^2}{n},$$

$$a_{12}=N\cos^2\chi+Q\sin^2\chi-(F-R)\sin\chi\cos\chi-K',$$

$$a_{21}=-N\sin^2\chi-Q\cos^2\chi-(F-R)\sin\chi\cos\chi+K'.$$

$$a_{22}=F\sin^2\chi+R\cos^2\chi-(N-Q)\sin\chi\cos\chi-\frac{K^2}{n}. \quad (72)$$

It may be seen from (71a,b) that the variables of these two equations are still not separated. But it can be proved from (72) and (48) that $a_{12}=a_{21}$. It means that (71a,b) are self-conjugate equations.

Apparently, in case the variables are completely separated, eqs.(71a,b) have to satisfy the following condition: $a_{12}=a_{21}=0$. If so, it can be written as $a_{12}+a_{21}=0$. Then we have

$$\tan 2\chi=\frac{N-Q}{F-R}. \quad (73)$$

Transforming the expressions of a_{11} and a_{22} in (72) by using (73), and let $U=a_{11}$, $V=a_{22}$, the we get

$$\frac{d}{ds}\left(n\frac{du}{ds}\right)=Uu,$$
$$\frac{d}{ds}\left(n\frac{dv}{ds}\right)=Vv; \quad (74a,b)$$

where

$$U=\frac{1}{2}(F+R)+\frac{1}{2}\sqrt{(F-R)^2+(N-Q)^2}-\frac{K^2}{n},$$
$$V=\frac{1}{2}(F+R)-\frac{1}{2}\sqrt{(F-R)^2+(N-Q)^2}-\frac{K^2}{n}. \quad (75)$$

From (70) and (73), we can derive the conditional relationship, which the coefficients n, K, F, R, N and Q must satisfy, when the variables are separated:

$$n[(N-Q)(F'-R')-(F-R)(N'-Q')]$$
$$=2K[(F-R)^2+(N-Q)^2]. \quad (76)$$

This is the orthogonal condition for systems with wide electron beam focusing. Systems, which satisfy the orthogonal condition (76), will be named orthogonal curvilinear "paraxial" systems.

For curvilinear electron optical systems satisfying the orthogonal condition, the image with point focusing or line focusing can be formed, as it has been pointed out by Sturrock [6].

(74a,b) are second order homogeneous linear differential equations of u and v respectively. Although in these equations a singular point exists at the cathode surface when $\varepsilon_S=0$. There are no essential difficulties in solving eqs.(74a,b), compared with the cathode lenses with axial symmetry. Thus, the general solutions u and v of eqs. (74a,b) can be expressed by two linear independent special solutions. Let u_α, u_β and v_α, v_β be the two sets of special solutions of eqs. (74a, b), which satisfy the following conditions:

$$u_\alpha(0)=v_\alpha(0)=0, \quad \sqrt{\varepsilon_S}u_\alpha'(0)=\sqrt{\varepsilon_S}v_\alpha'(0)=1;$$
$$u_\beta(0)=v_\beta(0)=1, \quad \sqrt{\varepsilon_S}u_\beta'(0)=\sqrt{\varepsilon_S}v_\beta'(0)=0. \quad (77)$$

In order to obtain an image of point focusing (stigmatic imaging), it is obviously necessary to make U and V in eqs.(74a,b) equal to each other. In that case, it becomes possible only when

$$F-R=N-Q=0. \quad (78)$$

Then at the image plane $s=s_i$, which corresponds to $\varepsilon_S=\varepsilon_{S1}$, we have

$$u_\alpha(\varepsilon_{S1},s_i)=v_\alpha(\varepsilon_{S1},s_i)=0,$$
$$u_\alpha'(\varepsilon_{S1},s_i)=v_\alpha'(\varepsilon_{S1},s_i). \quad (79)$$

The modulus of its magnification can be written as

$$M_u=M_v=M=\left|\frac{1}{u_\alpha'(\varepsilon_{S1},s_i)\sqrt{\phi(s_i)+\varepsilon_{S1}}}\right|. \quad (80)$$

It can be seen from (74a,b) and (76) that systems satisfying condition (78) will certainly satisfy the orthogonal condition given above. Condition (78) corresponds to the two relationships that follows:

$$\frac{\sqrt{\phi+\varepsilon_S}}{\rho^2}+\frac{\varphi_{33}-\varphi_{22}}{2\sqrt{\phi+\varepsilon_S}}+\frac{\varphi_2}{\rho\sqrt{\phi+\varepsilon_S}}-\sqrt{\frac{2e}{m_0}}\Big(B_{n3}$$
$$+\frac{B_b(s)\varphi_2+B_n(s)\varphi_3}{4\sqrt{\phi+\varepsilon_S}}\Big)=0,$$

$$\frac{\varphi_{23}}{\sqrt{\phi+\varepsilon_S}}-\frac{\varphi_3}{\rho\sqrt{\phi+\varepsilon_S}}-\sqrt{\frac{e}{2m_0}}\Big(B_{n2}-B_{b3}$$
$$-\frac{B_b(s)\varphi_3-B_n(s)\varphi_2}{2\sqrt{\phi+\varepsilon_S}}\Big)=0. \quad (81)$$

Condition (78) or (81) shows that the electron beam emitted from the cathode surface with $\varepsilon_S=\varepsilon_{S1}$ will have an ideal point focusing (stigmatic imaging), therefore the system will have properties similar to those of the electron optical system with axial symmetry.

If the system with wide electron beam focusing only satisfies the orthogonal condition (76), but does not satisfy condition (78), thus $U \neq V$, and the special solutions u_α, u_β and v_α, v_β are different. This means anisotropy in magnification, and its moduli in two directions in case of $\varepsilon_S = \varepsilon_{S1}$ will be

$$M_u = \left| \frac{1}{u_\alpha'(\varepsilon_{S1}, s_{1i})\sqrt{\phi(s_{1i}) + \varepsilon_{S1}}} \right|, \tag{82a}$$

$$M_v = \left| \frac{1}{v_\alpha'(\varepsilon_{S1}, s_{2i})\sqrt{\phi(s_{2i}) + \varepsilon_{S1}}} \right|. \tag{82b}$$

The positions s_{1i} and s_{2i} of focusing image for the electron beam do not coincide with each other even when the initial energy $e\varepsilon_S = e\varepsilon_{S1}$ is the same. It means that two ideal focusing images corresponding to $\varepsilon_S = \varepsilon_{S1}$ will be possible in a given orthogonal curvilinear system. A focusing segment in the binormal direction **b** at the location $s = s_{1i}$ is formed, and another focusing segment in the normal direction **n** at the location $s = s_{2i}$ at the same time. Thus, the difference between the two locations can be defined as the astigmatism of the system with wide electron beam focusing.

SOME PRACTICAL EXAMPLES FOR SYSTEMS WITH WIDE ELECTRON BEAM FOCUSING

1. System with Wide Electron Beam Focusing, in which the Principal Trajectory is a Linear Axis of Symmetry

When the principal trajectory is a linear axis of symmetry, $\frac{1}{\rho} = \frac{1}{\tau} = 0$. In consideration of axially symmetric fields, their potential on the axis takes the extreme value, and there is no difference between the normal and binormal directions.

Let $p = p_2 + ip_3$, $s = z$, $\varepsilon_s = \varepsilon_z$, $\phi(s) = \phi(z)$, $B_t(s) = B(z)$, we obtain

$$n = \sqrt{\phi(z) + \varepsilon_z}\,, \quad K = -\sqrt{\frac{e}{8m_0}}B(z), \quad N - Q = 0,$$

$$F = R = -\frac{1}{4}\frac{\phi''}{\sqrt{\phi(z) + \varepsilon_z}}. \tag{83}$$

Then

$$U = V = -\frac{1}{4}\frac{\phi''}{\sqrt{\phi(z) + \varepsilon_z}} - \frac{e}{8m_0}\frac{B^2(z)}{\sqrt{\phi(z) + \varepsilon_z}}. \tag{84}$$

It is obvious that the system of wide electron beam focusing with a linear axis of symmetry not only satisfies condition (76), but also satisfies condition (78). Thus, this system has the properties of ideal focusing (point focusing).

Let

$$u = u + iv = pe^{-i\chi}, \tag{85}$$

where

$$\chi' = \sqrt{\frac{e}{8m_0}}\frac{B(z)}{\sqrt{\phi(z) + \varepsilon_z}}.$$

eqs. (74a,b) will then have the following form:

$$[\phi(z) + \varepsilon_z]u'' + \frac{1}{2}\phi'(z)u' + \frac{1}{4}[(\phi''(z) + \frac{e}{2m_0}B^2(z)]u = 0. \tag{86}$$

It is just the linear differential equation of a system with wide electron beam focusing [8], where the principal trajectory is a straight axis having axial symmetry. Using eq.(86), we have solved a concentric spherical system of combined electrostatic and magnetic fields with wide electron beam focusing [1].

2. Electrostatic Cathode Lenses with Wide Electron Beam Focusing

For electrostatic cathode lenses with axial symmetry, the off-axis principal trajectory emitted from the cathode surface will be a plane curve which is normal to the cathode surface. Thus,

$$B = 0, \quad \frac{1}{\tau} = 0, \quad \varphi_{23} = 0,$$

then from (47) we have

$$n = \sqrt{\phi(s) + \varepsilon_s}\,, \qquad K = 0,$$

$$F = -\frac{3}{\rho^2}\sqrt{\phi(s) + \varepsilon_s} + \frac{\varphi_{22}}{2\sqrt{\phi(s) + \varepsilon_s}},$$

$$R = \frac{\varphi_{33}}{2\sqrt{\phi(s) + \varepsilon_s}}, \qquad N = Q = 0. \tag{87}$$

The coefficients in expressions (87) also satisfy the orthogonal condition (76). It is obvious that the variables u and v are also separated.

Because $\chi=0$, suppose that $u=p_z$, $v=p_3$; then the principal trajectory equation can be written as

$$\varphi_2=\frac{2[\phi(s)+\varepsilon_s]}{\rho}. \tag{88}$$

The curvilinear "paraxial" trajectory equations would be

$$\frac{d}{ds}\left(\sqrt{\phi(s)+\varepsilon_s}\frac{dp_2}{ds}\right)=\left\{-\frac{3\sqrt{\phi(s)+\varepsilon_s}}{\rho^2}+\frac{\varphi_{22}}{2\sqrt{\phi(s)+\varepsilon_s}}\right\}p_2,$$

$$\frac{d}{ds}\left(\sqrt{\phi(s)+\varepsilon_s}\frac{dp_3}{ds}\right)=\frac{\varphi_{33}}{2\sqrt{\phi(s)+\varepsilon_3}}p_3. \tag{89a,b}$$

The arc length s, which defines a coordinate curve of the principal trajectory, can be transformed into a form with cylindrical coordinates (r,z). From (88) and (89), we have the principal trajectory equation:

$$r''=\frac{1+r'^2}{2[\varphi(z,r)+\varepsilon_s]}\left(\frac{\partial\varphi}{\partial r}-r'\frac{\partial\varphi}{\partial z}\right), \tag{90}$$

and the curvilinear neighboring trajectory equation:

$$p_2''+F_1p_2'+F_2p_2=0,$$

$$p_3''+G_1p_3'+G_2p_3=0; \tag{91a,b}$$

where

$$F_1=G_1=\frac{1+r'^2}{2[\varphi(z,r)+\varepsilon_s]}\frac{\partial\varphi}{\partial z},$$

$$G_2=\frac{-(1+r'^2)}{2[\varphi(z,r)+\varepsilon_s]}\frac{1}{r}\frac{\partial\varphi}{\partial r},$$

$$F_2=\frac{3r''^2}{(1+r'^2)^2}+\frac{1}{2[\varphi(z,r)+\varepsilon_s]}\left(-\frac{\partial^2\varphi}{\partial r^2}+2r'\frac{\partial^2\varphi}{\partial r\partial z}-r'^2\frac{\partial^2\varphi}{\partial z^2}\right). \tag{92}$$

It is necessary to remind that the terms in the coefficients F_1, G_1, F_2, G_2, should be taken to be the values on the principal trajectory, and that the prime notations in the eqs.(90), (91a,b) and (92) denote derivatives with respect to the axial coordinate z.

For the electron beam emitted from the axial point of cathode surface, because the principal trajectory is actually the rotating symmetrical axis, we will have $r=r'=r''=0$, and

$$\left(\frac{1}{r}\frac{\partial\varphi}{\partial r}\right)_{r=0}=\left(\frac{\partial^2\varphi}{\partial r^2}\right)_{r=0}=-\frac{1}{2}\phi''(z). \tag{93}$$

Let $p=p_2+ip_3$, $\varepsilon_s=\varepsilon_z$, then eqs.(91a,b) take the following form:

$$p''+\frac{1}{2}\frac{\phi'}{\phi(z)+\varepsilon_z}p'+\frac{1}{4}\frac{\phi''}{\phi(z)+\varepsilon_z}p=0. \tag{94}$$

It is just the well-known "paraxial" equation in the case of electrostatic cathode lenses.

3. Two Dimensional Field of Plane Symmetry with Wide Electron Beam Focusing

Let the principal trajectory be an axis of symmetry for a two dimensional field with plane symmetry, then the orthogonal coordinate system (p_2,p_3,s) can be replaced by the Cartesian coordinate system (x,y,z), and the coordinate z, the axis of symmetry for the system, coincides with the principal trajectory. Therefore,

$$\frac{1}{\rho}=\frac{1}{\tau}=0.$$

For the two dimensional field, which is independent of the coordinate x, let $B_t(z)=B$, $\varepsilon_s=\varepsilon_z$, the "paraxial" trajectory equations (46a,b) can be written as

$$x'=-\sqrt{\frac{e}{2m_0}}\frac{1}{\sqrt{\phi+\varepsilon_z}}By+\frac{C}{\sqrt{\phi+\varepsilon_z}},$$

$$y''+\frac{\phi'}{2[\phi+\varepsilon_z]}y'+\left(\frac{\phi''}{2[\phi+\varepsilon_z]}+\frac{eB^2}{2m_0[\phi+\varepsilon_z]}\right)y=\sqrt{\frac{e}{2m_0}}C\frac{B}{\phi+\varepsilon_z}; \tag{95a,b}$$

where

$$C=\sqrt{\varepsilon_z}x_0'+\sqrt{\frac{e}{2m_0}}B_0y_0.$$

It will be seen from (95a,b) that this system does not satisfy the orthogonal condition.

For the case of the electrostatic field of plane symmetry with wide electron beam focusing, we obtain

$$x'=\frac{x_0'\sqrt{\varepsilon_z}}{\sqrt{\phi(z)+\varepsilon_z}},$$

$$y''+\frac{\phi'}{2[\phi(z)+\varepsilon_z]}y'+\frac{\phi''}{2[\phi(z)+\varepsilon_z]}y=0. \tag{96a,b}$$

It is obvious that the variables x and y are also separated. It is enough to prove that this kind of focusing system, which is actually a so-called cylindrical cathode lens, will focus only the plane electron beam located in the plane (y, z). In the direction perpendicular to the plane (y, z), there is no applied force of electric field. Thus, the solid electron beam emitted from the axial point of cathode surface, will be affected only in the direction perpendicular to the plane of symmetry. It will not have a point focusing as in the field with axial symmetry, but will have a line focusing in the direction perpendicular to the plane (y,z).

AN APPROACH TO THE TREATMENT OF DEFLECTION DEFOCUSING EFFECTS IN CATHODE-RAY TUBES USING THEORY OF WIDE ELECTRON BEAM FOCUSING

To describe the convergence defects of deflection fields at angles larger than 45°, Hutter [3] put forward a method to the theoretical treatment, based on the application of the theory of «focusing fields with arbitrary curved optical axes» . In his paper, a rotating system of coordinates u, v, w has been introduced, which rotates with respect to the Frenet local coordinate system, the potential φ and field vector **B** are expanded in a power series of u and v with coefficients that are functions of arc length w, the zero-order path conditions and the paraxial equations of pure electrostatic and pure magnetic deflection systems have been derived, and the aberrations of system have also been discussed. Hutter's paper presented a new approach to solve the theory of the deflection system at large deflection angles.

It must be pointed out that from the beginning of Hutter's paper the introduction of a rotating coordinate system with respect to Frenet local coordinate system seems to complicate the problem, and there is no clear relation in his paper between the zero-order path equation derived from the rotating coordinate system (see eq. (57) in ref. [3].) and the equation of central trajectory derived from the laboratory Cartesian coordinate system (see eq. (7) in ref. [3]).

In fact, one may extend the principal trajectory equations (45a,b) and the "paraxial" trajectory equations (46a,b) derived above to solve the problem of large angle deflection, if we assume $\varphi(s=0)\neq 0$, $\varepsilon_s=\varepsilon_0$, $\phi(s) \gg \varepsilon_0$. Therefore, problem with focusing of deflection system at large deflecting angle (in general, it is treated as a problem with narrow electron beam focusing [2, 4].) in a curvilinear coordinate system can be regarded as a special case thereof.

Now we illustrate with example for a pure magnetic deflection system with large angle deflection given by Hutter. In that case, $\phi=V_0=\text{const}$. From (45a,b)(46a,b), we have the principal trajectory equations:

$$\frac{1}{\rho}=\sqrt{\frac{e}{2m_0V_0}}B_b(s),$$

$$B_n(s)=0; \qquad (97a,b)$$

and the "paraxial" trajectory equations:

$$p_2''(s)-2\left(\frac{1}{\tau}-\sqrt{\frac{e}{8m_0V_0}}B_t(s)\right)p_3'(s)-\left[\left(\frac{1}{\tau^2}-\frac{1}{\rho^2}\right)-\sqrt{\frac{e}{2m_0V_0}}\left(\frac{1}{\tau}B_t(s)-B_{n3}\right)\right]p_2(s)$$

$$+\left(\frac{1}{\tau^2}\frac{d\tau}{ds}-\sqrt{\frac{e}{2m_0V_0}}B_{b3}\right)p_3(s)=0,$$

$$p_3''(s)+2\left(\frac{1}{\tau}-\sqrt{\frac{e}{8m_0V_0}}B_t(s)\right)p_2'(s)-\left[\frac{1}{\tau^2}-\sqrt{\frac{e}{2m_0V_0}}\left(\frac{1}{\tau}B_t(s)+B_{n3}\right)\right]p_3(s)$$

$$-\left(\frac{1}{\tau^2}\frac{d\tau}{ds}-\sqrt{\frac{e}{2m_0V_0}}B_{n2}\right)p_2(s)=0. \qquad (98a,b)$$

Transforming eq. (97a) into a form expressed by laboratory coordinate system, we obtain

$$\frac{d}{dz}\left(\frac{x'}{\sqrt{1+x'^2+y'^2}}\right)=\sqrt{\frac{e}{2m_0V_0}}(B_Y-y'Bz),$$

$$\frac{d}{dz}\left(\frac{y'}{\sqrt{1+x'^2+y'^2}}\right)=\sqrt{\frac{e}{2m_0V_0}}(x'B_z-B_x).$$

(99a,b)

It is enough to prove that eq. (97b) is automatically satisfied if eqs. (99a,b) is set up.

Similarly, for a pure electrostatic deflection system with large angle deflection, we have the principal trajectory equations:

$$\frac{1}{\rho}=\frac{\varphi_2}{2\phi},$$

$$\varphi_3=0; \qquad (100a,b)$$

and the paraxial trajectory equations:

$$p_2''+\frac{1}{2}\frac{\phi'}{\phi}p_2'-2\frac{1}{\tau}p_3'+\Big[\Big(\frac{1}{\rho^2}-\frac{1}{\tau^2}\Big)+\frac{\varphi_2}{\rho\phi}-\frac{\varphi_{22}}{2\phi}\Big]p_2+\Big(\frac{1}{\tau^2}\tau'-\frac{\phi'}{2\tau\phi}-\frac{\varphi_{23}}{2\phi}\Big)p_3=0,$$

$$p_3''+\frac{1}{2}\frac{\phi'}{\phi}p_3'+2\frac{1}{\tau}p_2'-\Big(\frac{1}{\tau^2}+\frac{\varphi_{33}}{2\phi}\Big)p_3-\Big(\frac{1}{\tau^2}\tau'-\frac{\phi'}{2\tau\phi}-\frac{\phi_{23}}{2\phi}\Big)p_2=0. \quad (101a,b)$$

Under the circumstances of the laboratory Cartesian coordinate system, the principal trajectory equation (100a) corresponds to the following equations:

$$\frac{d}{dz}\Big(\frac{\sqrt{\phi}}{\sqrt{1+x'^2+y'^2}}x'\Big)=\frac{\sqrt{1+x'^2+y'^2}}{2\sqrt{\phi}}\frac{\partial\phi}{\partial x},$$

$$\frac{d}{dz}\Big(\frac{\sqrt{\phi}}{\sqrt{1+x'^2+y'^2}}y'\Big)=\frac{\sqrt{1+x'^2+y'^2}}{2\sqrt{\phi}}\frac{\partial\phi}{\partial y}. \quad (102a,b)$$

We can also prove that eq.(100b) is automatically satisfied if eqs.(102a,b) are set up.

For computation, it is convenient to transform (98a,b) or (101a,b) into equations that are functions of z by using

$$\frac{dp}{ds}=\frac{dp}{dz}\frac{1}{\sqrt{1+x'^2+y'^2}},$$

$$\frac{d^2p}{ds^2}=\frac{d^2p}{dz^2}\frac{1}{1+x'^2+y'^2}-\frac{dp}{dz}\frac{x'x''+y'y''}{(1+x'^2+y'^2)^2}. \quad (103)$$

For example, transforming (98a,b) by (103) into a system of four first order equations, similar to eq.(65), where coefficients g_1, g_2, h_1, h_2, f_1, f_2, can be written as

$$g_1=\Big\{(1+x'^2+y'^2)\Big[\frac{1}{\tau^2}-\frac{1}{\rho^2}-\sqrt{\frac{e}{2m_0V_0}}\Big(\frac{1}{\tau}B_t(s)-B_{n3}\Big)\Big]\Big\}_{x=x_N,\ y=y_N,\ z=z_N}$$

$$g_2=\Big\{-(1+x'^2+y'^2)\Big(\frac{1}{\tau^2}\frac{d\tau}{ds}-\sqrt{\frac{e}{2m_0V_0}}B_{b3}\Big)\Big\}_{x=x_N,y=y_N,z=z_N}$$

$$h_1=\Big\{(1+x'^2+y'^2)\Big(\frac{1}{\tau^2}\frac{d\tau}{ds}-\sqrt{\frac{e}{2m_0V_0}}B_{n2}\Big)\Big\}_{x=x_N,\ y=y_N,\ z=z_N}$$

$$h_2=\Big\{(1+x'^2+y'^2)\Big[\frac{1}{\tau^2}-\sqrt{\frac{e}{2m_0V_0}}\Big(\frac{1}{\tau}B_t(s)+B_{n3}\Big)\Big]\Big\}_{x=x_N,\ y=y_N,\ z=z_N}$$

$$f_1=\Big(\frac{x'x''+y'y''}{1+x'^2+y'^2}\Big)_{x=x_N,\ y=y_N,\ z=z_N}$$

$$f_2=\Big\{\sqrt{1+x'^2+y'^2}\Big(\frac{2}{\tau}-\sqrt{\frac{e}{2m_0V_0}}B_t(s)\Big)\Big\}_{x=x_N,\ y=y_N,\ z=z_N} \quad (104)$$

Similarly, for eqs.(101a,b), we have

$$g_1=\Big\{-(1+x'^2+y'^2)\Big[\frac{1}{\rho^2}-\frac{1}{\tau^2}+\frac{\varphi_2}{\rho\phi}-\frac{\varphi_{22}}{2\phi}\Big]\Big\}_{x=x_N,\ y=y_N,\ z=z_N}$$

$$g_2=\Big\{-(1+x'^2+y'^2)\Big(\frac{1}{\tau^2}\tau'-\frac{\phi'}{2\tau\phi}-\frac{\varphi_{23}}{2\phi}\Big)\Big\}_{x=x_N,\ y=y_N,\ z=z_N}$$

$$h_1=-g_2,$$

$$h_2=\Big\{(1+x'^2+y'^2)\Big(\frac{1}{\tau^2}+\frac{\varphi_{33}}{2\phi}\Big)\Big\}_{x=x_N,\ y=y_N,\ z=z_N}$$

$$f_1=\Big\{\frac{x'x''+y'y''}{1+x'^2+y'^2}-\frac{1}{2}\frac{\phi'}{\phi}\sqrt{1+x'^2+y'^2}\Big\}_{x=x_N,\ y=y_N,\ z=z_N}$$

$$f_2=\Big\{2\frac{1}{\tau}\sqrt{1+x'^2+y'^2}\Big\}_{x=x_N,\ y=y_N,\ z=z_N} \quad (105)$$

By numerical method, one may determine a set of solutions of (65):

$$w_1^{(i)},\quad w_2^{(i)},\quad w_3^{(i)},\quad w_4^{(i)} \qquad (i=1,2,3,4)$$

which satisfies the following conditions:

i	1	2	3	4
$w_1^{(i)}=p_2^{(i)}$	1	0	0	0
$w_2^{(i)}=(p_2^{(i)})'$	0	1	0	0
$w_3^{(i)}=p_3^{(i)}$	0	0	1	0
$w_4^{(i)}=(p_3^{(i)})'$	0	0	0	1

(106)

The solutions of p_2, p_3 may be expressed by

$$p_2(z)=p_{20}w_1^{(1)}+p_{20}'w_1^{(2)}+p_{30}w_1^{(3)}+p_{30}'w_1^{(4)}$$

$$p_3(z)=p_{20}w_3^{(1)}+p_{20}'w_3^{(2)}+p_{30}w_3^{(3)}+p_{30}'w_3^{(4)} \qquad (107a,b)$$

where the prime denotes the derivatives to z, and

$$p_2(z_0)=p_{20},\quad p_2'(z_0)=p_{20}',$$
$$p_3(z_0)=p_{30},\quad p_3'(z_0)=p_{30}'.$$

It will be seen from (67) and (106) that a great difference exists in the assumption of the initial conditions for solving the two kinds of curvilinear paraxial trajectory equations.

SUMMARY

1. In the present paper, the author tries to develop his previous work. Based on the assumption that the electrostatic potential φ and magnetic field vector **B** are known functions in the laboratory Cartesian coordinate system, we have deduced the principal trajectory equation and the "paraxial" trajectory equations of systems with wide electron beam focusing by given mathematical expressions of fields along curved optical axis, we have also derived the orthogonal condition of a curvilinear "paraxial" system with wide electron beam focusing, Some practical examples of systems with wide' electron beam focusing satisfying the orthogonal condition are also given.

2. It has been found in the present paper that the zero order equations of trajectory derived above are just the same equations of principal trajectory deduced from the fundamental equation of electron motion. It seems that the principal trajectory is determined by two differential equations. But in fact, if we investigate the focusing of wide electron beam from a given electrostatic potential and magnetic field, only one of these two equations is required for determining the principal trajectory, the other equation is automatically satisfied.

3. For problems concerned with wide electron beam focusing, no matter whether it is a system with axial symmetry or a curvilinear "paraxial" system satisfying orthogonal condition, the principal trajectory as well as the curvilinear "paraxial" neighboring trajectories must take the same value for $\varepsilon_{\varepsilon_s}$, which is the initial energy of principal trajectory along the tangential direction. Only under such a condition will the system have properties of point focusing or line focusing.

4. This paper has given as a set of equations for solving the principal trajectory and "paraxial" trajectories with derivatives that are functions of laboratory Cartesian coordinate z, so that the given equations are convenient for computation.

5. The result of this paper can be extended to solve deflection systems with deflection angles larger than 45°, and the equations obtained for solving the principal trajectory equation and paraxial trajectories in the deflection systems appear to be simpler and more convenient than predecessors' work.

ACKNOWLEDGEMENT

The author wishes to thank Dr E.Munro from Imperial College for a review of the manuscript.

REFERENCES

1. Chou LW (Zhou Li-wei). (1979). Electron Optics of Concentric Electromagnetic Focusing Systems, in: Advances in Electronics and Electron Physics, L.Marton and C. Marton (ed), Academic Press, NY, **52**, 119-132.

2. Grinberg GA. (1948). Selected problems of Mathematical Theory of Electric and Magnetic Phenomena. Press of Academy of Sciences USSR, Moscow-Leningrad, 507-535. (in Russian)

3. Hutter RGE. (1970). Deflection defocusing effects in cathode-ray tubes at large deflection angles. IEEE ED. 17, 1022-1031.

4. Kas'yankov PP. (1956). Theory of Electromagnetic Systems with Curvilinear Axis. Press of Leningrad University, Leningrad, 50-57. (in Russian)

5. Spiegel MR. (1974). Schaum's Outline of Theory and Problems of Vector Analysis and an Introduction to Tensor Analysis. McGraw-Hill Book Co, Chapter 8.

6. Sturrock PA. (1955). Static and Dynamic Electron Optics. Cambridge at the University Press, Chapter 3.

7. Tsukkerman II. (1961). Electron Optics in Television. Pergamon Press, NY, Chapter 2.

8. Ximen Ji-ye. (1957). On the electron optical properties and aberration theory of a combined immersion objective. Acta Phys. Sinica. 13, 339-356. (in Chinese)

9. Zhou Li-wei. (1983). A generalized theory of wide electron beam focusing. Abstracts of 8th Symposium on Photoelectronic Image Devices. Imperial College, London 90.

10. Zhou Li-wei, Ai Ke-cong, Pan Shun-chen. (1983). On aberration theory of the combined electromagnetic focusing cathode lenses. Acta Phys. Sinica. 32, 376-392. (in Chinese)

Electron Optical Systems (pp. 63-73)
SEM Inc., AMF O'Hare (Chicago), IL 60666-0507, U.S.A.

0-931288-34-7/84$1.00+.05

RECENT DEVELOPMENTS IN NUMERICAL ELECTRON OPTICS

Erwin K. Kasper

Institut fuer Angewandte Physik
der Universitaet
Auf der Morgenstelle
74 Tuebingen, W. Germany
Phone: 07071/296747
292429

Abstract

The familiar methods for the numerical calculation of fields in electron optical devices are outlined briefly. For the solution of self-adjoint elliptic differential equations in orthogonal curvilinear coordinate systems a favourable ninepoint discretization is worked out which can be applied favourably e.g. to spherical mesh grids. The field calculation in magnetic deflection systems by means of an integral equation method is also highly advantageous. The methods for the field calculation can be still more improved by means of suitable hybrid procedures.

A second and shorter contribution is concerned with ray tracing and aberrations. Some favourable numerically stable new forms of the ray equation are derived and thereafter a new simple method for the determination of aberrations is outlined.

Key words: field calculation, ray tracing, aberrations, integral equation, hybrid method, Poisson's equation, discretization.

Introduction

Due to the rapid advances in computer technology it is possible to calculate now with reasonable effort and sufficient accuracy the properties of electron optical systems, and this is done very frequently, since the aid of a computer facilitates essentially the design of new devices. Generally such a computer-aided design consists of four subsequent and partly interdependent steps. These are the calculation of electric and magnetic fields from given source distributions and boundary conditions, the tracing of electron trajectories through these fields, the determination of paraxial properties and of aberrations from the obtained trajectories and finally the optimization of the design in question. The latter involves a repetition of the preceding steps under suitable variations of system parameters, until a relative minimum of the resulting aberration is found. In all of the above mentioned fields some progress was made during the last years, and it is the aim of this paper to present a short review of this progress and to give some outlook on further developments.

Field calculation

Generally three different methods of field calculation are in widespread use, these are the finite-element method (FEM), the finite-difference method (FDM) and the integral-equation method (IEM). Each of these has specific advantages and also disadvantages, so that none of them can be ignored completely or dealt with exclusively.

The finite-element method

The FEM consists in the dissection of the domain of solution into suitably chosen volume elements which must be irregular in the general case in order to fit to the boundaries. In the case of two-dimensional problems the mesh grid obtained in this way consists usually of irregular triangles with six elements joining together in each internal node.

It is familiar to derive the necessary equations for the values of the potential in the nodes from a variational principle: in each of the finite elements the stored electric or magnetic field energy must be minimized. In order to evaluate this principle, some assumptions about the required solution must be made. The simplest one is that the potential is a piecewise linear function. This version of the FEM has been introduced by Munro (1971, 1973) who applied it successfully to a large variety of electron optical devices, most recently to magnetic and electric deflection systems (Munro and Chu, 1982 a,b). A review of applications of Munro's method to the design of unconventional magnetic lenses is given by Mulvey (1982).

While the version of the FEM, described above is still frequently used, great improvements have been made in the past decade, mainly outside electron optics.Better approximations than piecewise linear functions for the potential are worked out (Silvester and Konrad, 1973), more generally curvilinear and even infinite elements are employed (Lencova and Lenc, 1982) and other than variational formulations are proposed. A review of recent improvements of the FEM is given in a volume edited by Chari and Silvester (1980). Even with respect to the still most familiar case of planar triangular grids progress was made, as Hermeline (1982) proposed a new and reliable method for the automatic generation of such grids under given boundary conditions and other constraints.

In spite of all these improvements the FEM has still severe disadvantages, the most serious one being the fact that the interpolation in irregular mesh grids is too complicated in order to ensure the continuity of the field strength and its derivatives along lines crossing orthogonally the mesh lines of the grid. This imposes strong restrictions on the ray tracing,as will be outlined in the "Ray Tracing and Aberrations" section.

The finite-difference method

In the applications of this method, the domain of solution is covered by a regular, usually square-shaped mesh grid. Most frequently this grid does not fit to the boundaries, so that many irregular nodes are to be considered. Theoretically it is no problem to derive discretization formulae for arbitrary irregular configurations, but in practice the consideration of many different irregular situations makes the method inconvenient. Nevertheless, the FDM has the advantage that - besides exceptional cases in electron guns - the grid is regular within the domain of the electron beam, so that the necessary calculation of the field-strength can be performed with sufficient accuracy.A review of these standard techniques has been given by Kasper (1982).

An interesting modification of the FDM has been proposed by Kang et al. (1981, 1983). In order to overcome the difficulties arising from the extreme differences of geometrical dimensions in electron guns with field emission cathodes, they proposed the use of an exponentially increasing spherical mesh grid which they named SCWIM (spherical coordinate with increasing mesh). Indeed, in this way a reasonable accuracy can be achieved. This method can be further generalized and improved, as will be outlined now.

We start from a general self-adjoint partial differential equation (PDE)

$$\frac{\partial}{\partial u}\left(p^2 v^\alpha \frac{\partial V}{\partial u}\right) + \frac{\partial}{\partial v}\left(p^2 v^\alpha \frac{\partial V}{\partial v}\right) + p^2 v^\alpha\left(\hat{q}\, V + s\right) = 0, \tag{1}$$

$p(u,v)$, $q(u,v)$ and $s(u,v)$ being non-singular analytic functions of the variables u and v, further $p > 0$, $\alpha \geq -1$, $v \geq 0$. The exponent α arises from a possible axial symmetry of the potential $V(u,v)$, for instance $\alpha = 1$ for rotationally symmetric fields and $\alpha = 2m + 1$ for multipole fields of order m. The product-relation

$$V(u,v) = U(u,v)/p(u,v) \tag{2}$$

reduces (1) to

$$\Delta_\alpha U = -g(u,v) := q(u,v)\cdot U - p\cdot s \tag{3}$$

with

$$\Delta_\alpha := \frac{\partial^2}{\partial u^2} + \frac{\partial^2}{\partial v^2} + \frac{\alpha}{v}\cdot\frac{\partial}{\partial v} \tag{4}$$

and

$$q(u,v) = \hat{q}(u,v) - p^{-1}\Delta_\alpha p \tag{5}$$

For the discretization of (3) very accurate nine-point-formulae have been derived (Kasper, 1976), but with respect to the transformation (2) a still more favourable form of the discretization can be found. If we introduce the coordinates $u = ih$, $v = kh$ (i,k integers) of the nodes and use corresponding subscripts ($U_{i,k} := U(ih, kh)$) then we can introduce favourably a new field

$$W_{i,k} := U_{i,k} + \frac{h^2}{12}\cdot g_{i,k} = P_{i,k}\left(V_{i,k} + \frac{h^2}{12}\left(q_{i,k} V_{i,k} + s_{i,k}\right)\right) \tag{6}$$

and have now for $k \geq 1$:

$$W_{i,k} = A_{-1,k} W_{i,k-1} + A_{1,k} W_{i,k+1} + A_{0,k} h^2 g_{i,k} + \sum_{j=-1}^{1} B_{ij}\left(W_{i-1,k+j} + W_{i+1,k+j}\right) + O(h^6) \tag{7}$$

with the coefficients

$$\gamma_k := \frac{3\alpha(\alpha-2)}{10(24k^2+\alpha^2-2\alpha-6)} \; ,$$

$$A_{\pm 1,k} = \frac{1}{5} \pm \frac{\alpha}{10k} + \gamma_k\left(1 \mp \frac{2\alpha+5}{6k}\right),$$

$$A_{0,k} = \frac{3}{10} - \gamma_k \; , \quad B_{0,k} = \frac{1}{5} - \gamma_k \, ,$$

$$B_{\pm 1,k} = \frac{1}{20} \pm \frac{\alpha}{40k} \mp \gamma_k \cdot \frac{\alpha+1}{12k} \; . \qquad (8)$$

On the axis of symmetry, that means for $k = v = 0$, instead of (7) and of (8) the formula

$$W_{i,0} = A_{1,0} W_{i,1} + \sum_{j=0}^{1} B_{j,0}(W_{i-1,j} + W_{i+1,j}) + h^2\left(A_{0,0}\, g_{0,0} + C\,(g_{0,0} - g_{0,1})\right) + O(h^6) \qquad (9)$$

with the coefficients

$$\beta := \frac{(1+\alpha)(6+\alpha)}{6(3+\alpha)} \; , \quad \gamma_0 := \frac{1}{2(2+\alpha-\beta)} \; ,$$

$$A_{0,0} = \gamma_0 \; , \quad A_{1,0} = 2\gamma_0(1+\alpha-\beta),$$

$$B_{0,0} = \gamma_0(1-\beta) \; , \quad B_{1,0} = \gamma_0 \beta \; ,$$

$$C = \gamma_0 \frac{\alpha(1+\alpha)}{6(3+\alpha)} \; . \qquad (10)$$

is to be employed. The derivation of this discretization cannot be presented here and will be published elsewhere.

A good example for the practical application of these formulae is the SCWIM grid. This can be considered as a conformal mapping of an originally cylindric grid with square-shaped meshes in the axial section. Let (z, r, ϕ) denote cylindric coordinates and (R, θ, ϕ) coaxial spherical coordinates with $z = R\cos\theta$, $r = R\sin\theta$, then we have

$$z + ir = R\exp(i\theta) = R_0 \exp(u + iv) \qquad (11)$$

with $R = R_0\exp(u)$ and $v \equiv \theta$. Transforming the rotationally symmetric Poisson equation $\nabla^2 V = -\varepsilon_0^{-1}\rho(R,\theta)$ into these coordinates, we find

$$\frac{\partial}{\partial u}\left(e^u \sin v \frac{\partial V}{\partial u}\right) + \frac{\partial}{\partial v}\left(e^u \sin v \frac{\partial V}{\partial v}\right) = \varepsilon_0^{-1} R_0^2\, e^{3u} \sin v \cdot \rho(R_0 e^u, v) \; . \qquad (12)$$

This can be cast into the form (1) with

$$\alpha = 1, \quad p^2 = e^u \frac{\sin v}{v}, \quad \hat{q} = 0 \; , \quad (v < \pi),$$

$$s = \varepsilon_0^{-1} R_0^2 e^{2u} \cdot \rho(R_0 e^u, v) \; ,$$

$$q = \frac{1}{4}\left(\frac{1}{\sin^2 v} - \frac{1}{v^2}\right) = \frac{1}{12}\left(1 + \frac{v^2}{5}\right) + O(v^4) \, . \qquad (13)$$

On the optical axis the coefficients p and q remain positive. By virtue of (11) the SCWIM-grid is transformed into an ordinary square shaped grid in u and v.

The discretization error is now of sixth order in the mesh width. Numerical tests have shown that in comparison with a five-point-discretization with the same mesh width a decrease of the error by a factor 10^{-5} can be achieved (Killes, to be published). But this is true only in the case of regular grids. Irregular meshes in the vicinity of boundaries will cause complications and a loss of accuracy. This can be circumvented by a combination of the FDM with the IEM, as will be obvious further below.

The integral-equation method

In the case of pure surface charge distributions with density $\sigma(\underline{r})$ the electrostatic potential satisfies

$$V(\underline{r}) = \int_B \frac{\sigma(\underline{r}')\, d^2 r'}{4\pi\varepsilon_0 |\underline{r} - \underline{r}'|} \; , \qquad (14)$$

where the integration is to be carried out over the boundary B. In the case of Dirichlet problems the boundary values of $V(\underline{r})$ are prescribed and (14) is then an integral equation for the unknown source distribution $\sigma(\underline{r})$. After its solution, (1) can be used in order to determine $V(\underline{r})$ in an arbitrary point $\underline{r}$. By means of an analogy-transfer also the magnetic field between unsaturated poles with very high permeability can be calculated. This version of the IEM, called the surface-charge method (SCM), is quite familiar in electron optics.

In order to perform the required integration, some simplifications are necessary. It is customary to dissect the surface into a large number of sufficiently small area elements in which $\sigma(\underline{r})$ can be considered as practically constant, then the integration can be carried out analytically and (14) is replaced by a linear system of equations for the piecewise constant, but unknown values of σ. The numerous publications concerning this matter differ only in details of the necessary discretization. Instead of discussing these items, we consider some recent developments.

Scherle (1983a) has applied the IEM to the calculation of magnetic fields in shielded deflection systems with toroidal and saddle coils, respectively. He was able to take into account the finite permeabilities of core materials and to solve the corresponding interface-condition problems. Hence his version of the IEM is far more general than the SCM outlined above. The starting point is the familiar separation of the magnetic field into the pure contribution $\underline{H}_0(\underline{r})$ from the coils, satisfying

$$\text{div}\, \underline{H}_0(\underline{r}) = 0, \quad \text{curl}\, \underline{H}_0(\underline{r}) = \underline{j}(\underline{r}) \qquad (15)$$

and hence determined from the theorem of Biot and Savart, and a gradient field grad $V(\underline{r})$ generated by the ferrite cores:

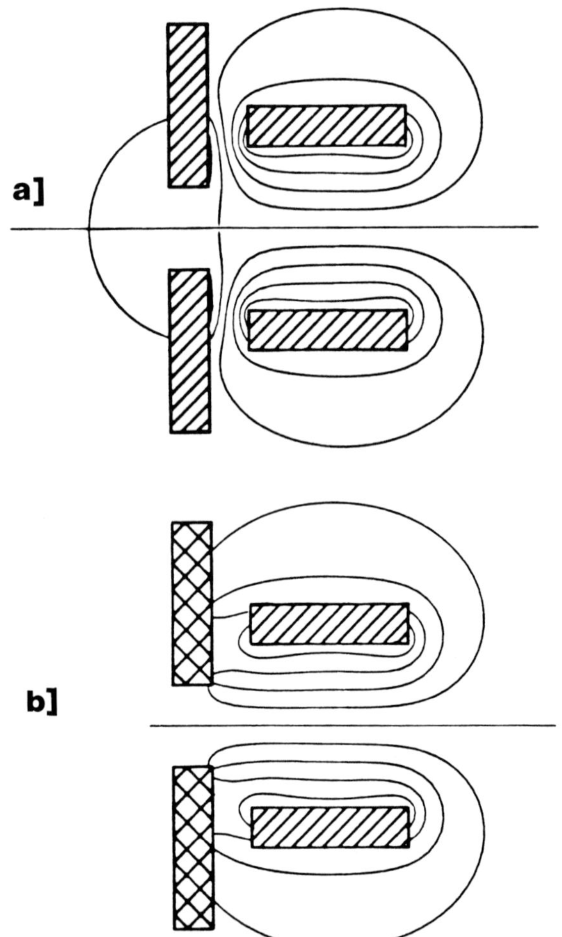

Figures 1a,b: equipotentials of the scalar magnetic pseudo-potential $V(\underline{r})$ in an axial section through a system of saddle coils, a ferromagnetic core and a shielding plate with a bore. For reasons of clearness the coils and their field $\underline{H}_0(\underline{r})$ are omitted.
(a) ferromagnetic shielding plate
(b) superconducting shielding plate
Courtesy of W. Scherle (1983b).

$$\underline{H}(\underline{r}) = \underline{H}_0(\underline{r}) + \text{grad } V(\underline{r}) \tag{16}$$

The latter contribution has its sources on the core surfaces and satisfies a Fredholm equation of second kind:

$$\frac{\mu+\mu_0}{2(\mu-\mu_0)} V(\underline{r}) + \int_B V(\underline{r}') \frac{\partial}{\partial n'} G(\underline{r},\underline{r}')\, d^2r' = -\int_B G(\underline{r},\underline{r}')\, \underline{n}'\cdot \underline{H}_0(\underline{r}')\, d^2r', \tag{17}$$

$\underline{n}'$ being the local surface normal in the directon towards the vacuum and $G = (4\pi|\underline{r}-\underline{r}'|)^{-1}$ Green's function of free space.

In the case of rotationally symmetric cores, even for unround current distributions $\underline{j}(\underline{r})$, eq.(17) can be reduced to a sequence of uncoupled one-dimensional Fredholm equations of second kind, as each term in (17) is expanded into a Fourier series with respect to the azimuth round the optic axis. After the decouplings, obtained in this way and the analytic integration over the azimuth, the remaining integration runs over the length s' of arc in the meridional section C through the boundaries. The resulting equations to be solved have the principal structure

$$\int_C K(s,s')\, y(s')ds' = U(s) + \lambda y(s), \tag{18}$$

U(s) being a given boundary value function, y(s) the required function, λ a fixed constant and K(s,s') a complicated kernel function with logarithmic singularity at s = s'.

Scherle (1983a) has used a cubic spline function for y(s) and in this way he obtained a high accuracy for y(s). In his thesis (1983b) he has developed a still more general version of the IEM for two different cores. The items cannot be described here, the plots of potentials, calculated in this way, are presented in fig. 1a,b.

In order to facilitate the integration over logarithmic singularities, Kasper (1983) has derived a new integration formula for functions with logarithmic singularities. This can be further generalized. Let $f_i(x)$, i = 1,2,3 be arbitrary regular functions, then we have

$$F(x) := f_1(x)\,\ln|x| + f_2(x)/x + f_3(x),$$

$$\int_{-h}^{h} F(x)dx = h\sum_{\mu=1}^{4} G_\mu\left(F(p_\mu h) + F(-p_\mu h)\right) + O(h^9) \tag{19}$$

with the abscissas and weight factors given by Basche (private communication, to be published)

P_1 = 0.0399 4596 2203
P_2 = 0.2801 7249 6204
P_3 = 0.6361 2394 4954
P_4 = 0.9223 6045 1138

G_1 = 0.1270 7679 2574
G_2 = 0.3267 4117 6078
G_3 = 0.3523 4912 8452
G_4 = 0.1938 3290 2896

The function y(s) was originally represented by a quadratic spline, i.e. a piecewise quadratic function with a continuous first derivative, in order to make eq. (19) applicable to the integration in eq. (18). This gives already highly accurate results. The details shall not be explained here. Instead of doing this, we present a still more improved method which can be found with very little effort.

We dissect the integration contour C into suitably chosen intervals, $s_0 < s_1 < s_2 \cdots < s_N$ and choose an expansion in terms of Legendre polynomials,

$$y(s) = \sum_{l=0}^{M} c_i^{(l)} P_l(t) \,, \quad s_{i-1} \leq s \leq s_i \tag{20}$$

with

$$h_i := (s_i - s_{i-1})/2 \,, \qquad \sigma_i := (s_i + s_{i-1})/2$$
$$t := (s - \sigma_i)/h_i \,, \qquad i = 1 \ldots N \tag{21}$$

Introducing the matrix elements

$$M_{ik}^{(l)} = h_k \int_{-1}^{1} K(\sigma_i, \sigma_k + t h_k) P_l(t) dt - \lambda P_l(0) \delta_{ik} \,, \quad (i,k = 1 \cdots N) \tag{22}$$

and the source terms $U_i := U(\sigma_i)$ we obtain the linear system of equations

$$\sum_{l=0}^{M} \sum_{k=1}^{N} M_{ik}^{(l)} c_k^{(l)} = U_i \,, \quad i = 1 \cdots N. \tag{23}$$

Provided that the coefficients $c_k^{(l)}$ for $l \neq 0$ are known, we can solve this for $c_1^{(0)} \ldots c_N^{(0)}$. By these coefficients the integral over the solution y(s) is determined: With the definition

$$Y(s) := \int_{s_0}^{s} y(s')ds' \,, \qquad Y_i := Y(s_i) \tag{24}$$

we have $Y_0 = 0$ and

$$Y_i = Y_{i-1} + 2h_i c_i^{(0)} \,, \quad i = 1 \ldots N \,. \tag{25}$$

Now we can determine $y_1 \ldots y_N$ and $y_1' \ldots y_N'$ by means of numeric differentiation. With these we obtain linear equations for the coefficients $c_i^{(l)}$ with $i = 1 \ldots N$ and $1 \leq l \leq M = 4$:

$$c_i^{(1)} = (6(y_i - y_{i-1}) + h_i(y_i' + y_{i-1}'))/10$$
$$c_i^{(3)} = (-(y_i - y_{i-1}) + h_i(y_i' + y_{i-1}'))/10$$
$$c_i^{(2)} = (10(y_i + y_{i-1} - 2c_i^{(0)}) - h_i(y_i' - y_{i-1}'))/14$$
$$c_i^{(4)} = (-3(y_i + y_{i-1} - 2c_i^{(0)}) + h_i(y_i' - y_{i-1}'))/14 \tag{26}$$

These are again substituted into (23). Starting with $c_k^{(l)} = 0$ for $l \neq 0$ in the first loop, the whole system of relations can be solved iteratively and converges rapidly. The accuracy is high if the function U(s) varies slowly. A modification for more rapidly varying functions U(s) is being investigated. Details will follow in a separate publication. In summary this method can be made highly accurate and efficient.

For truly three-dimensional problems the general situation is less favourable. It is not only the enlarged number of dimensions which causes problems, but - more than this - the fact that it is often impossible to map the given surfaces in a regular mesh grid in a two-dimensional parametric space. Concerning the surfaces there may arise similar difficulties like those in the FEM. A practicable approximation for the solution of Dirichlet problems in three dimensions has been published by Eupper (1982) who replaced the surface charge distribution by a set of suitably charged bars located behind the electrode surfaces. The charge distribution on these bars is determined in such a way that the potential assumes its prescribed boundary values at a sequence of control points on the electrode surfaces. The results for a field emission cathode with the shape of a hipped roof are presented in figs 2a,b.

Combinations of methods

In many practical cases a suitable combination of different field calculation methods can be more advantageous than the use of only one pure method. With respect to Dirichlet problems this question has been investigated in details by Schaefer (1982, 1983). His theory shall not be outlined here, since it is described in a separate paper at this conference (see Schaefer, this volume).

In this paper we shall consider first a combination between the IEM and the FDM, which is shown in fig. 3. We want to solve a Dirichlet problem for a domain G with boundary ∂G:

$$\nabla^2 V(\underline{r}) = -\rho(\underline{r})/\varepsilon_0 \,, \quad \underline{r} \in G \,, \tag{27}$$
$$V(\underline{r}) = \bar{V}(\underline{r}) \,, \quad \underline{r} \in \partial G \,. \tag{28}$$

The IEM is very favourable for the solution of Laplace's equation, even in the case of complicated boundaries, but is unsuitable for Poisson's equation, since the numerical evaluation of Coulomb integrals is a very time-consuming procedure. On the other hand the FDM is a very fast method for the solution of Poisson's equation, but useful only in the case of simple boundaries.

In this situation we propose to extend the regular mesh grid beyond the boundary ∂G and to extrapolate the space charge distribution continuously up to the new boundary $\partial \tilde{G}$. On $\partial \tilde{G}$ now simple guessed boundary values for a potential W are assumed and the PDE $\nabla^2 W = -\rho/\varepsilon_0$ is solved using the FDM. On ∂G we find then, of course, wrong boundary values $\bar{W}(\underline{r})$, but this is no misfortune. The true solution V can differ from the particular one (W) only by an appropriate solution U of Laplace's equation $\nabla^2 U = 0$. Hence we have to solve now

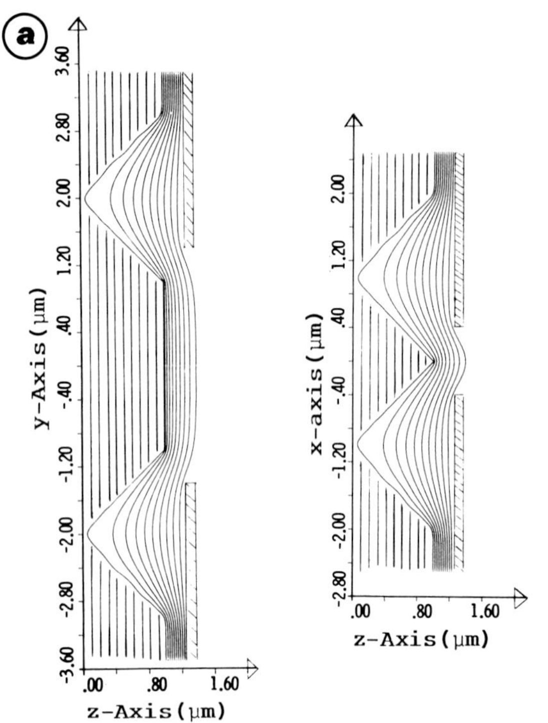

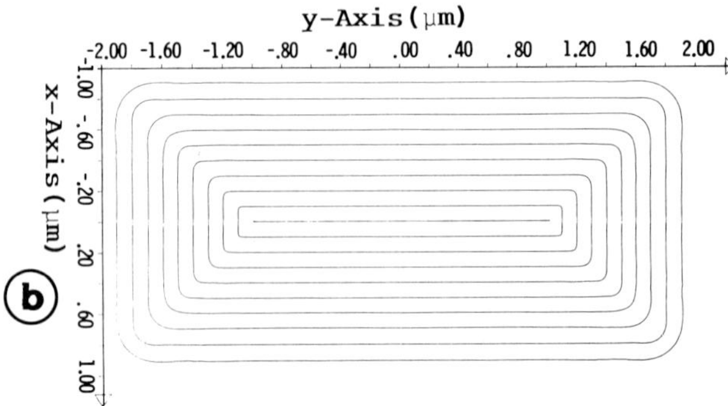

Figures 2a,b: equipotentials in different sections through a field electron emission source consisting of a cathode with a hipped roof and an anode with a rectangular bore.
Courtesy of M. Eupper (1982)

$$\nabla^2 U(\underline{r}) = 0 \ , \qquad \underline{r} \in G \tag{29}$$

$$U(\underline{r}) = \overline{V}(\underline{r}) - \overline{W}(\underline{r}) \ , \qquad \underline{r} \in \partial G \tag{30}$$

by means of the IEM in order to find $V = U + W$ as the correct total solution. In this way we have combined the advantages of both methods and circumvented their disadvantages.

This hybrid method shall be applied to the field calculation in electron guns with space charges, where it seems to be most useful; a corresponding publication is in preparation. In this context it is to be mentioned that the method outlined above is only one of the many possibilities to combine different procedures. For instance, in the vicinity of a spherical cathode, both the space charge distribution $\varrho(\underline{r})$ and the particular potential $W(\underline{r})$ can be expanded in series expansions containing Legendre polynomials, e.g.

$$W(R,\theta) = \frac{1}{R}\sum_{\ell=0}^{L} \phi_\ell(R) P_\ell(\cos\theta) \tag{31}$$

For the coefficient functions $\phi_\ell(R)$ a sequence of simple decoupled ordinary differential equations can be derived which are easily solved numerically. This procedure has been worked out by Weysser (1983a,b), who treated in this way an electron gun with a hemispherical tip cathode (fig. 4).

Finally we consider a simple combination between the FEM and the IEM, suitable for the field calculation in round magnetic lenses with saturation, as is shown in fig. 5. The FEM is advantageous for the field computation in nonlinear media, hence we discretize the axial section through the pole by a suitable triangular grid. Then using the FEM, we can calculate the values of the azimuthal vector potential $A(z,r)$ at the internal nodes, provided that the boundary values at the pole surfaces are known. Usually the latter are determined by the FEM in a grid extended to the practically field free space. This is, however, unfavourable; instead of this we better apply the IEM to the vacuum part of the field; then we have to solve the integral equation

$$\frac{1}{2}A(z,r) = \oint \left[A(z',r')\,\partial G_1/\partial n' - G_1(z,r;z',r')\left(\partial A(z',r')/\partial n'\right)_v \right] r'ds' + \mu_0 \iint G_1(z,r;z',r')\, j(z',r')\, r'dr'dz' \tag{32}$$

for the potential distribution on the pole surface. Here ds' denotes the length element of arc, $j(z,r)$ the azimuthal current density in the coils and $\partial/\partial n'$ the normal derivative in the direction towards the vacuum and on the vacuum side of the surface. The corresponding Green function is given by:

$$G_1(z,r;z',r') = \frac{1}{4\pi}\int_0^{2\pi} \frac{\cos\alpha \, d\alpha}{\sqrt{(z-z')^2 + r^2 + r'^2 - 2rr'\cos\alpha}} \tag{33}$$

This is an elliptic integral which can be calculated by well-known procedures. It has a logarithmic singularity, hence even the integration over the normal derivative by means of (19) is straightforward.

The solution of (32) requires the knowledge of $(\partial A/\partial n)_v$ on the vacuum side of the boundary. This derivative can be obtained in the following way. The FEM

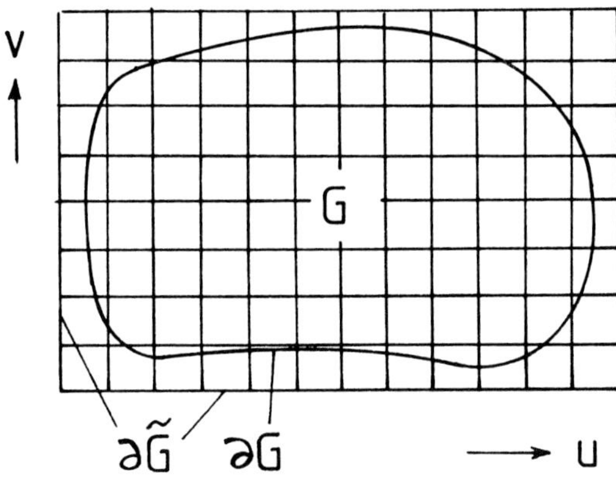

Figure 3: extension of a regular mesh grid beyond the curved boundary ∂G of a given domain G.

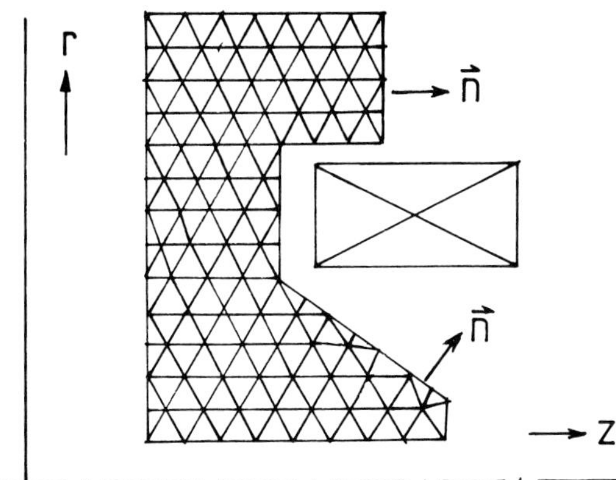

Figure 5: half axial section through an open round magnetic lens with a ferromagnetic core and a rectangular distribution of windings. Only the interior of the core is discretized by a triangular mesh grid.

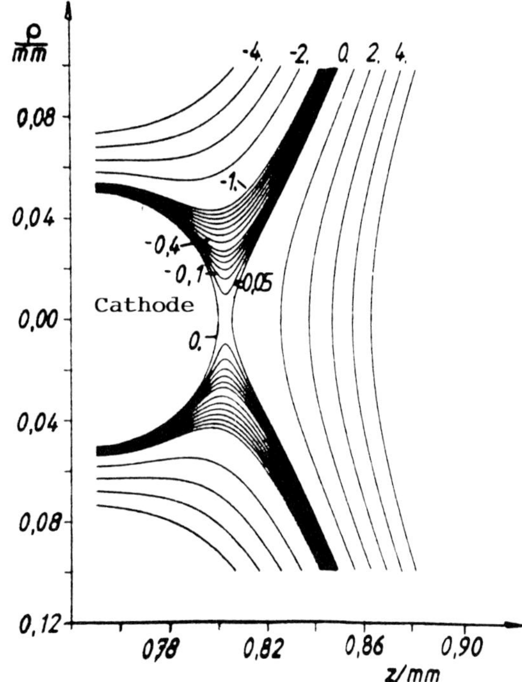

Figure 4: equipotentials in the vicinity of a hemispherical cathode. Due to the influence of space charge the potential is negative in front of the cathode. The cathode shank following to the left hand side, is not shown in this figure. Courtesy of R. Weysser (1983b)

can be formulated in such a manner that in each element the field strength $\underline{B}$ is calculated by means of numerical differentiation. Hence we know then $(\partial A/\partial n)_i$ on the inner side of the boundary. Moreover we can determine the reciprocal permeability by differentiation of the functional for the stored magnetic energy with respect to $B = |\underline{B}|$. But then $(\partial A/\partial n)_v$ is determined by the continuity of the tangential H-component:

$$\mu^{-1}\left[\partial(rA)/\partial n\right]_i = \mu_0^{-1}\left[\partial(rA)/\partial n\right]_v .$$

The whole system of equations has to be solved iteratively. This is no disadvantage, since for solving nonlinear problems iterative techniques are unavoidable in any case. Eventually the evaluation of $(\partial A/\partial n)_v$ requires an under-relaxation for reasons of stability; but this provides no principal problem. The computing time can be kept reasonably short, since the time-consuming calculation of the matrix elements for the solution of the integral equations needs to be carried out only once in the first loop. The practical investigations concerning this method are still going on and will be published later.

This technique removes the main disadvantage of the FEM, i.e. the inaccurate interpolations and differentiations in the vacuum part of the field. The essential gain is that now the field strength is truly a smooth analytic solution of the field equations with a relative error of about 10^{-3} or less. Then the ray tracing, especially for sufficiently large distances from the pole surfaces, provides no longer any problem.

Ray tracing and aberrations

The possibility and the quality of ray tracings depend strongly on the accuracy of the field calculation. If only the axial electrostatic potential $\phi(z)$, the axial flux density B(z) or other axial functions are computed by use of the FEM, then only the paraxial ray equations can be solved and this imposes the strong restriction to deal exclusively with the standard perturbation theory of aberrations. Probably this is the reason why the latter theory has been worked out in many details.

The physics of round magnetic lenses is developed in very many details and presented comprehensively in a volume edited by Hawkes (1982). Another comprehensive presentation of the aberration theory, also given by Hawkes (1980), deals with more general systems. Many items can also be found in the conference proceedings edited by Wollnik (1981). Here the quite numerous citations, to be found in these publications, shall not be repeated again. Some recent papers, not yet cited in the references given above, concern the aberration theory for microwave cavity lenses (Hawkes 1982/83 and 1983), for general systems with parasitic errors (Plies 1982a,b) and for planar cathode lenses (Ximen Ji-ye et al., 1983).

A quite general feature of the perturbation theory is its enormous complexity which forces many authors to use the aid of computer algebra languages; for instance, Hawkes applied the language CAMAL and Soma (1977) the language REDUCE. Usually the aberration coefficients are represented as integral expressions which assume such a length, that an outsider may have difficulties with them. Moreover, there are systems for which an aberration theory of third order is insufficient; examples are electrostatic filter lenses (Niemitz, 1980), deflection units at large deflection angles and annular systems like beta spectrometers of Siegbahn's type. Hence new concepts seem to be necessary.

Some ray equations

The accurate calculation of electron trajectories does not make any problem, provided that the electromagnetic fields can be computed accurately at arbitrary points of reference. Hence the effort is to be concentrated on the improvement of the field calculation programs. When this has been achieved, practically every kind of ray equation can be solved numerically. In Tuebingen we have good experience with a parametrization of the Lorentz equation in terms of a variable σ proportional to the relativistic proper time:

$$\ddot{\underline{r}}(\sigma) = \nabla\varphi(\underline{r}) + \underline{b}(\underline{r}) \times \dot{\underline{r}}(\sigma) \qquad (34)$$

Here dots denote differentiations with respect to σ , further

$$\varphi(\underline{r}) = \phi^{*}(\underline{r})/(2U^{*}) \qquad (35)$$

is a normalized relativistic acceleration potential and

$$\underline{b}(\underline{r}) = \sqrt{\frac{e}{2m_0 U^*}}\ \underline{B}(\underline{r}) \qquad (36)$$

a normalized magnetic field strength. The choice of the normalization constant U^* is free; very suitable is often the asymptotic acceleration potential $U^* = \phi^*(\infty)$.

The solution of (34) must satisfy the conservation law

$$\dot{\underline{r}}^2 = 2\varphi(\underline{r}) + \text{const} , \qquad (37)$$

the constant being determined uniquely by the bias of the electric potential and by the kinetic starting energy. The ray equation (34) is suitable for all systems with time-independent fields. Besides this applicability in numerical procedures, the particular choice of the parameter σ has the advantage that (34) is the simplest possible form of the general relativistic ray equation. This is the basis of the subsequent considerations.

Sometimes it is favourable to solve the ray equation not in the given form but in an incremental form yielding directly the shift $\underline{s}(\sigma)$ relative to another trajectory $\underline{r}(\sigma)$. The latter may be, for instance, the optical axis in a magnetic deflection prism. Small differences between neighbouring values of an arbitrary smooth function $f(\underline{r})$ can be avoided by numerical integration over its gradient:

$$\begin{aligned} f(\underline{r}) &:= f(\underline{r} + \underline{s}) - f(\underline{r}) \\ &= \underline{s}\cdot\int_0^1 \nabla f(\underline{r} + t\underline{s})\, dt . \end{aligned} \qquad (38)$$

This integration can be performed e.g. by means of Gauss quadrature formulae and gives always a reliable result, regardless how small $|\underline{s}|$ may be. In this sense we obtain from (34) quite correctly

$$\begin{aligned} \ddot{\underline{s}}(\sigma) &= \Delta(\nabla\varphi(\underline{r})) + \underline{b}(\underline{r}) \times \dot{\underline{s}} \\ &\quad + \Delta\underline{b}(\underline{r}) \times (\dot{\underline{r}} + \dot{\underline{s}}) . \end{aligned} \qquad (39)$$

This is valid for electrons having the same total energy. This ray equation can be generalized easily in order to include chromatic aberrations. The experiences made with (39) are still in the beginning; they will be published in a later paper.

Evaluation of aberration discs

By a solution of (34) and of (39) it is possible to determine accurately the endpoints of electron trajectories in a given recording plane, the initial conditions being still quite general. If we compute trajectories starting in one common object point with directions located on an aperture cone, the line connecting the obtained endpoints forms the contour of the corresponding aberration disc.This can be repeated for different aperture angles. A typical example is given in figs. 6a and 6b, which have been determined by Killes (1983, unpublished). These figures are the aberration discs for an axial object point in a double focusing magnetic sector field spectrometer. The effect of a small defocus is quite obvious. These contours have been obtained by a numerical solution of (34), numerical instabilities are not to be noticed.

A quite simple and practicable method for the quantitative evaluation of aberration discs has been developed by Scherle (1983). Let (x_k, y_k) denote the coordinates of individual trajectories in a given recording plane z = const, (x'_k, y'_k) the corresponding slopes and g_k suitable positive weight factors normalized to unit sum. Then

$$\bar{x} = \sum_{k=1}^{N} g_k x_k , \quad \bar{y} = \sum_{k=1}^{N} g_k y_k \tag{40}$$

are the coordinates of the center of the aberration disc; their deviations from the Gaussian image coordinates can be interpreted as weighted distortions. Furthermore, the second moments of the aberration disc are defined by

$$\begin{aligned} S_{xx} &= \sum g_k (x_k - \bar{x})^2 , \\ S_{xy} &= \sum g_k (x_k - \bar{x})(y_k - \bar{y}) , \\ S_{yy} &= \sum g_k (y_k - \bar{y})^2 . \end{aligned} \tag{41}$$

By these a variance ellipse is defined which approximates the cross section of the beam. A good approximation is obtained in many practical cases if the semiaxes of the ellipse are enlarged by a factor 2. Since the slopes of the trajectories are known, the local tangents are given. Hence, as far as it is sufficient to approximate the true trajectories by their tangents, for instance in a field free image space, the determination of the ellipse can be repeated for any plane z = const. In this way a defocus can be studied quite easily and rapidly. Fig. 7 demonstrates the approximation found for the spherical aberration of a round lens. The approximate circle of minimum radius is located near the familiar circle of least confusion.

Though this procedure is clearly an approximate one, it is certainly more convincing than the familiar practice to calculate a root-mean-square radius from roughly guessed radii for different aberration terms. If ever necessary, Scherle's theory can be further improved by taking into account the moments of third or higher order. It has one advantage which may be useful in the future: obviously the aberrations are minimized if after

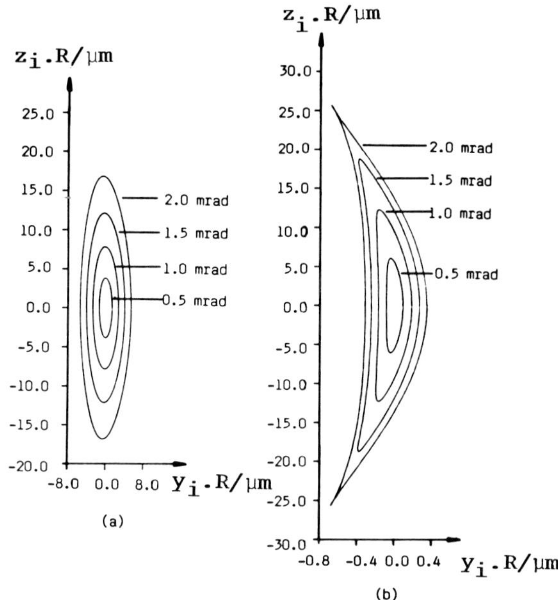

Figures 6a,b: examples for contour lines of aberration discs in a double focusing magnetic sector field with 90 degrees-deflection perpendicular to the z-direction. The rays start in an axial object point, the initial slopes form cones with the given semiaperture angles.
(a) configuration close to the best obtained focus
(b) defocus of 3 mm for a deflection radius 25 cm
Courtesy of P. Killes (unpublished work)

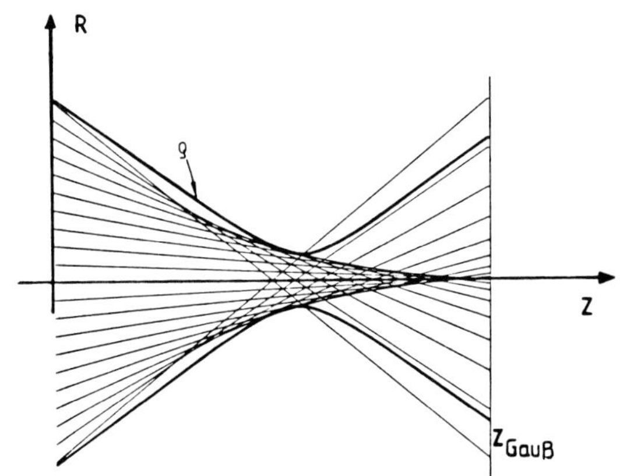

Figure 7: Scherle's approximation of the spherical aberration in a conventional round lens (the slopes are enlarged for reasons of distinctness). The hyperbola marked by ϱ represents the approximation. Courtesy of W. Scherle (1983b)

all allowed variations of system parameters the maximum diameter of all aberration discs has its smallest possible value. This gives a clear concept for optimization procedures. The practical application of this idea is still in the beginning, but seems to be straightforward.

In the case that not only the information about entire aberration discs is needed but also the knowledge of aberration coefficients, the latter are to be determined by suitable least-squares-fits with the endpoints (x_k, y_k), $(k = 1...N)$ of numerically calculated trajectories. Since this requires a very high accuracy, some care must be taken in the set-up of the corresponding routine. Such calculations were, for instance, performed by Niemitz (1980), who determined in this way the aberration coefficients for electrostatic filter lenses, even in fifth and higher orders. The research concerning the improvement of least-squares-fit techniques is going on. In summary there are already now some alternatives to the standard perturbation theory of aberrations.

References

Chari TVL, Silvester P. (1980). Finite Elements in Electrical and Magnetical Field Problems. Wiley, Chichester, New York.

Eupper M. (1982). Eine Methode zur Lösung des Dirichlet-Problems in drei Dimensionen und seine Anwendung auf einen neuartigen Elektronenstrahlerzeuger. Optik 62, 299-307.

Hawkes P.W. (1980). Methods of Computing Optical Properties and Combating Aberrations for Low-Intensity Beams, in Adv. Electr. Electron Phys. 13A, 45-157.

Hawkes P.W. (ed). (1982). Magnetic Electron Lenses, in: Topics in Current Physics 18. Springer; Berlin, Heidelberg, New York, ch. 1,3,4 and 5.

Hawkes P.W. (1982/83). Computer-aided calculation of the aberration coefficients of microwave cavity lenses. Part 1: Primary (second-order) aberrations. Optik 63, 129-156.

Hawkes P.W. (1983). Computer-aided calculation of the aberration coefficients of microwave cavity lenses. Part 2: Secondary (third-order) aberrations. Optik 65, 227-251.

Hermeline F. (1982). Triangulation automatique d'un polyedre en dimension N.R. A.I.R.O. Analyse numerique/Numerical Analysis 16, 211-242.

Kang NK, Orloff J, Swanson LW, Tuggle D. (1981). An improved method for numerical analysis of point electron and ion source optics. J. Vac. Sci. Technol. 19, 1077-1081.

Kang NK, Tuggle D, Swanson LW. (1983). A numerical analysis of the electric field and trajectories with and without the effect of space charge for a field electron source. Optik 63, 313-331.

Kasper E. (1976). On the numerical calculation of static multipole fields. Optik 46, 271-286.

Kasper E. (1982). Magnetic Field Calculation and the Determination of Electron Trajectories, in: Topics in current Physics 18, Magnetic Electron Lenses. P.W. Hawkes (ed), Springer; Berlin, Heidelberg, New York, 57-118.

Kasper E. (1983). On the solution of integral equations arising in electron optical field computations. Optik 64, 157-169.

Lencova B, Lenc M. (1982). Infinite elements for the computation of open lens structures. Electron Microscopy I (Physics), Hamburg, Proc. 10th Intl. Conf. on EM, Deutsche Gesselschaft fur Elektronenmikroskopie, Frankfurt, W.Germany, 317-318.

Mulvey T. (1982). Unconventional Lens Design, in Topics in Current Physics 18, Magnetic Electron Lenses. P.W. Hawkes (ed), Springer; Berlin, Heidelberg, New York, 359-412.

Munro E. (1971). Computer-aided design methods in electron optics.Dissertation, Cambridge.

Munro E. (1973). Computer-aided Design of Electron Lenses by the Finite Element Method, in Image Processing and Computer-aided Design in Electron Optics. P.W. Hawkes (ed), Academic Press, London, New York, 284-323.

Munro E, Chu HC. (1982a). Numerical analysis of electron beam lithography systems. Part I: Computation of fields in magnetic deflectors. Optik 60, 371-390.

Munro E, Chu HC. (1982b). Numerical analysis of electron beam lithography systems. Part II: Computation of fields in electrostatic deflectors. Optik 61, 1-16.

Niemitz P. (1980). Theoretische Untersuchung von elektrostatischen Drei-Elektroden-Filterlinsen. Dissertation, Tuebingen (W-Germany).

Plies E. (1982a). Berechnung zusammengesetzter elektronenoptischer Fokussier- und Ablenksysteme mit überlagerten Feldern. Teil I: Feldentwicklung und Bahngleichung. Siemens Forsch.- u. Entwickl.-Ber. 11, 83-90.

Plies E. (1982b). Berechnung zusammengesetzter elektronenoptischer Fokussier- und Ablenksysteme mit überlagerten Feldern. Teil II: Fundamentalbahnen und Bildfehler. Siemens Forsch.- u. Entwickl.-Ber. 11, 83-90.

Schaefer C. (1982). Methoden zur numerischen Lösung der Laplace-Gleichung bei komplizierten Randwertaufgaben in drei Dimensionen und ihre Anwendung auf Probleme der Elektronenoptik. Dissertation, Tuebingen (W.Germany).

Schaefer C. (1983). The application of the Alternating Procedure by H.A. Schwarz for computing three-dimensional electrostatic fields in electron-optical devices with complicated boundaries. Optik 65, 347-359.

Scherle W. (1983a). Eine Integralgleichungsmethode zur Berechnung magntischer Felder von Anordnungen mit Medien unterschiedlicher Permeabilität. Optik 63, 217-226.

Scherle W. (1983b). Berechnung von magnetischen Ablenksystemen. Dissertation, Tuebingen (W.Germany).

Silvester P, Konrad A. (1973). Axisymmetric Triangular Finite Elements for the Scalar Helmholtz Equation. Int. J. Num. Meth. Eng. 5, 481-497.

Soma T. (1977). Relativistic aberration formulas for combined electric-magnetic focusing-deflecting systems. Optik 49, 255-262.

Weysser R. (1983a). Feldberechnung in rotationssymmetrischen Elektronenstrahlerzeugern mit Spitzenkathode und Raumladungen. Optik 64, 143-156.

Weysser R. (1983b). Feldberechnung in rotationssymmetrischen Elektronenstrahlerzeugern mit Spitzenkathode unter Berücksichtigung von Raumladungen. Dissertation, Tuebingen (W.Germany).

Wollnik H. (ed), (1981). Charged Particle Optics. Proc. 1st Conf. Charged Particle Optics, Giessen (W.Germany) 1980. North Holland, Amsterdam; also Nucl.Inst.Meth. 187, parts IV and V.

Ximen Ji-ye, Zhou Li-wei, Ai Ke-cong (1983). Variational theory of aberrations in cathode lenses. Optik 66, 19-34.

Electron Optical Systems (pp. 75-84)
SEM Inc., AMF O'Hare (Chicago), IL 60666-0507, U.S.A.

0-931288-34-7/84$1.00+.05

SYNTHESIS OF ELECTRON LENSES

M. SZILAGYI

Department of Electrical and Computer Engineering,
University of Arizona, Tucson, AZ 85721
Phone:(602)621-6183

Abstract

This paper is a review of different approaches to one of the most ambitious goals of Electron and Ion Optics: to produce elements and systems with prescribed first-order properties and minimal aberrations. Synthesis of such elements is usually done in two steps: the first is a search for a field distribution with the given properties and the second is the reconstruction of electrodes (pole pieces) that would produce this field distribution. The first problem can be solved by the use of different techniques: Calculus of Variations, Dynamic Programming or Function Minimization. The second one is more complicated and requires a lot of ingenuity. Our novel approach takes a different course of action. It combines the two steps into one. Low-aberration field distributions are sought by dynamic programming or function minimization procedures in the form of continuous curves constructed of cubic splines. A very simple algorithm is used for the reconstruction of the electrodes or pole pieces. This approach combines the widely recognized advantages of our optimization techniques with a built-in accurate and effective reconstruction procedure. The final design is simplified on the basis of the requirements of manufacturability. High-quality electrostatic lenses have been designed by the use of this technique. Thus, electron and ion optical synthesis has been transformed from a dream to reality.

Key words: aberrations, chromatic aberration, electron lens, electrostatic lens, field distribution, ion lens, optimization, potential distribution, spherical aberration, synthesis.

Introduction

There has been a struggle to overcome the aberrations of electron lenses since the very beginnings of Electron and Ion Optics. The reduction of aberrations is of specific importance now because of the great interest in particle beam technologies. Electron and ion beam lithography and especially focused ion beams offer new possibilities for integrated circuit fabrication as well as new microanalytical capabilities.

Ion beams must be focused by electrostatic lenses to ensure the independence of the refraction of the charge-to-mass ratio of the ions. However, due to the fact that electrostatic fields can not be concentrated to narrow regions like magnetic fields, there is a widespread belief that the aberrations of electrostatic lenses are intrinsically much higher than those of magnetic lenses.

Scherzer (1936b) showed that it is impossible to correct all of the aberrations of axially symmetric electron lenses by another system of similar symmetry but it was he again who proposed several other ways to compensate them by introducing additional features: discontinuities in the field distributions, space charge, high-frequency fields or other types of symmetry (Septier, 1966). In this paper we shall not be concerned with compensation. Our topic is optimization i. e. search for such electron and ion lenses that would provide themselves (without additional compensating elements) the required optical properties with minimum aberrations.

There are two totally different approaches to optimization: analysis and synthesis.

If the method of analysis is used, the designer starts with a given simple set of electrodes or pole pieces. The design is gradually improved by analyzing the optical properties and changing the geometrical dimensions as well as the electric and magnetic parameters of the system. This process is repeated until it converges to an acceptable solution. Due to the infinite number of possible configurations, the procedure is extremely slow and tedious. It can yield the results quickly only if a reasonable guess of the answer is already available before the work starts or if the constraints are so severe that there is not very much choice left. The iteration procedure can be facilitated by automation (Chu and Munro, 1981).

Optimization by synthesis is based on the fact that the imaging field, its optical properties and aberrations are totally determined by the axial distribution of the field. Only the axial distribution and its derivatives appear in the paraxial ray equation and in the expressions of the aberration coefficients. Then, instead of analyzing a vast amount of different electrode and pole piece configurations, we can start with the criteria defining an optimum system as initial conditions and try to find the imaging field distribution (and subsequently the electrodes or pole pieces) that would produce it.

This is a complicated problem. Its solution has been keeping many researchers interested for an extended period of time. Before we start to review their work, let us first show the difference between the analytic and synthetic approaches by using a simple example.

This example is the ideal quadrupole lens. It consists of four identical infinite hyperbolic surfaces held at alternate positive and negative potentials. Since infinite surfaces can not be realized in practice, different approaches have been used to approximate the ideal quadrupole field. The analytic approach is to compensate the missing parts of the hyperbolic surfaces by changing the shapes of the remaining electrodes. The synthetic approach (Szilagyi, 1978b) starts with the ideal field distribution and tries to reproduce it by recognizing the fact that the most important characteristic of a quadrupole lens is not the number of its electrodes but the presence of exactly two mutually perpendicular symmetry planes. The solution then follows quite easily: a high number of simple electrodes held at suitably chosen potentials produces a much better approximation of the ideal quadrupole lens than any system of four sophisticated electrodes.

Let us now see how the problem of aberration reduction can be attacked by the synthetic approach.

Is Aberrationless Electron Optics Possible?

Contrary to a widespread belief, Scherzer's theorem does not exclude the possibility of aberrationless electron lenses. It only states that the spherical aberration coefficient can not change sign i. e. it can not be compensated by the spherical aberration of another axially symmetric electron lens. The situation is similar for the chromatic aberration. It is feasible, therefore, to search for a lens with zero aberrations.

Glaser (1940) calculated first the magnetic field distribution necessary to give zero spherical aberration. Unfortunately, the resulting field is not strong enough to produce a real image. It was established, however, that for the reduction of spherical aberration the second derivative of the axial distribution of the magnetic flux density must be positive throughout the lens. This led to a substantial improvement of beta-spectrographs (Siegbahn, 1946).

Although proofs of zero limit of the spherical aberration have been published (Kas'yankov, 1955), all the attempts to calculate a practically realizable field distribution with zero spherical aberration have failed (Crewe, 1977). Recknagel (1941) showed that electrostatic lenses free of aberrations can not form a real image either. Finally, Tretner (1959) calculated the lower limits of the spherical and chromatic aberration coefficients for both electrostatic and magnetic lenses.

Since the practical realization of aberrationless electron lenses seems to be impossible, our efforts must be concentrated on the reduction of aberrations.

Early Attempts of Synthesis

Synthesis of electron lenses with minimum aberrations has been attempted since the first decade of Electron Optics. Scherzer (1936a) and Glaser (1938) calculated the axial electrostatic and magnetic weak lens distributions, respectively, with the least spherical aberration.

The synthesis of real thick lenses is much more complicated. Kas'yankov (1952) was the first to try his wits at this formidable problem. He derived a set of high-order nonlinear differential equations the solution of which would minimize certain aberration integrals. Unfortunately, the initial conditions for these equations as well as practical methods of their numerical solution remained unclear.

Burfoot (1953) calculated a special set of quadrupole lenses of extremely complicated form to obtain a potential distribution for which the spherical aberration is corrected.

Septier (1966) mentions two other early approaches based on the general trajectory equation (P. Lapostolle) and the approximation of the ideal hyperbolic lens (R. Rudenberg and A. Septier), respectively. They produced comparatively weak lenses with not very low aberrations.

Calculus of Variations

Tretner (1954, 1959) realized that the Calculus of Variations is a natural approach to the problem of synthesis since the aberration coefficients can always be expressed as definite integrals of the form

$$C = \int_{z_o}^{z_i} F[r_1(z), r_1'(z), r_2(z), r_2'(z), V(z), V'(z), V''(z)]\,dz$$

$$= \int_{z_o}^{z_i} F(z)\,dz \qquad (1)$$

where F(z) may represent the integrand of a combination of aberration coefficients or just one coefficient, z_o and z_i are the object and image coordinates, respectively. The Calculus of Variations is well suited to minimize integrals. The difficulty is that the integrand F(z) is a complicated function of not only the unknown electrostatic and magnetic field (or potential) distributions [V(z)] and their derivatives [V'(z) and V"(z)] but also of two linearly independent solutions $r_1(z)$ and $r_2(z)$ of the paraxial ray equation

$$f[r(z),r'(z),r''(z),V(z),V'(z),V''(z)] = f(z)=0 \quad (2)$$

which is a second-order differential equation with coefficients depending on the same unknown field distributions and their derivatives. Though Tretner used the Calculus of Variations to find the limits of the chromatic and spherical aberration coefficients of electrostatic and magnetic lenses, he left the question of how to achieve these limits open. For special cases he managed to simplify the mathematical problem by a series of variable transformations but in a general case these simplifications usually do not work.

The first mathematically justified approach to synthesis with practical results was that of Moses. He first used Calculus of Variations for the optimization of quadrupoles (1970, 1971a, 1971b) but later successfully applied it to the design of axially symmetric magnetic lenses, too [(Moses, 1973) and (Rose and Moses, 1973)].

In order to determine the unknown functions V(z) and r(z) so that they minimize several aberration integrals simultaneously and at the same time satisfy the differential equation (2) together with additional constraints we first define the Lagrange multiplier $\mu(z)$ so that

$$\delta \int_{z_o}^{z_i} [F(z) + \mu(z)\, f(z)]\, dz = 0 \quad (3)$$

where δ means the first variation of the integral. Carrying out the variations of the integrand and integrating by parts we obtain two high-order nonlinear differential equations (the Euler-Lagrange-Poisson equations). Combining them with the paraxial ray equation (2), one has a system of three coupled nonlinear differential equations. An additional equation appears when the residues obtained from the partial integration are equated to zero. The initial conditions are determined by the constraints. As they may be given at different (even a priori unknown) points and the differential equations are usually very complicated, special numerical procedures are needed for their solution. Fortunately, for magnetic lenses the procedure yields a system of second-order differential equations and Moses was able to solve it. He actually managed to design a coma-free magnetic lens with low spherical aberration (Moses, 1973).

For electrostatic lenses the situation is much more complicated. The Euler-Lagrange-Poisson equations yield a system of fourth-order differential equations with complicated initial conditions.

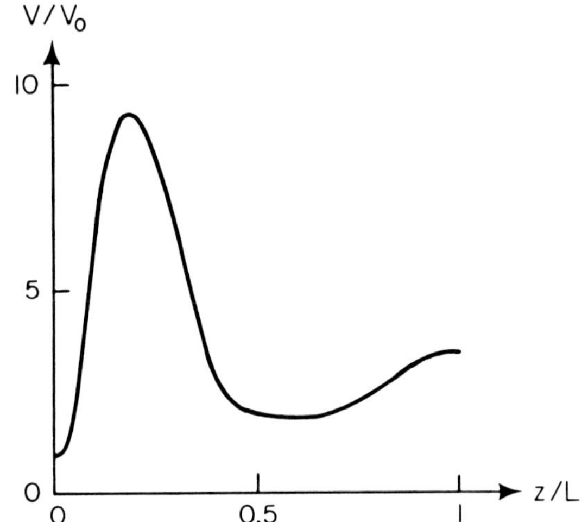

Fig. 1. Axial potential distribution for a high-performance electrostatic lens.

The solution of such a system is numerically intractable. Simpler procedures are needed for practical design.

A Simple Systematic Approach

If a mathematically justified approach yields such a complicated procedure, it is natural to try something much easier. The easiest way to search for field distributions that provide small aberrations is to investigate classes of parametrized analytical functions or simply try to produce certain distributions by the use of different curve fitting techniques (Szilagyi, 1983a). Using this simple approach in a systematic way we were able to find many electrostatic axial potential distributions that satisfy practical requirements for probe forming lenses and have very low spherical and chromatic aberrations. One of these distributions is shown in Fig. 1. The potential distribution and the axial dimensions are related to the object-side potential V_o and the effective length L of the lens, respectively. This distribution corresponds to a four-electrode electrostatic lens that has a spherical aberration coefficient C_{so} for infinite magnification, referred to the object space and related to the object-side focal length f_o equal to $C_{so}/f_o = 0.82$ and chromatic aberration coefficient for the same case equal to $C_{co}/f_o = 0.46$. For a six-electrode lens we could obtain even better results: $C_{so}/f_o = 0.63$ and $C_{co}/f_o = 0.29$, respectively.

If one compares these data with the aberrations of the best available electrostatic lenses, it is easy to see that a systematic search for good axial potential distributions can provide much better results than any analytical design. Even if we include the difficulties of the electrode reconstruction (see later), we can safely declare that general statements about the intrinsic inferiority of electrostatic lenses so common in the literature are certainly not justified.

This simple approach is very powerful because it produces good results without complex mathematical operations. One can find excellent imaging field distributions even with a personal computer.

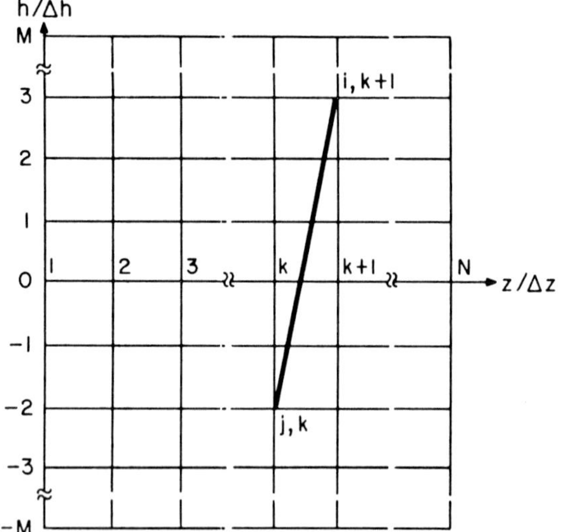

Fig. 2. Computational grid for the dynamic programming procedure.

However, if one really wishes to explore the vast universe of all feasible field distributions, mathematically justified and practically realizable optimization techniques are needed. Since the Calculus of Variations is too complicated for practical use, other approaches had to be considered.

Dynamic Programming

Our problem is to find the distribution V(z) that minimizes the aberration integral (1) simultaneously satisfying the differential equation (2) and the constraints imposed by practical requirements. The possible distributions constitute an infinite set even if the length of the lens as well as the maximum allowable field strength are both limited. We can reduce this infinite set to a finite but extremely large one in many different ways. For example, one can seek V(z) in the form of an n-degree polynomial or another parametrized function. In this case we have to determine the coefficients of the polynomial or the parameters of the function. As it was shown in the preceding section, it is a feasible alternative but without any hope to explore even a reasonable fraction of the different possibilities. Indeed, even if each coefficient of the polynomial can take only m different values, the number of possible variations is m^n which is an astronomical number for any pair of practical values of m and n. This brute force approach, therefore, does not satisfy our requirement of being mathematically justified. Is there any hope at all if e. g. n=10 and m=100?

Yes, there is not only hope but there are also practical ways of solution [(Szilagyi, 1977a), (Szilagyi, Yakowitz and Duff, 1984), (Szilagyi,1984b)].

Let us first see the Dynamic Programming approach (Szilagyi, 1977a). We start with a rectangular computational grid (Fig. 2) that defines the domain of existence for the field distribution. The field is limited by its maximum allowable value. Its axial extension is defined by the given effective length L of the lens. The effective length is divided into N equal regions. The unknown distribution V(z) or one of its derivatives is approximated by a straight line in each region. We shall denote this piecewise linear function by h(z). For a magnetic lens h(z) is the axial flux density distribution, for an electrostatic lens it is the electric field distribution. If one wishes to drastically reduce the amount of calculations, even a step-function model can be used where the field is constant in each region and its derivatives are calculated as differences at neighbouring regions. As we shall see later, the most effective approach is to use the highest derivative appearing in the aberration integral for h(z). In any case h(z) can take only 2M+1 different values at the boundaries of the regions (see Fig.2). Thus, our problem is reduced to that of finding N*(2M+1) intersection points of the computational grid that will provide the linear segments of the optimized field distribution.

The paraxial ray equation can be solved in each interval analytically or numerically. The continuity of the solution requires that the initial values of r and r' for each region be taken as equal to their terminal values in the preceding region. If N is a sufficiently large number, the aberration integral can be approximated by a finite sum and we can easily calculate the contribution of each region to this sum.

Let us now denote the initial and terminal values of h(z) in the kth interval by

$$h_k(z_k) = j\,\Delta h \tag{4}$$

and

$$h_k(z_{k+1}) = i\,\Delta h \tag{5}$$

respectively where Δh is the minimum amount of change in the value of the field distribution (see Fig. 2). Evidently, the solution of the paraxial ray equation as well as the contribution to the aberration sum will depend on the values of i, j and k.

Of course, we still have the astronomical number of $(2M+1)^N$ different possibilities. We must find a way to effectively search for the best possible solution.

Our multidimensional problem can be reduced to a one-dimensional multistage decision problem which can be solved step by step. Let us suppose that we have been able to find a potential distribution for which the value of the aberration integral between z_0 and z_k is minimal. We shall denote this optimized intermediate value of the integral by G_{jk}. How to proceed further? If F_{ijk} is the contribution of the kth region to the integral, then the minimized value $G_{i(k+1)}$ of the aberration integral between z_0 and z_{k+1} is given by the recursive relationship

$$G_{i(k+1)} = \min_j (G_{jk} + F_{ijk}) \tag{6}$$

where j is a variable and the optimization procedure is aimed at finding that particular optimal value of j for which the value of the sum in parentheses is minimal. As we have already calculated everything needed to obtain F_{ijk} for the given value of i the procedure yields both the optimized value of j and the new value of $G_{i(k+1)}$. Since we started with the assumption that the optimum distribution leading to the point (j,k) was known, now one further portion of the optimum distribution has been found.

We start at k=1 with $G_{j1}=0$ which expresses the fact that in the field-free region beyond the lens the contribution of the aberration integral is zero. Therefore, the search in the first region is reduced to the comparison of the different F_{ijk} values. For each i we find the corresponding j_{opt} value that minimizes F_{ijk}. We proceed region by region recursively toward the object space using Eq. (6) for the determination of the optimum values of j for each k and i until we reach the end of the computational grid (k=N+1). Note that at each region we only have to evaluate $(2M+1)^2$ possibilities. As a result of $N*(2M+1)^2$ evaluations we shall have 2M+1 different optimized axial field distributions (with different terminal values of i) together with the corresponding particle trajectories. Each distribution can uniquely be traced back by following the optimum values of j for all the pairs of the i and k values. One chooses the particular distribution that best satisfies the practical requirements of the given design. For example, if the field must vanish at the boundary of the lens and h(z) means the field distribution, then naturally i=0 must be chosen at the terminal point. When this is not required but we would like the image to be located inside the lens, we can make N a variable and stop the calculations at the pair of values i and N where r=0.

The most important advantages of the dynamic programming procedure are as follows:

1. No initial guess of the result is required. The algorithm searches through the entire problem space in an efficient way without the requirement of any initial assumptions.

2. A global search is provided with only $N*(2M+1)^2$ evaluations.

3. In general, rich patterns of optimized distributions are produced. They contain vast numbers of distributions as sub-solutions of the original problem with different initial and terminal conditions.

4. The procedure is simplified by any additional constraint that reduces the number of possible choices at a particular stage.

5. The procedure can be directly applied to any symmetry.

This method has been successfully used for electron lens design. Numerous interesting configurations have been found both for magnetic (Szilagyi, 1977b) and electrostatic (Szilagyi, 1978a, 1978c and 1983b) lenses. A typical case is shown in Fig. 3. This is a 50*60 computational grid for the design of accelerating electrostatic immersion lenses (Szilagyi, 1978a). Here h(z) is proportional to the axial electrostatic field distribution. All the acceptable solutions start and end with approximately zero fields and represent two-cylinder immersion lenses with the second cylinder having a smaller diameter and a higher potential.

As usual, there are also some limitations:

1. The procedure is aimed at subsequent nodes of the computational grid and not at the a priori unknown terminal node. As a consequence, it may happen that a "non-optimal" distribution provides a solution sooner than the "optimal" one.

2. Due to the discrete nature of the procedure, storage and manipulation of large arrays of data are required.

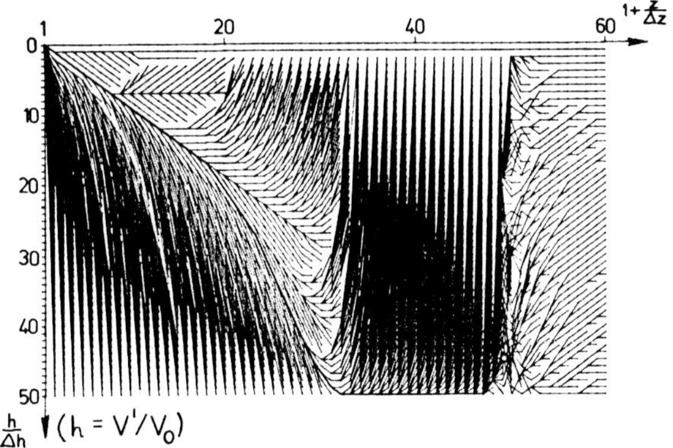

Fig. 3. Optimized electrostatic field distributions obtained by dynamic programming.

The procedure can be improved by any of the following suggestions:

1. Aim the optimization procedure at the given terminal condition. Then for an arbitrary node of the computational grid the predecessor node is chosen to minimize the value of the aberration integral not at the given node but at an a priori chosen terminating point.

2. Maintain the original algorithm but make the value of Δz a variable to ensure a smooth distribution even for small grids.

3. Replace the computational grid in the h-z plane by a grid in the phase plane r-r' of the paraxial ray.

4. Avoid the difficulties connected with the discrete character of the method by the utilization of differential dynamic programming.

5. Try something totally different. This will be discussed in the next section.

Function Minimization

A very effective optimization procedure was recently proposed by Szilagyi, Yakowitz and Duff (1984). We again divide the axial length of the lens into N regions and in each region represent the unknown axial distribution V(z) by a simple polynomial expression. The simpler this expression is, the faster and easier the procedure becomes. We require the continuity of V(z) and its lower derivatives. This requirement imposes some relationships between the coefficients of the polynomials at neighbouring regions. We always define the requirements of continuity in such a way that the set of coefficients at the highest-degree terms remain free. Then our problem is reduced to that of finding N coefficients satisfying the paraxial ray equation (2) and the constraints of the problem and minimizing the aberration integral (1).

Let us start with an arbitrary set of the free coefficients. The paraxial equation is solved with these coefficients for each interval and the corresponding aberration coefficients are evaluated numerically. Naturally, the aberration coefficients will be too high and the constraints will not be met. Now we add a penalty reflecting the violation of the constraints to the value of the objective function constructed from the aberration coefficients. This sum is the new objective function we are trying to minimize.

The minimization can be done simply by applying some nonlinear programming technique (e.g. a quasi-Newton algorithm) to the new objective function of the N coefficients (function minimization). The result is a set of coefficients that minimize the aberrations with the simultaneous satisfaction of the constraints.

Though this procedure requires an initial set of coefficients to be chosen by the user, it is extremely fast and effective. Almost any combination of the initial data converges to a good solution in seconds of CPU time. We were able to design some very good electrostatic lenses by the use of this technique.

Reconstruction of the Electrodes (Pole Pieces)

When the optimized axial electrostatic or magnetic scalar potential distribution has been found, our next task is to find the electrode (pole piece) configuration that will produce the optimized distribution. Since the potential produced by an axially symmetric system at an arbitrary point in space is uniquely determined by its axial distribution, from theoretical point of view the electrode reconstruction does not constitute a problem.

The power series expansion of the potential, however, requires the axial distribution to be a 2(n-1) times differentiable function of the axial coordinate z where n is the number of terms used in the power series. Unfortunately, for a good convergence a large value of n is needed. If the axial distribution is given as a numerical data set, or even if it is given in the form of a complicated analytical function, the higher derivatives must be produced by numerical techniques which are extremely inaccurate. This difficulty can be avoided by using the charge density method (Hawkes, 1981) but it requires excessive computer time for the reconstruction of the potential distribution at a considerable distance from the axis.

The axial potential distribution given as a discrete data set can be replaced by a continuous, n times differentiable function with minimum interpolation noise (Szilagyi, 1980). The computational feasibility of this approach, however, is also questionable.

The difficulties of electrode (pole piece) reconstruction, therefore, constituted a common weakness for all methods of electron and ion optical synthesis. Some researchers have even expressed scepticism about the very possibility that the problem of electron optical synthesis can ever be solved (Kasper, 1981).

We were able to achieve quite acceptable accuracy of reconstruction by fitting the numerical data sets with smooth cubic spline curves (Szilagyi, 1983a). Of course, inside the regions the cubic splines do not have any derivatives higher than the third. However, the difficulty is that the higher derivatives are undefined at the boundaries of the regions. Nevertheless, with some insight and experience one can always reconstruct the electrodes. Fortunately, in most cases the complicated curved equipotential surfaces can be replaced by simpler ones with straight boundaries. As a result we were able to design lenses with electron optical properties very close to those of the optimized axial potential distributions. A four-electrode electrostatic lens based on the axial potential distribution given in Fig.1 is shown in Fig. 4. This lens has very good optical properties in a wide range of the parameters. Its aberrations are quite low even at large working distances. If the electrodes are held at the relative potentials shown in the figure, the spherical aberration coefficient is equal to C_{so}/f_o=0.92 and the chromatic aberration coefficient has a value of C_{co}/f_o=0.55.

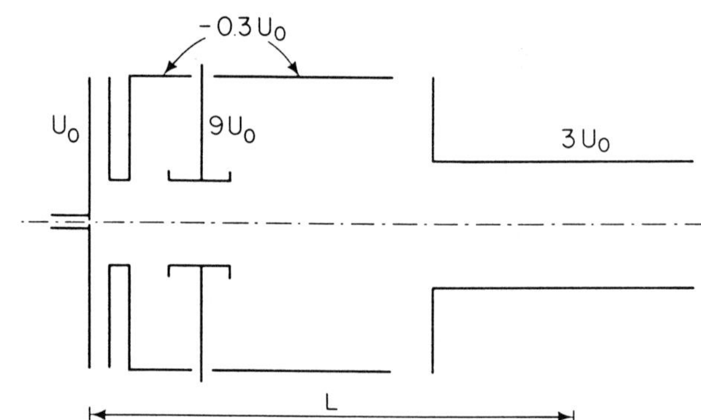

Fig. 4. Electrostatic lens designed on the basis of the potential distribution shown in Fig. 1.

The success of utilization of the cubic spline curve fitting technique has led us to the solution of the synthesis problem.

Solution of the Problem of Electron Lens Synthesis

The difficulties of electrode (pole piece) reconstruction were connected with the fact that the reconstruction procedure was totally separated from the optimization algorithm. When we search for an optimum axial field distribution we do not care about the problem of off-axis expansion. This is wrong: <u>the reconstruction process must be an integral part of the optimization procedure</u>. Instead of replacing the discrete axial distribution data sequence by a continuous curve in a complicated way or trying to accomplish the off-axis expansion by some other sophisticated techniques, <u>we must seek the solution directly in the form of a continuous piecewise cubic spline function</u>. The simple expression

$$v(r,z) = V(z) - \frac{r^2}{4} V''(z) \qquad (7)$$

will then be used for the potential distribution in the entire space. The reconstruction of the electrodes or pole pieces thus becomes an almost trivial task (Szilagyi, 1984a).

This approach allows us to combine the evident advantages of both the dynamic programming and the function minimization procedures with an easy, fast and accurate reconstruction technique.

In the following we shall briefly outline our method of electron optical synthesis (Szilagyi, 1984b).

The axial electrostatic or magnetic scalar potential distribution will be represented by a piecewise cubic function. As usual, the axial length of the lens is divided into N equal regions. The unknown distribution V(z) is sought in the form of

$$V_k(z) = A_k + B_k(z - z_k) + C_k(z - z_k)^2 + D_k(z - z_k)^3 \tag{8}$$

for each region (k=1,2,...,N) where z_k is the axial coordinate of the starting point of the kth region. The coefficients A_k, B_k, C_k and D_k are different for each region, therefore V(z), V'(z) and V"(z) have different expressions for each region. It is very easy, however, to ensure the continuity of these functions by requiring the satisfaction of the following relationships between the coefficients:

$$A_{k+1} = A_k + B_k \Delta z + C_k (\Delta z)^2 + D_k (\Delta z)^3 \tag{9}$$

$$B_{k+1} = B_k + 2 C_k \Delta z + 3 D_k (\Delta z)^2 \tag{10}$$

and

$$C_{k+1} = C_k + 3 D_k \Delta z \tag{11}$$

where

$$\Delta z = L/N = z_{k+1} - z_k \tag{12}$$

Let us formulate the constraints now. Naturally, the fields must be practically realizable. Therefore, the magnitudes of the potential and its derivatives must be limited:

$$V_I \le V(z) \le V_{II} \tag{13}$$

$$|V'(z)| \le V'_I \tag{14}$$

and

$$|V''(z)| \le V''_I \tag{15}$$

where V_I, V_{II}, V'_I and V''_I are a priori given numbers. It is usually required that the fields vanish at both the object and the image:

$$V'(z_o) = V'(z_i) = 0 \tag{16}$$

The particle trajectories must be focused toward the optical axis but should not cross it inside the optical element. In addition, we must provide some working distance beyond the lens. Therefore, we must require

$$r(z) > 0 \tag{17}$$

and

$$r'(z_o) > 0 \tag{18}$$

Depending on the particular problem, the constraints may be different from these.

If we relate the potential distribution to V_o then

$$A_1 = 1 \tag{19}$$

Conditions (16) yield

$$B_1 = 0 \tag{20}$$

and

$$B_N + 2 C_N \Delta z + 3 D_N (\Delta z)^2 = 0 \tag{21}$$

The above constraints leave only the coefficients C_1 and D_k (k=1,2,...,N-1) free. For a given set of coefficients the paraxial ray equation can easily be solved and the aberration integrals evaluated numerically. Therefore, the problem of searching for a function in the infinite set of different possibilities is reduced to that of finding these N coefficients so that they satisfy the paraxial equation and our constraints and at the same time minimize the aberration integrals.

This is an N-dimensional constrained optimization problem. We can solve it either by the dynamic programming or the function minimization approach with the fundamental advantage over the original versions of these methods that the optimized potential distribution will now be available in the form of a continuous curve with only three derivatives instead of a digital data set.

If the <u>dynamic programming</u> procedure is used, we shall utilize the fact that for the piecewise cubic lens model the second derivative of the axial potential distribution is a linear function of the axial coordinate within each region:

$$V''_k(z) = 2 C_k + 6 D_k (z - z_k) \tag{22}$$

The distribution of the second derivative V"(z) is then given by a series of continuous linear segments. We simplify the search by restricting these linear segments to those connecting the nodes of the discrete computational grid i. e. we only allow 2M+1 different discrete values of the second derivative at the boundaries of the regions. We have to substitute V" for h in equations (4) and (5).Then the coefficients are expressed through i and j by

$$C_k = j\,\Delta V''/2 \tag{23}$$

and

$$D_k = \frac{i-j}{6\Delta z}\,\Delta V'' \tag{24}$$

with

$$\Delta V'' = V''/M_I \tag{25}$$

The rest of the procedure is identical to that outlined above in the description of the dynamic programming method.

We have modified our <u>function minimization</u> procedure, too, to provide a continuous, piecewise cubic function for the optimized axial potential distribution in the form of Eq. (8), the coefficients of which are determined by the procedure.

Dynamic programming and function minimization very well complement each other: while the first provides a global search in a discrete domain the second is very fast and accurate. Both give quite satisfactory results but in extraordinarily difficult cases the synthesis procedure may combine them by starting with dynamic programming and refining the optimized axial potential distributions by the function minimization procedure.

The reconstruction of the electrodes (pole pieces) is extremely simple now because we already know not only the axial distribution of the potential but also its continuous second derivative. The third derivative is not continuous but it does not appear anywhere so it cannot cause any problem. The fourth and higher derivatives are zero everywhere except the boundaries of the regions where they are still undefined. However, we are not using the spline function for curve fitting now. Therefore, it is justified to assume that these higher derivatives not appearing anywhere in the expressions of the focusing properties and aberrations will not affect the potential distribution either. This is why the simple expression (7) can be used for the potential distribution in the entire space. By the use of this formula we are able to reproduce the equipotential surfaces that will provide the same functions V(z), V'(z) and V"(z) and thus the same focusing properties and aberrations as the original theoretical distribution obtained from the optimization procedure.

First we examine the potential distribution and determine its inflection points. The number of electrodes (pole pieces) is always one more than the number of inflection points that separate the neighbouring electrodes. Then we substitute the suitably chosen values of the electrostatic or magnetic scalar equipotentials into Eq. (7) for each electrode and obtain simple relationships r=r(z) for the shapes of the electrodes or pole pieces. If the value of V" is negative between two inflection points, the electrode (pole piece) potential is chosen slightly higher than the maximum value of V(z) in the given region. If V" is positive, the electrode (pole piece) potential is lower than the minimum value of V(z) in the region.

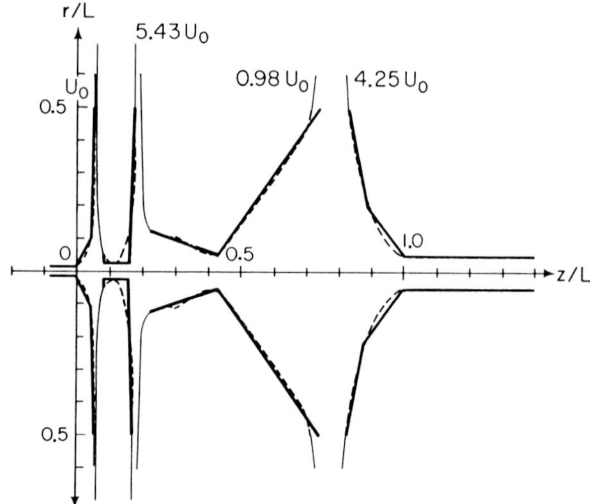

Fig. 5. Four-electrode optimized electrostatic lens designed by the author's new method of synthesis. The electrode shapes are gradually simplified to plane surfaces.

The maximum value of r must be limited to a realistic size, usually to half of the length of the system. Since the value of r rapidly approaches infinity at the inflection points, parts of the electrodes (pole pieces) must be omitted. This automatically ensures a certain minimum distance between them which is necessary to avoid electric breakdown in case of electrostatic lenses.

A very important remark is due here. When the final electrode shapes are chosen, we actually depart from the optimized potential distribution by omitting parts of the electrodes. During this process we can also simplify the electrode shapes and replace the complicated curved boundaries by easily manufacturable simple straight surfaces. The accuracy of the reconstruction will only slightly be affected by these actions but on the other hand the practical value of the method is tremendously increased by them.

As an example, the schematics of a four-electrode electrostatic lens designed by the function minimization method of synthesis is shown in Fig. 5. The dimensions of the lens are given in units of the effective length L (the distance between the axial points beyond which the potentials are practically constant at both sides of the lens). The electrode potentials are related to the potential of the first electrode at the object side. The optimization procedure first yields (1) the infinitely long electrodes, parts of which are shown as thin lines in the figure. As the next step (2), the electrodes are cut to reasonable sizes and parts of them are totally omitted in order to avoid breakdown (broken lines). The final simple electrode shapes (3) are shown in thick continuous lines.

The axial potential distributions of the reconstructed lenses were re-calculated by the charge density method and the optical properties (locations of the cardinal elements and values of the aberration coefficients) were determined by ray tracing and numerical integration. The results for the three systems are compared in the following table:

System #	Focal plane z_o/L	Principal plane P_o/L	Focal length f_o/L	Spherical aberr. C_{so}/L	Chromatic aberr. C_{co}/L
1.	-0.029	0.182	0.212	0.190	0.169
2.	-0.025	0.187	0.212	0.224	0.175
3.	-0.018	0.194	0.212	0.235	0.177

As we can see, even the optical properties of the simplest lens are excellent and not very much different from those of the optimized theoretical potential distribution. At a very low maximum-to-minimum electrode potential ratio (U_{max}/U_{min}=5.54) we have C_{so}/f_o= 1.11 and C_{co}/f_o = 0.83. The negative value of the location of the focal plane provides comfortable working distances when the lens is used in the probe forming mode (the zero point coincides with the center of the entrance aperture).

We were also able to design a three-electrode electrostatic lens with U_{max}/U_{min}= 8.86, C_{so}/f_o = 0.95 and C_{co}/f_o = 0.72 (Szilagyi, 1984a) as well as a five-electrode einzel-lens with U_{max}/U_{min}=6.75, C_{so}/f_o= 1.03 and C_{co}/f_o= 0.76 (Szilagyi, 1984b).

Conclusion

Electron and ion optical synthesis is a very real problem of Particle Beam Optics. The final goal is to be able to design an entire optical column automatically on the basis of given properties. Many years of work by a number of different researchers has eventually culminated in our solution outlined above. It provides a powerful practical tool for the synthesis of electron and ion lenses. The reconstruction of electrodes (pole pieces) is built in the optimization procedure and does not require any extra effort. Our design examples not only prove the effectiveness of this approach but also give encouragement to the possibility of building very high performance electrostatic lenses so much needed for ion beam lithography.

Acknowledgment

This work is supported by the National Science Foundation - Solid State and Microstructures Engineering - Grant # ECS 8317485.

References

Burfoot JC. (1953). Correction of electrostatic lenses by departure from rotational symmetry. Proc. Phys. Soc. B66, 775-792.

Chu HC, Munro E. (1981). Computerized optimization of electron-beam lithography systems. J. Vac. Sci. Technol. 19(4), 1053-1057.

Crewe AV. (1977). Ideal lenses and the Scherzer theorem. Ultramicroscopy 2, 281-284.

Glaser W. (1938). Die kurze Magnetlinse von kleinstem Öffnungsfehler. Z. Phys. 109, 700-721.

Glaser W. (1940). Über ein von sphärischer Aberration freies Magnetfeld. Z. Phys. 116, 19-33.

Hawkes PW. (1981). Some uses of computers in electron optics. J. Phys. E: Sci. Instrum. 14, 1353-1367.

Kasper E. (1981). Numerical design of electron lenses. Nuclear Instruments and Methods 187(1), 175-180.

Kas'yankov PP. (1952). On the problem of calculating an electron lens with given conditions on the third-order aberrations. (In Russian). Zh. Tekhn. Fiz. 22, 80-83.

Kas'yankov PP. (1955). On electron lenses with arbitrarily small spherical aberration. (In Russian). Zh. Tekhn. Fiz. 25, 1639-1648.

Moses RW. (1970). Minimum aperture aberrations of quadrupole lens systems. Rev. Sci. Instrum. 41, 729-740.

Moses RW. (1971a). Minimum chromatic aberrations of magnetic quadrupole objective lenses. Rev. Sci Instrum. 42, 828-831.

Moses RW. (1971b). Quadrupole-octopole aberration correctors with minimum fields. Rev. Sci. Instrum. 42, 832-839.

Moses RW. (1973). Lens optimization by direct application of the Calculus of Variations, in: Image Processing and Computer-Aided Design in Electron Optics, P. W. Hawkes (ed), Academic Press, London and NY, 250-272.

Recknagel A. (1941). Über die sphärische Aberration bei elektronenoptischer Abbildung. Z. Phys. 117, 67-73.

Rose H, Moses RW. (1973). Minimaler Öffnungsfehler magnetischer Rund- und Zylinderlinsen bei feldfreiem Objektraum. Optik 37, 316-336.

Scherzer O. (1936a). Die schwache elektrische Einzellinse geringster sphärischer Aberration. Z. Physik 101, 23-26.

Scherzer O. (1936b). Über einige Fehler von Elektronenlinsen. Z. Physik 101, 593-603.

Septier A. (1966). The struggle to overcome spherical aberration in Electron Optics, in: Advances in Optical and Electron Microscopy, R. Barer and V. E. Cosslett (eds), Academic Press, London and NY, 1, 204-274.

Siegbahn K. (1946). A magnetic lens of special field form for beta- and gamma-ray investigations; designs and applications. Phil. Mag. 37, 162-184.

Szilagyi M. (1977a). A new approach to electron optical optimization. Optik 48, 215-224.

Szilagyi M. (1977b). A dynamic programming search for magnetic field distributions with minimum spherical aberration. Optik 49, 223-246.

Szilagyi M. (1978a). A dynamic programming search for electrostatic immersion lenses with minimum spherical aberration. Optik 50, 35-51.

Szilagyi M. (1978b). A solution to the problem of the ideal quadrupole lens. Optik 50, 121-128.

Szilagyi M. (1978c). Reduction of aberrations by new optimization techniques. Proc. Ninth International Congress on Electron Microscopy, Canadian Microscopical Society, Toronto, 1, 30-31.

Szilagyi M. (1980). Generation of a continuous potential distribution from digital data for electron optical synthesis. Electron Microscopy, Proc. 7th European Congress, The Hague, 7th Europ. Congr. on EM Foundation, Leiden, The Netherlands, 1, 62-63.

Szilagyi M. (1983a). Improvement of electrostatic lenses for ion beam lithography. J. Vac. Sci. Technol. B1(4), 1137-1140.

Szilagyi M. (1983b). Dynamic programming minimization of lens aberrations for ion beam lithography, in: Applied Simulation and Modelling, Proc. ASM'83 San Francisco, Acta Press, Anaheim, Calgary and Zurich, 41-44.

Szilagyi M. (1984a). Reconstruction of electrodes and pole pieces from optimized axial field distributions of electron and ion optical systems. Appl. Phys. Lett. (in press).

Szilagyi M. (1984b). Electron optical synthesis and optimization. Proc. IEEE, 45, 499-501.

Szilagyi M, Yakowitz SJ, Duff MO. (1984). Procedure for electron and ion lens optimization. Appl. Phys. Lett. 44, 7-9.

Tretner W. (1954). Die untere Grenze des Öffnungsfehlers bei magnetischen Elektronenlinsen. Optik 11, 312-326.

Tretner W. (1959). Existenzbereiche rotationssym metrischer Elektronenlinsen. Optik 16, 155-184.

Discussion at the Conference

This work has generated an unusually great interest at the Conference. Very active and lengthy discussions took place both during its presentation and at the special workshop on electron lens design. Although it is not possible to describe the entire debate, a summary of the most important items is presented in the following:

1. Synthesis is necessary because different applications require different designs. Therefore, it is not sufficient to catalogue some simple cases but a general design method is needed.

2. The author's method must be used with additional simplification of the electrodes (pole pieces). The strength of the method is that it is able to find the locations and potentials of the electrodes. Their shapes do not influence the optical properties very much.

3. The simplification of the electrode shapes naturally introduces higher derivatives into the axial potential distribution. However, this does not constitute any real problem since our aim is not an exercise in pure mathematics but practical design.

4. To obtain more general designs it is convenient to define the electrode shapes and potentials in dimensionless units. The unit of potential is determined by the given energy of the particles either in the object or the image space. Then the minimum value of the effective length is chosen from the requirement that in order to avoid breakdown the electric field must not exceed a certain maximum value.

Electron Optical Systems (pp. 85-89)
SEM Inc., AMF O'Hare (Chicago), IL 60666-0507, U.S.A.

0-931288-34-7/84$1.00+.05

THE APPLICATION OF THE ALTERNATING PROCEDURE OF H.A. SCHWARZ FOR COMPUTING THREE-DIMENSIONAL ELECTROSTATIC FIELDS IN ELECTRON-OPTICAL SYSTEMS

Christoph H. Schaefer*

Lehrstuhl für Theoretische Elektronenphysik
Institut für Angewandte Physik
University Tubingen
7400 Tubingen
West Germany

Abstract

The numerical solution of three-dimensional electrostatic field problems in electron-optical devices with complicated boundaries requires the use of very large and fast computers. A numerical counterpart of the Alternating Procedure by H.A.Schwarz (SAP) makes efficient use of both storage space and time available on the computer and therefore represents a useful new method in this context. In this method, the domain of solution is decomposed into independent subdomains where the appropriate individual methods of solution may be chosen. Correspondingly, the original boundary value problem is decomposed into a sequence of boundary value problems in these subdomains. SAP requires no explicit influence matrix and uses little storage space. Time requirements depend essentially on the methods applied in the respective subregions. SAP is particularly useful in obtaining analytical formulas for field distributions in "large" regions of empty space (such as in fringe field regions). Numerical results obtained with a computer program treating an electrostatic quadrupole lens problem show that a high degree of accuracy can be obtained with this method.

Key words: Alternating Procedure, three-dimensional, electrostatic field, potential theory, numerical, quadrupole lens, fringe field, subregion.

* Present address: IBM T. J. Watson Research Center, Yorktown Heights, N.Y. 10598, USA. Phone no.: 914-945-2046

Introduction

In the computer design of electron-optical devices, the Laplace equation $\Delta U=0$ arises when electrostatic and magnetostatic fields are to be calculated. The complexity of boundaries encountered in realistic devices requires the use of numerical methods for solving the corresponding boundary value problems. Three numerical methods in common use are the Finite-Difference Method (FDM) [2,4], the Finite-Element Method (FEM) [12], and the Integral Equation Method (IEM) [1;6].

We shall be concerned in particular with electrostatic field computation in complicated devices *not* possessing rotational symmetry. Moreover, since we want to consider fringe field problems, we must allow for domains of definition D for the potential U which are unbounded. An example for this type of problem is furnished by the electrostatic quadrupole lens pictured in fig. 1, consisting of four cylindrical electrodes arranged symmetrically as shown but held at the respective potentials $+V_0$ and $-V_0$. Though D is bounded by a grounded shielding in the x- and y-directions, it is assumed to be unbounded in the z-direction, and a fringe field must be taken into account.

Applied to this class of problems, each of the three methods mentioned gives rise to certain difficulties. In the following, we investigate briefly the pros and cons of each method. The applicability of each method is closely related to the geometry of the domain it is to be used in.

An FDM is adequate for "small" bounded domains enclosed by irregular boundary surfaces (perhaps containing edges and corners). For "large" domains, the dimension of the associated linear system grows too large, requiring too much storage space and computer time during iterative solution. Code is simple and short. The desired field is obtained from stored potential data using interpolation techniques. This requires much storage space but produces the field at little time expenditure. Smoothness of the field depends on meshwidth and interpolation formulae in use. For rectangular domains of reasonable size, fast solvers may be quite useful. (An account of fast direct methods in rectangular 2-d regions is found in [5]. Many of these methods "have a natural extension to three dimensions".)

Most of these considerations equally apply to the FEM. However, the FEM is more flexible both regarding domain size and irregularity of boundaries, at the cost of more complex code. (For a comparison of the FDM and the FEM with respect to electron-optical applications, see [7].)

An IEM is adequate for treating an exterior or mixed interior and exterior boundary value problem (e.g. any electrostatic field problem in unbounded space arising from potential values being posed on arbitrarily positioned electrodes). The associated code is flexible. An IEM is inadequate if the boundary surfaces are "large" in total surface area, as the dimension of the associated linear system grows too large and the matrix of coefficients

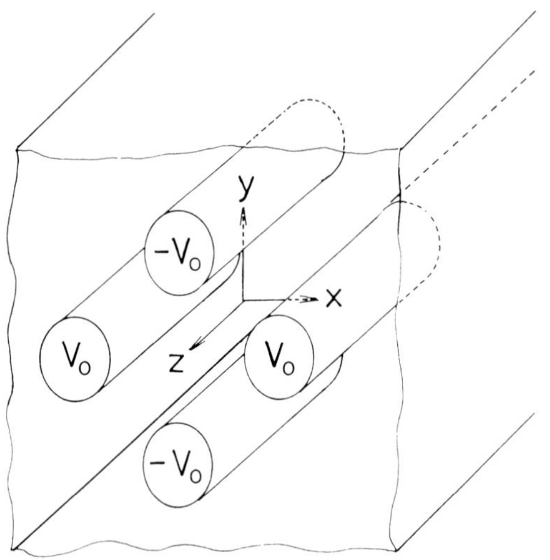

Fig. 1: Electrostatic quadrupole lens with cylindrical electrodes and shielding

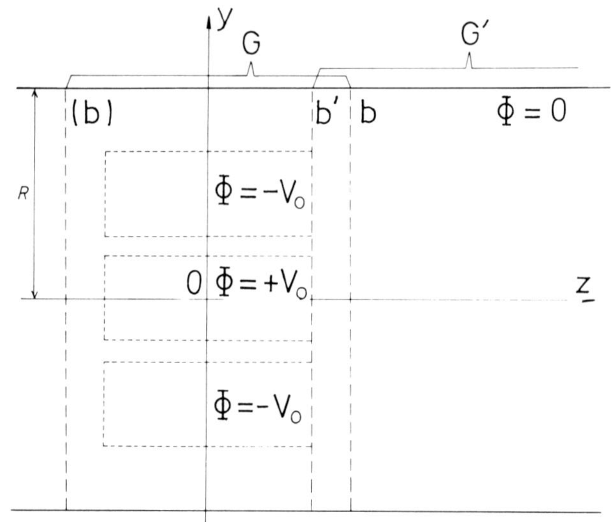

Fig. 2: Cross-section $x=0$ through the electrostatic quadrupole lens. Definition of surfaces b,b' and subregions G,G'.

is dense. The resulting potential is defined continuously and is "smooth", and the field and derivatives of higher order may be computed analytically. However, potential and field computation are expensive on computer time, and accuracy deteriorates near boundaries.

In practical 3-d field problems, different methods may seem appropriate to use in different parts of the domain D. Moreover, a certain advantage of one method may seem desirable in a certain region of D (such as obtaining an analytical solution). The method we are presenting allows one to decompose D into a finite number of subdomains. In each subdomain, the appropriate method of solving the Dirichlet problem associated with the Laplace equation is chosen. Instead of solving the original Dirichlet problem, a sequence of Dirichlet problems with different boundary conditions is solved in each subdomain. Besides combining the advantages of several methods, the proposed method has two notable properties: it requires storing values of functions only on surfaces, not in volumes (which is also true for the IEM) but requires no storage space for a matrix of coefficients. In fact, this matrix is never explicitly set up.

This method originally was conceived as a theoretical tool in potential theory by H.A.Schwarz in 1870 [11] and is referred to as "Alternating Procedure" (SAP). Its numerical counterpart has been employed by F.Lenz (private communication) since the early 1950's, and more recently by Kern [8] and the author [9,10].

In the first section of this paper, SAP is illustrated and discussed in the physical context of a typical fringe field problem in three dimensions. Only the case of two subdomains is treated there. The following section defines the precise algorithm to proceed by in the case of $n \geq 2$ subdomains. (The associated proofs may be found in [9,10].) The results of an application of SAP to the electrostatic-field problem associated with a quadrupole lens are reported in a final section.

Fringe field determination through SAP

As an example, suppose we were to determine the electrostatic field induced in the quadrupole lens pictured in fig. 1. A longitudinal cross-section $x=0$ through that system is shown in fig. 2. The dotted lines (-----) represent the projection of the contours of the electrodes. While the electrodes have a finite length of $2L$, the shielding is assumed to extend infinitely in the $\pm z$-directions. The shielding is of square cross-section, and its sides are perpendicular to the x- and y-axes, respectively, each having a distance R to the z-axis. Let D denote the domain of solution of the associated Dirichlet problem (DP; the boundary value problem of Dirichlet boundary conditions being imposed on a solution of the Laplace equation is referred to as DP. Dirichlet boundary conditions prescribe the values of the potential on the boundary ∂D of D.).

We artificially introduce two plane surfaces b' and b which are perpendicular to the z-axis (indicated by dotted lines (-.-.-.-) in fig. 2). This enables us to define two overlapping subregions of D: G, a bounded region enclosing the electrodes, and G', an unbounded fringe field region. From the figure it should be clear that G extends from boundary b (left) to boundary b (right), and that G' extends from b' to infinity. We are assuming for simplicity that that problem is symmetric with respect to the x,y-plane ($z=0$).

In G', the desired potential Φ may be represented analytically using the following series:

$$\Phi(x,y,z) = \sum_{m,n=1}^{\infty} A_{mn} \cos\left[(2m-1)\frac{\pi}{2}\frac{x}{R}\right] \cos\left[(2n-1)\frac{\pi}{2}\frac{y}{R}\right] \cdot$$

$$\cdot \exp\left(-\sqrt{(2m-1)^2+(2n-1)^2}\,\frac{\pi}{2}\frac{z-L}{R}\right) \quad (1)$$

If the potential values $\Phi(x,y,L)$ ($|x|, |y| < R$) were known, the coefficients A_{mn} could readily be determined by evaluating the iterated integrals

$$A_{mn} = \frac{4}{R^2}\int_0^R\int_0^R \Phi(x,y,L) \cos\left[(2m-1)\frac{\pi}{2}\frac{x}{R}\right] \cdot$$

$$\cdot \cos\left[(2n-1)\frac{\pi}{2}\frac{y}{R}\right] dx\,dy \quad (m,n \geq 1) \quad (2)$$

More generally, we may prescribe on b' arbitrary smooth boundary values $V(x,y)$ (which exhibit the required symmetry) and obtain a series solution (1) to the associated Dirichlet problem by substituting these values $V(x,y)$ for $\Phi(x,y,L)$ in (2).

Correspondingly, let us assume that we are capable of solving the Dirichlet problem in G associated with arbitrary boundary values posed on b.

Consider the following alternating sequence of Dirichlet problems posed in G and G':

1) Solve the DP associated with zero boundary values on b and the originally imposed boundary values on all other boundaries of G. Denote the solution by $U^{(0)}$.

2) Restrict $U^{(0)}$ to b', denote these values defined on b' by $U^{(0)}_{b'}$. Let $V^{(0)}$ be the solution of the DP in G' which is defined by the values $U^{(0)}_{b'}$ on b' and the original (vanishing) values on the remainder of $\partial G'$, the boundary of G'.

3) Restrict $V^{(0)}$ to the surface b, denote these values by $V^{(0)}_b$. Let $U^{(1)}$ be the solution of the DP in G which is defined by the values $V^{(0)}_b$ on b and zero boundary values on the remainder of ∂G, the boundary of G.

4) Restrict $U^{(1)}$ to b', denote these values by $U^{(1)}_{b'}$. Let $V^{(1)}$ be the solution of the DP in G' defined by the values $U^{(1)}_{b'}$ on b' and zero boundary values on the remainder of $\partial G'$. ...

Continuing in this fashion yields two sequences $(U^{(i)})_{i=1,2,...}$, $(V^{(i)})_{i=1,2,...}$ of harmonic functions defined in G and G', respectively. The series

$$\sum_{i=0}^{\infty} U^{(i)}(x,y,z) \quad and \quad \sum_{i=0}^{\infty} V^{(i)}(x,y,z)$$

converge uniformly in G and G', respectively, and their sums are equal to the desired potential values $\Phi(x,y,z)$ in the respective regions. The rate of convergence is one of a geometric series $\Sigma_i q^i$ with $0<q<1$; q becomes smaller as the overlap becomes larger and vice versa. For exact solutions $U^{(i)}$ and $V^{(i)}$, this is the assertion of the Alternating Procedure. (For a proof, see [3].) A precise description of how the procedure is performed discretely (on the computer) will be given in the next section. This includes a sufficient condition of convergence in the discrete case.

SAP prescribes a repeated performance of the following four steps: Solve a DP in G; evaluate the solution on $b' \subset G$; solve a DP in G'; evaluate the solution on $b \subset G'$. Following this procedure on the computer, only one of the two subroutines for solving a DP in G and G' is kept in the high-speed store at a time. For SAP to proceed after the i-th iteration in G, there is no need to evaluate and store all values of the solution $U^{(i)}$ in G; only the values $U^{(i)}_{b'}$ are of immediate interest. The same is true for $V^{(i)}$ and G'; i is any current value of the iteration parameter. Keeping stored, say, the sum

$$\sum_{k=0}^{i} U^{(k)}_{b'},$$

after a sufficient number M of iterations,

$$\Phi \cong \sum_{k=0}^{M} U^{(k)}_{b'} \ ;$$

from these approximated values of Φ on b', Φ can later be determined anywhere in G' (and hence in G). Analytical solutions like the functions $V^{(i)}$ in G' are particularly convenient; they allow the determination of the values $V^{(i)}_b$ without computing any other values of $V^{(i)}$ in G'. This means that very little storage space is required to obtain the values $V^{(i)}_b$ on b corresponding to the solution $V^{(i)}$ of the i-th DP in G'. Comparing SAP to conventional methods, the reduction in storage space stems from the fact that potential values need not be computed and stored in volumes but only on surfaces (b and b'). In this respect, SAP resembles the IEM. However, no matrix of coefficients for the unknown values of Φ on these surfaces is explicitly set up; the Alternating Procedure is merely followed, employing the two subroutines for solving the respective Dirichlet problems in G and G'.

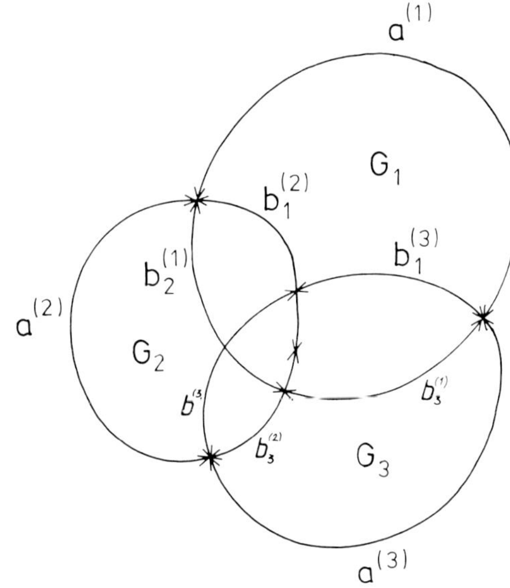

Fig. 3: Decomposition of a 2-d domain into three subregions G_1, G_2, G_3. Definition of boundary segments $a^{(i)}$ and $b_j^{(i)}$.

The other major advantage of SAP is that the domain of definition D is divided into subregions where individually, the appropriate methods of solution can be applied. In the fringe field region, for example, an analytical solution seems appropriate since the region is unbounded and its geometry is simple. On the other hand, in a bounded region such as G containing a complicated device, a flexible discrete method like the FDM seems most suitable. SAP tells us how to join the separate solutions "harmonically".

All these arguments apply equally well to the case of $n \geq 2$ subregions, treated in the next section.

The general Alternating Procedure

In this section, it is shown how a given domain D is properly decomposed into n overlapping subdomains and how SAP is performed step by step, making use of existing methods of solution in the individual subdomains. A condition is stated which each method must satisfy for SAP to converge. No proofs are issued in this paper; they may be found in [9] or [10]. No analytical estimate of the approximation error of SAP is offered by this author. It will depend on the accuracy of the individual methods in use.

Consider decomposing an open and connected region (domain) D in 3-space into $n \geq 2$ overlapping subdomains G_i $(i=1,...,n)$:

$$D = \bigcup_{i=1}^{n} G_i$$

(fig. 3 illustrates a two-dimensional case where n=3). Let ∂D denote the boundary of D. Each boundary $\Gamma_i = \partial G_i$ of G_i consists of two complementary parts

$$a^{(i)} := \Gamma_i \cap \partial D \quad and \quad b^{(i)} := \Gamma_i \cap D; \quad \Gamma_i = a^{(i)} \cup b^{(i)} \ .$$

(Perhaps $a^{(i)} = \emptyset$ but $b^{(i)} \neq \emptyset$.) By $\bar{b}^{(i)}$ we refer to the closure $\bar{b}^{(i)} = b^{(i)} \cup \partial b^{(i)}$ of an open surface segment $b^{(i)}$. Each surface $b^{(i)}$ is divided into n−1 open surface segments

$$b_1^{(i)}, b_2^{(i)}, ..., b_{i-1}^{(i)}, b_{i+1}^{(i)}, ..., b_n^{(i)}$$

such that for each $j=1,...,i-1,i+1,...,n$, $b_j^{(i)}$ is contained in the interior of the region G_j:

$$b_j^{(i)} \subset G_j \, .$$

More precisely,

$$\bar{b}^{(i)} = \bigcup_{j=1;j\neq i}^{n} \bar{b}_j^{(i)} \quad but \quad b_j^{(i)} \cap b_k^{(i)} = \emptyset \quad if \quad j\neq k$$

$$(j,k=1,...,i-1,i+1,...,n)$$

Of course, this division of each $b^{(i)}$ generally is not unique and is only possible if the following vital condition is satisfied:

Overlap condition. For every subdomain G_i $(i=1,...,n)$, that part $b^{(i)}$ of its boundary surface Γ_i which is located in the interior of the original domain D must satisfy the following condition: every point of $b^{(i)}$ must lie within at least one of the other subdomains G_j $(j\neq i)$.

Next, consider distributing a finite number of mesh points P_k in some uniform manner over the complete set of (open!) boundary surface segments $b_j^{(i)}$ $(i=1,...,n;\ j=1,...,i-1,i+1,...,n)$. With these points we associate an array of variables x_k used in SAP. The array x_k constitutes most of the storage space required by SAP. Certain subsets of this array will sometimes be referred to as "the values associated with" one or the other surface segment.

Assume that for each subdomain G_i, a numerical method (implemented as a computer subroutine) is available which will solve a given DP in G_i. Any such method will linearly associate with a complete set of boundary values x_l (corresponding to a complete set of mesh points P_l located on Γ_i), any required set of interior values x_m (corresponding to points P_m located in the interior of G_i). If the discrete set of boundary values x_l approximates some set of boundary values continuously defined on Γ_i, the values x_m should approximate the potential values $U(P_m)$ of the associated exact solution U .

For convergence of SAP, only one condition is required of each of the numerical methods employed in the respective subdomains: it must be subject to the discrete counterpart of the fundamental maximum-minimum principle for harmonic functions. It is stated below.

Discrete maximum-minimum principle: if $M_1 \leq x_l \leq M_2$ at every boundary point $P_l \in \Gamma_i$, solving the DP in G_i numerically leads to values $M_1 < x_m < M_2$ at any $P_m \in G_i$. It has been assumed that: 1)$M_1 < M_2$, and that $M_1 \leq 0 \leq M_2$ if G_i is unbounded; 2)the boundary values x_l are not constant.

As in the case of n=2 subregions, the general Alternating Procedure prescribes moving from subdomain to subdomain in a defined sequence. In each subdomain, a current set of (discrete) boundary values defines a DP, this DP is solved, and the solution is evaluated at certain mesh points on surface segments lying within the current subdomain. The following initial and general steps define this procedure.

By $\psi_j^{(i)}$denote the array of variables x_k associated with a surface segment $b_j^{(i)}(i=1,...,n;\ j=1,...,i-1,i+1,...,n)$. Let G_{i_0} be any subregion where after the completion of SAP, the solution is required $(1 \leq i_0 \leq n)$. For each $j=1,...,i_0-1,i_0+1,...,n$, let $\phi_j^{(i_0)}$ denote an independent array of variables associated with the set of mesh points located on $b_j^{(i_0)}$.

Initial procedure

Set the values associated with all surfaces $b^{(i)}$ $(i=1,...,n)$ equal to zero. Let the parameter i run from 1 to n in steps of 1. For each i, carry out the following step completely (before increasing i):

Solve the DP in G_i. The boundary values on $a^{(i)}$ are the values inherited from those prescribed on ∂D, the boundary values on $b_j^{(i)}$ $(j=1,...,i-1,i+1,...,n)$ are those currently stored in connection with $b_j^{(i)}$. Restrict the obtained solution ψ to all surfaces $b_i^{(j)} \subset G_i$ $(j=1,...,i-1,i+1,...,n)$ and store these values $\psi_i^{(j)}$ in connection with $b_i^{(j)}$. If $i\neq i_0$, let $\phi_i^{(i_0)} := 0$.

General procedure

For $i=1,2,...,n,1,2,...,n,1,2,...$ perform the following step:

Solve the DP in G_i . The boundary values on $a^{(i)}$ are zero, the boundary values on $b_j^{(i)}$ $(j=1,...,i-1,i+1,...,n)$ are those currently stored in connection with $b_j^{(i)}$. Restrict the obtained solution ψ to all surfaces $b_i^{(j)} \subset G_i$ $(j=1,...,i-1,i+1,...,n)$ and replace the old values associated with $b_i^{(j)}$ by the new values $\psi_i^{(j)}$. If $i\neq i_0$, update $\phi_i^{(i_0)} := \phi_i^{(i_0)} + \psi_i^{(i_0)}$. When the values $\psi_i^{(i_0)}$ have become uniformly "small", stop the procedure.

The solution in G_{i_0} may now be obtained by solving a final DP in G_{i_0} with the original boundary values posed on $a^{(i_0)}$ and the values $\phi_j^{(i_0)}(j=1,...,i_0-1,i_0+1,...,n)$ posed on $b_j^{(i_0)}$. Consider any quantity representing a linear functional operating on the potential in G_{i_0} (such as the value of this potential at any specific interior point or any coefficient in some series representation of this potential). Such quantities may replace the arrays $\phi_j^{(i_0)}$ and make solving the final DP in G_{i_0} unnecessary if these quantities were the only information required with respect to this subdomain.

Numerical results obtained with SAP

Using SAP, a program has been developed which computes the electrostatic potential in a quadrupole lens as shown in fig. 1. The domain of solution D has been decomposed into n=6 overlapping subdomains, four of which are rectangular and have been treated using analytical (series) solutions. The FDM serves to solve the respective Dirichlet problems in each of the remaining two. More details of this program, including some plots of equipotential lines in various sections across D, may be found in [10].

Some figures which may serve to estimate the accuracy of the computed potential are presented in table 1.

Table 1: Maximum error of computed potential

Run no.		1	2	3
$\Delta_r(\Delta_a)$ in	T_1	.00071(.00005)	.00019(.00002)	.00014(.00001)
	T_2	.00072(.00017)	.00019(.00006)	.00014(.00005)
	T_3	.00096(.00033)	.00039(.00010)	.00025(.00007)
	T_4	.00164(.00070)	.00064(.00026)	.00045(.00017)
No. of iterations		23	25	24
Time (h:min)		0:19	2:47	10:08

The columns of this table correspond to various runs of the program with differing degrees of precision. In lines two through five we find the maximum relative error Δ_r (in parentheses: the maximum absolute error Δ_a) within four plane square regions of test T_1-T_4 in the fringe field area of the lens:

$$T_i = \{(x,y,z) \mid z = z_0; \mid x \mid , \mid y \mid \leq r_i\}.$$

(See fig. 1 for the orientation of the coordinate system. The origin is located at the center of the lens.) Let a denote the distance of the electrodes from the optical axis. z_0 has been chosen such that:

$$\frac{(z_0 - L)}{a} \cong 1.2$$

The numbers r_i are defined as follows:

$$r_1 = \frac{a}{4}\ ; r_2 = \frac{a}{2}\ ; r_3 = \frac{3a}{4}\ ; r_4 = \frac{11a}{12}\ .$$

The absolute error Δ_a is given with respect to unit potential values ± 1 on the electrodes. The relative error at a point $P \epsilon D$ is defined as

$$\Delta_r(P) = \frac{\Delta_a(P)}{\mid \Phi(P) \mid}.$$

In the last two lines are reported the number of iterations performed, i.e. the number of times a DP is solved in *each* subdomain (with the boundary values uniformly growing smaller each iteration), and the computer time required on a Sperry-Univac 1100/80 computer. With respect to these time requirements, we would like to emphasize that a high speed store of 480 KBytes was used and no peripheral storage space was required. We note that even with respect to complicated electrostatic field problems, SAP could be used on smaller computers where storage space is sparse while computer time is cheap.

Acknowledgements

The author wishes to thank Prof. Dr. F. Lenz for introducing him to the basic ideas of this work and for his continuing support during the work. Thanks are due to the Zentrum für Datenverarbeitung der Universität Tübingen for making available to the author substantial amounts of computing time on their Univac 1100.

References

[1] Christiansen S, Hansen EB. (1978). Numerical solution of boundary value problems through integral equations. ZAMM 58, T14-T25.

[2] Collatz L. (1960). The Numerical Treatment of Differential Equations (Die Grundlagen der mathematischen Wissenschaften Band 60). Springer-Verlag, Berlin - Heidelberg - New York.

[3] Courant R, Hilbert D. (1968). Methoden der mathematischen Physik II (Heidelberger Taschenbücher Band 31). Springer-Verlag, Berlin - Heidelberg - New York. 264-266.

[4] Forsythe GE, Wasow WR. (1960). Finite Difference Methods for Partial Differential Equations. John Wiley & Sons, Inc., New York - London.

[5] Hockney RW. (1970). The Potential Calculation and Some Applications (Methods in Computatonal Physics Vol. 9 Plasma Physics). Academic Press, New York - London. 136-164.

[6] Jaswon MA, Symm GT. (1977). Integral Equation Methods in Potential Theory and Elastostatics. Academic Press, New York - London.

[7] Kasper E, Lenz F. (1980). Numerical methods in geometrical electron optics, in: ELectron Microsc. **1** (ed. by P. Brederoo and G. Boom) North Holland, Amsterdam, 10–15.

[8] Kern D. (1978). Theoretische Untersuchungen an rotationssymmetrischen Strahlerzeugungssystemen mit Feldemissionsquelle. Dissertation, Universität Tübingen, 7400 Tübingen, W-Germany.

[9] Schaefer CH. (1982). Methoden zur numerischen Loesung der Laplacegleichung bei komplizierten Randwertaufgaben in drei Dimensionen und ihre Anwendung auf Probleme der Elektronenoptik. Dissertation, Universität Tübingen, 7400 Tübingen, W-Germany.

[10] Schaefer CH. (1983). The application of the Alternating Procedure by H.A.Schwarz for computing three-dimensional electrostatic fields in electron-optical devices with complicated boundaries. Optik 65, No. 4, 347-359.

[11] Schwarz HA. (1870). Gesammelte mathematische Abhandlungen, Band II 133.

[12] Schwarz HR. (1980). Methode der finiten Elemente (Teubner Studienbücher Band 47). B.G.Teubner, Stuttgart.

Electron Optical Systems (pp. 91-96)
SEM Inc., AMF O'Hare (Chicago), IL 60666-0507, U.S.A.

0-931288-34-7/84$1.00+.05

CONSTRAINED OPTIMIZATION DESIGN OF AN ELECTRON OPTICAL SYSTEM

Chang-xin GU[1,2] and Li-ying SHAN[2]

[1]Present address:
Department of Physics
State University of New York at Stony Brook
Stony Brook, New York 11794

[2]Modern Physics Institute
Fudan University
Shanghai, China

Abstract

An electron optical system can be optimized using the "simplex method" or "complex method". By these methods, the final structure of an electron optical system, for example, an extended field lens (EFL), can be searched with a criterion of minimum objective parameter (in the present case, the coefficient of spherical aberration). Because there is no constraint in the simplex method, the constrained optimization method (the complex method) described in this paper is better than the simplex method in the design of electron optical systems. In the simplex method as well as the complex method, it is not necessary to know the explicit functional relation between the objective function and the searching parameters; and the variations of aberration coefficient with respect to some machining tolerance can be easily obtained. Therefore, comparing with other optimization methods, the simplex method and complex method have significant advantage in the optimization design of electron optical systems.

KEY WORDS: Electron optical system design, Constrained optimization design, Complex method.

Address for correspondence:
Chang-xin GU
Modern Physics Institute
Fudan University, Shanghai
China 201903 Phone no.: 480906-175

Introduction

The computer-aided design (CAD) method for electron optical systems has been considerably developed since the 1960's. However, there are some limitations in the application of the CAD method in electron optical systems; this CAD method can only be used to calculate certain electron optics characteristic parameters from given boundary conditions, such as geometric structure and electrical parameters[2]. In recent years, the question of how to determine the optimal structure and the corresponding electrical parameters of an electron optical system from given electron optics characteristic parameters, i.e. the optimization design of electron optical systems, has gradually received more attention.

The optimization design method is a method which can minimize the objective function (e.g. an aberration coefficient of the electron optical system) under certain constraint conditions, from which the optimal structure can be obtained. In this method, some suitable geometric and electrical parameters are chosen as the search parameters.

The optimization design is an objective which has been sought by electron optics researchers for a long time. Some optimization design methods[5,6,8-11] for electron optical systems have been suggested since the 1970's. Using the dynamical programming method of Szilagyi[9-11] and the variational method of Rose[8], the potential and/or magnetic field distribution along the electron optical system's symmetry axis producing a minimum spherical aberration could be searched. However, the final optimal structure cannot be obtained by these methods. In order to determine the optimal structure, it is necessary to carry out experimental analog studies conforming with the potential distribution along the symmetry axis obtained by these methods. The "Simplex Method", an optimization design method in electron optics as suggested by Gu and Chen[1] in 1982, can overcome the disadvantages of both the dynamical programming and the variational methods. The optimal structure can be directly obtained with a criterion of minimum spherical aberration coefficient using the simplex method. However, there are no constraint conditions in the simplex method, so there must be an on-line control of the search parameters during the optimization process. This makes automatic search difficult. Hence, although the simplex method is a usable method for electron optical system design, it still suffers some shortcomings.

The "Complex Method" proposed in this paper is an important improvement in the optimization design method of electron optical systems.[7] It is a multidimensional "constrained-extreme-value problem". Not only can the optimal structure be searched and corresponding parameters be obtained automatically, but one can also be certain that the results will satisfy the constraint conditions.

Principle

In the optimization design of electron optical systems, the electron optics characteristic parameters are chosen as the optimization objective function (or "error function"). It is defined as the sum of weighted squares (or the square root of the sum of weighted squares) of various electron optical aberration coefficients. Some of the aberration coefficients $f_1, f_2, \ldots, f_m$(e.g. spherical aberration, central chromatic aberration, axial astigmatism, etc.) will be used as components of the objective function subject to the concrete requirements. Each kind of aberration coefficient $f_i(i=1,2,\ldots,m)$ is regarded as a function of certain geometric parameters (e.g., the diameter or length of an electrode cylinder, the gap between two electrodes, the shape of the magnetic pole pieces, etc.) and electrical parameters (e.g. the electrode potentials, the currents driving the magnetic field, etc.). These parameters will be taken as the search parameters expressed as $x_1, x_2, \ldots, x_n$. Thus, the objective function is defined as follows:

$$f(\vec{x}) = \sum_{i=1}^{m} (W_i f_i(\vec{x}))^2 \qquad (1)$$

or

$$f(\vec{x}) = [\sum_{i=1}^{m} (W_i f_i(\vec{x}))^2]^{\frac{1}{2}} \qquad (2)$$

where $\vec{x}$ is an independent vector-argument in the n-dimensional space:

$$\vec{x} = (x_1, x_2, \ldots, x_n)^T \qquad (3)$$

where T refers to transposition and $x_1, x_2, \ldots, x_n$ are n unknown search arguments. The $\vec{x}$-domain is R^n. Each W_i is a weighted factor, $0 < W_i \leq 1$, $i=1,2,\ldots,m$, determined according to the design requirements.

Mathematically, the optimization problem is an extreme value problem of the objective function $f(\vec{x})$ defined in the space of $\vec{x} \in R^n$. Its mathematical programming form is:

$$\begin{cases} \min\limits_{\vec{x} \in R^n} f(\vec{x}) & (4) \\ a_i \leq x_i \leq b_i \quad i=1,2,\ldots,n \text{ explicit constraints} \\ g_j(\vec{x}) \leq 0 \quad j=n+1,\ldots,m \text{ implicit constraints} \end{cases}$$

This problem is a nonlinear programming problem.

In the design of electron optical systems, the limitation on a part of the search parameters, e.g., the size of an electrode cylinder and magnetic pole, the values of electrode potential and driving current etc., will be taken as the explicit constraint conditions. The limitation on the electron optical characteristic parameters, which are implicit functions of the search arguments, can be taken as the implicit constraint conditions. As it is known, the spherical aberration coefficient can be used as a criterion for the performance of an electron lens under a fixed focal length condition. However, the focal length of the lens is an implicit function of the search arguments. Usually, the expression of the functional relations between the focal length and the search arguments cannot be written in a closed form. Only if the iterative calculation of the potential or the magnetic field is finished, the focal length of the lens can be evaluated in accordance with a certain geometrical structure and electrical parameters. Therefore, the limitation on the focal length can be taken as an implicit constraint condition in the optimization design of electron optical systems.

The complex method described in this paper is used as a constrained optimization design method of electron optical systems, i.e., an electron optics constrained extreme value problem. The complex used in this method is a polyhedron which has k vertices in the n-dimensional space: $\{\vec{x}^{(1)}, \vec{x}^{(2)}, \ldots, \vec{x}^{(k)}\}$, $k \geq n+2$, usually we take $k=2n$. In order to avoid degeneracy, k should be taken at a high value.

The complex method is summarized as follows:

A. The complex method iterative process:

(i) Give an initial feasible point $\vec{x}^{(1)}$. Feasible means that this point is defined in the $\vec{x}$-domain and satisfies the constraint conditions.

(ii) Starting with $\vec{x}^{(1)}$, set up an initial complex $\{\vec{x}^{(1)}, \vec{x}^{(2)}, \ldots \vec{x}^{(k)}\}$. The vertices of the complex must be limited in the feasible set.

(iii) Calculate the worst point $\vec{x}^{(h)}$ and the best point $\vec{x}^{(\ell)}$ among the vertices of the complex, i.e. the point of maximum and minimum value of $f(\vec{x})$:

$$f(\vec{x}^{(h)}) = \max_i f(\vec{x}^{(i)}) \qquad (5)$$

$$f(\vec{x}^{(\ell)}) = \min_i f(\vec{x}^{(i)}) \qquad (6)$$

(iv) Calculate $\bar{x}$ - the centroid of the complex excluding $\vec{x}^{(h)}$, i.e. calculate the vectorial mean:

$$\bar{x} = \frac{1}{k-1} \sum_{i \neq h}^{k} \vec{x}^{(i)} \qquad (7)$$

(v) Calculate $\vec{x}^{(r)}$ - the reflection point of $\vec{x}^{(h)}$ with respect to $\bar{x}$, i.e.

$$\vec{x}^{(r)} = \bar{x} + \alpha(\bar{x} - \vec{x}^{(h)}) \qquad (8)$$

where $\alpha > 1$; generally, $\alpha = 1.3$.

(vi) Check whether the $\vec{x}^{(r)}$ is a feasible point, if not, readjust $\vec{x}^{(r)}$ to become a feasible point.

(vii) Take $\vec{x}^{(r)}$ instead of $\vec{x}^{(h)}$, if it cannot improve the worst point position, $\vec{x}^{(r)}$ should be constricted towards $\bar{x}$ until it satisfies the requirement.

The terminate condition of the complex method is

$$f(\vec{x}^{(h)}) - f(\vec{x}^{(\ell)}) \leq \varepsilon . \qquad (9)$$

The condition (9) should be satisfied q times consecutively, where ε is the accuracy the objective function is to be computed with.

The complex method calculation flow diagram is shown in Fig. 1.

B. The method of checking and readjusting the feasible points

In this method, the non-feasible points are readjusted to become feasible points by means of known feasible points: $\vec{d}^{(1)}, \vec{d}^{(2)}, \ldots, \vec{d}^{(s)}$. This process is shown as follows:

(i) If $\vec{x}$ cannot satisfy the explicit constraints, it should be adjusted with a small displacement δ_i such that

$$\text{if } x_i > b_i, \text{ then } x_i = b_i - \delta_i \ ;$$

$$\text{if } x_i < a_i, \text{ then } x_i = a_i + \delta_i \ ;$$

the value of δ_i depends on practical requirements of the electron optical systems.

(ii) If $\vec{x}$ cannot satisfy the implicit constraints, it should be constricted towards the centroid of the set of points $\vec{d}^{(1)}, \vec{d}^{(2)}, \ldots, \vec{d}^{(s)}$:

$$\vec{x}^{(N)} = \frac{\vec{x} + \vec{x}^{(c)}}{2} \qquad (10)$$

where

$$\vec{x}^{(c)} = \frac{1}{S} \sum_{i=1}^{S} \vec{d}^{(i)} . \qquad (11)$$

(iii) Repeat step (i) and recheck if $\vec{x}$ satisfies the explicit and implicit constraints until $\vec{x}$ is a feasible point.

The flow diagram for this method is shown in Fig. 2.

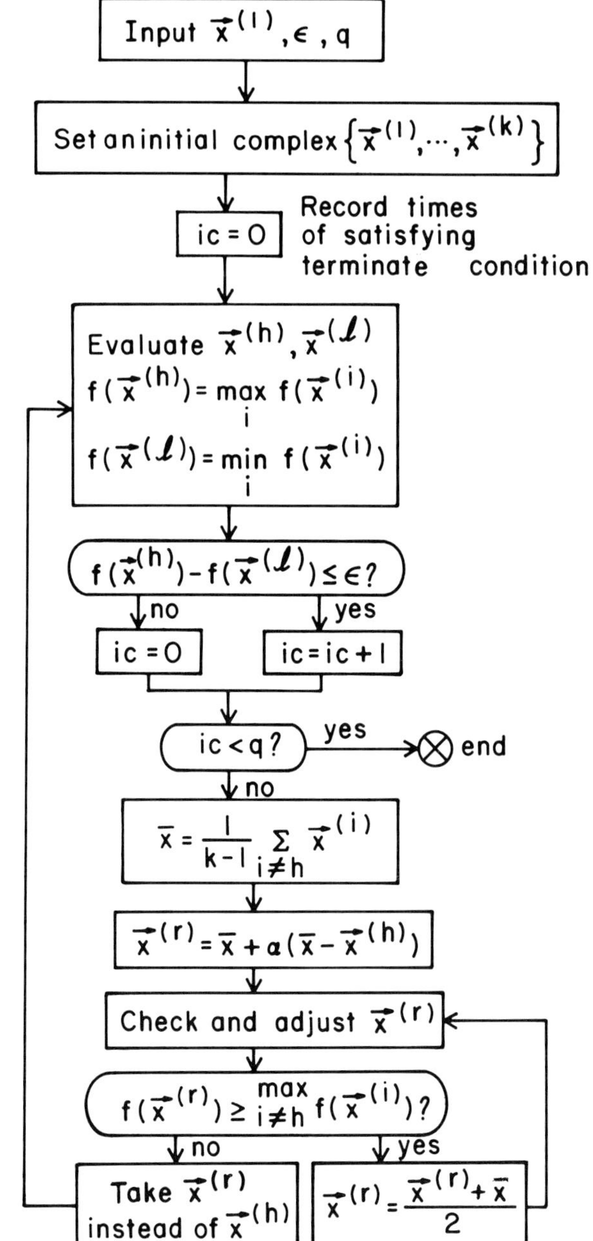

Fig. 1. Flow diagram for the complex method.

An Example and Results

An Extended Field Lens (EFL) is taken as an example for optimization design using the complex method. The computed results for EFL using the CAD and simplex method have already been obtained before.[1,3,4] The 4EFL structure (it consists of 4 equidiameter electrodes) is shown in Fig. 3.

In this paper, the spherical aberration is taken as an objective function, the length and potential of electrodes, and the gap between the electrodes are taken as the search arguments.

Owing to the limitation of computer capacity at our university, the method of successive optimal search in two dimensional space is used in the present case. We change the pair of search arguments: $L_3, S_3; L_2, L_3$ and V_2, V_3 respectively in each searching process.

The explicit constraints are

$$\begin{aligned} 15.0 \leq L_2 &\leq 25.0 \quad (\text{mm}) \\ 1.0 \leq L_3 &\leq 5.0 \quad (\text{mm}) \\ 1.0 \leq S_3 &\leq 3.0 \quad (\text{mm}) \\ 6.0 \leq V_2 &\leq 8.0 \quad (\text{kV}) \\ 9.0 \leq V_3 &\leq 12.0 \quad (\text{kV}) \end{aligned} \qquad (12)$$

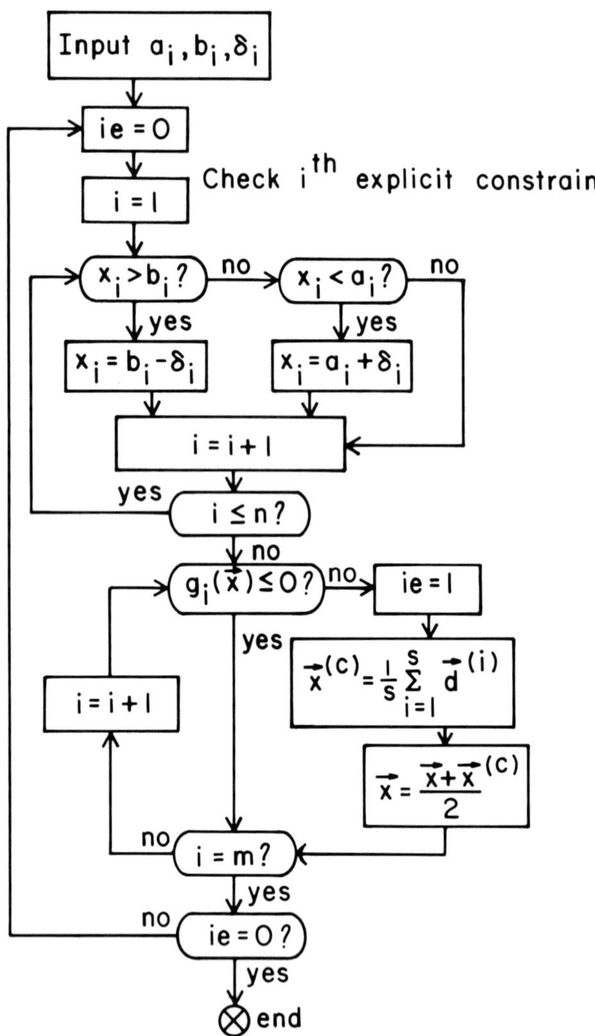

Fig. 2. Flow diagram for checking and adjusting the feasible point.

The implicit constraint is

$$F_2(\vec{x}) - 50.5 \leq 0 , \qquad 13)$$

where F_2 is the image space focal length. Let $\varepsilon = 0.5$, $\delta_1=\delta_2=\delta_3= 0.5$ (mm), $\delta_4=\delta_5=0.2$ kV.

Because the 4EFL is often used as a main lens in electron optical systems, the potential on the electrodes and the electron velocity in the space of 4EFL are rather high. Therefore, the space charge effect can be neglected. The evaluation of the electric field in the space of 4EFL can be reduced to a boundary value problem of a rotational symmetrical Laplace equation.

The finite difference method with successive overrelaxation was used in the numerical calculation of the Laplace equation. The selected value of relaxation factor ω depends on the number of mesh nodal points and the iterative calculation order. For some optimum value ω between 1 and 2, the rate of

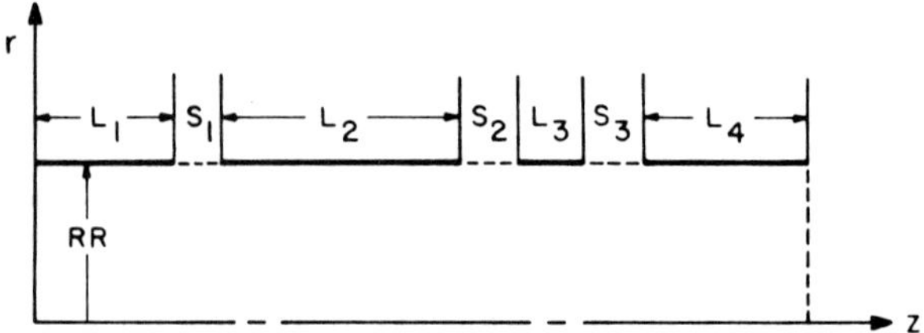

Fig. 3. 4EFL (Extended Field Lens) structure.

convergence can be improved. The auto-selected value of the ω factor is used in our calculations.

The famous Scherzer formula is used for the calculation of the spherical aberration coefficient.

$$C_s = \frac{1}{16\sqrt{\phi_0}} \int_{z_0}^{Z_n} \phi^{-3/2} \left[\frac{5}{4}\phi''^2 + \frac{5}{24}\frac{\phi'^4}{\phi^2} + \frac{14}{3}\frac{\phi'^3}{\phi} \cdot \frac{r_1'}{r_1} - \frac{3}{2}\phi'^2\left(\frac{r_1'}{r_1}\right)^2\right] r_1^4 \, dz \qquad (14)$$

where $z = z_0$ is the objective plane, $z = z_n$ is the image plane, r_1 is a paraxial-ray emitted from the objective point on the axis with 45^0 initial emission angle, ϕ_0 is the potential at the objective point, ϕ' and ϕ'' are first and second derivatives of the axial potential with respect to z respectively.

The calculation of the electron trajectory makes use of the Picht equation:

$$\rho'' + \frac{3}{16}\left(\frac{\phi'}{\phi}\right)^2 \rho = 0 , \qquad (15)$$

where

$$\rho = r\,\phi^{1/4} .$$

The method of the parallel trajectory is used for computing the electron optical characteristic parameters: principal point, focal point and focal length.

The Fox-Goodwin formula shown as follows is used for the numerical calculation of Eq. (15)

$$\left(1-\frac{1}{12}a^2 g_{n+1}\right)\rho_{n+1} = \left(2+\frac{5}{6}a^2 g_n\right)\rho_n - \left(1-\frac{1}{12}a^2 g_{n-1}\right)\rho_{n-1}$$

where

$$g_i = -\frac{3}{16}\left(\frac{\phi_i'}{\phi_i}\right)^2 , \quad \phi_i' = \frac{\phi_{i+1} - \phi_{i-1}}{2a} ,$$

and a is an axial step. The results computed using the CAD method are used to form the initial complex.

The computed results using the complex method are listed in Table 1. For the purpose of comparison, the computed results using the simplex method and the CAD method are also shown in this table.

Table 1. The Main Computed Results

Fixed Parameters: RR = 4.5 mm L_1 = 13.0 mm S_1 = 2.0 mm S_2 = 2.0 mm L_4 = 13.0 mm V_4 = 30 kV

Method	No	L_2 mm	L_3 mm	S_3 mm	V_1 kV	V_2 kV	V_3 kV	F_2 mm	C_s	Remarks
Complex	1	20.0	4.00	3.00	10.0	7.0	10.0	50.11	763.2	The initial complex
	2	"	4.26	2.50	"	"	"	50.35	795.9	L_3,S_3 searching process
	3	"	2.86	2.50	"	"	"	46.41	802.0	"
	4	"	3.36	2.94	"	"	"	48.77	728.5	L_3,S_3 optimal point
	5	20.0	3.50	2.94	"	"	"	48.81	734.1	L_3,L_2 initial value
	6	15.5	2.943	2.938	"	"	"	47.75	457.2	L_3,L_2 optimal point
	7	15.5	2.943	2.938	10.0	6.0	11.0	41.39	408.6	V_2V_3 initial value
	8	15.5	2.94	2.94	10.0	6.9	11.8	50.10	371.9	V_2,V_3 optimal point
Simplex	9	20.0	4.00	3.00	10.0	7.0	10.0	50.11	763.2	L_3S_3 initial value
	10	"	3.54	2.80	"	"	"	48.78	730.4	L_3S_3 optimal point
	11	"	2.56	4.60	"	"	"	49.72	664.7	S_3 too large
	12	16.5	4.24	2.80	"	"	"	49.90	549.9	L_3L_2 initial value
	13	16.5	4.00	2.80	"	"	"	49.81	541.8	L_3L_2 optimal point V_2,V_3 initial value
	14	"	"	"	"	5.85	11.3	40.76	393.2	V_2V_3 optimal point
CAD	15	20.0	4.0	3.2	10.0	7.0	10.0	50.61	826.9	CAD optimal result
	16	14.5	9.5	2.0	8.0	6.0	8.0	46.94	1152.5	" worst result

Discussion and Conclusions

A. From the computed results listed in Table 1, it can be seen that the computed results using the complex method are consistent with the computed results using the simplex method; and the former are better than the latter. Using the complex method, the relative value of the coefficient of spherical aberration decreases from 826.9 to 371.9 while maintaining the same focal length.

B. The computed results using the simplex method show that the spherical aberration coefficient C_s reduces as the gap S_3 increases (see No. 11 in Table 1). Of course, this is consistent with the electron optics principle: the distribution of potential along the axis is extended as the gap S_3 increases. This consequence is beneficial to the reduction of the spherical aberration coefficient. But, if the gap S_3 is too large, the external electric and magnetic field will interfere with the electric field in the lens, and the image focal length will increase, and the design requirement cannot be satisfied. Since there are no explicit and implicit constraints in the simplex method, the aforementioned consequence is hard to avoid. However, using the complex method there is no such situation, so the constrained optimization design method (the complex method) is better than the simplex method in the design of electron optical systems.

C. In the complex method as well as simplex method, variations of C_s, the spherical aberration coefficient, caused by small changes of the search parameters due to machining tolerances are readily available from the computed data in the search process. It is quite accurate and easy to obtain in comparison with the calculation of the spherical aberration coefficient by using theoretical aberration coefficient formulas.

D. In this paper, the computed results show that the choice of the search arguments, the calculation of the objective function and the determination of the constraints are reasonable. In the simplex method and the complex method, it is not necessary to know the explicit functional relation between the objective function and the searching parameters; and the variations of aberration coefficients with respect to some machining tolerance can be easily obtained. Therefore, comparing with other optimization methods, the simplex method and complex method have significant advantage in the optimization design of electron optical systems. The complex method provides an effective mathematical method for the optimization design of electron optical systems. This method has a wide-range application in the field of high resolution electron beam technique.

Acknowledgments

The authors would like to acknowledge Prof. Z.Y. Hua for his kind advice and Prof. Y.H. Kao for his valuable suggestions; and to thank Dr. K.M. Chen for many beneficial conversations and Mr. N.Q. Chen for his assistance. One of the authors (Gu Chang-Xin) is a recipient of the Lee Hysan Fellowship through the Committee on Educational Exchange with China. The financial support by Mr. Lee Hysan is gratefully acknowledged.

References

1. Gu CX and Chen NQ. (1982). The optimization design of an electron-optical system. The Second National Electron Optics Conference,(China), Abstract. Journal of Electronics (China) 12,41-47 (1984).
2. Hawkes PW. (1973). Computer-aided design in electron optics, Computer Aided Design, 5, 200-214.
3. Hua ZY and Gu CX. (1977). The Calculation of Electrostatic Field by Using Finite Element Method. Fudan Univ., Shanghai, China, 35-55.
4. Hua ZY and Gu CX (1978). The computer analysis of 4 Extended Field Lens. The proceedings of the annual Conf. of Shanghai Electronics Society, Shanghai, China, Abstract. 22.
5. Moses RW. (1971). Minimum Chromatic aberrations of magnetic quadrupole objective lenses. Rev. Sci. Instrum. 42, 828-831.
6. Munro E (1975). Design and optimization of magnetic lenses and deflection systems for electron beams. Journal of Vacuum Science and Technology. 12, 1146-1150.
7. Richardson JA, Kuester JL. (1973). The Complex Method for Constrained Optimization. Communications of the ACM. 16, 487-489.
8. Rose H, Moses RW. (1973). Minimaler öffnungsfehler magnetischer rund-und zylinderlinsen bei feldfreiem objektraum. Optik. 37, 316-336.
9. Szilagyi M.(1977). A new approach to electron optical optimization. Optik. 48, 215-224.
10. Szilagyi M.(1977). A dynamic programming search for magnetic field distribution with minimum spherical aberration. Optik. 49, 223-246.
11. Szilagyi M.(1978). A dynamic programming search for electrostatic immersion lenses with minimum spherical aberration. Optik. 50, 35-51.

Electron Optical Systems (pp. 97-108)
SEM Inc., AMF O'Hare (Chicago), IL 60666-0507, U.S.A.

0-931288-34-7/84$1.00+.05

PROGRESS IN THE THEORY OF ELECTRON-BEAM DEFLECTION

E.F. Ritz, Jr.

Tektronix, Inc.
P.O. Box 500
Mail Station 50-493
Beaverton, Oregon 97077
Phone: (503) 627-5435

Abstract

Analysis, simulation, and design of electron-beam-deflection systems are reviewed in light of the current state of theoretical understanding. A brief review of the physical principles is followed by a detailed discussion of electrostatic, magnetostatic, mixed-field, traveling-wave, and scan-expansion systems. Each methodology is examined from a triple perspective: calculation of electromagnetic fields, calculation of electron trajectories, and calculation of the ensemble of trajectories forming the beam. Applications discussed include deflectors for television displays, lithography, scanning microscopes, and CRT oscillography. Developments of the last ten years are stressed, thereby supplementing and updating the author's previous review on this subject.

In field calculation, recent developments in the use of numerical methods on computers dominate. These methods include finite-difference, finite-element, and charge-density or integral-equation techniques. In trajectory calculations, increasing use of numerical integration as well as improvements and extensions of the aberration theory are found. In treatment of the beam bundle, the growing sophistication of numerical deflected-beam models has lead to increased use of aberration figures, current-density plots, and phase-space methods.

Key Words: electron-beam deflection, electrostatic deflection, magnetostatic deflection, mixed-field deflection, traveling-wave deflection, scan-expansion lenses, numerical field calculation, electron-trajectory calculation, description of electron beams, aberration theory of deflection.

Introduction

Scope and Organization

In this review, we shall consider the present theoretical understanding of electron-beam deflection, concentrating on the work of the last ten years. Thus, the discussion overlaps and updates the coverage of this author's previous review (Ritz, 1979). Hutter (1974) also reviews progress in electron-beam deflection, covering about 20 years prior to 1974. Both of these previous reviews have extensive lists of references and are recommended reading as an introduction to the field. No textbook exists that adequately serves this purpose.

In the theory of electron-beam deflection, we are faced with three fundamental problems. First, we must solve the Maxwell equations for the electric and magnetic fields produced by the deflection system. Second, we must solve the Lorentz equation of motion for the trajectories of the individual electrons in these fields. Third, we must use the solutions to the first and second problems to calculate the collective behavior of the ensemble of electrons forming the beam. In general, these problems cannot be solved analytically. Therefore, simplified analytical models, approximations such as perturbation analysis, or numerical simulation on a computer are used. The various subfields of our subject differ greatly in the degree of success obtained with each of these methods.

We shall consider our subject under the headings of electrostatic, magnetostatic, mixed-field, traveling-wave, and scan-expansion systems. However, for future reference, we shall first briefly review the physical principles and basic viewpoints needed for all of these topics.

Calculation of Electromagnetic Fields

We now review the relevant portions of electromagnetic theory, using MKSA formulas and units. Our problem is to solve the Maxwell equations

$$\nabla \times \mathbf{E} = -\frac{\partial \mathbf{B}}{\partial t} \tag{1}$$

$$\nabla \times \mathbf{H} = \mathbf{J} + \frac{\partial \mathbf{D}}{\partial t} \tag{2}$$

$$\nabla \cdot \mathbf{D} = \varrho \tag{3}$$

$$\nabla \cdot \mathbf{B} = 0 \tag{4}$$

for electric intensity **E**, electric displacement **D**, magnetic induction **B**, and magnetic intensity **H**, given the physical structure of the deflector and the electrical sources that give rise to the electric current density **J** and the electric charge density ϱ. In addition, **J** and ϱ must satisfy the equation of continuity

$$\nabla\cdot\mathbf{J} + \frac{\partial\varrho}{\partial t} = 0 \tag{5}$$

Finally, we shall generally assume isotropic, homogeneous media with

$$\mathbf{D} = \epsilon\mathbf{E} \tag{6}$$

$$\mathbf{B} = \mu\mathbf{H} \tag{7}$$

in which the permittivity ϵ and the permeability μ are constant within a material (with vacuum values ϵ_0 and μ_0 as usual).

It is frequently convenient to employ the electric scalar potential Φ and the magnetic vector potential **A** in place of the field vectors **E**, **B** to which they are related by

$$\mathbf{E} = -\nabla\Phi - \frac{\partial\mathbf{A}}{\partial t} \tag{8}$$

$$\mathbf{B} = \nabla\times\mathbf{A} \tag{9}$$

In the general time-dependent case, the electromagnetic potentials satisfy the coupled wave equations

$$\nabla^2\mathbf{A} - \frac{1}{v_p^2}\frac{\partial^2\mathbf{A}}{\partial t^2} = \mu\mathbf{J} \tag{10}$$

$$\nabla^2\Phi - \frac{1}{v_p^2}\frac{\partial^2\Phi}{\partial t^2} = -\frac{\varrho}{\epsilon} \tag{11}$$

in which v_p is the phase velocity

$$v_p = \frac{1}{\sqrt{\mu\epsilon}} \tag{12}$$

and the Lorentz gauge condition

$$\frac{1}{v_p^2}\frac{\partial\Phi}{\partial t} + \nabla\cdot\mathbf{A} = 0 \tag{13}$$

applies. The solutions of the wave equations represent waves of **E** and **B** propagating with phase velocity v_p.

In the static-field case, eqs. (10) - (11) reduce to

$$\nabla^2\mathbf{A} = -\mu\mathbf{J} \tag{14}$$

$$\nabla^2\Phi = -\frac{\varrho}{\epsilon} \tag{15}$$

If in addition we are in a source-free region, then **J** and ϱ vanish so that

$$\nabla^2\Phi = 0 \tag{16}$$

$$\mathbf{E} = -\nabla\Phi \tag{17}$$

$$\nabla^2\Psi = 0 \tag{18}$$

$$\mathbf{H} = -\nabla\Psi \tag{19}$$

where Ψ is the magnetic scalar potential now permitted by the vanishing of $\nabla\times\mathbf{H}$. This Laplace formulation of the problem is particularly useful when the potentials or currents on conductors are given and can be used to obtain boundary conditions on Φ or Ψ.

In most problems of electron-beam deflection, we can neglect the wave aspects of the fields and solve the simpler static Laplace equations, (16) and (18), rather than the full wave equations. This *static-field* approximation assumes, in effect, that the driving fields travel from the input terminals to all parts of the deflector in a time that is short compared with the time in which the input signal changes appreciably. That is, for a sinusoidal input of frequency f, we must have

$$2\pi f t_w << 1 \tag{20}$$

in which t_w is the transit time required for the wave to reach the farthest part of the deflector. This condition is violated in traveling-wave deflectors for high-speed deflection. Note that in vacuum, t_w will be about 33 psec for every centimeter of travel.

So far, we have formulated the field problem in terms of partial differential equations relating the fields and their sources. One can also express the field as an integral over the sources. In the static case of electric charges on the surfaces of conductors with specified potentials, the integral in question is

$$\Phi(\mathbf{x}_2) = \frac{1}{4\pi\epsilon_0}\iint_S \frac{\sigma(\mathbf{x}_1)}{R}\, dS_1 \tag{21}$$

in which σ is the density of surface charge, S represents the surface (or surfaces) of the conductor(s), and $R = |\mathbf{x}_2 - \mathbf{x}_1|$. When the point of observation $\mathbf{x}_2$ lies on one of the conductors, then Φ must be the constant potential of that conductor. Thus, (21) can be regarded as an integral equation for the unknown charge density $\sigma(\mathbf{x}_1)$ on the conductors. Once this density is determined, the potential at any point can be calculated from (21). For the general static case, we must add to (21) integrals over the volume density ϱ and over the surface dipole layers (e.g., Stratton, 1941).

The general time-dependent problem can be treated similarly. If we assume a sinusoidal dependence $\exp(i\omega t)$ for the fields, currents, and charges, then

$$\mathbf{A}(\mathbf{x}_2) = \frac{\mu}{4\pi}\iiint_V \frac{\mathbf{J}(\mathbf{x}_1)e^{-ikR}}{R}\, dV_1 \tag{22}$$

$$\Phi(\mathbf{x}_2) = \frac{1}{4\pi\epsilon}\iint_S \frac{\sigma(\mathbf{x}_1)e^{-ikR}}{R}\, dS_1 \tag{23}$$

in which V is the volume containing the currents and $k = \omega(\mu\epsilon)^{1/2} = 2\pi/\lambda$, with λ the wavelength. We shall not pursue this formulation here because it has yet to be applied in numerical field calculations for traveling-wave deflectors.

We can give a third formulation of the field problem: the variational formulation. We define a functional F to describe the fields. For static electric or magnetic fields with scalar potentials Φ or Ψ, the functionals are

$$F = \iiint \frac{1}{2}\epsilon(\nabla\Phi)^2 dV$$
$$F = \iiint \frac{1}{2}\mu(\nabla\Psi)^2 dV \qquad (24)$$

Equations (24) are the energies stored in the fields, expressed as integrals over the volume V containing the fields.

The potentials Φ and Ψ are determined by the condition that the variation of F vanish with respect to variations of the fields, thus minimizing the energy:

$$\delta F = 0 \qquad (25)$$

The resulting Euler-Lagrange equations reproduce (16) and (18).

Each of the three formulations of the field problem is the basis of a corresponding technique for numerical solution of that problem on a computer. The partial differential equations (10) - (11) lead to the finite-difference method (FDM). The integral equations (21) - (23) lead to the integral-equation or boundary-element method (IEM). The variational equations (24) - (25) lead to the finite-element method (FEM). Rather than discuss these methods here, we shall introduce them as appropriate in later sections. For a general view of these methods, see Kasper (1982) on magnetic-field calculation. Much of his discussion also applies to electrostatic fields. Schaefer (1983) gives a helpful comparison of these numerical techniques as well as methods of combining numerical and analytical solutions.

Calculation of Electron Trajectories

Electron trajectories in an electromagnetic field are governed by the Lorentz equation of motion for the position vector $\mathbf{r}(t)$:

$$\frac{d}{dt}\left[\frac{1}{\sqrt{1-\beta^2}}\frac{d\mathbf{r}}{dt}\right] = -\eta\left[\mathbf{E} + \frac{d\mathbf{r}}{dt}\times\mathbf{B}\right] \qquad (26)$$

in which $\beta = |d\mathbf{r}/dt|/c$, where c is the speed of light in vacuum and η is the ratio of electronic charge e to rest mass m. In the literature, the nonrelativistic limit

$$\ddot{\mathbf{r}} = -\eta[\mathbf{E} + \dot{\mathbf{r}}\times\mathbf{B}] \qquad (27)$$

is usually employed, and we shall also do so for simplicity. It is sometimes convenient to replace the independent variable t with arc length s along the trajectory or with z, the coordinate in the direction of the beam.

For numerical calculation, it is usually convenient to replace (27) with two equivalent first-order equations:

$$\mathbf{v} = \dot{\mathbf{r}} \qquad (28)$$

$$\dot{\mathbf{v}} = -\eta[\mathbf{E} + \mathbf{v}\times\mathbf{B}] \qquad (29)$$

in which we have defined the second variable $\mathbf{v}$. It is also useful to think of the electron as moving in the *phase space* defined by the six-vector $(\mathbf{r}, m\mathbf{v})$ or, less rigorously, $(\mathbf{r},\mathbf{v})$.

In principle, the fields $\mathbf{E}$ and $\mathbf{B}$ should include both the fields of the deflector as discussed in *Calculation of Electromagnetic Fields* and the fields due to the electrons in the beam. However, the latter fields have not as yet been incorporated successfully into calculations of beam deflection, although there has been much progress in the study of such effects in electron guns using the average-field or space-charge approximation (e.g., Hauke, 1977). Therefore, we shall not discuss the self-fields of the beam.

The trajectory equations (26) or (27) or (28) - (29) must be solved subject to initial conditions $\mathbf{r}(0) = \mathbf{r}_0$, $\dot{\mathbf{r}}(0) = \dot{\mathbf{r}}_0$ on the starting position and velocity of the electron in question. Each electron in the beam has a different set of initial parameters, although only slightly different. In what follows, we shall assume that the beam when undeflected is directed along the z axis of a rectangular coordinate system and travels in the positive-z direction. Thus, all electrons in the beam will have approximately the same value of $\dot{z}_0$. We take the term "beam" to mean that the transverse velocities $\dot{x}_0$ and $\dot{y}_0$ are small compared with $\dot{z}_0$.

In general, the trajectory equations cannot be solved analytically. To circumvent this difficulty, perturbation methods called *aberration theory* have been developed. In the classical formulation (e.g., Glaser, 1949,1952; Haantjes and Lubben, 1957, 1959; and Kaashoek, 1968), the Lagrangian function is expanded in a four-dimensional power series with dimensions x, y, x', y'. The initial conditions are specified by giving the total energy (or beam potential), the position and slope x_s, y_s, x_s', x_s' *at the screen* for the undeflected ray, and the position $z = z_0$ of the starting plane. Here, primes indicate z derivatives. The deflecting fields and certain of their derivatives must be supplied along the z axis.

The resulting equations of motion are solved by successive approximations. The result is a power series for the landing position of a ray at the screen or target plane. The first correction to the landing point of the undeflected ray is the Gaussian deflection, which is proportional to the deflection potential or deflection current. The next correction consists of the so-called third-order aberrations, which fall into three classes. Terms that are cubic in the x and y Gaussian deflections but independent of x_s, y_s, x_s', y_s' produce *raster distortion*. Terms that are quadratic in the Gaussian deflections but linear in the ray parameters produce *curvature of the field* and *astigmatism*. Terms that are linear in the Gaussian deflection but quadratic in the ray parameters produce *coma*. Higher corrections are rarely used because of their number and complexity. In all cases, the coefficients of the power series are complicated integrals of Gaussian deflections and field functions on the z axis.

The aberration theory has recently been improved (e.g., Chu and Munro, 1982a) in connection with deflection systems for scanning microscopy and electron-beam lithography. We shall consider these improvements under *Mixed-Field Deflection*.

It is important to note that the aberration theory of deflection is a *narrow-angle theory*. In the third-order approximation, the theory is limited to deflection angles of 20° or 25°. The series expansions employed in the derivation of the theory fail altogether to converge for angles of 45° or greater. To avoid this limit, some authors have employed curved optical axes for their perturbation expansions. However, such theories are even more unwieldy than the narrow-angle versions. The paper of Hutter (1970) expounds this method and also has a good review of earlier work. In all cases, these perturbation approximations apply only to static fields.

The failure to obtain analytical solutions and the limitations of aberration theories have led to the development of numerical methods for use on computers. The problem is the solution of the three coupled second-order equations (27) or the six first-order equations (28) - (29). These are linear ordinary differential equations for which several methods of solution exist. A recent review of this topic appears in Kasper (1982).

In the discussion above, we have tacitly assumed that we can neglect any changes in the deflection fields that occur during the passage of a particular electron through the deflector. For this assumption to be valid, a condition like (20) must apply. That is, for a sinusoidal input to the deflector, we must have

$$2\pi f t_e << 1 \qquad (30)$$

in which f is the frequency and t_e is the transit time for a single deflector. If this is true, we can use the instantaneous values of the fields at the time the electron enters to calculate the entire trajectory. Thus, *two* conditions, (20) and (30), must be satisfied for the static-deflection approximation to hold.

Calculation of Beam Properties

After we have calculated a single trajectory in the deflecting fields, we are far from knowing how the entire beam behaves. If we think of each electron as a point in the phase space $(x, y, z, \dot{x}, \dot{y}, \dot{z})$, then the beam is represented at any instant by a cloud of points in the same six-dimensional space (we assume non-interacting electrons). It is clearly impossible to calculate the paths of all these electrons through phase space, hence the importance of the methods used to determine the properties of the beam and of the final spot by less drastic means. Since most of these methods involve small subsets of the electrons forming the beam, it is essential to remember the full complexity of the distribution in phase space; otherwise, it is easy to draw unjustified conclusions.

Under the conditions of static deflection, a cloud of points representing a group of electrons moves with time so that the volume of phase space occupied remains constant (Liouville's theorem); however, the shape of that volume can alter significantly. Thus, if the spread in velocity increases, the spread in position must decrease and vice versa. Consequently, considerable knowledge of the beam can be obtained by following the evolution of the *boundary* of the volume. To aid in visualization, the volume can be projected onto the three phase planes $(x,\dot{x})$, $(y,\dot{y})$, $(z,\dot{z})$. Where one or more of these phase planes are uncoupled in the equations of motion, Liouville invariance can be applied to the projection of the total volume onto the uncoupled plane or planes and the remaining volume. See the reviews by Lejeune and Aubert (1980) and by Crawford and Brody (1966) for an introduction to these ideas.

If the beam is approximately monoenergetic, a useful description of the beam is the amount of charge passing through the point (x,y) in the transverse plane z = constant per unit time and per unit volume $dxdydx'dy'$ in the transverse *trace space* (x,y,x',y'). This quantity is variously named and symbolized. We shall call it *brightness* and denote it by $R(x,y,x',y',z)$ after the original German term *Richtstrahlfunktion*. Note that Liouville invariance does not generally apply in trace space.

The brightness function can be measured (e.g., Lejeune and Aubert, 1980; Lauer, 1982) or simulated on a computer (e.g., Hauke, 1977) for some electron guns. Then the brightness in the exit plane $z=z_0$ of the gun can be transformed into the target or screen plane $z=z_1$ by using the transformation functions $x_1=x_1(x_0,y_0,x_0',y_0')$, etc. that connect the initial with the final ray parameters. These functions are obtained from trajectory calculations for the deflectors. The initial and final brightnesses are obtained from the relation

$$R(x_1,y_1,x_1',y_1',z_1)dx_1dy_1dx_1'dy_1' = R(x_0,y_0,x_0',y_0',z_0)Jdx_0dy_0dx_0'dy_0' \qquad (31)$$

in which J is the Jacobian determinant of the parameter transformations.

The current density in any plane $z=z_1$ can now be found by integrating the left side of (31) over x_1' and y_1'. This has actually been done by Wang (1967a) for some special cases using the full Jacobian formalism. This was possible because Wang obtained analytical transformation functions using aberration theory from which the Jacobian was calculated analytically.

However, for computer simulations this cannot be done directly. Consequently, it is more common to calculate a large number of trajectories by numerical integration and to plot their landing points to give an impression of the deflected spot. The initial parameters of these trajectories are chosen to represent the density distribution of electrons in phase space. Examples of this approach include Wang (1967b), Lucchesi and Carpenter (1979), Kanaya and Baba (1980), and Baba and Kanaya (1981). (Calculations of this kind require careful interpretation because the actual density in phase space is frequently approximated using more or less drastic assumptions.)

A simpler and much older method of characterizing the deflected spot uses *aberration figures*. In this method, the undeflected beam is represented by a hollow cone having base radius r and height L (in the z direction). This cone is traced out by a rotating ray starting at an azimuthal angle φ in the base. The aberration figure is the closed curve described at the screen by the landing point of the deflected generating ray. The values of r and L can be varied to explore the behavior of various parts of the brightness distribution. These aberration figures can be calculated (e.g., Wang, 1967b; Kasper and Scherle, 1982; Kanaya and Baba, 1980) and also measured using suitable apparatus (e.g., Friend, 1951). A disadvantage is the limitation to meridional rays (point-focused beam).

Electrostatic Deflection

Introduction

Most of the published work on this topic after the review of Ritz (1979) appears to be confined to electrostatic multipole deflectors. This review, together with that of Hutter (1974), covers earlier work. Great improvements in the calculation of electrostatic fields in these systems have been made since 1979. The level of sophistication now approaches that already attained in the study of magnetostatic deflecting fields. The new work has been done in connection with narrow-angle deflectors for electron-beam lithography and scanning electron microscopy.

There is nothing new of interest in the calculation of trajectories by numerical integration; however, the aberration theory is being applied extensively, and several improvements have been made. As these improvements are general enough to apply to systems with mixed electric and magnetic fields for both focusing and deflecting the beam, they will be considered in the section on mixed-field deflection.

In the study of deflection aberrations of the beam, accurate determination of the fields has permitted wider use of the coefficients from the aberration theory. These have been used to construct aberration figures and spot profiles generated from a multitude of individual landing points of deflected rays. Also, progress involving heavy use of the aberration theory has been reported in the dynamic correction of beam aberrations with stigmators.

Calculation of the Electrostatic Field in Multipole Deflectors

Multipole deflectors are of two types, which we shall call surface multipoles and rod multipoles. Both types consist of longitudinal electrodes arranged on a surface of revolution, usually a cylinder. However, surface electrodes are thin and conform to the surface of revolution, while rod electrodes are usually cylindrical in cross section with their *axes* lying on the surface of revolution. Appropriate deflection potentials are applied to each electrode through a resistor chain or by separate voltage supplies. Such deflectors may provide one or two axes of deflection. If both x and y deflection are provided, the deflector is often referred to as an electrostatic *yoke.* In the following account of field calculations for these deflectors, we follow the treatment of Munro and Chu (1982b).

The potential Φ for surface multipoles can be written as a Fourier series

$$\Phi(r,\varphi,z) = \sum_{m=1}^{\infty} [U_m(r,z)\cos m\varphi + V_m(r,z)\sin m\varphi], \quad m \text{ odd} \tag{32}$$

in the azimuthal angle φ (cylindrical coordinates r,φ,z). The terms in U_m give the horizontal deflection and those in V_m give the vertical deflection. When substituted into the Laplace equation (16), this Fourier expansion gives the *reduced* Laplace equation

$$\frac{\partial^2 U_m}{\partial r^2} + \frac{1}{r}\frac{\partial U_m}{\partial r} + \frac{\partial^2 U_m}{\partial z^2} - \frac{m^2 U_m}{r^2} = 0 \tag{33}$$

which is satisfied by both U_m and V_m.

On the surface $r = R_d(z)$ of the deflector, U_m and V_m take the values

$$U_m(R_d,z) = \frac{1}{\pi}\int_0^{2\pi} \Phi(R_d,\varphi,z)\cos m\varphi d\varphi \tag{34}$$

$$V_m(R_d,z) = \frac{1}{\pi}\int_0^{2\pi} \Phi(R_d,\varphi,z)\sin m\varphi d\varphi \tag{35}$$

which give the required Dirichlet conditions for the solution of (33) for U_m and V_m. (This assumes that the variation of potential in the gaps between electrodes can be neglected or approximated in some manner, e.g., linearly.)

The *multipole potentials* U_m and V_m can be expanded about the z axis in power series

$$U_1(r,z) = -f_1(z)r + \frac{1}{8}f_1''(z)r^3 - \ldots \tag{36}$$

$$U_3(r,z) = -f_3(z)r^3 + \frac{1}{16}f_3''(z)r^5 - \ldots \tag{37}$$

and so on. Consequently, all of the U_m and V_m vanish for $r=0$ and vary as r^m near the axis. That is, near the axis we can neglect the terms beyond some finite value of m as being small corrections and truncate the infinite series (32). At large distances from the deflector, the multipole potentials must remain finite. Thus, the three-dimensional problem is reduced to a finite sum of two-dimensional problems.

In general, the reduced Laplace equation must be solved by numerical computation. Use of the finite-element method has been reported by Munro and Chu (1982b). As in magnetic deflection, one could also use the finite-difference method. We shall later discuss such a solution for the full three-dimensional field. The functional to be minimized has the form

$$F_m = \iint \frac{1}{2}\epsilon_0 \left[\left(\frac{\partial U_m}{\partial r}\right)^2 + \left(\frac{\partial U_m}{\partial z}\right)^2 + \left(\frac{mU_m}{r}\right)^2\right] \pi r dr dz \tag{38}$$

for U_m with a similar expression for V_m. The integration is over the part of the (r,z) plane with $r \geq 0$ and enclosed within an outer boundary (at a large distance from the deflector) on which $U_m = V_m = 0$. References describing the details of the finite-element minimization procedure were given under *Calculation of Electromagnetic Fields*, above.

The method just described is inappropriate for rod-multipole deflectors because the boundary potentials are not defined on a surface of revolution about the z axis. Consequently, the calculation remains a true three-dimensional problem. To solve it, Munro and Chu (1982b) introduce the "charge-density method," a special case of the integral-equation approach.

Each electrode surface is subdivided into many small patches or subelectrodes, the j-th one of which carries a uniform surface-charge density σ_j. The potential is then expressed as the integral (21) over the surfaces of the electrodes. This integral becomes a summation over the subelectrodes. The potential at the center of subelectrode i resulting from all the N σ_j is

$$\Phi_i = \sum_{j=1}^{N} P_{ij}\sigma_j \tag{39}$$

in which the coefficients P_{ij} are

$$P_{ij} = \frac{1}{4\pi\epsilon_0}\iint_{S_j} \frac{dS_j}{R_{ij}} \tag{40}$$

where $R_{ij} = |\mathbf{x}_i - \mathbf{x}_j|$ and S_j is the area of subelectrode j. Equation (39) represents N equations in N unknowns for the charge densities σ_j. The Φ_i are all known from the given electrode potentials and the P_{ij} are calculated from the integral (40). These equations are solved numerically by any suitable standard method. Once the σ_j are known, the potential anywhere can be calculated by performing the integral (21) for each subelectrode and then summing over all the subelectrodes.

Munro and Chu (1982b) also describe the calculation of an effective capacitance for these deflectors using both the finite-element and integral-equation results. The calculation rests on the relationship between stored energy and capacitance.

Baba and Kanaya (1981) have calculated the field of a quadrupole deflector having noncylindrical rod electrodes of finite length. They use the finite-difference method with successive over-relaxation to calculate the full three-dimensional field. The computational mesh is defined by horizontal and vertical lines in the (r,z) planes and by circles and radial lines in the (r,φ) planes.

In comparing the finite-element, finite-difference, and integral-equation methods, it is well to note that, according to Kasper (1982), the finite-element method suffers from a loss of accuracy in rotationally symmetric systems near the axis. This is a result of the linear approximation to the potential that is ordinarily employed in each finite element. The problem can be corrected by using a higher-order approximation, but at the expense of increased complexity. Although a study of this problem for deflection fields has not been published, it appears that the finite-element method should be used cautiously.

Calculation of Electron Trajectories and Beam Properties

These calculations have been carried out in the framework of aberration theory in the papers seen. Numerical integration of trajectories has not been used. No doubt this is because the applications for these deflectors are in narrow-angle systems, for which the third-order theory is generally adequate. Aberration theory also has the advantage of giving the landing positions as power series in the initial ray parameters, which permits economical calculation of numerous landing points to simulate the deflected spot.

Chu and Munro (1982a) formulate extensive improvements of third-order theory to permit calculations for any combination of electric or magnetic, focusing or deflection fields. Kanaya and Baba (1980) and Baba and Kanaya (1981) display numerous examples of both aberration figures and spot profiles constructed from many individual landing points. (The first of these two papers deals with a sequential system of parallel-plate deflectors.) Kanaya and Baba also describe the use of dynamic focusing and an octupole stigmator to correct for deflection aberrations.

In closing, note that the multipole potentials in $m=1$ and $m=3$ completely characterize the fields for purposes of third-order aberration theory. These first two harmonics contain all terms depending on r or r^3. These are precisely the terms needed to specify the field profiles on axis that appear in the integrals for the aberration coefficients.

Magnetostatic Deflection

Introduction

As was the case in the earlier review (Ritz, 1979), the bulk of the work in the theory of deflection continues to be done in magnetostatic deflection, and the largest portion of that effort relates to color television. However, the work in magnetostatic deflection for electron-beam lithography is increasing.

In field calculation, the finite-difference methods that were previously the standard have been supplemented with finite-element and integral-equation calculations. In the latter method, the requirement of infinite permeability for the core has been relaxed, permitting investigation of the effects of finite permeability on the deflecting fields. Also, the field calculations have been used to calculate the stored energy and inductance for magnetic yokes.

In trajectory calculation, both numerical integration and aberration theory continue to supply useful results. Substantial advances in the theory of self-converging inline color TV have been reported using both techniques, including a new method for reducing the residual misconvergence in these systems. The improvements in aberration theory alluded to above apply also to magnetostatic deflection, although we again defer discussion to *Mixed-Field Deflection*.

In the calculation of beam and spot properties, detailed simulations have been carried out for examples of self-converging inline systems using fields calculated by finite differences and trajectory end points determined by a combination of numerical integration and interpolation. The results agree very well with experiment. Several studies have produced aberration figures for deflection in magnetic yokes. A method for combining a minimum number of directly integrated trajectories with interpolating functions of a theoretically desirable form has been described. This technique permits large numbers of landing points to be calculated from a small number of suitably chosen sample trajectories.

Finally, some progress has been reported on the problem of synthesis in the design of magnetostatic deflectors. As opposed to the problem of analysis of given structures to determine their behavior, the problem of synthesis requires the determination of structures that will produce a desired behavior. In general, such a structure may not exist. The problem of synthesis is then one of optimization. The reported work addresses both the existence and the optimization problems.

Calculation of Fields

In the work reviewed by Ritz (1979), the technique of finite differences dominated the numerical calculation of fields in magnetostatic deflectors when permeable cores were present. The finite-element method was just being introduced. In recent years, the fashion has shifted somewhat in the direction of integral-equation methods, beginning with Fye (1979) and Ţugulea *et al.* (1979). These two papers illustrate divergent approaches, however. Fye reduces the dimension of the problem from three to two by Fourier analysis in the azimuthal angle φ as above, expressing the density of surface current in a Fourier series. Ţugulea and his co-workers solve the full three-dimensional problem, as do Munro and Chu (1982b) in the analogous electrostatic case. Kasper and Scherle (1982) follow Fye in this respect, but employ a variant procedure in which rings of magnetic charge replace the permeable surfaces and currents. In Scherle (1983a,b), the permeability of the core need not be infinite. He provides numerous field profiles for saddle and toroidal yokes having values of μ/μ_0 ranging from 10^5 to 1. Above 10^3, he finds the effects of further increases hardly noticeable.

The finite-difference method continues to be used.

Yokota *et al.* (1979) apply the FDM to a yoke for 110°, 18-inch diagonal deflection for a color-TV application. This yoke has toroidal vertical and saddle horizontal windings. The toroid is wound on the surface of the core while the saddle is detached. They employ the method of Nomura (1971) in dealing with this detached winding. Equipotential plots, field profiles, and computer-program organization are given. However, they encounter unexplained difficulties with some of their trajectory calculations. Ximen and Chen (1980) also employ the FDM in the simulation of a toroidal yoke on a cylinder. In all cases, these authors first decompose the scalar potential Ψ and the current distribution into Fourier series in azimuth φ as described, for example, in the review by Ritz (1979) and then solve the resulting two-dimensional problems. Thus, nothing fundamentally new is being reported here.

Munro and Chu (1982a) describe the application of the finite-element method to the fields of saddle yokes with or without the presence of permeable elements. As with the FDM, two-dimensional multipole potentials are calculated. The FEM permits the use of finite permeability for the core. A valuable feature of this paper is a comparison of the finite-element solution for a particular yoke with the analytic solution as calculated by the Biot-Savart law from the currents (no core present). An accuracy of 1% to 2% is reported.

An interesting analytic method is described by Dasgupta (1983b). He assumes that in each cross section $z = \text{constant}$, the fields behave as do those of an infinite cylindrical yoke of the same cross section. The effect of the end turns in saddle and toroidal yokes is more difficult to calculate. Dasgupta ignores the effect of the core on the end turns in saddle windings and neglects toroidal end turns altogether. The resulting fields are checked against fields calculated by the method of Fye (1979). There is a good resemblance between the two but errors are substantial, on the order of 20% in field strength. This accuracy is probably inadequate for most purposes. Also, the basic calculation uses the Biot-Savart law and is quite involved, requiring the use of hypergeometric and beta functions.

In another paper, Dasgupta (1983a) analyzes the effect of finite winding thickness on the field. This is an important question, because ordinarily it is assumed that the windings can be represented by sheet currents. He uses analytical calculations for infinite-cylinder yokes to show that for thick windings the higher harmonics of the field have smaller amplitudes than in the thin-winding case. Also, since the size of the error depends on m, there is no "average" position at which the approximate sheet current can be placed. The effect is more severe for detached than for surface windings.

Calculation of Energy Storage and Inductance

Ritz (1980) describes the calculation of energy storage and inductance for magnetic yokes, using fields calculated by the FDM for the angular harmonics. He shows that each harmonic of the field contributes independently to the stored energy and inductance. The calculation rests on the relation

$$W = \frac{1}{2} L I^2 \tag{41}$$

amongst inductance L, current I, and energy W stored in the field of the yoke. The stored energy is calculated by integrating the energy density $\frac{1}{2}\mu|\mathbf{H}|^2$ over the entire volume containing the field. This leads to two methods: a surface integral in the (r,z) plane and a contour integral in that plane. The second follows by use of Green's first identity of potential theory. The theory is used to compare energy storage in saddle and toroidal yokes. [Note that the sign of his Eq. (7) is reversed.]

Dasgupta (1982) calculates the inductance directly by means of flux linkages for saddle yokes only. His result is a special case of Ritz's second (contour integral) method. The fields used are calculated by the FDM.

Scherle (1983b) also calculates the stored energy but uses fields calculated by the IEM. His method permits the core permeability to be finite, and he gives a very instructive graph showing the effect of permeability on the total stored energy for saddle and toroidal yokes. The energy increases until $\mu/\mu_0 \approx 500$, after which little change occurs.

Calculation of Trajectories

Both numerical-integration and aberration-theory calculations of trajectories are well established in magnetostatic deflection. Consequently, applications rather than newly introduced methods are currently interesting. The important areas of study are misconvergence in self-converging color inline CRTs and deflection aberrations in systems for electron-beam lithography. As noted previously, the latter topic is considered under *Mixed-Field Deflection*. Good accounts of the misconvergence problem, based on third-order aberration theory, are given by Heijnemans *et al.* (1980) and by Hutter (1979). We summarize them here.

An inline color CRT contains three electron guns: red, blue, and green. The green gun is aimed along the z axis while the other two lie at equal distances either side of green in the horizontal (x,z) plane. The two outer beams are aimed so that all three undeflected beams land at the center of the screen. The common landing point must also be maintained as the beams are deflected. Departures from this condition are termed *misconvergence*. In *self-convergent* inline CRTs, convergence is achieved by proper design of the yoke, gun, and sometimes magnetic pole pieces rather than with auxiliary convergence coils driven by dynamic currents that depend on the deflection currents, as in delta-gun CRTs.

Self-convergence is quite difficult to achieve. Ordinary deflection yokes with so-called homogeneous fields cause the three beams to cross and diverge before reaching the screen. This defect is partially corrected by giving the horizontal deflecting field a pin-cushion shape and the vertical field a barrel shape. This is done by introducing a large third-harmonic component into the angular distribution of winding current for each axis of deflection. The strengths are adjusted to remove *horizontal* red-blue misconvergence at the ends of the x and y axes on the screen.

In the third-order aberration theory, this also removes *horizontal* red-blue misconvergence in the corners but leaves a *vertical* red-blue misconvergence due to *astigmatism* error. There is also *coma* error, which causes the green beam to land inward from the average position of the red and blue spots.

Astigmatism is affected most strongly by the amount of third harmonic at the yoke *exit* while coma is affected most strongly at the *entrance* and with the opposite sign. Consequently, by having the amount of third harmonic in the windings change sign from front to rear, both astigmatic and comatic misconvergence can be removed at the ends of the x and y axes, although the vertical astigmatic misconvergence between red and blue remains in the corners. The desired winding distribution is achieved with nonradial (nonmeridional) turns. All of these conclusions follow from use of the third-order aberration theory.

There is another widely used method of correcting for the coma error. Permeable pole pieces called *shunts* are placed at the exits of the two outer guns (red and blue) to reduce their deflection slightly, and another pole piece called an *enhancer* is placed at the exit of the center (green) gun to increase its deflection slightly. Because the effect depends linearly on the strength of the yoke field, it exactly corrects everywhere for coma error, which depends linearly on the yoke currents. Formerly, the pole pieces were designed empirically, but recently Fye and Grinberg (1980) have calculated the effects of these field controllers using the integral-equation method.

Ando *et al.* (1977) and Nakamura *et al.* (1982) have asserted and demonstrated, respectively, that the residual vertical red-blue misconvergence in the corners can be reduced or even eliminated by displacing the horizontal winding toward the gun with respect to the vertical winding. The demonstration is accomplished with the third-order aberration theory as formulated by Kaashoek (1968).

Calculation of Beam Properties

Severe aberration of the deflected spot is an unwanted consequence of self-convergent deflection fields. The yoke produces a vertical line focus everywhere on the screen for a cone of rays focused at the center of the screen. The diameter of the base of the cone is the distance between the two outer beams at the exit of the gun. The deflected image of the smaller cone approximating the green (center) beam should also be a (shorter) vertical line. However, because practical beams have spherical aberration and hence are focused (undeflected) for the circle of least confusion, the resultant spot is actually quite different. There is a bright horizontal line or cigar with triangular regions of flare above and below and with a large elliptical halo surrounding the core region. (See Yoshida *et al.*, 1974 and Hutter, 1979 for qualitative explanations of these phenomena.)

Lucchesi and Carpenter (1979) have simulated both the undeflected and deflected spots by computer for a self-converging inline yoke. This is the most elaborate and complete spot simulation in electron optics that is known to this reviewer. They begin with a computer program that computes positions and slopes for 123 trajectories at the exit of the electron gun, including the effects of space charge and thermal velocities. These trajectories, calculated originally in the (r,z) plane, are repeated at 100 equally spaced angles in the (r,φ) transverse plane. The resulting 12,300 rays are projected through the yoke fields with the aid of numerical integration.

However, Lucchesi and Carpenter do not integrate *all* these rays, which would be impossibly expensive and time consuming. Instead, they calculate a much smaller number, taking twelve different azimuths, four different radii, and six different slopes, plus the central ray, for a total of 289 trajectories. The landing points for all of the other rays are found by linear interpolation between the directly calculated rays.

The result is a picture of the deflected spot that agrees remarkably well with photographs of actual spots. The main discrepancy is a reduction of size, which the authors ascribe to the neglect of space charge in the deflecting fields. This reviewer suggests that their apparent neglect of skew rays may also contribute to the discrepancy.

The principal defect of their method is the still large number of trajectories to be calculated. This number can be reduced by an order of magnitude using a technique described by Ritz (1981). The x and y positions at the screen are expressed as Fourier series in the azimuthal angle φ for a ray lying on the surface of a cone having base radius r and length L (undeflected). The ordinary aberrations of astigmatism and coma require only the terms in φ and 2φ for their representation. Ritz shows that only 16 trajectories with various values of r, φ, and L need be calculated to determine all the necessary Fourier coefficients. The angular dependence is fitted using the fast Fourier transform. The dependences on r and L are fitted by least squares.

Once the coefficients are known, any landing point can be interpolated between the sample values. In this way, landing points become easy, fast, and inexpensive to calculate in great numbers. Thus, aberration figures and the current distribution in the spot become much more accessible.

This method is actually a special case derived from a six-dimensional power series expansion in a phase-space perturbation theory of wide-angle deflection. Typically, powers of order 2 or 3 (Fourier components through 2φ or 3φ) suffice for excellent accuracy of fitting. If the power series formulation were used, least-squares fitting to the samples in r and L would be unnecessary and the number of samples required would be reduced further for a given accuracy.

The Problem of Synthesis

Suppose that we ask what *winding density* is required to produce a desired set of *harmonics* of the winding, determined for example by the procedure described above for achieving a self-convergent inline yoke. At first glance, this question appears to be a trivial matter of adding up the individual harmonics of the winding according to their required amplitudes.

However, in practical yokes the mechanical constraints on the placement of the wires may make it impossible to produce the harmonics required, and only those harmonics. We must therefore synthesize a winding that comes as close as possible to the desired one. This inverse problem is exceedingly difficult and has not been addressed previously in any effective way.

The three papers of Vassell (1981a,b,c) address the question of windability for winding densities with a finite number of adjustable parameters, discrete densities, and densities with an infinite number of parameters. These papers are difficult though elegant. In essence, Vassell transforms the winding variable from azimuthal angle φ to $\xi = \cos^2\varphi$ and then applies some results of the branch of mathematics concerned with the theory of power moments of distributions.

Each winding density is characterized by its normalized harmonic ratios

$$h_{2p+1} = \frac{A_{2p+1}}{A_1} \tag{42}$$

in which A_{2p+1}, $p = 1, 2, 3, \ldots$ is the amplitude of the $2p+1$ harmonic of the winding. For a given set of constraints on the winding, the set of feasible windings is represented by a subregion of the space with coordinates

$(h_3, h_5, h_7, \ldots)$. These regions of feasibility are determined using the moment theory. Since in practice, one specifies a small finite number of harmonic amplitudes, the space in question is usually of low dimensionality. In any case, these papers deserve wider notice than they have received so far.

Dasgupta (1983c) addresses the same problem as Vassell, but limits his study to toroidal yokes having radial (meridional) windings. Like Vassell, Dasgupta derives conditions on the values of the harmonics permitted by the winding constraints, but his approach and results are far less general. He also proposes to circumvent these restrictions by adding turns at properly chosen angles.

Another approach to determining a structure that will have the desired electron-optical behavior is taken by Chu and Munro (1982b). They consider a set of m functions f_i depending on a set of n parameters. Each function is the product of a weight and an individual aberration. A *defect function* or *merit function* equal to the sum of the squares of the m functions f_i is to be minimized by varying the n parameters.

This is done using the *damped least squares* method, according to which the partial derivatives of the f_i^2 with respect to the parameters are used to obtain an improved value of the defect function. A damping factor is used at each step to stabilize the process and to speed convergence. The calculations are done by computer. As an example, they optimize a pure magnetic focusing and deflection system suitable for electron-beam lithography, although a pure deflection system could presumably be treated in the same manner.

Mixed-Field Deflection

Introduction

Under this topic, we shall consider deflecting systems composed of electric or magnetic fields for focusing or deflection in any number or combination. Historically, mixed-field devices have been used typically in imaging tubes such as vidicons and in such devices as scan converters. However, the current field of greatest activity is in electron-beam lithography and scanning microscopy.

This section is organized somewhat differently from the previous sections. Because we have already discussed the calculation of electrostatic and magnetostatic deflecting fields individually, we need not do so again here. Likewise, as the calculation of fields for focusing lenses is widely discussed in the literature of electron optics, there is no point in repeating that discussion here. Consequently, we shall proceed under the combined topic of trajectory and beam calculation.

Calculation of Trajectories and Beam Properties

Because the typical electron-beam lithography or scanning-microscope deflecting system is a narrow-angle system, the aberration theory of deflection is appropriate and is heavily used. However, the previous formulations of aberration theory are intended for single- or double-axis deflectors and not for multiple sequential deflectors or for deflectors with overlapping focusing fields.

This circumstance has prompted many recent improvements of the third-order aberration theory. The most comprehensive and accessible treatment is that of Chu and Munro (1982a). We summarize here the principal features of their formulation and compare it with the traditional formulation.

Chu and Munro permit any combination of focusing and deflecting fields; these may overlap. The electron source may have a finite size. Both electric and magnetic fields can be used.

The deflection currents are expressed in complex form $I = I_x + iI_y$ and the deflection potentials as $V = V_x + iV_y$, with complex conjugates $\overline{I}$ and $\overline{V}$. The same currents and potentials are applied to all deflectors in a multiple deflector. The complex current and complex potential appear explicitly in the ultimate aberration expansions. In the conventional theory, they are concealed in the Gaussian deflections.

The theory uses the complex position $w = x + iy$ and complex slope $w' = x' + iy'$ as well as their complex conjugates $\overline{w}$ and $\overline{w}'$. The ray parameters are expressed as $w = w_0$, $w' = s_0$, I, V in some initial plane $z = z_0$. Contrast this with the conventional method of specifying position and slope of the undeflected ray *at the screen.*

The fields and trajectories are expanded in powers of w, $\overline{w}$, w', $\overline{w}'$, I, $\overline{I}$, V, and $\overline{V}$ as appropriate. However, instead of using these expansions to construct the Lagrangian and Euler-Lagrange equations, Chu and Munro use the Lorentz equation (27) directly (after changing variables from t to z). The resulting equations of motion are solved by successive approximations. In first order, the two Gaussian deflections in x and y are replaced by four principal rays. Each principal ray is specified by w_0, s_0, I, and V, of which one parameter is unity and the rest zero for a given ray. These principal rays are then used to obtain third-order corrections.

The final aberrations are expressed in terms of the following complex quantities at the image plane: aperture angle s_i, Gaussian spot size w_{ig}, magnetic deflection vector w_{im}, and electrostatic deflection vector w_{ie}. These are simply related to w_0, s_0, I, and V by the four principal rays. In general, there are 59 complex coefficients. When the Gaussian spot size w_{ig} can be neglected, there are only 27.

The 27 aberration coefficients for the latter case are given in terms of two general integration functions $F(w_1, w_2, \overline{w}_3)$ and $G(\overline{w}_1, \overline{w}_2, \overline{w}_3)$ in which w_1, w_2, and w_3 are dummy arguments. These functions are complicated integrals that require computer evaluation.

The effect of these improvements is to increase generality, simplify the use of symmetry arguments in eliminating terms, and simplify algebraic bookkeeping and computer programing.

Many other authors have applied similar ideas to various systems too numerous to mention here individually. These include Goto and Soma (1977); Ohiwa (1978, 1979); Kuroda (1980); Lencova (1981); Hosokawa *et al.* (1981); Ximen and Li (1982); Li and Ximen (1982); and Chen *et al.* (1983). The paper of Chu and Munro (1982b) on optimization was reviewed above under *Magnetostatic Deflection.*

Traveling-Wave Deflection

Assessment

There are no recent publications reporting anything new of particular note on the theory of traveling-wave deflectors since the review by Ritz (1979). This writer still retains his opinion that further progress requires use of computer simulations of the full wave fields as described in the *Introduction* of the present paper. Such simulations have been done for other high-frequency structures but not yet for traveling-wave deflectors.

Scan Expansion

Introduction

Only a few papers have appeared on the theory of scan expansion since the previous review by Ritz (1979). These papers fall into three categories: mesh scan expansion; meshless expansion with rotationally symmetrical lenses; and meshless expansion with quadrupole lenses. The techniques of analysis used include geometrical optics, field calculations in two and three dimensions using numerical methods, and third-order aberration theory.

Calculation of Fields

The FDM is now being used in the calculation of fields for scan-expansion systems. Hawes and Nelson (1978) employ this technique to calculate the field for a rotationally symmetric mesh post-deflection lens. This is a two-dimensional calculation in the (r,z) plane. Franzen (1983) describes a program that combines three-dimensional FDM (Liebman) calculations with Fourier expansion methods to calculate the fields in a complex, nonrotationally symmetric meshless lens of quadrupole type. The Fourier-Bessel expansion is used to provide boundary values for the FDM relaxation calculation. Further details are given in Janko *et al.* (1983) and in Hawken *et al.* (1983).

Haley *et al.* (1979) describe the calculation of fields for a rotationally symmetric expansion lens using the method of moments, which is an integral-equation technique. Their work appears to be two-dimensional, although this is not explicitly stated. Their lens is used as a projection lens to magnify a storage mesh illuminated by a flood-beam of electrons, and hence is not a true scan-expansion lens operating on the deflected rays.

Calculation of Trajectories and Beam Properties

Hawes and Nelson (1978) have calculated rays for their mesh lens using both a thin-lens geometrical-optics method and numerical ray-tracing in a field determined by the FDM. The calculated magnifications agreed within 2%.

Franzen (1983) calculates trajectories in his meshless quadrupole lens numerically and uses the results to study and characterize the distortions of the raster. His program uses gradients of potentials calculated both numerically and from the Fourier series he uses near the outer boundary of his fields. Franzen also mentions simulations of beam defocusing, although no details are given.

Lenz (1979) presents a third-order theory of distortion in rotationally symmetric mesh post-deflection expansion lenses. The potential near the axis is obtained analytically. Then the equations of motion are solved successively in first- and third-order approximations to find the landing points at the viewing screen. No account is taken of the effects of individual mesh holes.

Conclusion

In his review of 1979, the present author predicted successfully that application of the numerical methods then being used in magnetostatic deflection to problems in electrostatic deflection and scan expansion would produce useful advances in those areas.

However, having perhaps exaggerated the relative merits of numerical simulations over aberration theory, he failed to perceive the renascence of aberration theory that was even then occurring in electron-beam lithography and scanning microscopy. Ironically, this new interest is partly due to the accurate field calculations made possible by numerical methods and computers. Adding embarrassment, the classical aberration theory again displayed its power of explanation in the theory of self-convergent inline deflection, even though exact calculations eluded its grasp.

When his previous review went to press, the author was unsure whether numerical simulations were sufficiently accurate to calculate beam properties. That question has now been answered resoundingly in the affirmative, and we can expect wider use of such calculations.

Regrettably, he was all too correct when asserting that further progress in traveling-wave deflection must await the application of numerical methods. Those methods have not been applied, and we are still waiting.

In the study of scan expansion, conventional methods continue in use for computer simulation of rotationally symmetrical expansion systems. For complicated systems of the quadrupole type, angular Fourier expansions have been combined with three-dimensional finite-difference methods to deal with an intrinsically three-dimensional problem. This parallels the earlier developments in magnetic-yoke theory discussed in the author's last review.

An aberration theory has been developed for scan expansion, but has not yet been applied. As many scan-expansion systems have small angles of deflection, the combination of aberration theory with numerically calculated fields might prove fruitful.

With the exception of traveling-wave deflection, the work described in this review generally shows increasingly sophisticated computer-aided simulation of increasingly complex deflection systems. Aberration theory and numerical calculation of fields and trajectories have been improved. Full three-dimensional field calculations are becoming feasible. The output of space-charge models of electron guns has been combined with deflector models to produce excellent depictions of deflected spots, although space-charge effects in the deflectors themselves have yet to be treated successfully. Thus, many of the long-standing theoretical problems of electron-beam deflection have either been solved or have solutions in view.

With this increasing power of *analytical* techniques and understanding (including numerical analysis), attention is beginning to shift to the *synthetic* or design problem in electron-beam deflection. This reviewer believes that several influences tend in this direction.

First, the power and accuracy of numerical simulations are usually gained at the expense of analytical and intuitive understanding. Each simulation is a special case, much like a single experiment, and inference of general behavior requires many special cases to be studied in a manner almost empirical.

Second, although in some cases the use of aberration theory provides analytical insight, this theory is very complicated even for simple deflection systems. This complication greatly reduces its usefulness in aiding understanding.

Third, the complicated mixed focus and deflection systems now used in lithography and scanning microscopy have so many parameters that it is difficult to comprehend the design possibilities and limitations of a given system.

There are thus two challenges for future work in this field. First, momentum must be maintained in attacking the remaining analytical problems. Second, means must be found to assimilate and use effectively the mass of results being obtained as a result of successes in analytical theory and computer simulation.

Several of the papers reviewed show some progress in meeting the second challenge, but much remains to be done to prevent our current successes from burying us.

References

Ando K, Hirota R, Akatsu M, Maruyama K. (1977). New self-convergence yoke and picture tube system with 110° in-line feature. *IEEE Trans. Consumer Electron.* **CE-23**, 375-381.

Baba N, Kanaya K. (1981). Correction of the distorted electron beam spot deflected by electrostatic crossed deflection fields. *J. Phys. E:* **14**, 183-193.

Chen ZW, Qiu PY, Wang JK. (1983). The optical properties of "swinging objective lens" in a combined magnetic lens and deflection system with superimposed field. *Optik (Stuttgart)* **64**, 341-347.

Chu HC, Munro E. (1982a). Numerical analysis of electron beam lithography systems. Part III: Calculation of the optical properties of electron focusing systems and dual-channel deflection systems with combined magnetic and electrostatic fields. *Optik (Stuttgart)* **61**, 121-145.

Chu HC, Munro E. (1982b). Numerical analysis of electron beam lithography systems. Part IV: Computerized optimization of the electron beam lithography systems using the damped least squares method. *Optik (Stuttgart)* **61**, 213-236.

Crawford CK, Brody MD. (1966). Phase space representations of charged particle beams, in: *Proc. Second Int'l. Conf. on Electron and Ion Beam Science and Technology, Metallurgical Society Conferences* **51**, Bakish R (ed.), American Institute of Mining, Metallurgical, and Petroleum Engineers, New York, 53-73.

Dasgupta BB. (1982). Calculation of inductance of the horizontal coil of a magnetic-deflection yoke. *IEEE Trans. Consumer Electron.* **CE-28**, 455-462.

Dasgupta BB. (1983a). Effect of finite coil thickness in a magnetic deflection system. *J. Appl. Phys.* **54**, 1626-1627.

Dasgupta BB. (1983b). An analytic method for calculating the magnetic field due to a deflection yoke. *RCA Review* **44**, 404-423.

Dasgupta BB. (1983c). Harmonic analysis of a planar-wound toroidal deflection coil. *IEEE Trans. Consumer Electron.* **CE 29**, 508-515.

Franzen N. (1983). A quadrupole scan-expansion accelerating lens system for oscilloscopes, in: *1983 SID Int'l. Sym. Digest of Tech. Papers,* Lewis Winner, Coral Gables, Florida, 120-121.

Friend AW. (1951). Deflection and convergence in color kinescopes. *Proc. IRE* **39**, 1249-1263.

Fye DM. (1979). An integral equation method for the analysis of magnetic deflection yokes. *J. Appl. Phys.* **50**, 17-22.

Fye DM, Grinberg EL. (1980). Magnetic field controller analysis using an integral equation technique. *IEEE Trans. Magn.* **MAG-16**, 1265-1271.

Glaser W. (1949). Über die Theorie der elektrischen und magnetischen Ablenkung von Elektronen strahlbündeln und ein ihr angepasstes Störungsverfahren. *Ann. der Phys. (Leipzig)* **6** Folge, Bd. 4, Heft 7, 391-408.

Glaser W. (1952). *Grundlagen der Elektronen optik.* Springer-Verlag, Vienna, 467-486.

Goto E, Soma T. (1977). MOL (moving objective lens) formulation of deflective aberration free system. *Optik (Stuttgart)* **48**, 255-270.

Haantjes J, Lubben GJ. (1957). Errors of magnetic deflection I. *Philips Res. Rep.* **12**, 46-68.

Haantjes J, Lubben GJ. (1959). Errors of magnetic deflection II. *Philips Res. Rep.* **14**, 65-97.

Haley JF, Carnahan PE, Balint FA, Pallas EJ. (1979). A projection storage CRT. *Proc. SID* **20**, 61-64.

Hauke R. (1977). Theoretische Untersuchungen rotations symmetrischer Elektronenstrahlerzeugungs-systeme unter Berücksichtigung von Raumladung. Dissertation, Univ. of Tübingen, 1-54.

Hawes JD, Nelson G. (1978). Mesh PDA lenses in cathode-ray tubes. *Proc. SID* **19**, 17-21.

Hawken K, Franzen N, Sonneborn J. (1983). A novel high-performance CRT with meshless scan expansion, in: *Proc. 3rd Int'l. Display Res. Conf.* (Japan Display '83), Society for Information Display, Los Angeles, and Institute of Television Engineers of Japan, Tokyo, 136-139.

Heijnemans WAL, Nieuwendijk JAM, Vink NG. (1980). The deflection coils of the 30AX colour-picture system. *Philips Tech. Rev.* **39**, 154-171.

Hosokawa T, Takeuchi Y, Ohshiba H. (1981). A multistage deflector theory. *Optik (Stuttgart)* **58**, 241-250.

Hutter RGE. (1970). Deflection defocusing effects in cathode-ray tubes at large deflection angles. *IEEE Trans. Electron Devices* **ED-17**, 1022-1031.

Hutter RGE. (1974). The deflection of electron beams, in: *Advances in Image Pickup and Display,* vol. **1**, Kazan B (ed.), Academic Press, NY, 163-224.

Hutter RGE (1979) Electron optics in CRT, in: *Society for Information Display and University of Illinois 1979 Seminar Lecture Notes (SID Int'l Sym.),* vol. **1**, Society for Information Display, Los Angeles, 42-68.

Janko B, Franzen N, Sonneborn J. (1983). CRT architecture with meshless scan expansion, in: *1983 SID Int'l. Sym. Digest of Tech. Papers,* Lewis Winner, Coral Gables, Florida, 118-119.

Kaashoek J. (1968). A study of magnetic deflection errors. *Philips Res. Rep. Suppl.* **11**, 1-114.

Kanaya K, Baba N. (1980). A method of correcting the distorted spot shape of a deflected electron probe by means of dynamic focusing and stigmator. *J. Phys. E:* **13**, 415-426.

Kasper E. (1982). Magnetic field calculation and the determination of electron trajectories, in: *Magnetic Electron Lenses,* Hawkes PW (ed.), Springer-Verlag, Berlin, 57-118.

Kasper E, Scherle W. (1982). On the analytical calculation of fields in electron optical devices. *Optik (Stuttgart)* **60**, 339-352.

Kuroda K. (1980). Simplified method for calculating deflective aberration of electron optical system with two deflectors and lens. *Optik (Stuttgart)* **57**, 251-258.

Lauer R. (1982). Characteristics of triode electron guns, in: *Advances in Optical and Electron Microscopy*, vol. **8**, Barer R and Cosslett VE (eds.), Academic Press, London, 137-206.

Lejeune C, Aubert J. (1980). Emittance and brightness: definitions and measurements, in: *Adv. Electron. Electron Phys. Suppl.* **13A**, Septier A (ed.), Academic Press, 159-259.

Lencova B. (1981). Deflection aberrations of multi-stage deflection systems. *Optik (Stuttgart)* **58**, 25-35.

Lenz F. (1979). Third-order distortion of axisymmetric electrostatic post-deflection lenses. *Proc. SID* **20**, 65-70.

Li Y, Ximen JY. (1982). On the relativistic aberration theory of a combined focusing-deflection system with multi-stage deflectors. *Optik (Stuttgart)* **61**, 315-332.

Lucchesi BF, Carpenter ME. (1979). "Pictures" of deflected electron spot from a computer. *IEEE Trans. Consumer Electron.* **CE-25**, 468-474.

Munro E, Chu HC. (1982a). Numerical analysis of electron beam lithography systems. Part I: Computation of fields in magnetic deflectors. *Optik (Stuttgart)* **60**, 371-390.

Munro E, Chu HC. (1982b). Numerical analysis of electron beam lithography systems. Part II: Computation of fields in electrostatic deflectors. *Optik (Stuttgart)* **61**, 1-16.

Nakamura Y, Ichida K, Sugii Y, Saito T. (1982). Application of aberration theory to the deflection yoke design of a color picture tube, in: *1982 SID Int'l. Sym. Digest of Tech. Papers,* Lewis Winner, Coral Gables, Florida, 64-65.

Nomura T. (1971). Calculation of magnetic field produced by distributed current on revolving surfaces. *Elec. Eng. Jpn.* **91**, 147-155.

Ohiwa H. (1978). Design of electron-beam scanning system using the moving objective lens. *J. Vac. Sci. Technol.* **15**, 849-852.

Ohiwa H. (1979). Moving objective lens and the Fraunhofer condition for pre-deflection. *Optik (Stuttgart)* **53**, 63-68.

Ritz EF. (1979). Recent advances in electron beam deflection. *Adv. Electron. Electron Phys.* **59**, 299-357.

Ritz EF. (1980). Computer calculation of field energy storage and inductance in magnetic deflection yokes by multipole field analysis, in: *1980 SID Int'l. Sym. Digest Tech. Papers,* Lewis Winner, Coral Gables, Florida, 52-53.

Ritz EF. (1981). Analysis of deflection defocusing in large-angle CRTs, in: *1981 SID Int'l. Sym. Digest Tech. Papers,* Lewis Winner, Coral Gables, Florida, 126-127.

Schaefer CH. (1983). The application of the alternating procedure by H.A. Schwartz for computing three-dimensional electrostatic fields in electron-optical devices with complicated boundaries. *Optik (Stuttgart)* **65**, 347-359.

Scherle W. (1983a). Eine Integralgleichungsmethode zur Berechnung magnetischer Felder von Anordnungen mit Medien unterschiedlicher Permeabilität. *Optik (Stuttgart)* **63**, 217-226.

Scherle W. (1983b). Berechnung von magnetischen Ablenksystemen. Dissertation, Univ. of Tübingen, 1-103.

Stratton JA. (1941). *Electromagnetic Theory.* McGraw Hill, NY, 192.

Ţugulea A, Nemoianu C, Măries VA, Panaite D. (1979). Determination of magnetic field in magnetic deflection coils. *Rev. Roum. Sci. Techn-Electrotechn. et Energ.* **24**, 403-417.

Vassell MO. (1981a). Curve fitting the winding densities for magnetic coils: I. Densities with a finite number of parameters. *J. Phys. D:* **14**, 523-541.

Vassell MO. (1981b). Curve fitting the winding densities for magnetic coils: II. Discrete densities. *J. Phys. D.* **14**, 543-557.

Vassell MO. (1981c). Curve fitting the winding densities for magnetic coils: III. Densities with an infinite number of parameters. *J. Phys. D:* **14**, 559-573.

Wang CCT. (1967a). Coma-dominated aberration in magnetically deflected electron beams. *J. Appl. Phys.* **38**, 3991-3994.

Wang CCT. (1967b). Spot distortion caused by magnetic deflection of an electron beam. *J. Appl. Phys.* **38**, 4938-4944.

Ximen JY, Chen WX. (1980). On the analysis and calculation of static electromagnetic multipole fields. *Optik (Stuttgart)* **57**, 259-286.

Ximen JY, Li Y. (1982). Dynamic aberration correction in a combined scanning electron beam system. *Optik (Stuttgart)* **62**, 287-297.

Yokota Y, Li T, Toyofuku T, Tabatake K. (1979). The calculation of the deflection magnetic field and the electron-beam trajectory for color television. *IEEE Trans. Consumer Electron.* **CE-25**, 468-474.

Yoshida S, Ohkoshi A, Miyaoka S. (1974). 25-V inch 114 degree Trinitron color picture tube and associated new developments. *IEEE Trans. Broadcast Telev. Receivers* **BTR-20**, 193-200.

Electron Optical Systems (pp. 109-114)
SEM Inc., AMF O'Hare (Chicago), IL 60666-0507, U.S.A.

0-931288-34-7/84$1.00+.05

ELECTRON BEAM DEFLECTION WITHOUT OFF-AXIS ABERRATIONS

H.C. Pfeiffer* and M.A. Sturans

IBM General Technology Division
East Fishkill, Route 52
Hopewell Junction, NY 12533

Abstract

A novel focusing/deflection system for high accuracy, high throughput E-beam lithography, denoted as Variable Axis Immersion Lens (VAIL), has been successfully demonstrated. The main attributes of this system include: 1)Perpendicular landing at all points of a deflection field > (10 x 10 mm), 2)Elimination of transverse chromatic aberration, 3)High resolution (< 0.2μm edge slope) over the entire deflection field, 4)Elimination of eddy current effects in the target area, and 5)Total magnetic shielding of the target from external fields.

Key Words: Variable axis lens, Immersion lens, Transverse chromatic aberration, Telecentric lens, Parallel deflection, Predeflection, Eddy currents, Normal landing, Ferrite shielding, Dynamic focus

*Address for correspondence:
H.C. Pfeiffer
IBM General Technology Division
D/88g B-500 Zip 90K
East Fishkill, Route 52
Hopewell Junction, NY 12533
Phone no.: (914) 894-4129

Introduction

The Variable Axis Lens (VAL) was demonstrated [7,8] to have very desirable properties for sub-micron large field electron lithography systems. By electronically shifting the electron optical axis in synchronism with the deflected beam the resolution remains essentially constant over a 10 x 10mm^2 field since all deflection aberrations are eliminated. Practical implementation of the system for production machines, however, was hampered by eddy current generation in the target area by the dynamic field of the correction yoke, stray field effects from the X-Y table and general complexity and instability of required driving electronics. All these problems have been alleviated in the Variable Axis Immersion Lens (VAIL) configuration. This has been accomplished by replacing the lower pole piece of the projection lens with a solid ferrite disc. This configuration creates an asymmetrical lens with zero bore in the lower pole piece, with the target plane located slightly above the pole piece. This eliminates the need for the lower field correction yoke and two dynamic focus coils, hence there are no dynamic fields in the target area. In addition, the solid pole piece acts as a shield for any extraneous fields which might otherwise affect beam stability.

Description

The variable axis concept was first discussed by Ohiwa et al.[2] It was denoted as MOL (Moving Objective Lens) but was not reduced to practice due to unsatisfactory field matching[1] and prohibitive design constraints.[3] The VAL theory of operation and experimental results have been published by Pfeiffer et al.[7,8] It consists of a telecentric lens, double deflection yokes and a projection lens capable of moving its axis in synchronism with the deflected beam. See Fig. 1. The variable axis capability is established by the superposition of field correcting yokes (see Fig. 2) and the lens field, providing the condition:

$$B_x (Z)_{yoke} = 1/2 \frac{dB_{lens} (Z)}{dZ} X_o \qquad (1)$$

where X_o is the axis displacement and is directly proportional to the deflection yoke current. The double pre-deflection yokes are precisely balanced in sensitivity so as to provide a deflection which is always parallel to the geometrical axis of the beam. Since both yokes are outside the magnetic fields of adjacent lenses, they no longer have to be air core. This allows for yoke designs with improved sensitivity and speed. As was shown in Ref. 7, the field correcting yoke over-compensates the radial field component of the lens field resulting in the field configuration shown in Fig. 3. This condition is corrected by introducing dynamic focus coils which provide a field to counter this effect, such that:

$$B_{dyn}(Z) = -1/4\, B_L''(X)\, X_o^2 \qquad (2)$$

In the VAL configuration two field correction yokes and four dynamic focus coils symmetrically positioned in the upper and lower pole pieces are required to obtain the aforementioned conditions. As can be seen from Fig. 1 the close proximity of the lower correction yoke and dynamic focus coils to the target impose difficult design and material restriction on the sample holder and X-Y table since any bulk conductive material will generate detrimental eddy currents and stray fields from any magnetic material will likewise adversely affect the beam.

The VAIL configuration eliminates the dynamic and stray field target problems and in addition simplifies mechanical alignment and driving electronics requirements. The concept can be easily deduced from Fig. 4. If the excitation of the VAL lens is increased until the focal plane coincides with the geometrical center of the lens, then theoretically a slab of infinitely high permeability material can be inserted in the center of the lens and the magnetic field configuration will remain unchanged. By introducing a zero bore lower pole piece and "immersing" the target inside the lens we have accomplished the following:

1. Eliminated one field correction yoke.
2. Eliminated two dynamic focus coils.
3. Eliminated dynamic fields from the target area.
4. Shielded the target from stray fields.

The implementation of such a lens requires that a means be provided for introducing and removing samples from the target area. The practical design, as shown in Fig. 5, provides a gap above the lower solid pole piece sufficiently far removed so that its fringing fields have no effect in the target area. We have found that for our lens geometries a 2.5 cm gap presents no adverse effects on the beam over a 10 x 10mm^2 field. The fact that there is no dynamic field influence on the target can be seen from Hall probe measurements (Fig. 6) of the dynamic correction yoke field which has essentially zero field strength in the 2.5 cm gap where the target and its holder are located. The same figure also illustrates field matching between the correction yoke and the transverse field of the lens. The extent to which the condition of Eq. (1) is satisfied can be seen in the same figure. The small difference between the yoke field and the transverse field of the lens is symmetrically distributed in both directions so that the integral over the difference is vanishingly small.

Figure 7 illustrates the entire VAIL configuration. It consists of two lenses which image the object plane into the target plane acting as a telecentric system. The upper lens is of the conventional type with finite bore pole pieces, while the lower, projection lens, has a zero bore lower pole piece and utilized as an immersion lens. Deflection is achieved by means of two composite predeflection yokes which shift the beam parallel to itself.
The variable axis projection lens contains only one field correction yoke and two dynamic correction coils. The predeflection and excitation of the field correction yoke takes place in synchronism so that the predeflected beam always enters the variable axis projection lens on its electron optically displaced axis. The focused spot is located a short distance above the lower pole piece, and the beam landing is always perpendicular to target plane.

Telecentricity of the beam entering the projection lens is required to assure that the object point will not move when electrons undergo energy changes, and provide complete elimination of transverse chromatic aberration.

A dynamic stigmator coil assembly, located in the collimating lens, permits the correction for axial astigmatism. It consists of two quadrupoles rotated by 45° to each other in the same plane.

Experimental Results

In order to demonstrate that VAIL eliminates transverse chromatic aberration, the following experiment was performed. The beam was first aligned on the geometrical axis of the lens and observed to have no positional shift due to changes in lens excitation or accelerating potential. The double predeflection yokes were then excited to deflect the beam to a corner of a 10 x 10mm^2 field (i.e., 7.07mm displacement). Here the beam profiles, in both X and Y were observed by the well known method of traversing wires in both directions, collecting the transmitted signal and looking at its derivative.[5,6] The accelerating potential was first changed by 100 volts and then 200 volts, each time the beam profile was stored and finally a composite picture photographed, see Fig. 8. Although the beam defocussing effect is very evident (as expected) the position of the beam did not change, proving the elimination of transverse chromatic aberration.

Since the VAL virtually eliminates the transverse chromatic aberration (which was the dominant aberration) and coma, the system parameters, particularly the beam semi-angel α, can now be chosen such that the spherical aberration ($\sim \alpha^3$), the axial chromatic aberration ($\sim \alpha$), and the interaction blur (α^{-1}),

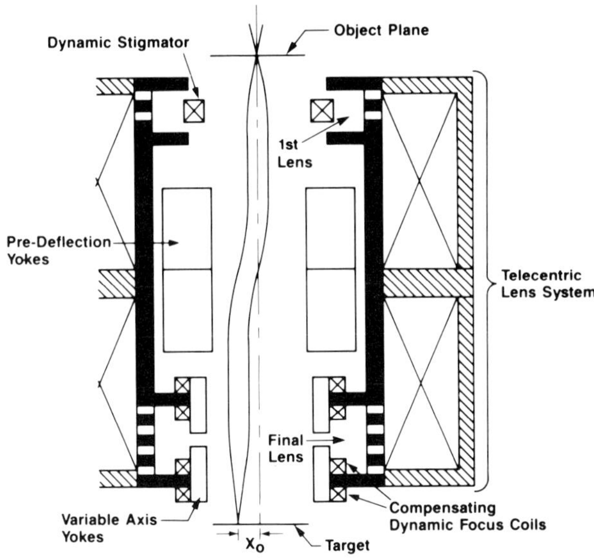

Fig. 1 Variable Axis Lens (VAL) diagram.

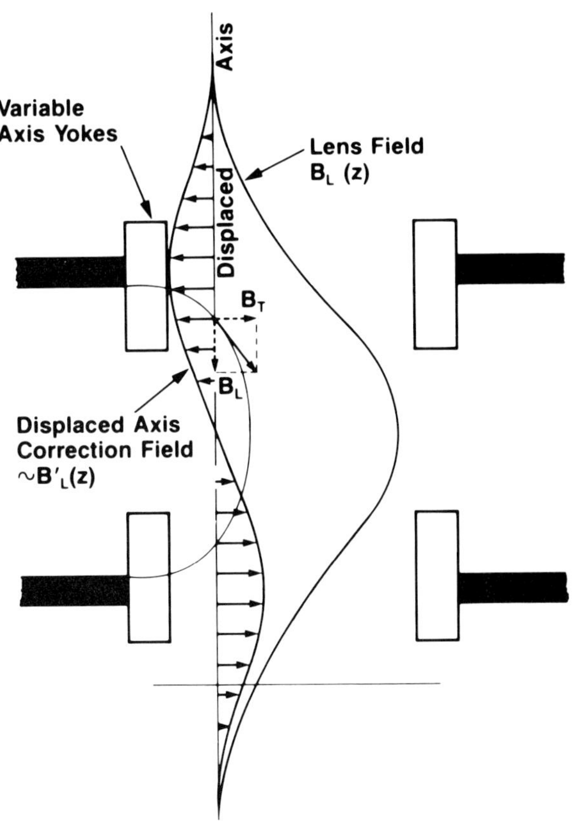

Fig. 2 Variable axis field correction concept.

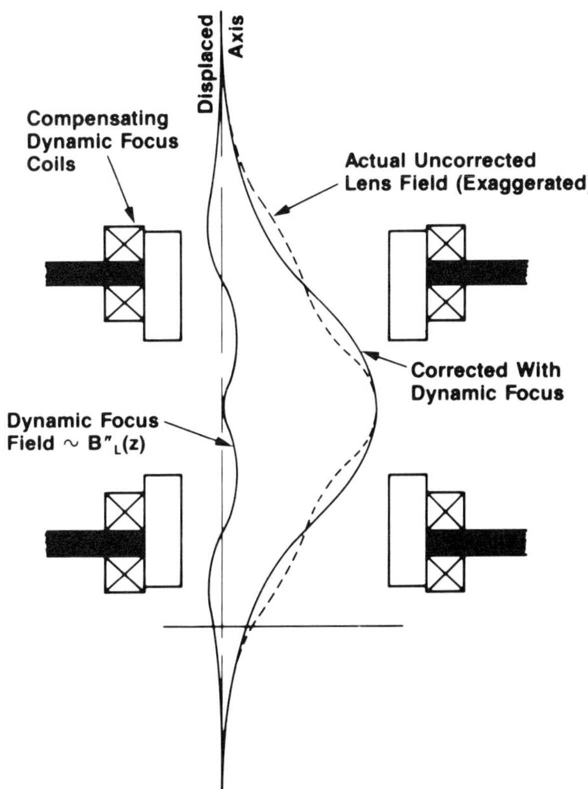

Fig. 3 Variable axis dynamic focus correction concept.

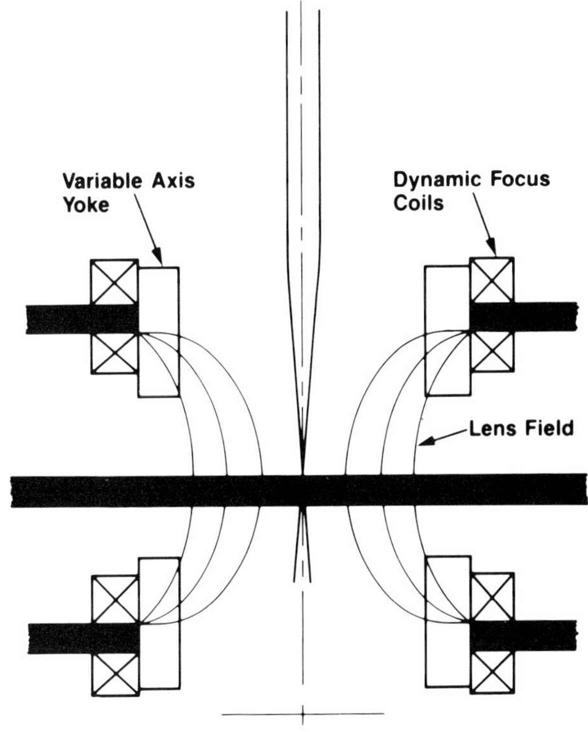

Fig. 4 Immersion lens concept.

are balanced and their geometrical sum is minimized. Experimental results in Fig. 9 indicate that the optimal α for our VAL geometry occurs at 7.2 mrad. It is also evident that the edge slope deteriorates with increased current density, as evidenced by the two sets of data, one for 50 amps/cm^2 and the other for 14 amps/cm^2. It can also be noted that the edge slope for the lower current density reaches a minimum at smaller α. Since the filament temperature is kept constant, the improvement in resolution for smaller beam current due to chromatic aberration is less significant than that due to electron interaction, hence the optimum has shifted to 6.5 mrad. For a more detailed explanation of this effect see Ref. 4.

Figure 10 illustrates edge slope of $<0.2\mu m$ with current density of 14 amps/cm^2. Similar photographs were taken at nine points of a 10 x 10mm^2 deflection field (see Fig. 11). From the composite photograph of all nine locations it can be seen that the edge slope remains essentially constant over the 10 x 10mm^2 field. Dynamic focus and stigmatism are of course compensated at each location. All are directly proportional to the square of the distance from the geometrical axis.

Summary

When looking at requirements placed on E-beam lithography systems for the future, submicron resolution over $\geq$ 10 x 10mm^2 fields, greater field stitching accuracy (vertical beam landing) and improved overlay capability (elimination of deflection aberrations) are commonly heard as key words. The variable axis lens in theory provides these necessary improvements and the VAIL construction allows for its practical implementation.

Elimination of transverse chromatic aberration has always been of prime concern in the development of IBM's high throughput E-beam lithography tools. A summary of these improvements, evidenced by manufacturing systems EL1 and EL3, are shown in Fig. 12. The EL1 system with a single yoke provided 0.5μm resolution over 5 x 5mm^2 field. The EL3 with double deflection yokes extended the field coverage to 10 x 10mm^2 with the same 0.5μm resolution. VAIL technology maintains 0.2μm resolution over 10 x 10mm^2 field and allows for further expansion of the field size.

Acknowledgements

The authors gratefully acknowledge the contribution of Dr. G. O. Langner who participated in the early conception of VAIL technology, Dr. W. Stickel for helpful discussions, and R. F. Bush for performing various experiments for system evaluation.

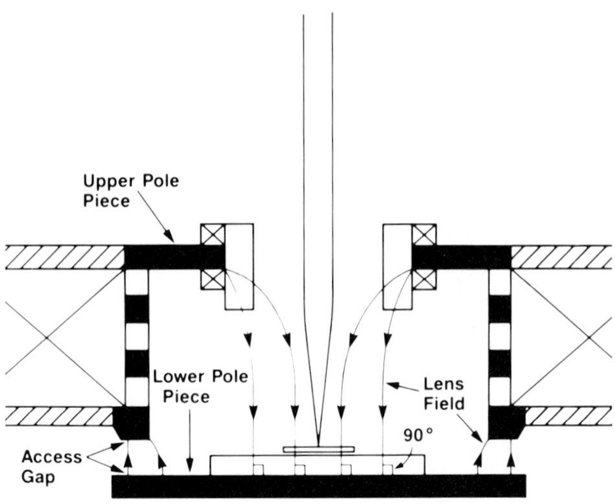

Fig. 5 Immersion lens practical design.

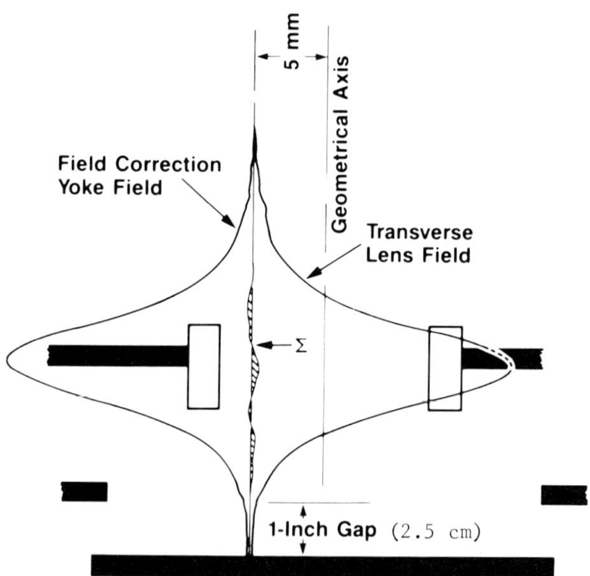

Fig. 6 Variable Axis Immersion Lens (VAIL) compensating field Hall probe measurements.

References

(1) Ohiwa H. (1978) "Design of electron-beam scanning systems using the moving objective lens", J. Vac. Sci. Technol. 15, 849-852.

(2) Ohiwa H, Goto E, Ono A. (1971) "Elimination of third-order aberrations in electron-beam scanning systems", Electron. Commun. Japan Sect. Vol. 54-B, No. 12, 1971, 44-51.

(3) Pearce-Percy HT, Spicer DF. (1980) "Integration of trajectory equations for deflection and focusing systems avoiding paraxial type approximations", Microcircuit Engineering, edited by Ahmed, H., and Nixon, W. C., Cambridge University, London, p. 535-545.

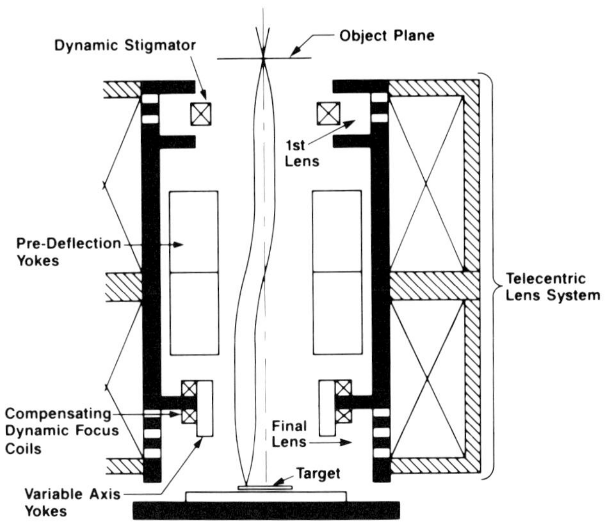

Fig. 7 Variable Axis Immersion Lens (VAIL) diagram.

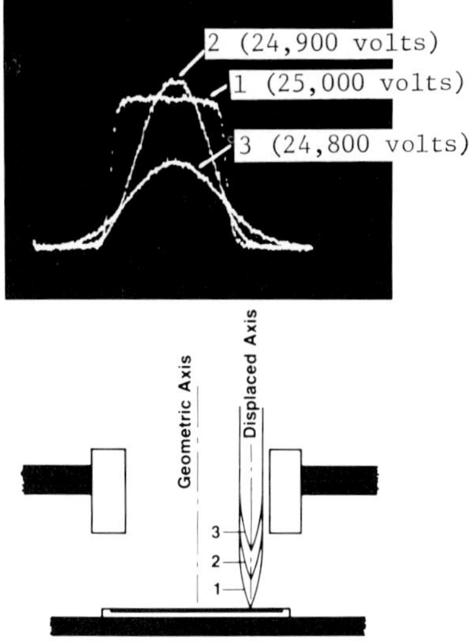

Fig. 8 Elimination of transverse chromatic aberration.

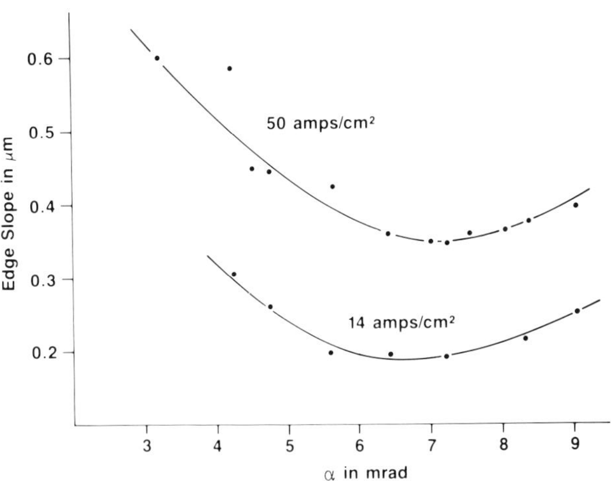

Fig. 9 Optimization of alpha.

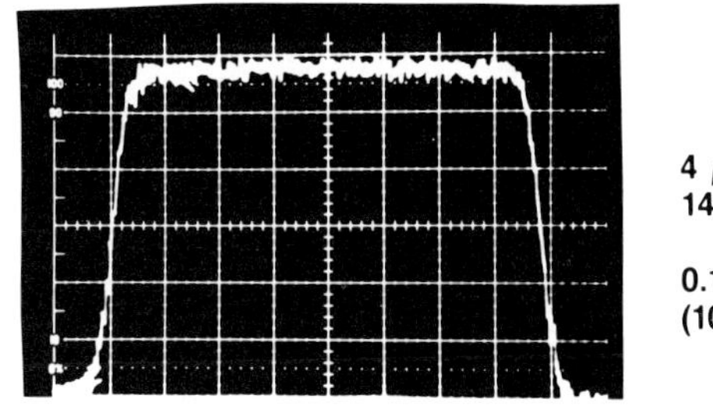

Fig. 10 Edge slope measurement.

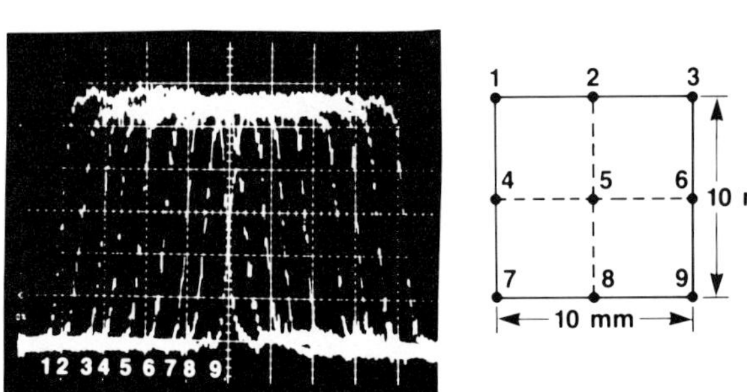

Fig. 11 Resolution over 10 x 10mm^2 field.

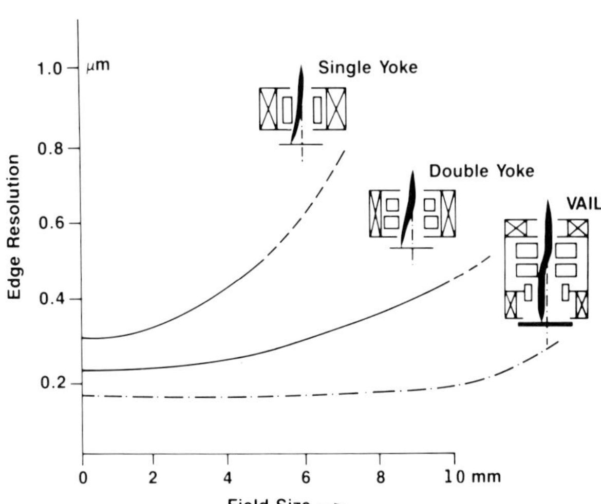

Fig. 12 Comparison of the edge resolution over the deflection field for three projection/deflection system concepts.

(4) Pfeiffer HC. (1972) "Basic limitations of probe forming systems due to electron-electron interaction", Scanning Electron Microsc. 1972:113-117.

(5) Pfeiffer HC. (1975) "New imaging and deflection concept for probeforming microfabrication systems", J. Vac. Sci. Technol. 12 (6), 1170-1173.

(6) Pfeiffer HC. (1979) "Recent advances in electron-beam lithography for the high-volume production of VLSI devices", IEEE Trans. Electr. Dev. ED-26 (4), 663-674.

(7) Pfeiffer HC, Langner GO. (1981) "Advanced deflection concept for large area, high resolution e-beam lithography", J. Vac. Sci. Technol. 19 (4), 1058-1063.

(8) Pfeiffer HC, Langner GO, Sturans MA. (1981) "Variable axis lens for electron beams", Appl. Phys. Lett. 39 (9), 775-776.

Electron Optical Systems (pp. 115-126)
SEM Inc., AMF O'Hare (Chicago), IL 60666-0507, U.S.A.

0-931288-34-7/84$1.00+.05

COMPUTER PROGRAMS FOR ANALYZING CERTAIN CLASSES OF 3-D ELECTROSTATIC FIELDS WITH TWO PLANES OF SYMMETRY

Norm Franzen

Tektronix, Inc.
P.O. Box 500
Beaverton, OR 97077
Phone No. (503) 627-5136

Abstract

At Tektronix, two classes of non-rotationally symmetric electrostatic lenses which lead to significant three-dimensional-field problems are currently under investigation. Extensive interactive computer programs have been developed to study these lens structures.

The first class is referred to as Klemperer *lipped* lenses; the accelerating quadrupole being the most important. Klemperer lenses are formed from shaped concentric cylindrical electrodes and employ two or three independent voltages. These lenses are predominantly of quadrupole type, but possess higher multipoles to correct for geometry distortion.

The second class of lenses, referred to as *wafer-lenses,* consists of lenses formed from spaced sequences of metal wafer electrodes with varying apertures and two planes of symmetry. Typical lenses formed in this manner include quadrupoles, octopoles, slot-lenses, stigmators and solid quadrupoles with exit field shaping.

The primary method of calculation used here is three-dimensional relaxation in either a rectangular or cylindrical coordinate mesh representing one quadrant of the field. The three-dimensional relaxation is followed by multipole decomposition of the core field using Fast Fourier Transforms (FFTs). Numerous graphical methods are used for program diagnostics and to display lens distortions. The electrodes are described by simple shape parameters and the computer program does the complete analysis of the boundary conditions. Program setup time for a complex lens can be less than one half hour. Large fields can be segmented if necessary.

Key Words: Electron Optics, electrostatic quadrupoles relaxation methods, meshless scan expansion, Klemperer lenses, wafer lenses, Fourier-Bessel expansion, Electron gun.

Introduction

With ever greater frequency, designers of electron optical systems are employing electrostatic and magnetostatic fields which cannot be reduced to two-dimensional field problems. Most deflection systems are of this type.[13,15] The newer in-line gun systems for color raster displays employ non-rotationally symmetric optics.[6,11] Beginning with Martin and Deschamps,[12] designers of electron guns for oscilloscopes, have begun to exploit the advantages of quadrupole optics.[2,3,5,8,14] In each of the problem areas covered in the referenced material, field calculations are confronted that stress the capabilities of modern high-speed computers. Not being reducible to two dimensions or amenable to classical analysis, such three-dimensional problems would not have been attempted at all just a few short years ago.

The most frequently used method for two-dimensional problems has been the finite difference method (FDM) or relaxation technique. To most researchers this method did not seem feasible in three dimensions since as many as 10,000 to 100,000 mesh nodes might be required to represent the field. However in 1977 Odenthal,[14], developed a three-dimensional program using relaxation methods to study the *Box-lens.* Roughly 30,000 mesh nodes were used and the relaxation process converged in 90 CPU seconds on a CYBER 73 system. The success of the method rested in the fact that the electrodes were situated in coordinate planes.

This article will show that there are other quite large classes of non-rotationally symmetric lens structures that can be analyzed by finite difference methods. In fact the ease of use and setup procedure for the study of these structures is vastly more simple than for most of the newer methods. The setup time for some rather complex lens configurations discussed below can be measured in minutes rather than hours or days.

The FDM techniques employed at Tektronix cannot be readily adapted to the treatment of misaligned structures, however the method referred to as the charge density or *Boundary Element Method (BEM)* seems to be most suited for such calculations. A. Wexler and his students,[10,18], have combined this method with techniques for smooth modeling of surfaces. This approach has the virtue of having to store only the sources on two-dimensional surfaces as well as yielding efficient means to perform the required surface integrals. In Wexler's approach the concept of *surface spline* or *Coons-patch,*[1], is employed to represent the subelectrodes and interconnect them in a smooth fashion. The charge is not assumed to be constant over such patches but takes a form convenient for performing the surface integrals by Gaussian quadrature. The accidental singularity when $|r-r'|=0$ in $\int \varrho(r')/|r-r'|da$ is given special attention. Since the surfaces and integrals are setup in parametric form, the electrodes can be tilted in space by simple transformations without complications. The charge distribution is determined by a variational principle which is a form of the *Method of Moments* as discussed by R. Harrington.[4] As is the case for the *Finite Element Method* (FEM), the Boundary Element Method requires the inversion of a *large matrix.* Usually the number of subelectrodes is much less for the BEM than the number of elements required for the FEM approach. Also the *elements* are confined to surfaces so the subdivision process is much simpler.

Since it is our primary purpose to demonstrate the utility of the Finite Difference Method (FDM) in certain significant classes of electrostatic field problems with two planes of symmetry, the class of structures called Klemperer lenses is briefly reviewed in the following paragraph.

Klemperer Lenses of Type I & II

In the second edition of his textbook *Electron Optics,*[9], published in 1953, Klemperer proposed two quadrupole type accelerating lenses as shown in figures 1a and 1b. Both lenses have been called MSE lenses, an abbreviation for *meshless scan expansion.* MSE lenses are being used to displace the dome-mesh scan expansion lens commonly employed in electron guns used in high performance oscilloscopes. The Type I lens is also called the *In-line* lens for obvious reasons. The Type II lens is referred to as the *Low-Voltage Profile lens,* or LVMSE. In each case the entrant electrode on the left is operated at gun potential, say 2 kV, and the exit electrode is at screen potential, say 10–20 kV. Both Type I and Type II lenses have played an important role in the development of the quadrupole gun design displayed in figures 2a and 2b.* These figures show the action of the four quadrupoles (Q1,Q2,Q3, and Q4) in the formation of the beam, and the development of the scan from the deflection system. Quadrupoles Q1 and Q2 are adjusted to regain focus with change in drive, while Q3 and Q4 operate in a static mode and together constitute the scan-expansion and accelerating system. Further details of the operation of the gun is given in references [2,3,5,8]. The physical implementation of the quadrupoles Q1, Q2, and Q3 is in the form of *wafer-lens* as illustrated later (figures 11-17). The quadrupole MSE lens, Q4 is illustrated in figure 1c, and is a Type II Klemperer lens. The action of the lens in the vertical plane and horizontal plane is shown in figures 3a, 3b respectively. These fields were calculated using a three-dimensional cylindrical coordinate mesh. Most of the development work to determine the shape of the contoured inner electrode displayed in figure 1c, was accomplished using the Type I, In-line Klemperer lens.

Analysis of the Type I Lens

The field problem posed by the In-line lens corresponds to the classic solution of Laplace's equation in a cylindrical cavity with specified boundary values. The potential is given by the Fourier-Bessel expansion:

**Janko B, U.S. Pat. No. 4,188,563, Feb. 1980; Cathode-ray tube having an electron lens system including a meshless-scan-expansion post-deflection accelerating lens.*

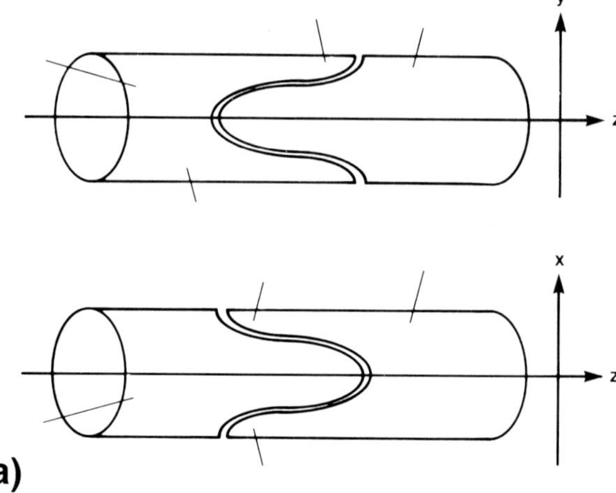

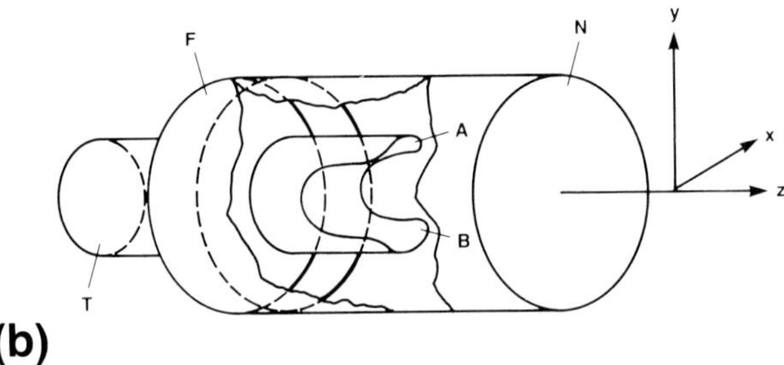

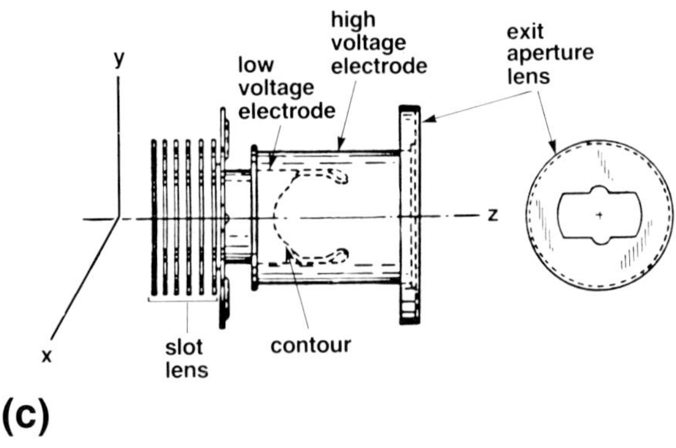

Figure 1. a) Type I Klemperer lens (in line). b) Type II Klemperer lens (LVMSE). c) Accelerating quadrupole Type II lens as used in the Tektronix quadrupole CRT. (Note slot and exit aperture lenses.)

$$V(r,\theta,z) = \sum_{k=0}^{\infty}\sum_{m=0}^{\infty} C_{mk}\, I_{2m}(\hat{k}r)\cos(2m\theta)\cos(\hat{k}z) \tag{1}$$

where for convenience we reflect the structure in figure 1a, about its exit plane $z=L/2$, and assume the symmetry conditions:

$$V(\theta,L-z) = V(\theta,z) = V(\pi-\theta,z). \tag{2}$$

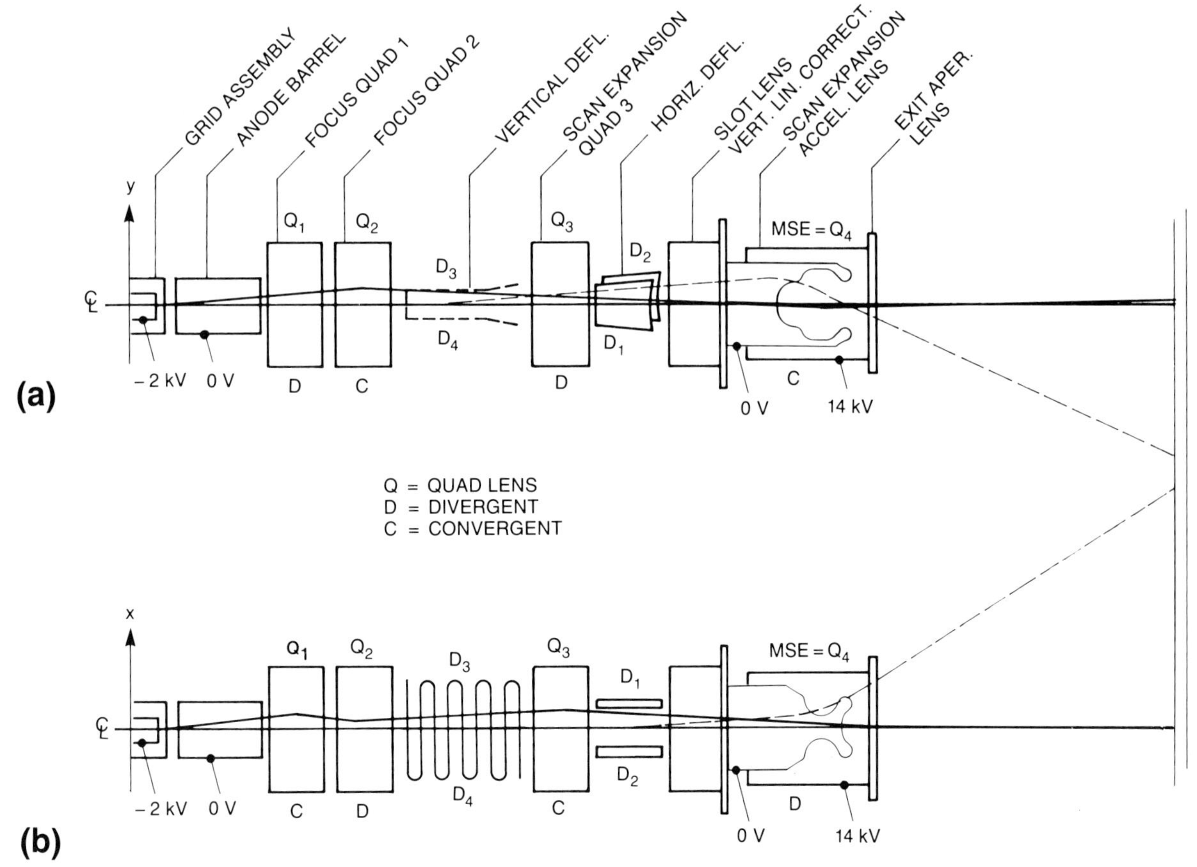

Figure 2. a) Side view of the quadrupole CRT showing the gun's vertical plane of symmetry. b) Top view of the quadrupole CRT showing the horizontal plane of symmetry.

Also $\hat{k} = (2\pi/L)$ k, and $I_{2m}(\hat{k}r)$ is to be interpreted as the Modified Bessel function of order 2m when k≠0, and r^{2m} otherwise. The Fourier coefficients C_{mk} are then given by the surface integrals:

$$C_{mk} = \frac{1}{N_{mk}} \iint V(\theta,z) \cos(2m\theta) \cos(\hat{k}z)\, d\theta dz, \tag{3}$$

where $N_{mk} = 4\pi^2 I_{2m}(\hat{k}R)$, k≠0 (see reference[7]).

Clearly the practical difficulty is to efficiently evaluate these surface integrals for a general class of contours on the surface of the cylinder. What shape of electrodes will be required to eliminate geometry distortion and provide good linearity of the display is generally unknown at the outset. For example, a sinusoid scribed on the surface of the cylinder as a lens shape, yielded a very pincushioned display. The family of curves on the cylinder that has proven to be convenient to work with is made up of concatenations of straight line segments tangent to segments of circular arcs. Such a typical curve is shown in figure 4. A member of this family of curves is described by a list of triads (A,D,R), where A is the angle a line segment of length D makes with the axis of the cylinder (Z-axis), and R is the radius of the circular segment to which it is tangent. These curves have continuous first derivatives and are easy to use.

Evaluating the C_{mk} by Contour Integration

Referring to figure 5a, one quadrant of the surface of a cylinder is divided into 3 regions using contours C_1 and C_2. These regions are assigned voltages V_1, V_2, and V_3 respectively. It is clear that by using Green's Theorem in the plane, the surface integrals (3) can be reduced to contour integrals of the form:

$$M^i_{mk} = \frac{2}{\hat{k}R} \int_{C_i} \sin(\hat{k}z) \cos(2\hat{m}y) \frac{dy}{ds} ds \quad i = 1,2 \tag{4}$$

where $y = R\theta$, $\hat{m} = m/R$ and k≠ 0. The special case k = 0 is treated similarly. This results in:

$$C_{mk} = \frac{1}{N_{mk}} \left[(V_1 - V_2) M^1_{mk} + (V_2 - V_3) M^2_{mk} \right]. \tag{5}$$

The only contribution to the integrals is along C_1 and C_2. These integrations can be performed efficiently by the computer for this general family of contours.

An independent feature on the profile is needed to control horizontal linearity. The four lobes on the profile shown in figure 2 function in this capacity. That the vertical scan crosses through the axis in this plane is most advantageous.

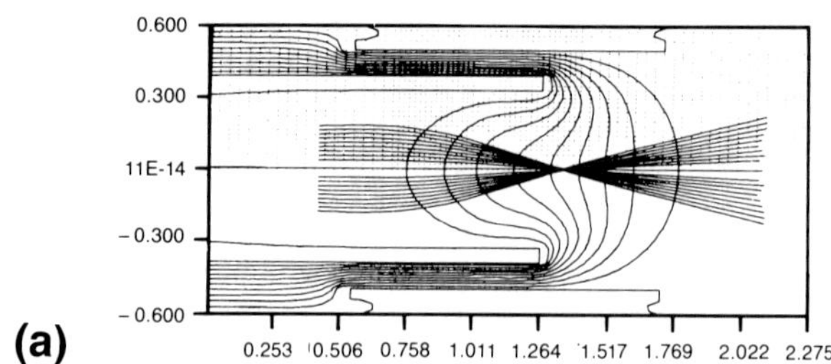

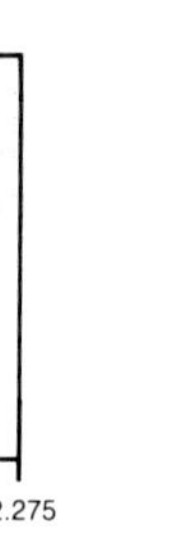

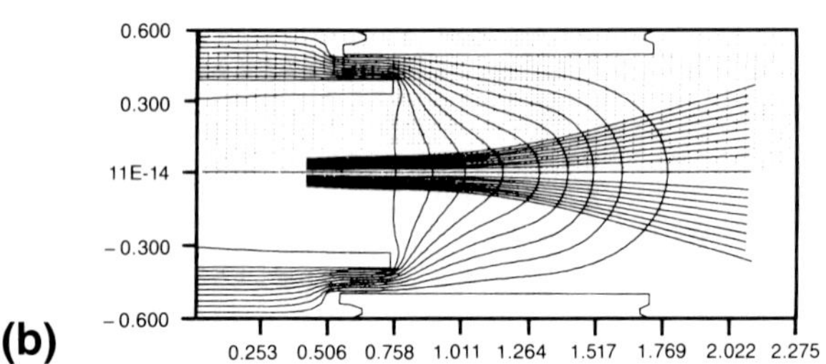

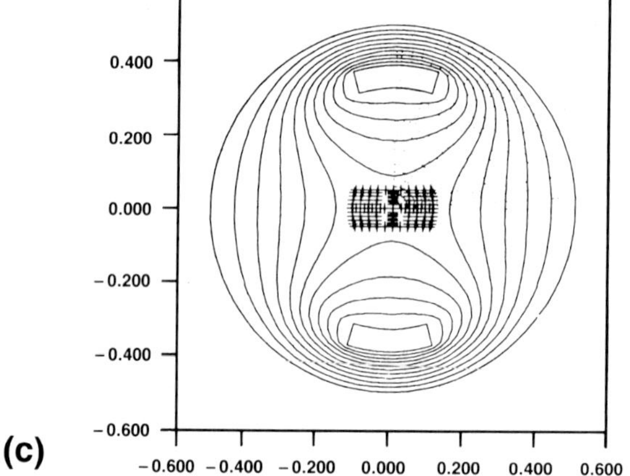

Figure 3. a) Vertical section of a Type II Klemperer lens showing equipotentials and a family of trajectories in the YZ-plane. The exit aperture lens is not included in the calculation. b) Horizontal section of the same lens showing the typical quadrupole characteristic. c) A cross section normal to the Z-axis showing development of the scan. When followed to the CRT face, this scan results in the display shown in Figure 8.

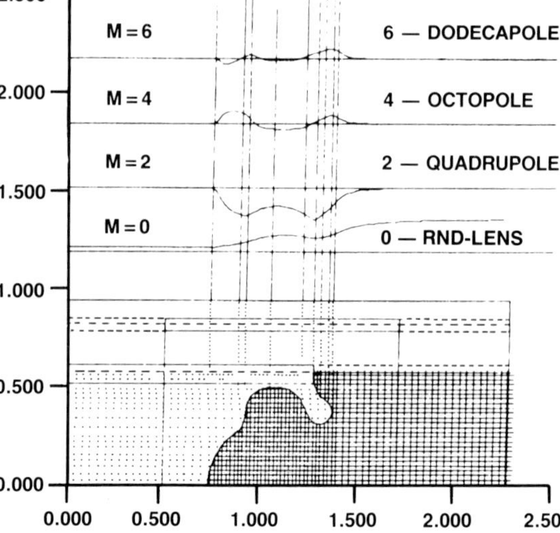

Figure 4. Computer graphics showing the contoured inner electrode, relaxation mesh, and moment functions up to cos(6θ).

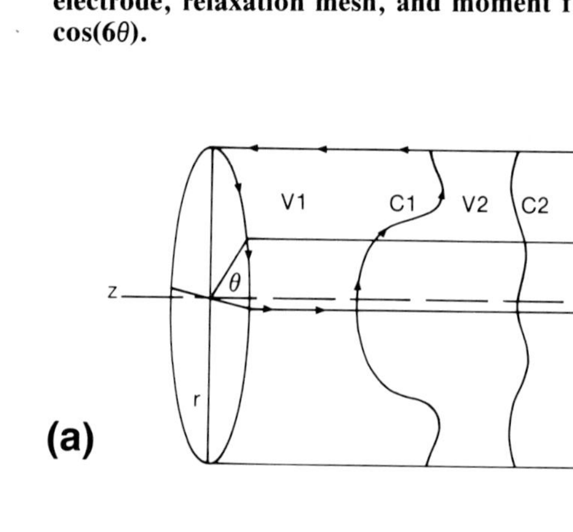

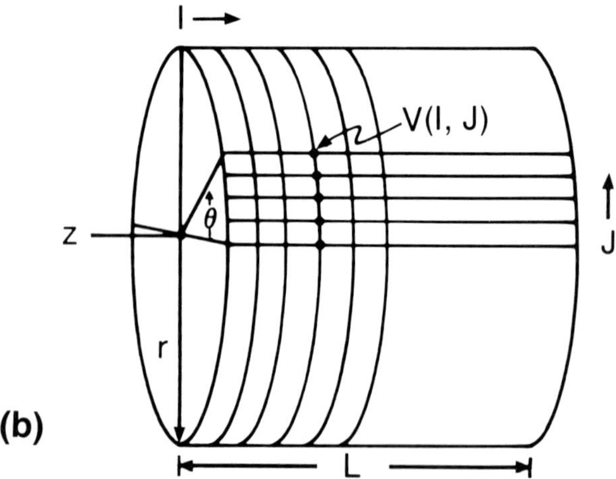

Figure 5. a) Cylindrical surface divided into regions V1, V2 and V3 by contours C1 and C2. A typical contour path of integration, indicated by the arrows, is used to calculate Fourier coefficients. b) Discrete potentials V(I,J) are passed to Fast Fourier Transforms to produce Fourier coefficients.

Evaluating C_{mk} with Fast-Fourier Transforms

A second method of evaluating the C_{mk} that can be used in combination with relaxation methods has proved to be very useful. If $V(z,\theta)$ denotes the boundary potentials on the cylinder, then:

$$A_m(z) = \int V(z,\theta)\cos(m\theta)\,d\theta \qquad (6)$$

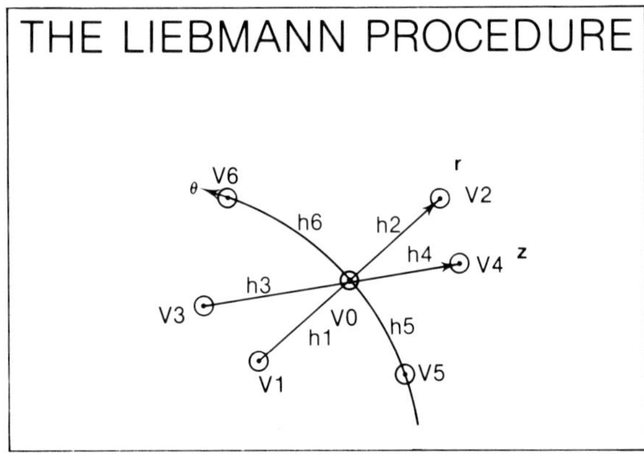

Figure 6. Relaxation formula (8) is applied to a cylindrical mesh. The central node 0 is surrounded by six nearest neighbors 1,2-6 in the r, θ, z coordinate mesh.

is the m-th Moment Function. The Fourier coefficients are then:

$$C_{mk} = \frac{1}{N_{mk}} \int A_{2m}(z) \cos(\hat{k}z)\, dz. \qquad (7)$$

Let the continuous boundary values $V(z,\theta)$ be approximated by the discrete values V(I,J), as indicated in figure 5b. For each fixed z_i, the integrals (6) may be evaluated by a call to a Fast-Fourier Transform. After building the discrete functions $A_m(i)$, they are used to evaluate the integrals (7), by further calls to an FFT, for each m. This process for an array of size 100×50, requires about 1 CPU second on a CYBER 170/760.

As indicated above, if the potentials V(I,J), have been determined by any method, relaxation for instance, then the potentials and gradient of the field inside this radius can be determined by evaluating the Fourier series, or respectively, the differentiated series.

Figure 4 illustrates a typical lens profile of Type II, together with the moment functions $A_m(z)$, for $m = 0,2,4,6$. It can be shown that even the multipole corresponding to $m = 8$ has a small but detectable effect on the display.

Analysis of Klemperer Lenses of Type II

It is obvious that the Fourier methods alone will not suffice to determine the fields in the Type II lens. The method employed is relaxation in a three-dimensional cylindrical mesh, the potential matrix V(I,J,K) representing one quadrant of the field. A typical potential matrix may be of dimension V(88,25,35), or 77,000 mesh points. The method depends very heavily on having the electrodes lie in coordinate surfaces. The relaxation process proceeds along coordinate cylinders beginning at the center axis. Boundary analysis also follows the same sequence. The computer graph displayed in figure 4 shows that the mesh density can be varied along each axis as most convenient. The crosshatch shows the mesh distribution as well as giving a convincing demonstration that the program has determined the appropriate relaxation weights. This diagnostic plot shows that for each point sufficiently near the contour, the program has calculated the correct displacements required to perform the relaxation at this point.

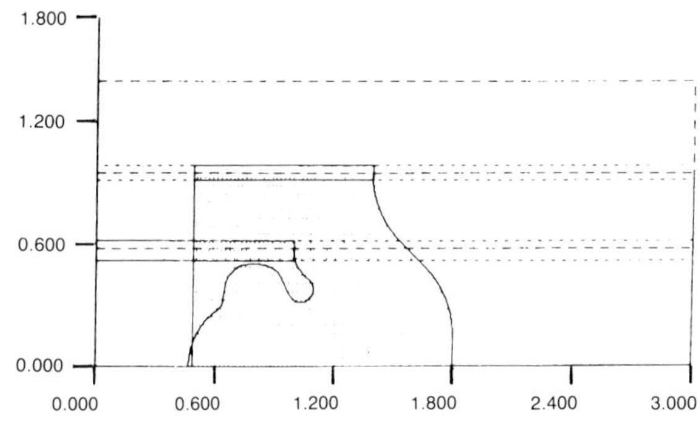

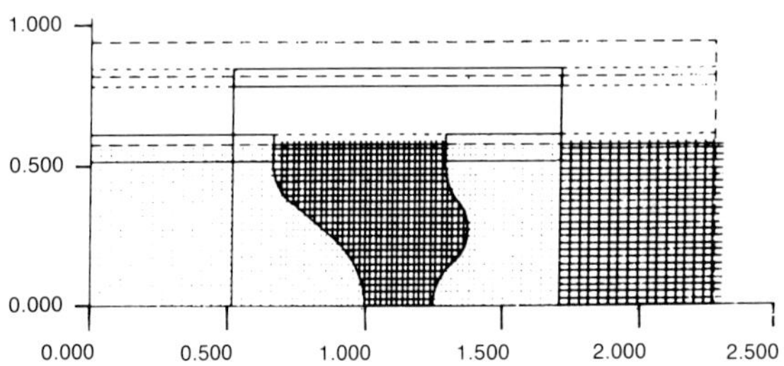

Figure 7. Computer graphics showing software package flexibility. Crosshatch diagnostic shows correct boundary analysis and requires only one CPU second of computer time.

Using the notation indicated in figure 6, the relaxation formula in use becomes:

$$V_0 = \frac{\sum_{i=1}^{6} w_i V_i}{\sum_{i=1}^{6} w_i} \qquad (8)$$

where the weights, $w_1, w_2, \ldots, w_6$, are given by:

(r)

$$w_1 = \frac{1}{h_1(h_1 + h_2)} (1 - h_2/2r) \qquad ,(r \neq 0)$$

$$w_2 = \frac{1}{h_2(h_1 + h_2)} (1 - h_1/2r) \qquad ,(r \neq 0)$$

(z)

$$w_3 = \frac{1}{h_3(h_3 + h_4)}$$

$$w_4 = \frac{1}{h_4(h_3 + h_4)}$$

(θ)

$$w_5 = \frac{1}{h_5(h_5 + h_6)}$$

$$w_6 = \frac{1}{h_6(h_5 + h_6)}$$

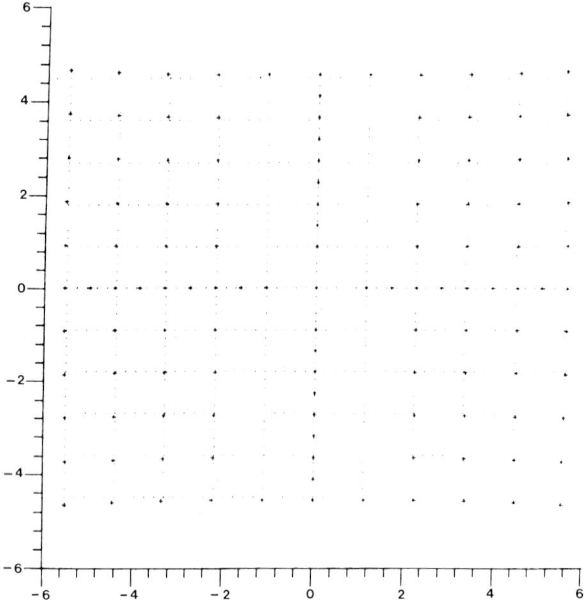

Figure 8. Calculated raster (small +) overlayed by an *ideal* raster (small dot array) is employed to determine lens distortions in the MSE lens. Some pincushion distortion is evident since the exit aperture effect was not included in the calculation.

A modified formula must be used when $r = 0$. Southwell[17] provides the classic reference for relaxation methods in two dimensions.

The process of determining the displacements h_1, h_2, ... h_6, is the heart of the computer program and has been fully automated requiring no operator intervention. This fact is responsible for the great ease of calculating fields using this technique.

Analysis of Boundary Conditions

Once contours C_1 and C_2 (see figure 5a) have been defined and the mesh distribution specified, the program builds electrode *templates*. A template is an array, NCODE(I,J), containing coded information pertaining to the mesh points which lie in the mesh cylinders falling within the diameter or an electrode. The coded information instructs the relaxation process denoting whether a given point needs *special* or *regular* relaxation weights and whether it is inside or outside the electrode. The program uses an efficient procedure to decide inside-outside issues, for instance to decide if a given mesh node is in region 1 or not. Use of templates makes it possible to store only small tables of *special* relaxation weights; the next special weight to be required always being the next one in the list. Features that make significant reduction in memory requirements are of fundamental importance in the design of a three-dimensional field program.

As shown in figure 7, shaped electrodes can be formed on two concentric diameters. The boundary analysis procedure is very efficient and requires only 1 CPU second on a CYBER 170/760. The short time required for these calculations makes the SETUP procedure for a lens very interactive and the designer can display numerous diagnostic plots and all internally generated quantities at the graphics terminal. The diagnostic plots in figures 4 and 7 are driven by the same subroutine that drives the relaxation process, thus demonstrating that the procedure is performing correctly.

The SETUP program *primes* the relaxation process so that the only calculation that has to be performed during the relaxation stage is the simple arithmetic in equation (8). Since relaxation may take from 15 minutes to an hour of CPU time on a CYBER 170/760, the relaxation process is done in batch mode and not in prime time, to reduce costs. Memory requirements to calculate the Type II lens is 270K octal words, of which 220K words is the large potential matrix. The program is so arranged that during relaxation the potential matrix, the templates and required weight tables are all in high-speed memory together with a minimal amount of FORTRAN code to drive the process. The program does not have to use the disk and hence is not swapped out frequently.

It is expected that if the Fourier series were used to propagate the potentials from the inner electrode surface to the center axis periodically as the relaxation continues, the convergence time could be reduced by 40 to 50 percent. This process has been partially implemented but is as yet untested.

Lens Distortion Analysis

Since the Klemperer lens contains significant multipoles up to $\cos(8\theta)$, it is clear that 3rd order or even 5th order aberration theory will not suffice to model the geometry distortions of these lenses. The principal method employed to study these distortions is to perform the numerical integration of the equations of motion for a rectilinear family of rays emanating from the deflection centers. Once the deflection centers for the gun are specified, the sequencing of the trajectory calculations to generate the display in figure 8, is fully automated. The rays nearest the center axis on both the X and Y axes are used to define the linear central region from which an *ideal* undistorted raster is generated. In figure 8, the ideal raster is represented by the array of small dots. The calculated raster, determined by the landing positions of the rectilinear family, is then overlaid onto the *ideal* raster. These landing positions are denoted by the heavier crosses. Both geometry and non-linearity distortions can then be seen at a glance. Defocus also can be calculated automatically. This method takes much of the labor out of lens distortion analysis.

The gradients of the field are generated either by a somewhat involved interpolation process in the cylindrical mesh or by employing the Fourier-Bessel series. Since the interpolation near the center axis may not be sufficiently well behaved, the designer can specify a diameter inside of which the program will calculate using the series. Use of the series also allows deeper insight into the effects of various multipoles on the display. Multipoles to be used in the calculation can be specified. Figure 9a shows the dot array resulting from use of only the round-lens and quadrupole terms in the series. When the octopole term is included, the dot array in figure 9b is displayed. This technique together with the moment functions shown in figure 4 provide a way to see what various regions on the profile are contributing to the display.

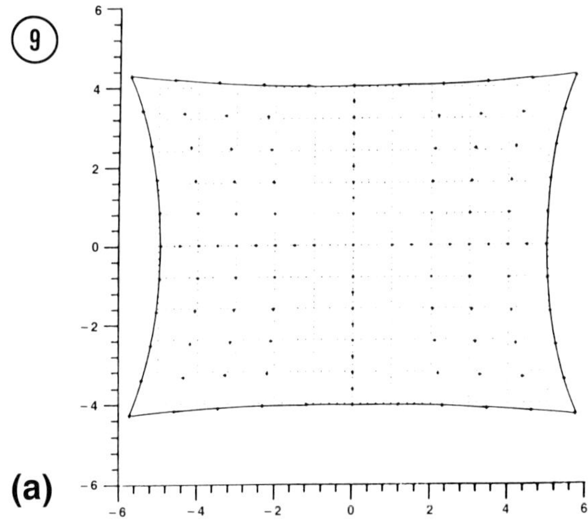

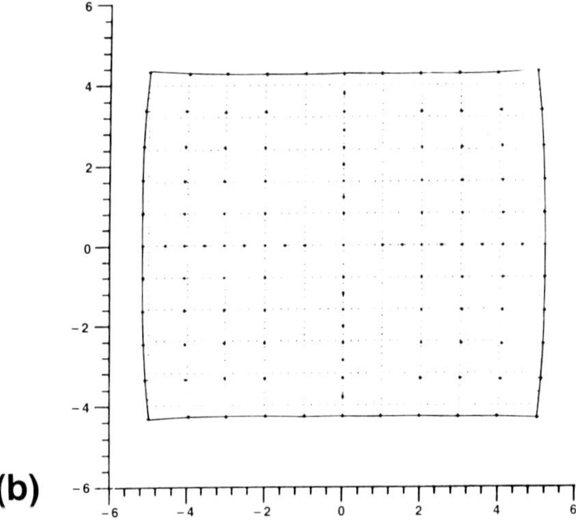

Figure 9. a) Calculated display using only round lens and quadrupole multipoles. b) Calculated display adding the octopole term to the raster calculation.

Lens Performance

Figure 8 shows the quality of the MSE lens field before correction by the exit-aperture lens. The exit-aperture lens shown in figure 1c is a weak lens which removes the residual pincushion in the horizontal lines. This lens was designed empirically. Computer code has been written to model the exit aperture lenses but it remains untested at this time. Figures 8 and 16c illustrate the effect of the *slot-lens* whose function is to correct vertical linearity. Vertical linearity before correction by the slot-lens is shown in figure 10b as a function of the ratio between cathode potential and screen potential. Displayed in figure 10a, calculated focal lengths agree well within the range of experimental error with measured data. See reference [3,5] for details of the MSE quadrupole system performance.

The slot-lens is one example of what is called a *wafer-lens*. The description of a general software package to analyze such structures follows.

Wafer-Lens Analysis

Implementation of quadrupoles Q_1, Q_2, Q_3 and the *slot-lens* in the Quadrupole CRT take the form of sequences of flat wafer electrodes. These wafers contain various apertures with two planes of symmetry. Lenses formed in this manner can be accurately aligned and are inexpensive. Early modeling of quadrupoles was done using the transformations noted by Martin and Deschamps[12], for *ideal* quadrupoles. These quadrupole models have limited usage and cannot be used to study aberrations nor to accurately relate structural dimensions to the required operating voltages. Considerable octopole moment is intentionally introduced into the Q_3 lens to compensate for a defocusing effect in the MSE lens. This defocusing effect cannot be treated by any known approximate model. On the other hand, the action of the slot-lens can be treated approximately by assuming the slots are of infinite extent in the horizontal direction. Field calculation is then reduced to a two-dimensional problem and can be calculated by existing programs. However, the finite width of the slot apertures in the XZ-plane produces significant unwanted lens effects in the horizontal plane. These effects can only be determined by a three-dimensional field program.

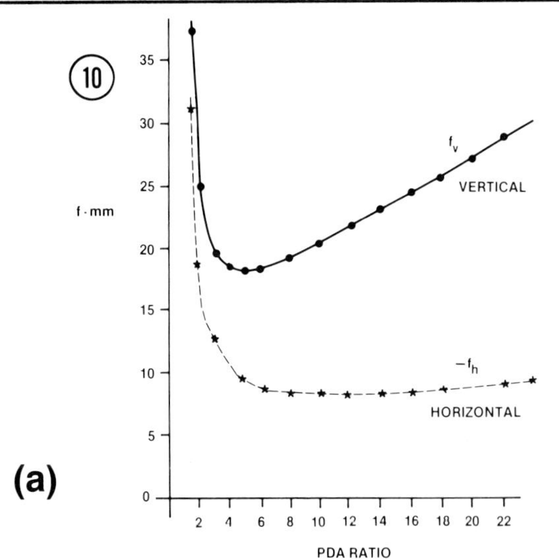

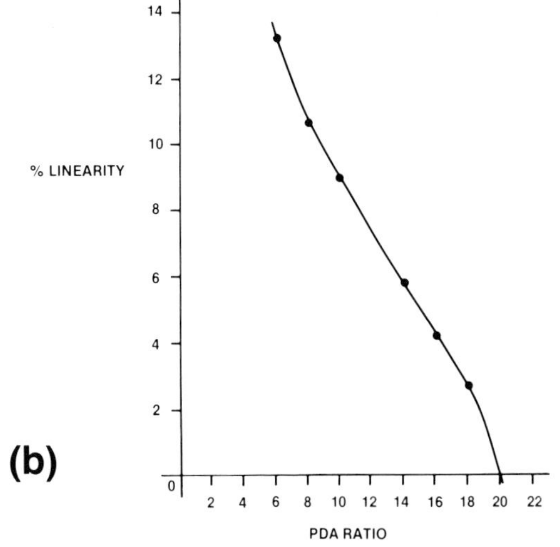

Figure 10. a) Horizontal and vertical focal lengths are shown as a function of PDA ratio (ratio of cathode to screen potential). b) MSE vertical lens non-linearity with slot lens turned off.

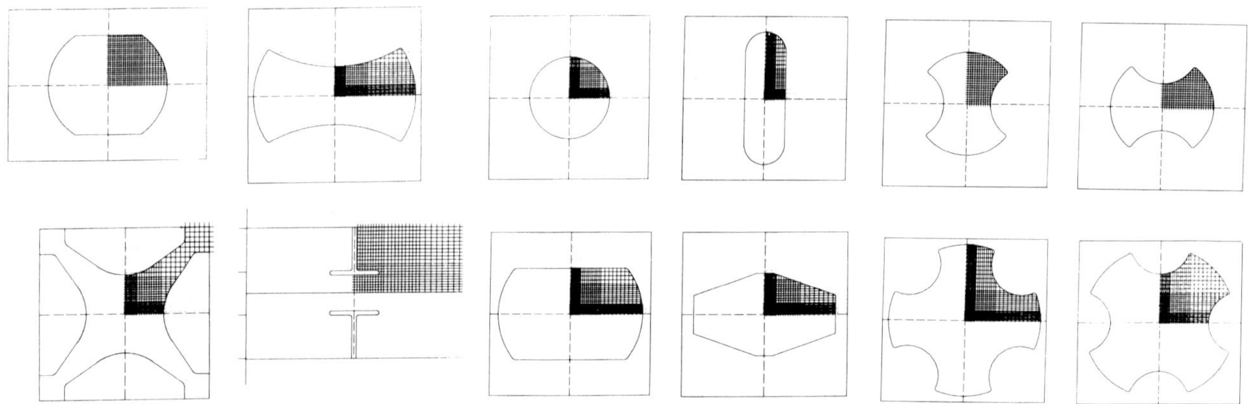

Figure 11. Various aperture types that can be generated with the wafer software program.

After developing a program to study Type II Klemperer lenses, it was realized that the same software techniques could be used to analyze a large class of wafer-lenses. Such a program has been developed at Tektronix and is essentially complete. Typical apertures that can be treated by the program are illustrated in figure 11. These wafer electrodes are not assumed to be thin structures and can even have more than one voltage assigned as in the *solid-quadrupole* discussed below. It is also possible to use the program to study a class of segmented-plate electrostatic deflectors by forcing a ground plane in the XZ-plane. The fringe fields and interaction with the shields can be investigated. This discussion, however, is restricted to three examples: the Q_3 lens shown in figures 12–15, the slot-lens of figure 16, and the *solid-quadrupole* pair of figure 17.

General Program Features

Most of the apertures in figure 11 can be described by three to five parameters. Besides the *general* element, there are 13 aperture types that can be defined very simply by a minimal set of parameters. The *general* element allows the CRT designer to create many complex aperture shapes. These apertures are again formed from concatenations of straight line segments, tangent to segments of circular arcs.

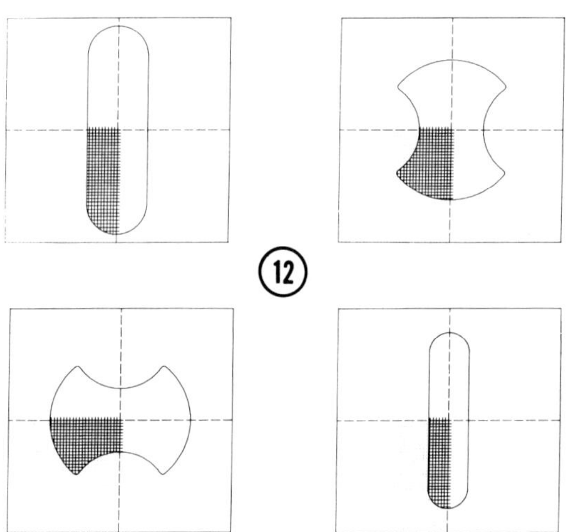

Figure 12. Menu of apertures used in the Q3 lens.

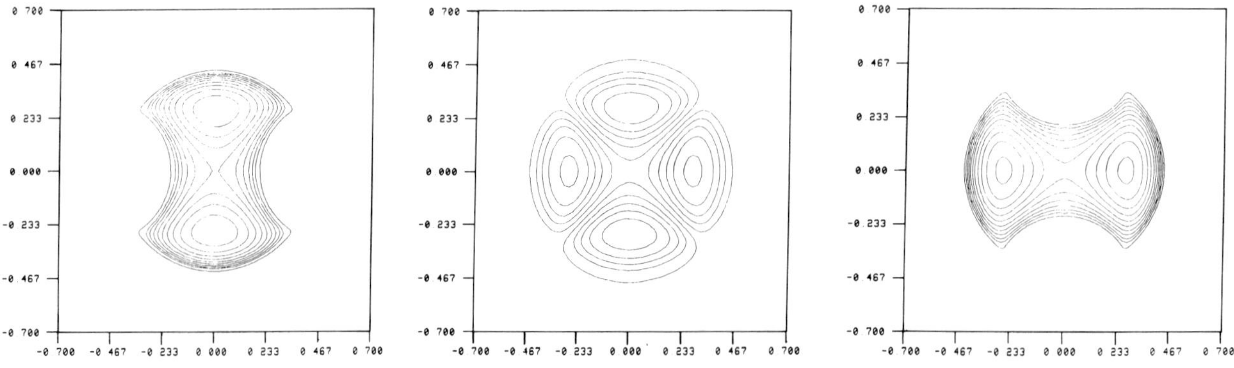

Figure 13. Typical equipotential plots in various cross sections of the Q3 lens. Calculations were performed using relaxation in a Cartesian coordinate mesh.

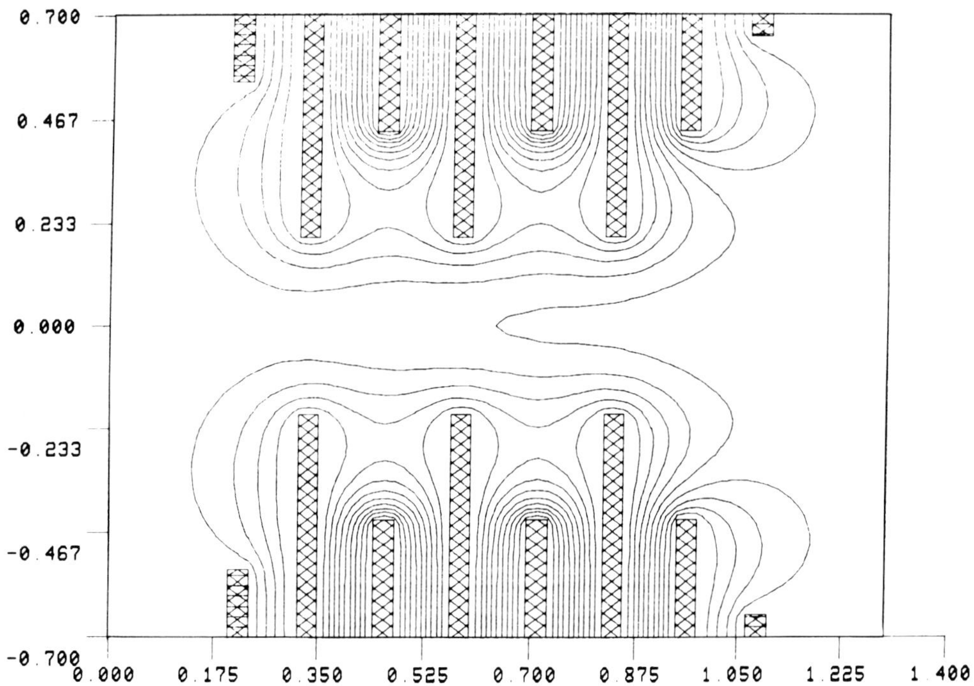

Figure 14. Equipotentials are displayed in a vertical cross section of the Q3 lens.

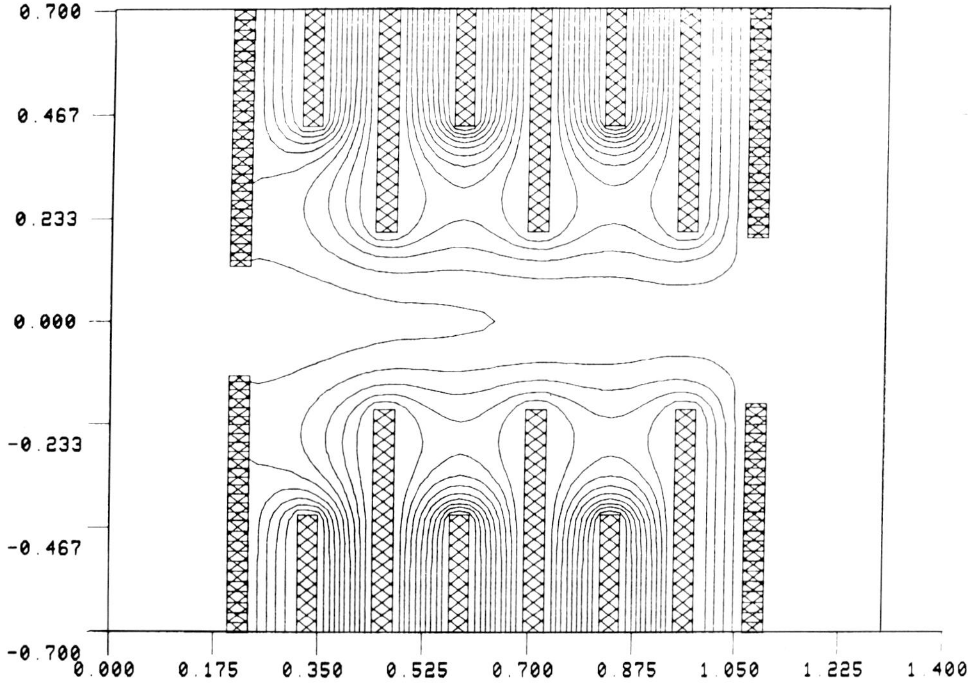

Figure 15. Horizontal cross section of the Q3 lens displaying equipotentials.

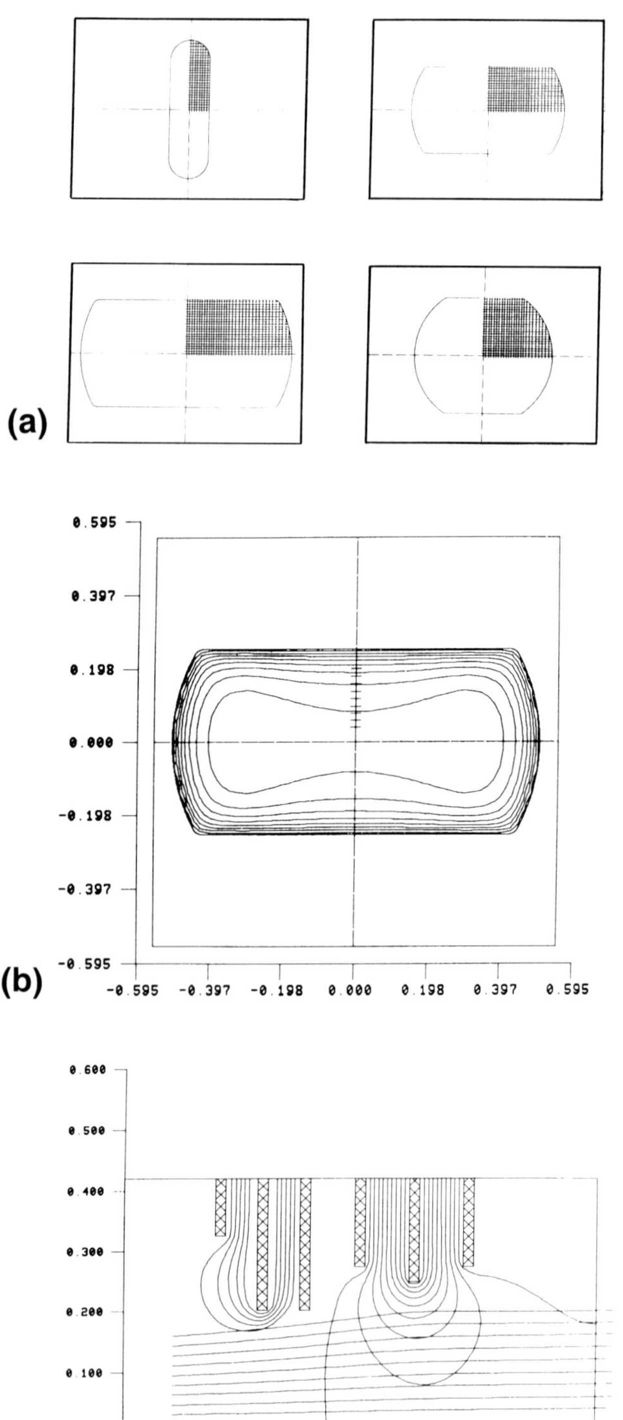

Figure 16. a) Menu of apertures for the slot lens. b) Equipotentials in the plane of the slot wafer (second from right in 16c). Crosses show where the trajectories pierce the plane. c) Vertical cross section of the slot lens showing the action of the field on the vertical scan as it passes the slot wafer.

Three special elements exist: vacuum space, Neumann boundary, and a voltage plane. For a given problem the designer defines a *menu* of apertures from which elements can be selected to form a lens system. The lens configuration is defined simply by specifying a sequence of indices which point to the desired member of the *menu*. The same aperture can be used repeatedly with different voltages. For each element, thickness, voltages and number of Z-planes is specified. This information essentially defines a lens configuration. Once the XY-mesh spacing has been given the program analyzes boundary conditions and computes all the required relaxation weights. This happens automatically without operator intervention and provides many diagnostic plots, if desired. The designer might for instance wish to insert an octopole element into a quadrupole to compensate for unwanted spot distortions in the system.

The Q3 Lens

Figure 12 shows the *menu* of apertures used for the Q3 lens. This example is actually a two times scale up of the lens used in the quadrupole MSE gun. Figure 13 displays the equipotential contours in typical apertures and in a plane midway between them. Figures 14 and 15, respectively, illustrate the YZ and XZ cross-sections of the Q_3 lens. Fields are calculated by relaxation in a three-dimensional, Cartesian coordinate mesh. It was somewhat surprising to find that the relaxation process converged in approximately 1/5 to 1/7 the time for the typical MSE lens field calculations. The relaxation time for the Q_3 lens was 359 CPU seconds on a CYBER 170/760. One quadrant of the field was represented by a matrix of size V(35,35,77). The maximum residual dropped to 0.00005 volts in 121 iterations through the matrix.

Useful Techniques

The Schwarz-Alternating-Procedure (SAP)[16] can be very useful for solving complex boundary value problems. A form of this procedure is used at Tektronix to calculate the fields in *wafer-lenses*. Slightly more than half of the total field is kept in the high speed memory of the computer at any given time. The remainder resides on the disk. A communicating overlap region is mapped back and forth in memory as the two segments of the field are swapped in and out of high speed memory. For the Q_3 lens, for instance, the total field had 77 Z-planes. Segment 1 contained 45 Z-planes. The overlap region consisted of 13 Z-planes. Hence segment two also contained $(77-45)+13=45$ mesh planes. Each segment was relaxed five times before being swapped out again. The overlap region is relaxed twice as often as the rest of the field.

Voltage Scaling

Voltage scaling has been an often used procedure. In calculating three-dimensional fields this process becomes imperative since the relaxation times are fairly long. At this time the Tektronix program is set up to handle voltage scaling for three independent voltages. If two independent solutions of the fields have been calculated, then any other set of voltages can be obtained by linear super-position. Since voltage scaling requires only about one to two CPU seconds on a CYBER 170/760, interactive voltage changes can be made on the lens followed by immediate display of the new equipotentials.

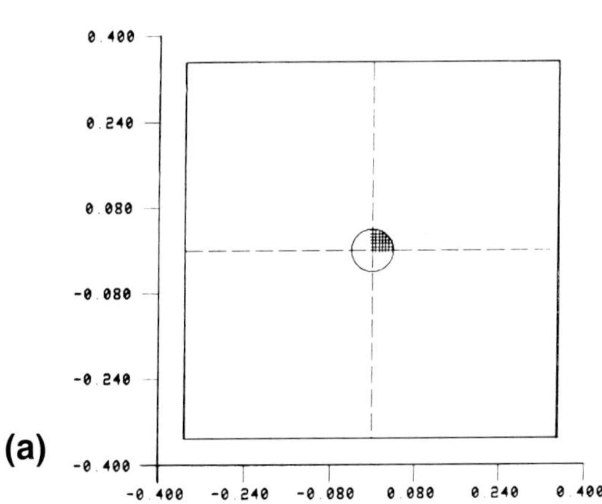

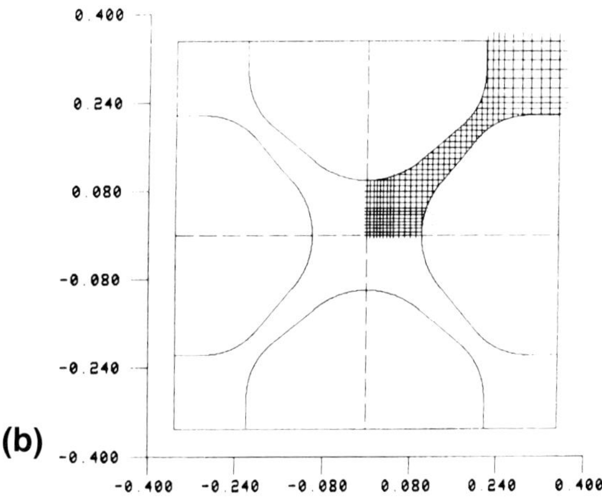

Figure 17. Solid quadrupole pair. a) Isolation shield. b) Wafer cross section. c) Vertical cross section and a pair of trajectories. d) Equipotentials in cross section of the lens.

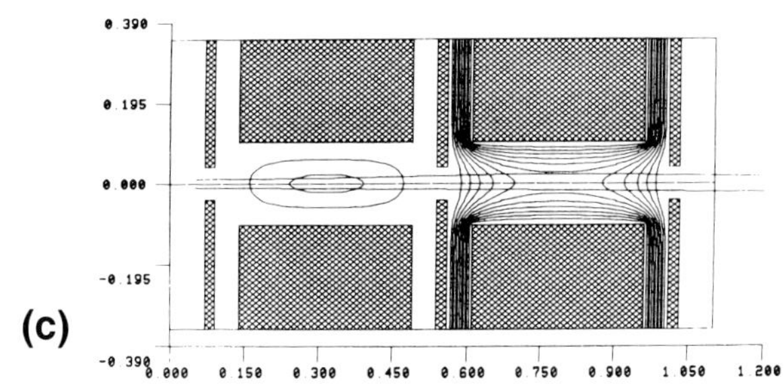

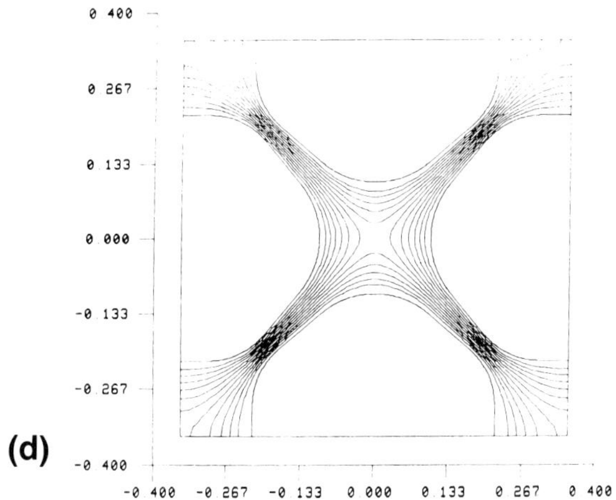

Initializing With Old Potentials

Calculation times can drop to about one third by initializing the relaxation using the potentials from a similar investigation. This is generally the case if the geometry of the lens has changed only slightly or a spacing has changed. If the lens design is progressing incrementally, calculations can be fairly inexpensive.

The Slot Lens

In figure 16c the wafer second to the last is called the *slot* wafer. It operates at a potential of about 1 kV below its neighbors. Non-linear distribution of the equipotentials near the wafer are shown in two aspects in figures 16b and 16c. The primary effect of the lens is to change the slope of the rays in the vertical scan as they pass the slot wafer. The net result is to improve the vertical linearity of the system. The program can display graphically the positions of trajectories as they penetrate any Z-plane (see figure 16b). For this illustration the slot field was represented by a potential matrix of size V(40,35,76) and was constructed with field segments of length 45 and 44, and an overlap of 13 planes. The relaxation process required 17.5 minutes of CPU time on the CYBER to reduce the maximum residual to 0.00013 volts. A total of 500 passes through the matrix was required.

A Solid Quadrupole Pair

Figure 17 is an example using *thick* electrodes which also employ two voltages for a single element. Two trajectories are displayed in figure 17c. This field was calculated using a potential matrix of size V(35,35,79). Both field segments were of length 45 and an overlap region consisting of 11 Z-planes. Total iterations was 90 to bring the maximum residual down to 0.00003 volts. 233 CPU seconds were required for the calculations.

Conclusions

For the class of lenses discussed herein the developmental setup effort by any other known method would be very labourious, if possible at all. On the other hand, some of the examples above can be setup in less than one hour using the software programs and techniques presented here. The software methods used in the analysis of wafer-lenses could be extended to allow for non-symmetric apertures or wafer displacements normal to the Z-axis. Clearly, all four quadrants of the fields would have to be represented and relaxation times would be quite long. Misalignments of the MSE structures could best be handled by the Boundary Element Method. It may well be possible to automate setting up the BEM procedure for a broad class of structures. The

BEM method could more easily handle general *tilts* and misalignments. Application of the Fast Fourier Transform can be easily used to decompose the central field into its multipole moments, thus giving more insight into the fields.

References

[1] Coons SA, (1974) "Surface patches and B-spline curves in computer aided geometric design," Barnhill and Riesenfeld, Eds. *Academic Press.*

[2] Franzen N, (1983) "A quadrupole scan expansion accelerating lens system for oscilloscopes," *Society for Information Display Technical Digest,* Vol. 14, pp. 120–121, *Lewis Winner,* Coral Gables, Florida.

[3] Franzen N and Janko B, (1983) "Meshless scan expansion CRT improves trace quality," *Electronic Imaging,* Vol. 2-7, pp. 48–53, *Morgan-Grampian,* Boston Mass.

[4] Harrington R, (1967) "Matrix methods for field problems," *Proc. of IEEE,* Vol. 55, No. 2, pp. 136–149.

[5] Hawken K, Franzen N and Sonneborn J, (1983) "A novel high-performance CRT with meshless scan expansion," *Proc. of 3rd International Display Research Conf.,* Kobe Japan, pp. 136–139.

[6] Hosokoshi K, Ashizaki S and Suzuki H, (1983) "Improved OLF in-line gun system," *Proc. of 3rd International Display Research Conf.,* Kobe Japan, pp. 272–275.

[7] Jackson D, (1975) Classical Electrodynamics, 2nd Ed., *Wiley,* pp. 102.

[8] Janko B, Franzen N, Sonneborn J, (1983) "CRT architecture with meshless scan expansion," *Society for Information Display Digest,* Vol. 14, pp. 118–119, *Lewis Winner,* Coral Gables, Florida.

[9] Klemperer O, (1953) Electron Optics, pp.297–298, *Cambridge Univ. Press.,* 2nd Ed.

[10] Lean M and Wexler A, (1982) "Accurate field computation with the boundary element method," *IEEE Trans. on Mag,* Vol. MAG-18, pp. 331-335.

[11] MacGregor DM, (1983) "Computer-aided design of color picture tubes with a three-dimensional model,"*IEEE Conf. Proceedings of International Conference on Consumer Electronics,* Des Plaines, Ill., pp.130.

[12] Martin A and Deschamps J, (1971) "A short-length rectangular oscilloscope tube with high deflection sensitivity by an original technique," *Proc. Society for Information Display,* Vol. 12/1, pp. 16–21.

[13] Munro E and Chu HC, (1982) "Numerical analysis of electron beam lithography systems," Part I, *Optik 60, No. 4* and Part II, *Optik 61, No. 1, Wissenschaftliche Verlagsgesellshaft mbH,* Stuttgart.

[14] Odenthal C, (1977) "A box-shaped scan-expansion lens for an oscilloscope CRT," *Society for Information Display Digest,* Lewis Winner, NY, pp. 134–135.

[15] Ritz E, (1979) "Recent Advances in Electron Beam Deflection," *Advances in Electronics and Electron Physics,* Vol. 49, *Academic Press,* pp. 299-356.

[16] Schaefer C, (1983) "The application of the alternating procedure, by H.A. Schwarz, for computing three-dimensional electrostatic fields in electron-optical devices with complicated boundaries," *Optik 65,* No. 4, pp. 347-359.

[17] Southwell R, (1940) Relaxation Methods in Engineering Science, *Oxford U. Press.*

[18] Yildir Y and Wexler A, (1983) "MANDEP-A FEM/BEM data preparation package," *IEEE Trans. on Magnetics,* Vol. MAG-19, No. 6, pp.2562.

Acknowledgement

The author wishes to give special thanks to Doctor Friedrich Lenz, Professor, University of Tubingen, West Germany, for the excellent courses in Electron Optics which he delivered at Tektronix over the years. His presence is always a great inspiration. Many thanks also to the members of the Display Device Engineering Group at Tektronix for their support of this work.

Electron Optical Systems (pp. 127-135)
SEM Inc., AMF O'Hare (Chicago), IL 60666-0507, U.S.A.

0-931288-34-7/84$1.00+.05

OPTIMIZATION OF STROBOSCOPIC ELECTRON DEFLECTION SYSTEMS

H. Rose* and J. Zach†

Center for Laboratories and Research
New York State Department of Health
Albany, NY 12201
and
†Institut für Angewandte Physik
Technische Hochschule Darmstadt
D-6100 Darmstadt, FRG

Abstract

Highly corrected electron-beam blanking systems (EBBS) are required for analyzing fast periodic processes at submicron spatial resolutions by stroboscopic methods. The deleterious degradation of the probe during the blanking operation can be avoided by a new straight-vision deflection system. Chopping of the beam is performed within this system by deflecting it across a knife edge. Calculations demonstrate that this system should be able to generate almost rectangular beam pulses with rise times of a few picoseconds and spot sizes smaller than 0.5 μm. The proposed EBBS consists of two wedge-shaped plate capacitors located symmetrically about the midplane of a rotationally symmetric double lens. Time-of-flight effects are largely compensated by driving the two capacitors as a traveling-wave structure to yield resonance deflection.

KEY WORDS: Electron-beam blanking systems, electron stroboscopy, resonance deflection, aberration correction, oblique plate capacitor, time-of-flight effects, temporal resolution, spatial resolution.

*Address for correspondence:
H. Rose, Inst. fur Angew. Physik
T.H. Darmstadt, D-6100 Darmstadt, FRG
Phone No.: 6151-162481

Introduction

Electron-beam blanking systems (EBBS) are increasingly used as basic devices in electron lithography and electron-beam testing systems and for measurement of time-dependent, spatially localized processes in thin specimens. For each application the EBBS must meet different requirements with respect to pulse width, repetition rate, beam current, and accelerating voltage. Pulsed electron beams can be generated by (a) modulation of the Wehnelt voltage at the electron gun [7,12,15], (b) rotationally symmetric cavities [9] or filter lenses [10], or (c) beam deflection across a chopping aperture or a knife edge [3,4,6,8,11]. Owing to its simplicity the last approach is widely employed in stroboscopically operating scanning electron microscopes (SEM). These instruments are increasingly used for functional testing of integrated circuits by voltage-contrast measurements. This method enables one to visualize periodically changing potentials, either for a constant phase position (by scanning the beam over the surface of the specimen) or as a function of phase (at a fixed position in the circuit) [1,2].

High time resolution necessitates electron pulses with very short rise times. Furthermore, to prevent falsification of the signal, both the position and the size of the scanning spot at the specimen should be as unaffected by the blanking operation as possible. Investigation of emitter-coupled logic chips and GaAs circuits in the near future, for example, will require almost rectangular pulses of widths between 50 ps and 5 ns and stable spot sizes smaller than 0.5 μm [6]. These requirements cannot be fulfilled with presently used deflection blanking systems consisting of a single capacitor. Therefore we shall investigate in this paper EBBS composed of more than one capacitor together with static magnetic round lenses. These lenses may belong to the SEM or may be incorporated separately into the EBBS. Furthermore we shall impose no restrictions on the geometry of the deflection system except that its fields can be considered as two-dimensional in the region of the electron beam. Magnetic deflection coils will not be considered because they are not suitable for obtaining very fast deflections.

Quasistatic Field Approximation

The propagation of electrons in fast time-varying deflection fields has received little attention in theoretical electron optics. However, the standard stationary treatment is no longer applicable when the fields change significantly during the passage of the electrons through an EBBS. In this case time-of-flight effects influence the electron path and must therefore be taken into account.

Let us restrict our considerations to signals whose dominant frequencies are small compared to c/L where c is the velocity of light and L is the length of the EBBS in the direction of the beam. With this assumption the electric fields can be considered as quasistatic. This approximation assumes that the magnetic vector potential $\vec{A}$ does not explicitly depend on the time variable t, so that $\partial\vec{A}/\partial t = 0$, which means that the effects of the magnetic fields produced by the time-varying electric fields are negligibly small. As a result both the electric field $\vec{E}$ and the magnetic flux density $\vec{B}$ outside the region of the conductors and currents can be obtained from scalar potentials $\phi = \phi(\vec{r},t)$ and $\psi = \psi(\vec{r})$ respectively:

$$\vec{E} = -\text{grad}\ \phi, \qquad \vec{B} = -\text{grad}\ \psi. \tag{1}$$

Assume further that the capacitors can be considered as infinitely extended in the y direction perpendicular to the straight optic axis, which is chosen as the z axis of the Cartesian coordinate system. Because the electric signal applied to the capacitors travels in the y direction and because the diameter of the electron beam is small compared to the wavelengths of the plane-wave components of the signal, the electric potential in the region of the beam can be written as:

$$\phi(\vec{r},t) = \Phi_D(x,z)g(t) + \Phi + \delta\Phi . \tag{2}$$

Here Φ is the constant axial potential at a large distance from the deflection system; $\delta\Phi$ is the deviation due to the energy spread of the emitted electrons; and:

$$\Phi_D(x,z) = \Delta + \Phi_1 x - \Delta'' x^2/2 - \Phi_1'' x^3/6 + \ldots \tag{3}$$

is the static potential of a two-dimensional capacitor. Dashes represent derivatives with respect to the z coordinate; $\Phi_1 = \Phi_1(z)$ is the strength of the dipole field, while $\Delta=\Delta(z)$ represents the deviation of the axial potential from the anode potential . This deviation vanishes only if the plates of the capacitors are arranged symmetrically about the plane $x = o$ and are excited antisymmetrically, so that one plate is at the potential ϕ_D=U and the other at ϕ_D = -U with respect to ground. A nonrealistic example of such a capacitor, consisting of two inclined plates, is shown in Fig. 1. The plates intersect at the line $x = o$, $z = z_0$ and enclose an angle 2θ. In the case of antisymmetric excitation the potential is

$$\begin{aligned}\Phi_D &= (U/\theta)\arctan[x/(z-z_0)] \approx \\ &(U/\theta)[x/(z-z_0) - x^3/3(z-z_0)^3 + ..]\end{aligned} \tag{4}$$

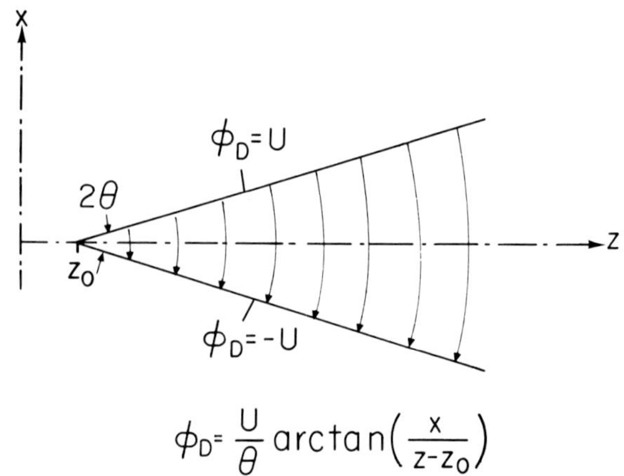

Fig. 1. Electric-field distribution in the region between two inclined half-planes at potentials ϕ_D= U and ϕ_D = -U.

The lines of constant electric field are circles about the origin $z = z_0$, $x = o$. If we neglect the fringing fields, relation (4) also describes approximately the electric field within a real capacitor with inclined plates, provided their extension is large compared to the maximum gap width.

In the expansion (3) the terms which comprise the axial potential and its derivatives represent the potential expansion of a cylinder lens. Hence any asymmetrically excited capacitor can be considered as a linear superposition of a cylinder lens and a deflection field whose potential is antisymmetric with respect to the midplane $x = o$. The time-varying cylinder lens acts like a buncher.

Upon introduction of the complex off-axial coordinates:

$$w = x + iy, \ \bar{w} = x - iy \tag{5}$$

the power series expansion of the scalar magnetic potential of the round lenses has the form:

$$\psi = \Psi - (w\bar{w}/4)\Psi'' + \ldots \tag{6}$$

where $-\Psi' = B_0$ is the magnetic flux density along the symmetry axis.

Equations of Motion

The equations for describing electron motion in time-dependent fields can be obtained either from the Lagrangian, as outlined by Sturrock [13], or by employing Newton's law together with the Lorenzian expression for the electromagnetic force acting on the electrons. Assuming $\vec{A} = 0$ with nonrelativistic electron velocities, we eventually arrive after some manipulations at the following equations:

$$m(w'/t')' = e\left\{2t'\frac{\partial\phi}{\partial\bar{w}} + 2i\frac{\partial\psi}{\partial\bar{w}} - iw'\frac{\partial\psi}{\partial z}\right\}, \tag{7}$$

$$E' = \frac{m}{2}\left(\frac{1+w'\bar{w}'}{t'^2}\right)' = e\left\{\frac{\partial\phi}{\partial z} + 2\ \mathrm{Re}\left(\bar{w}'\frac{\partial\phi}{\partial\bar{w}}\right)\right\}. \tag{8}$$

Here e, m, and $1/t' = v_z$ denote the charge, mass, and axial velocity component of the electron respectively. Its kinetic energy $E = E_z + E_w$ consists of an axial component:

$$E_z = mv_z^2/2 \quad (9)$$

and a transverse component:

$$E_w = mw'\bar{w}'/2t'^2. \quad (10)$$

The solution w = w(z), t = t(z), of the nonlinear coupled differential equations (7,8) is an implicit representation of the electron position as a function of time. The right-hand side of the temporal equation (8) is a total differential with respect to z only if the electric potential does not depend on time. Then the z integration can be performed readily to yield a constant total energy. This well-known result, however, no longer holds true in the case of a time-varying potential.

The system (7,8) of nonlinear differential equations can be solved approximately by means of well-established perturbation methods. For this purpose we write

$$t(z) = T(z) + \tau(z) , \quad (11)$$

where T denotes the time taken by an electron traveling along the axis when the deflection system is switched off. Accordingly τ is the time difference for an electron not moving along this straight axis to reach the same plane. In a zeroth-order approximation we derive from (8):

$$T' = (m/2e\Phi)^{\frac{1}{2}} = \text{const.} , \quad \tau^{(0)} = 0, \quad (12)$$

which corresponds to $E^{(0)} = E_z^{(0)} = E_o = e\Phi = mv_o^2/2$. Here v_o and E_o are the initial average velocity and the mean kinetic energy respectively of the electrons below the anode. Within the frame of our approximation procedure we consider $w, \bar{w}, w', \bar{w}', \tau, \Phi_1, \delta\Phi$ and Δ to be small quantities.

Dynamic Gaussian Dioptrics

The first-order equations are obtained from (7,8) by retaining only those terms which are linear in the expansion parameters. The resulting dynamic paraxial equations

$$w^{(1)''} - 2i\chi' w^{(1)'} - i\chi'' w^{(1)} = g(T)D(z) , \quad (13)$$

$$E^{(1)} = E_z^{(1)} = -m\tau^{(1)'}/T'^3 = e\int_{z_i}^{z}\Delta' g\,dz + e\delta\Phi \quad (14)$$

take account to a first approximation of time-of-flight effects, since they replace the local distributions $D(z) = \Phi_1/2\Phi$ and Δ of the deflection field by the effective fields $g(T)D$ and $g(T)\Delta$ respectively, as seen by an electron traveling along the axis with the initial axial velocity v_o. The angle:

$$\chi = (e/8m\Phi)^{\frac{1}{2}}\int_{z_i}^{z} B_o\,dz \quad (15)$$

describes the Larmor rotation of an electron starting from the initial plane z_i. Also:

$$T = (z - z_i)/v_o \quad (16)$$

is the time an electron needs to travel from this plane along the axis to plane z.

The first-order time equation (14) convincingly demonstrates that the so-called temporal aberrations [5] are closely connected with changes in the kinetic energy of the electrons. This energy deviation is caused by the energy spread of the emitted electrons and by the time-varying part of the electric potential. Expressing the time T in the time function by relation (16), we find from (14) that, within the frame of the first-order approximation, the kinetic energy lost or gained by the electron when passing through the asymmetrically driven capacitor is proportional to the convolution of the time function and the static axial field $-\Delta'$. To survey this energy change we assume a step pulse:

$$g(T) = \begin{cases} 0 \text{ for } T \le T_o = (z_o - z_i)/v_o \\ 1 \text{ for } T > T_o \end{cases} \quad (17)$$

This choice implies that the capacitor is switched on instantaneously when the electron is at plane z_o. With this relation the integration in (14) can be performed readily, yielding:

$$E^{(1)} = \begin{cases} 0 \text{ for } z \le z_o \\ e(\Delta - \Delta_o) \text{ for } z > z_o \end{cases} \quad (18)$$

where $\Delta_o = \Delta(z_o)$. In the opposite case where the capacitor is switched off, we find:

$$E^{(1)} = \begin{cases} e\Delta \text{ for } z \le z_o \\ e\Delta_o \text{ for } z > z_o . \end{cases} \quad (19)$$

The comparison shows that at a large distance from the capacitor, where $\Delta = o$, the electron has lost the kinetic energy $e\Delta_o$ in the former case, while it gains the same amount if the capacitor is off. This difference in behavior occurs because the kinetic energy of the electron remains unchanged when the potential jumps. In the former case the electron loses kinetic energy when it leaves the capacitor because the potential Δ decreases in the exit region slowly toward zero. When the capacitor is switched off abruptly, the electron retains the kinetic energy which it has gained by entering regions of steadily higher potential within the capacitor. In practice the kinetic energy is changed significantly only when the transit time of the electron through the blanking system is large compared to the rise time of the pulse during the passage. No such first-order energy change occurs when the capacitor is driven in such a way that the potential in the midplane remains equal to the constant anode potential.

It is convenient to express the Gaussian rays in the rotating coordinate system:

$$u = we^{-i\chi} . \quad (20)$$

Then the general solution of the paraxial path equation (13) has the form:

$$u^{(1)} = u_C u_\gamma + \omega_A u_\alpha + u_\delta . \quad (21)$$

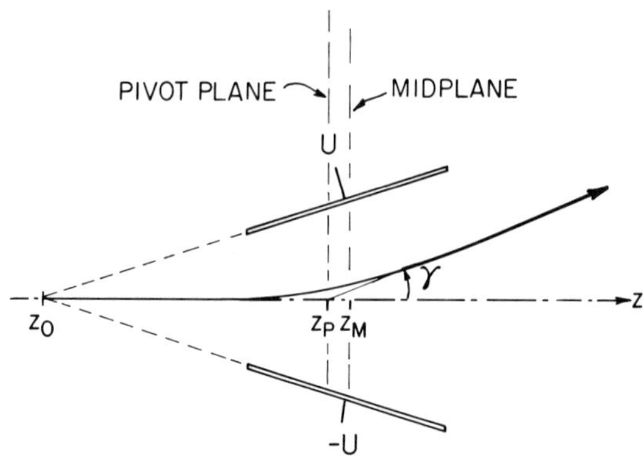

Fig. 2. Beam deflection by a wedge-shaped plate capacitor; γ denotes the deflection angle.

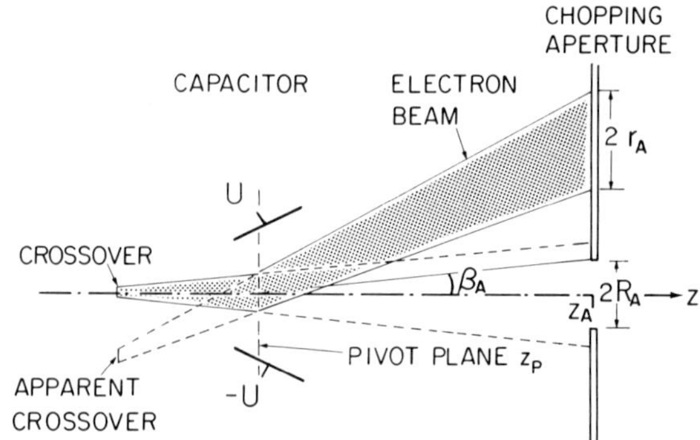

Fig. 3. Scheme of electron-beam blanking by a single plate capacitor placed between the crossover and the effective (final) chopping aperture.

The term

$$u_\delta = u_\alpha\int_{-\infty}^{z} gDe^{-i\chi}u_\gamma dz - u_\gamma\int_{-\infty}^{z} gDe^{-i\chi}u_\alpha dz \qquad (22)$$

represents the deflection of the beam. The real functions u_α and u_γ are linearly independent solutions of the homogeneous part (D = 0) of (13) which satisfy the boundary conditions:

$$u_\gamma(z_A) = u_{\gamma A} = u_\alpha(z_C) = 0, \qquad (23)$$
$$u_\gamma(z_C) = u_\alpha{}'(z_A) = 1 \quad .$$

Here z_C and z_A denote either the locations of the crossover and the final beam-limiting aperture respectively or the locations of their intermediate images; and $|u_C|$ and $u_{\alpha A}\ |\omega_A| = |u_A|$ are the radii of the crossover and of the final beam-limiting aperture at the corresponding planes.

To survey the action of a dipole field on the beam we evaluate (22) in the absence of overlapping round lenses and for a single capacitor with inclined plane plates, as shown in Fig. 2. In this case we have χ = o and g = 1. Using the expansion (4) for the deflection potential together with (3) and neglecting the fringing fields, we find:

$$D = \pm\, U/[2(z-z_0)\theta\Phi] \qquad (24)$$

The sign is plus when z_0 is on the left side of the capacitor and minus when z_0 is on the right. If we insert this expression into (22), the integration can be performed analytically to yield the following expression for the deflection in the region below the capacitor:

$$u_\delta = w_\delta = (z-z_p)\gamma \qquad (25)$$

where

$$\gamma = (U/2\theta\Phi)\ln(1+l/s) \qquad (26)$$

is the deflection angle;

$$z_p = z_0 \pm l/\ln(1+l/s) \qquad (27)$$

denotes the location of the pivot plane, s is the distance between z_0 and the capacitor, and l is the length of the capacitor. Here the sign is plus when the intersection z_0 of the extended plates is in front of the capacitor, viewed in the direction of the axis; the sign is minus in the opposite case. For a wedge-shaped capacitor the pivot plane z_P does not coincide with the midplane z_M, as in the case of a parallel plate capacitor, but is shifted toward the intersection point z_0.

Degradation of Spatial Resolution by Blanking

Beam pulses are generated by deflecting the beam across the hole of a chopping diaphragm or across a knife edge. In most instruments the beam-limiting diaphragm in the final lens is used for chopping. The EBBS is usually placed above or below the first lens, because the virtual image of this diaphragm is very small in this region. The final image, located at a plane z_A below the EBBS is an effective chopping aperture, as electrons which do not pass through this aperture are removed from the beam by the final diaphragm.

The blanking operation is generally connected with an apparent movement of the crossover, as shown in Fig. 3. The corresponding displacement $u_{\delta S} = u_\delta(z_S)$ of the probe at the specimen plane z_S causes an enlargement of the probe size and hence a degradation of spatial resolution. We define the degradation:

$$\delta = |u_{\delta S}|/(u_{\gamma S}|u_C|) = |u_C|^{-1}\int_{-\infty}^{z_S} I(z)dz \qquad (28)$$

as the ratio of beam displacement to the radius $u_{\gamma S}|u_C|$ of the real probe at the specimen plane. Here:

$$I(z) = g(T)De^{-i\chi}u_\alpha \quad . \qquad (29)$$

The spot remains at rest during blanking only if the degradation vanishes. This occurs, for example, if the integrand (29) on the right-hand side of (28) is antisymmetric with respect to a certain plane of the EBBS. It has often been stated that the degradation would vanish if the center of gravity of the deflection field D were placed at an intermediate image of the crossover. However, this conjecture holds true only for the static situation (g = 1) or for an infinitely short deflection structure. In the real case of an extended EBBS time-of-flight effects may cause a substantial degradation when the deflection voltage changes significantly during the transit time of the electrons. This behavior becomes more obvious

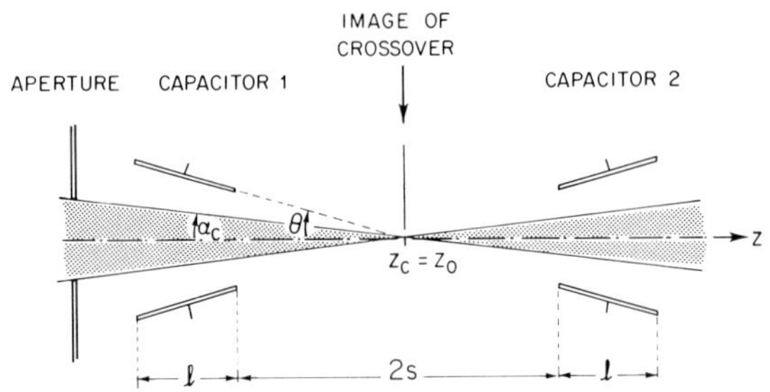

Fig. 4. Deflection structure consisting of two wedge-shaped plate capacitors positioned symmetrically about an image of the crossover.

if we consider that electrons within the capacitor will pivot on different planes when the deflection voltage is changed abruptly, depending on their instantaneous axial positions.

In the case of a fast time-varying deflection voltage the integrand (29) cannot be made antisymmetric with respect to the midplane z_M of the deflection structure. Fortunately the degradation also vanishes if the integrand fulfills the conditions:

$$I(z+b) = -I(z) \quad , \quad 0 \leq z_M - z \leq b \, , \tag{30}$$

where $b+l$ is the total length of the deflection system. This requirement can be met by placing two wedge-shaped capacitors separated by a distance 2s symmetrically about an intermediate image $z_C = z_M$ of the crossover. Such an arrangement is depicted schematically in Fig. 4. The plates are inclined in such a way that their elongations intersect each other at the crossover ($z_O = z_C$). In this case the factor $Du_\alpha \exp(-i\chi)$ is antisymmetric with respect to the midplane and constant within each capacitor. This behavior follows if we use expression (24) together with $u_\alpha = z - z_O$ and remember that χ is constant in the absence of magnetic fields at the deflection system. Hence condition (30) is fulfilled if the time function has the property:

$$g[T(z+b)] = g[T(z)] \tag{31}$$

where $b = 2s + l$.

For this purpose we must delay the deflection signal at the second capacitor by the time $T_d = b/v_o$ with respect to the first capacitor. This delay time corresponds to the transit time of the electrons through each half of the deflection system. The delay time can be adjusted conveniently by connecting the capacitors by a wave guide of proper length. Then we obtain a so-called traveling-wave deflection structure, which has the property that each electron passes conjugate segments of the two capacitors at equal phases of the deflection voltage. As a result all electrons seem to pivot on the midplane, regardless of their initial position when the signal is switched on or off. This behavior is demonstrated more convincingly in Fig. 5, where the numbers indicate the positions of various electrons during passage through the first and second capacitor when the signal is switched on.

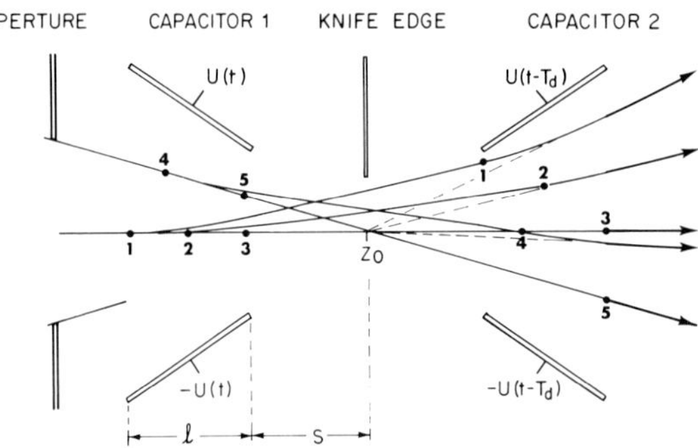

Fig. 5. Scheme of resonance beam deflection, employing the deflection structure shown in Fig. 4. The signal at the second plate capacitor is delayed by the time $T_d = (l + 2s)/v_O$, where v_o is the velocity of the electrons. The numbers represent the two positions of each different electron as it passes through the first and second capacitor when the deflection voltage is switched on. The dashed lines indicate the asymptotes of the emergent electron trajectories.

Temporal Resolution

Plate capacitors can be used for stroboscopically operating EBBS only if the transit time of the electrons through the deflection field is small compared to the pulse repetition rate. For accelerating voltages higher than about 1 kV and capacitor lengths in the centimeter range, the repetition frequencies must be lower than 100 MHz. Higher frequencies in the GHz region will not be considered here because they require the use of traveling-wave structures or reentrant cavities [14]. Nevertheless the rise times of the electric pulses may be in the picosecond range, depending on the shape of the pulses, which need not be sinusoidal [6].

Chopping of the beam is usually performed by a dc-biased deflector, so that the beam is normally off. The beam pulse is generated by dropping the beam into the chopping aperture, as sketched in Fig. 3. The velocity of the deflection at this plane is obtained from (19) by partial differentiation with respect to T to yield:

$$\dot{u}_{\delta A} = u_{\alpha A} \int_{\infty}^{z_A} D\dot{g} e^{-i\chi} u_\gamma dz \quad . \tag{32}$$

This relation shows that the deflection velocity of the beam is proportional to the convolution of the velocity of the driving voltage and the deflection field times a factor $u_\gamma \exp(-i\chi)$. This convolution is a function of the position of the electron at a given time. To clarify this statement let us consider a single parallel-plate capacitor which is driven by a step pulse (17). Its derivative with respect to time is $\dot{g} = v_o\, \delta(z - z_o)$, where z_o is the position of the electron when the pulse is switched on, and $\delta(z)$ is the delta function. If we insert this relation into (32), assume $\chi = o$, and remember that in the absence of round lenses the fundamental rays are straight trajectories, we eventually obtain:

$$\dot{u}_{\delta A} = (z_A - z_O) v_o D(z_o) \quad . \tag{33}$$

If we further suppose that the throw $z_A - z_P$ is large compared to the capacitor length l, and to $z_P - z_0$, we find for the angular deflection velocity

$$\dot{\gamma} = \gamma F(z_o)/\tau_t = (Uv_o/d\Phi)F(z_o) \ . \qquad (34)$$

Here 2U and d are the voltage and the distance between the capacitor plates respectively; $\tau_t = l/v_o$ is the transit time of the electrons through the capacitor; and F(z) is the normalized distribution of the deflection field along the z axis. If chopping is performed at an intermediate image of the crossover the fundamental rays u_γ and u_α must be exchanged in (32). However this exchange does not alter relation (34). If we neglect the fringing fields, F represents a box function which is unity inside and zero outside the capacitor, and γ is the maximum deflection angle. Hence in the case of a step pulse the deflection velocity conceived as a function of time differs from zero only during the period τ_t.

Next consider the other extreme situation, where the driving voltage is a slowly varying function during the transit time τ_t. In this case the resulting angular deflection velocity:

$$\dot{\gamma} = \gamma \dot{g} \qquad (35)$$

varies synchronously with the driving voltage.

We define the time resolution limit τ_r as the rise time of the probe current at the specimen plane from zero to maximum intensity. This time is roughly given by the relation:

$$\tau_r = 2\beta_A/\dot{\gamma}_C \qquad (36)$$

provided that the angular deflection velocity $\dot{\gamma}_C$ does not change significantly while the beam is traversing the knife edge at the chopping aperture. The angle:

$$\beta_A = [R_A + r_A - |R_A - r_A|]/2(z_A - z_P) \qquad (37)$$

in Fig. 3 is determined by the smaller of two radii, either the beam radius r_A at the chopping plane or the radius R_A of the chopping aperture. Relation (36) states that the temporal resolution $1/\tau_r$ will be improved as the angle decreases and as the angular deflection velocity increases. Expressions (34) and (35) show that a high deflection velocity can be achieved if the rise time of the driving voltage is in the range of the transit time τ_t and if the electric field strength is as high as possible.

Small angles β_A can be obtained in two ways: by placing the EBBS in the region of the gun where the effective chopping aperture is very small; or by blanking the beam across a knife edge placed at the strongly demagnified intermediate image of the crossover above the final lens. The use of a field-emission gun further decreases the size of the crossover and hence the angle β_A.

Correction of Blanking Aberrations

The EBBS generates time-varying aberrations at the specimen plane which degrade the probe. In practice the main degradation stems from the displacement of the scanning spot. This zeroth-order aberration cannot be eliminated or significantly reduced by the stigmator used for astigmatism control. Although this stigmator can always be adjusted to produce a circular spot, its resulting diameter will not be appreciably smaller than the original displacement, contrary to statements found in the literature [e.g. 3,14].

To achieve high spatial resolution the beam should be pivoted on an intermediate image of the crossover. Then the displacement vanishes, as has been discussed in detail in the section on degradation of spatial resolution. The residual blurring of the probe arises primarily from two effects:

(a) The total energy of the electrons is not conserved in time-varying fields. Accordingly the energy width of the beam is broadened and further increases the chromatic aberration.

(b) The deflected beam intersects the subsequent round lenses at increasingly larger off-axial distances. This combination of the large deflection, with the defects of the final lens generates strong aberrations at the specimen plane. The direct aberrations of the deflection fields at this plane are negligibly small if the crossover at the EBBS is appreciably demagnified by the subsequent round lenses.

Diminution of chromatic aberration

The chromatic aberrations are ultimately connected with the delay time:

$$\tau = \int_{z_i}^{z} (t' - 1/v_o)\,dz \ . \qquad (38)$$

This temporal deviation occurs when the electrons change their flight directions (directional delay) and/or when their total energy is altered by either a time-varying field or the emission process (chromatic delay). Within the frame of the first-order approximation the delay time $\tau^{(1)}$ (14) is entirely chromatic and is produced by the axial component of the time-varying field and the initial energy spread. The corresponding energy change depends on the axial position of the electron at a given point of time or, from a different point of view, on the phase of the time-varying signal. Hence the overall effect of the different phases is a broadening of the energy width of the beam.

Each asymmetrically driven capacitor acts as a combination of a pure deflection element and a cylinder lens. Primarily the time-dependence of this lens causes a significant energy-broadening of the beam when the pulse rise times are of the order of the transit time of the electrons. The cylinder lens also introduces a defocus and an astigmatism at the specimen plane, because such a lens refracts the rays toward its midplane.

These deleterious defects can be avoided if the capacitor plates are excited antisymmetrically so that the midplane remains at anode potential ($\Delta = 0$). This is achieved in practice (e.g. in television sets) by using two perfectly symmetrical electrical supply units with the same midpoint. Assuming this driving mode ($\Delta = 0$), we obtain for the remaining primary chromatic aberration the expression:

$$u_{cS}^{(2)} = (\delta\Phi/\Phi)C_c^{(0)} \ . \qquad (39)$$

Here:

$$C_c^{(0)} = u_{\gamma S} \int_{-\infty}^{z_S} u_\alpha \{ u_\delta [\chi'^2 - i(\chi' u_\delta^2)'/2u_\delta^2] - Dge^{-i\chi} - (z-z_i) g' De^{-i\chi}/2 \} dz \qquad (40)$$

defines the constant of the zeroth-order chromatic aberration or, from another viewpoint, of the dispersion. Within the frame of our approximation this zeroth-order chromatic aberration is a second-rank aberration: it is bilinear in the chromatic deviation and the deflection field yet independent of the boundary (Seidel) parameters of the beam. The first term of the dispersion (40) results from the beam deflection and the unavoidable chromatic aberration of the subsequent round lenses. When the displacement is compensated, this term is found to be $\gamma C_c/M$, where C_c is the chromatic aberration constant of the final lens and $M = u_{\gamma S}$ is the magnification of the subsequent round lenses. The second term in (40) simply represents the displacement of a quasistatic deflection field. This term is zero when the pivot plane coincides with an intermediate image of the crossover. The third term, which contains the derivative of the time function g(T), allows for the fact that electrons whose velocities differ from the average axial velocity v_o intersect a given plane at slightly different times. During each time interval the deflection field at this plane changes somewhat, so the electrons are deflected differently.

The contribution of the last two terms to the dispersion is proportional to the magnification of the crossover produced by the lenses beyond the EBBS. Hence in the case M << 1 and $\gamma \neq 0$ the dispersion (40) is primarily determined by the first term, which represents a combination aberration. To avoid this most deleterious aberration, chopping must be performed within a so-called omega deflection system, which produces a bent ray-path with parallel directions for the incident and emergent electrons. Such a straight-vision system represents the electron-optical analog of the well-known Dove prism in light optics. In such a system all but the last term in (40) vanish.

Secondary temporal deviation

The delay time $\tau^{(1)}$ obtained in the first step of the approximation procedure (14) is caused entirely by the initial energy spread in the case $\Delta = 0$. The secondary temporal deviation $\tau^{(2)}$ is derived from (8) in the second step of the approximation to yield:

$$\frac{\tau^{(2)\prime}}{T'} = -\frac{E_z^{(2)}}{E_o} = \frac{E_w^{(2)}}{E_o} + \frac{E_p^{(2)}}{E_o} - \frac{E_t^{(2)}}{E_o}. \qquad (41)$$

Here:

$$E_w^{(2)}/E_o = w^{(1)\prime}\bar{w}^{(1)\prime}/2 \qquad (42)$$

represents the normalized off-axial kinetic energy in the paraxial approximation, and:

$$E_p^{(2)}/E_o = -g(T)Dx^{(1)} \qquad (43)$$

is the change of the potential energy if the electron is at an off-axial position $x^{(1)}$ within the deflection field. These two spatial terms do not vanish in the static case g = 1. They account for the different flight times of monoenergetic electrons traveling in different directions to reach the same plane. The third term:

$$E_t^{(2)}/E_o = -(3/2)(\delta\Phi/\Phi)^2 - \int_{z_i}^{z} Dg'x^{(1)}dz \qquad (44)$$

is chromatic, since it results from the change of the total energy. Here the first term on the right-hand side arises from the initial energy spread, while the second term accounts for the energy gained or lost by the electron at a fixed position when the electric field changes its strength. The chromatic aberrations generated by the secondary time deviation $\tau^{(2)}$ are of third rank and are negligibly small in the case of a straight-vision EBBS.

Third-order geometric aberrations

Apart from the chromatic aberration the spot size is limited by the third-order geometric aberrations. These aberrations consist of the third-order deflection aberrations of the EBBS, the third-order aberrations of the magnetic round lenses, and a combination aberration resulting from the beam deflection and the defects of the round lenses. The combination aberration has the form of an axial coma and may become quite large in the absence of a beam-limiting aperture at the last lens. The third-order geometric aberrations generated by the EBBS are negligibly small if the crossover there is demagnified by the subsequent lenses. In the case of the straight-vision EBBS or if a beam-limiting aperture is placed at the last round lens, the probe size is affected only by the third-order spherical aberration and the chromatic aberration of this lens and by the diffraction at the effective beam-limiting aperture.

Outline of a High Performance Blanking System

Chopping of the beam at an intermediate image of the crossover is most feasible in the region above the final lens, where the crossover is strongly demagnified. However, beam-blanking in this region by a single capacitor strongly displaces the probe. This degradation can be avoided by incorporating two capacitors which are driven in such a way that the virtual crossover between them remains at rest during the blanking operation, yet the emanating rays are strongly deflected away from the axis, as demonstrated in Fig. 5. The beam-limiting diaphragm should therefore not be placed in the final lens, because this diaphragm also chops the beam and unduly increases the rise time of the beam pulse. If the diaphragm is removed, however, the strong deflection, in combination with the unavoidable chromatic and spherical aberrations of the subsequent round lenses, produces a strong dispersion and a large axial coma, which significantly deform the probe during blanking.

The highly corrected EBBS outlined in Fig. 6 avoids these disadvantages. This Ω-deflection system consists of two identical round lenses and two oblique plate capacitors, one located above, the other below the round lenses. The pivot planes of the capacitors are at focal planes of the round lenses. An intermediate image of the crossover is located at the symmetry plane midway between these lenses. At this plane is placed a knife edge which chops the beam. The elongations of the plates intersect at intermediate images at equal distances above and below the midplane.

Because the system incorporates two round lenses, the second capacitor compensates for the deflection and some aberrations of the first capacitor, provided that the second capacitor is driven with the appropriate time delay. To guarantee that the deflection vanishes below the EBBS the focal length $f_1 = f_2 = f$ of the two lenses must be sufficiently large to prevent overlap of their magnetic fields with the electric deflection fields. Then the course of the Gaussian rays in front of the final lens is not affected by the blanking procedure.

The capacitors must be rotated with respect to each other about the optic axis by the angle of the Larmor rotation of the magnetic round lenses. This rotation vanishes if the coil currents have opposite directions for the two lenses.

Installation of the proposed EBBS in a SEM increases its column length. The extent of this increase depends on the required demagnification of the crossover at the chopping plane z_C. Since the axis of the deflected beam is parallel to the optic axis between the two round lenses, we have $\beta_A = r_C/f$, where r_C is the radius of the demagnified crossover.

To survey the performance of the proposed EBBS let us presuppose that the currents are sufficiently low that broadening of the probe by the Boersch effect can be neglected. In this case angles $\beta_A \simeq 10^{-5}$ are attainable. If we further assume driving pulses with rise times in the range of the transit time of the electrons through a single capacitor and if we use the formula (36) together with (34) and (37), we find that it should be possible to generate beam pulses whose rise times are smaller than 1 ps.

In the case of low beam voltages the magnetic round lenses between the capacitors can be replaced with electrostatic round lenses. This substitution simplifies the mechanical setup considerably, because the two electrostatic lenses can be designed as immersion lenses. Since these lenses must be identical, their fields can be generated by three spatially separated coaxial cylinders where the potentials of the outer cylinders are at ground. Then the two immersion lenses resemble a thick electrostatic einzel lens with an extremely extended inner electrode.

The chopping aperture (or knife edge) at the center of this lens must be kept at the potential of the central electrode. Since the focal length of the two lenses is large compared to both the diameter of the cylinders and their separation distances, the thin-lens approximation can be used for the actual design.

During the blanking operation the probe remains roughly at rest at the specimen plane, apart from a small residual chromatic displacement. The time-dependent dispersion constant $C_c^{(0)}$ of this zeroth-order chromatic aberration is always smaller than $2M\gamma l\,(1 + s/l)^2$ for the proposed system (Fig. 6). As an example assume a capacitor length l = 1 cm a separation distance 2s = 6 cm, a maximum deflection angle $\gamma \simeq 10^{-2}$, and a final magnification M = 0.1 to yield $C_c^{(0)} \simeq$ 30 μm. Even at low beam

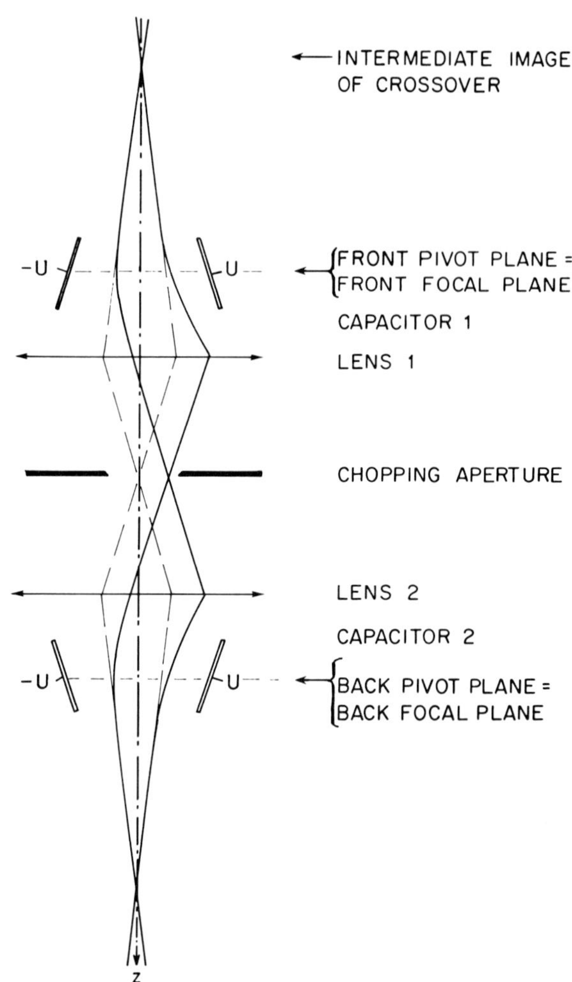

Fig. 6. Ray path in a symmetric straight-vision EBBS with an intermediate image of the crossover at the chopping aperture in the cases U = 0 (- - -) and U ≠ 0 (——).

voltages of several kV the maximum probe displacement during blanking will remain less than about 30nm. This displacement is small compared with the spot sizes used in electron-beam testing systems.

Conclusions

High-resolution electron-beam blanking can be achieved by using two oblique plate capacitors which are driven in such a way that they form a traveling-wave structure. Chopping of the beam is performed either by placing a relatively simple deflection structure (Fig. 4) at the first intermediate image of the gun crossover or by incorporating a more sophisticated straight-vision EBBS (Fig. 6) in the region of a strongly demagnified image of the crossover near the final lens of the SEM. The latter system, consisting of two capacitors and two identical round lenses, yields the highest spatial and temporal resolution, but its incorporation into the SEM extends the column length. Both blanking systems avoid probe movement, even in the case of very short pulses, because time-of-flight effects are compensated. The proposed blanking systems should considerably surpass the performance of EBBS now in use.

Acknowledgements

We want to thank Drs. H. Pfeiffer and D. Kern (IBM) for making us aware at the conference of an existing high speed parallel-plate blanker which compensates to a certain extent for time-of-flight effects. (Kuo HP, Foster J, Haase W, Kelly J, Oliver BM. (1983). A high speed blanker for a 300 MHz lithography system, in: Proc. 10th Intl. Conf. on Electron and Ion Beam Science and Technology, R. Bakish (ed.) Electrochemical Soc. Inc., Pennington, NJ Vol. 1, 78-91).

References

[1] Feuerbaum HP. (1979). VLSI testing using the electron probe, Scanning Electron Microsc. 1979; I:285-296.

[2] Feuerbaum HP, Otto J. (1978). Beam chopper for subnanosecond pulses in scanning electron microscopy. Sci. Instr. 11, 529-532.

[3] Fujioka H, Ura K. (1983). Electron beam blanking systems. Scanning 5, 3-13.

[4] Gopinath A, Hill MS. (1977). Deflection beam chopping in SEM. J. Phys. E: Sci. Instr. 10, 229-236.

[5] Hawkes PW. (1983). Computer-aided calculation of the aberration coefficients of microwave cavity lenses. Optik 63, 129-156.

[6] Lischke B, Plies E, Schmitt R. (1983). Resolution limits in stroboscopic electron beam instruments, Scanning Electron Microsc. 1983; III:1177-1185.

[7] Lukianov AE, Spivak GV. (1966). Electron mirror microscopy of transient in semiconductor diodes, in: Proc. 6th Int. Cong. Electron Microsc. Kyoto, Vol. I, Maruzen Company Ltd., Tokyo,611-612.

[8] Menzel E, Kubalek E. (1979). Electron beam chopping systems in the SEM, Scanning Electron Microsc. 1979; I:305-318.

[9] Oldfield LC. (1976). A rotationally symmetric electron beam chopper for picosecond pulses. J. Phys. E: Sci. Instr. 9, 455-463.

[10] Plies E. (1982). Proposal for an electron beam blanking system with monochromator effect, in: Proc. 10th Int. Congr. Electron Microsc. Hamburg, Vol. I, Deutsche Gesellschaft für Elektronenmikroskopie,Frankfurt,319-320.

[11] Plows GS, Nixon WC. (1968). Stroboscopic scanning electron microscopy. J. Phys. E: Sci. Instr. 1, 595-600.

[12] Szentesi OI. (1972). Stroboscopic electron mirror microscopy at frequencies up to 100 MHz. J. Phys. E: Sci. Instr. 5, 563-567.

[13] Sturrock PA. (1955). Static and Dynamic Electron Optics, University Press, Cambridge, 148-185.

[14] Ura K, Fujioka H, Hosakawa T. (1978). Electron optical design of picosecond pulse stroboscopic SEM, Scanning Electron Microsc. 1978; I:747-753.

[15] Weinfeld M, Bouchoule A. (1976). Electron gun for generation of subnanosecond electron packets at very high repetition rate. Rev. Sci. Instr. 47, 412-417.

Electron Optical Systems (pp. 137-147)
SEM Inc., AMF O'Hare (Chicago), IL 60666-0507, U.S.A.

0-931288-34-7/84$1.00+.05

FIELD EMISSION SOURCE OPTICS

L. W. Swanson

Department of Applied Physics and
Electrical Engineering
Oregon Graduate Center
19600 N.W. Walker Road
Beaverton, Oregon 97006

Phone no.: (503) 645-1121

Abstract

The use of field emission processes as a means of generating high brightness sources of both electrons and ions is becoming of increasing practical use in electron/ion optical systems. A review is given of the source optics and emission characteristics of a zirconium oxide coated, <100> oriented, tungsten (ZrO/W(100)), thermal field emitter electron (TFE) source and a liquid metal ion source (LMIS). The primary interest at present is the use of these high field sources in microprobe focusing columns where beam sizes of < 0.5 μm and target current densities of ~ 1 to 10 A/cm^2 (in the case of a LMIS) and 10^2 to 10^4 A/cm^2 (in the case of a ZrO/W(100) TFE) are required. For predicting the performance of field emission sources in microprobe columns source properties such as beam energy spread (ΔE), angular intensity (I') and virtual source size (d_v) are of interest. For the LMIS experimental values of ΔE, I' and d_v are found to depend on particle charge to mass ratio and total source current. Similarly for the ZrO/W(100) TFE experimental values of ΔE increase with I'. A computer program, known as SCWIM, has been designed to model point-to-plane source geometries and is used to model such emitter parameters as d_v and I'.

Key words: Field Emission, Liquid Metal Ion Source, Field Ionization, High Brightness.

Introduction

High field sources for the formation of both electrons and ions are becoming of increasing importance in optical systems used to form high current density, focused beams of size < 0.5 μm. A field emission mechanism is common to both ion (e.g., gas phase field ionization and solid or liquid phase field evaporation) and electron formation where a conically shaped emitter of radius r < 1 μm is typically employed to achieve the high field strengths. Since field emission involves a quantum mechanical tunnelling process it can theoretically occur at T = 0 K. Nevertheless for various reasons the emitters usually operate at elevated temperature thereby resulting in a thermal-field (TF) operating mode.

For field electron emission analytical expressions governing the relationships among cathode electron energy distribution, current density, work function ϕ, surface electric field F and temperature T have been developed [13]. In contrast, for ion formation via field evaporation or field ionization the theoretical model is less well developed in terms of the current density $J_c(F,\phi,T)$ and full width at half maximum (FWHM) of the total energy distribution $\Delta E(F,\phi,T)$ relationships. Nevertheless important source optic characteristics such as $\Delta E(F,T)$, the angular intensity $I' = dI/d\Omega$ (where Ω is the solid angle of emission) and the virtual source size d_v can be either measured experimentally [1,8,9] or obtained from computer analysis of emitter/extractor electrode geometry [5]. In this study we shall examine these source optical parameters for two promising, high brightness emitters: the zirconium oxide/W(100) TF electron (TFE) source for electrons and the liquid metal ion source (LMIS). The results of a computer program SCWIM [3,4] especially designed to compute trajectories and d_v values for point-to-plane electrode geometries will be discussed.

Source Optics

The source optical characteristics of primary importance for the formation of high current density focused beams using high field sources are summarized in Table 1. Figure 1 depicts the typical geometry and trajectory relationships for the hemisphere on cone type high field source. The angular magnification m is given by

List of Symbols

B_A - apparent axial brightness

C_c - chromatic aberration coefficient

D_s - spherical aberration coefficient

d - focused beam size

d_v - virtual source diameter

D(w) - transmission coefficient

E_F - Fermi level

ΔE - full width at half maximum of energy distribution

F - electric field

f - facet radius

h_θ - angular mesh size

I' - angular intensity

J_A - apparent current density

J_c - cathode current density

J(E) - current density per unit energy

m - angular magnification

m' - linear magnification

N - no. radial mesh points

r - emitter radius or radial distance

r_g - radius of Gaussian image disk

r_s - radius of emission disk on cathode

R_m - maximum mesh radius

R_o - minimum mesh radius

T - temperature

V - beam voltage

W - normal energy

α - beam angle

ε - energy relative to Fermi level

ε_o - permittivity of free space

λ - deBroglie wavelength

ϕ - work function

ρ - charge density

θ - emission angle

$$m = \frac{d\alpha}{d\theta} \tag{1}$$

and d_v can be determined from knowledge of the trajectories. In addition d_v can be determined experimentally, although such experiments are difficult to carry out. If m is known then the following relationship between I' and the cathode current density J_c

$$I' \simeq \frac{J_c r^2}{m^2}, \tag{2}$$

can be derived for small values of α where r is the emitter radius. Usually I' is easily obtained experimentally whereas the predicted value of J_c requires knowledge of the surface electric field and the theoretical emission relationship $J_c(F,\phi,T)$.

The radius of a disk on the cathode r_s is related to a corresponding Gaussian image disk r_g by

$$r_g = M' r_s \tag{3}$$

where M' is the linear magnification. From these fundamental parameters other source characteristics such as apparent current density

$$J_A = 4I'\alpha^2/d_v^2$$

and apparent axial brightness

$$B_A = J_A/\pi\alpha^2 \tag{4}$$

can be derived.

For a single lens system the focused beam size d is given by

$$d = M[d_v^2 + (\tfrac{1}{2}C_s\alpha^3)^2 + (C_c\alpha\frac{\Delta E}{eV})^2 + (0.6\frac{\lambda}{\alpha})^2]^{1/2} \tag{5}$$

where the spherical and chromatic aberration coefficients C_s and C_c are referred to the object side, M is the column magnification and λ is the particle wave length. The importance of ΔE and I' can be illustrated for the focused beam case when d_v and chromatic aberration disks dominate the aberration terms. In this case it can be shown that the beam current I_b is given by:

$$I_b = (\frac{d^2}{M^2} - d_v^2)(\frac{eV}{\Delta E C_c})^2 I'\pi \tag{6}$$

Thus for a specified value of d the beam current is proportional to the source parameter $I'/\Delta E^2$. In addition, I_b is proportional to $d^2/M^2 - d_v^2$. This simple example illustrates the importance of the source parameters I', ΔE and d_v.

Computation of Electric Potential and Trajectories

In order to obtain the field required for high current density at convenient voltages, the emitter is usually much smaller in comparison to the size of other electrodes. This difference can be as large as 10^7, as in the case of a field ion emitter, and is normally in excess of 10^3. To complicate matters the emitter shape may be non-hemispherical depending on the operating condition. For example a faceted shape frequently occurs in the case of TF emission due to field build up [4]. To deal with this problem a special coordinate system was employed known as SCWIM (spherical coordinate with increasing mesh size) [3,4].

Figure 2 is a schematic representation of the SCWIM model. The radial mesh size increases from the emitter outward according to a geometrical series with a term ratio $(1 - h_\theta)^{-1}$, where h_θ is the angular mesh size, in radians. If one

TABLE 1

Summary of source characteristics and their effect on focused beam optics. Optics column magnification, chromatic aberration, source angular magnification, emitter apex radius and aperture half angle are M, C_c, m, r and α, respectively.

Source Characteristics	Optics Function	Quantitative Effect
Angular Intensity $I'(F,\phi,T)$	Determines Beam Current	$I'\pi\alpha^2$
Virtual Source Size $d_v(\alpha,r,F)$	Determines Minimum Beam Size	Md_v
Beam Energy Spread $\Delta E(F,\phi,T)$	Determines Minimum Chromatic Aberration Disk	$MC_c\alpha \frac{\Delta E}{eV}$
Emitter Current Density J	Determines I'	Jr^2/m^2

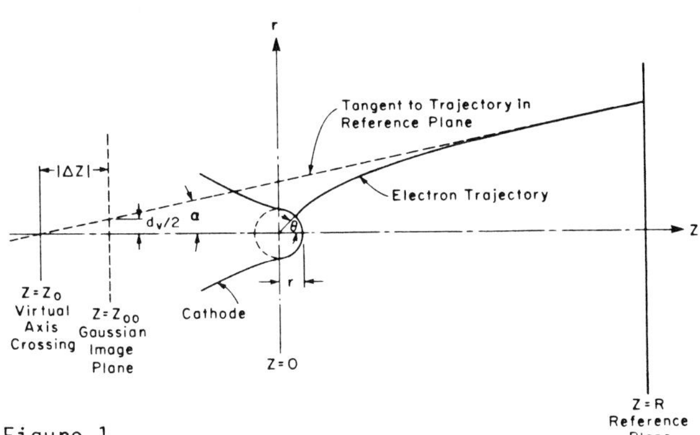

Figure 1.

Illustration of a typical trajectory and position of virtual axis crossing relative to Gaussian image plane.

takes h_θ on both sides to be the same, then a quasi-equidistant central difference scheme occurs (i.e., $h_s = h_w = h_e$ in Figure 2) which leads to a truncation error between that of nonequidistance and equidistance schemes, i.e., approximately proportional to h_θ^3. In order to realize this scheme, the radial mesh interval h_n (along a radius vector) is:

$$h_n = h_o/(1 - h_\theta)^n \tag{7}$$

where n is the radial mesh point number. The first and also the smallest radial mesh interval is determined by

$$h_o = R_o/(\frac{1}{h_\theta} - 1) \tag{8}$$

where R_o is the minimum radius which is most conveniently chosen as the apex radius for a spherical emitter while for a faceted emitter R_o can be chosen as the radius of the facet. From R_o and the maximum dimension of the gun R_m (i.e. the center of the spherical coordinate system to the most distant point), the required number of mesh points along a radius vector can be calculated by

$$N = \frac{\ln(R_o/R_m)}{\ln(1 - h_\theta)} - 1 \tag{9}$$

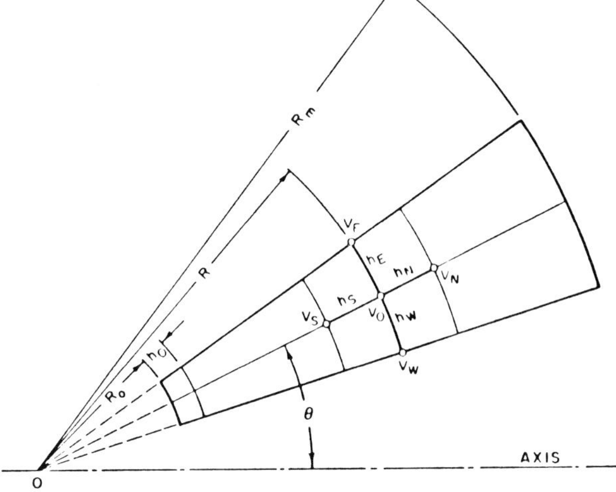

Figure 2.

Schematic representation of the mesh arrangement used in the SCWIM method. The minimum and maximum electrode dimensions are R_o and R_m respectively.

Figure 3 shows the dependence of N on R_m/R_o for various values of h_θ. One notices that N increases very slowly with increasing R_m/R_o. For example, when R_m/R_o increases by four orders of magnitude (from 10^2 to 10^6), N increases only by a factor of 3 for $h_\theta = 0.1$. Generally, the number of radial mesh points approximately doubles when the ratio of geometrical sizes increases by two orders of magnitude. Because of this unique feature of SCWIM one is able to treat the entire gun at once without unreasonable demand

TABLE 2

Calculated emitter characteristics for the faceted ZrO/W(100) TFE source where ϕ = 2.8 eV, T = 1800 K. The minimum value of d_v and corresponding aperture half angle α_m are indicated for each value of r.

r(μm)	d_v (nm)	α_m (mr)	I' (mA/sr)	B (10^9 A/cm^2 sr)	J_c (A/cm^2)	F(V/Å)
0.3	7.0	5.5	0.72	1.9	2.8×10^4	0.11
0.5	8.8	3.5	0.98	1.6	1.5×10^4	0.97
1.0	12.0	2.5	0.97	0.86	4.0×10^3	0.070
2.0	16.5	2.0	1.06	0.46	1.2×10^3	0.049
3.0	19.5	1.0	1.46	0.49	7.7×10^2	0.041

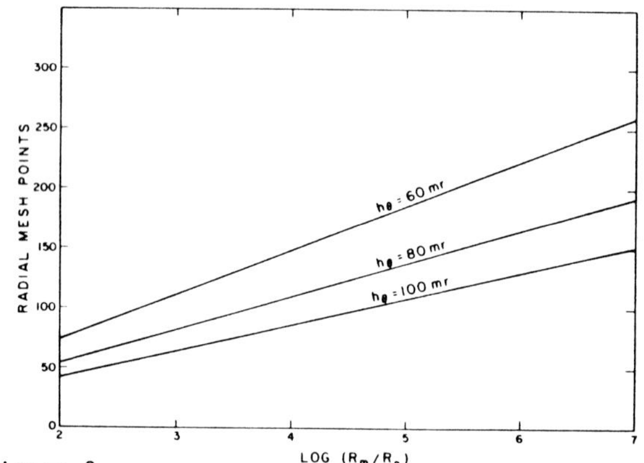

Figure 3.

Plots of the number of radial mesh points N vs the electrode geometric ratio R_m/R_o for the SCWIM method. The angular mesh sizes h_θ are indicated.

on computer storage capacity. At the same time, the inaccuracy and complexity connected with the "successive magnification" approach are eliminated. In areas remote from the emitter the mesh size appears too large. However, the fields in such remote areas are fairly weak, therefore, the mesh arrangement in SCWIM matches the features of field emission guns very well.

By the use of Eq.(7), the axisymmetric Poisson equation in spherical coordinates, which is given by

$$\frac{\partial^2 V}{\partial r^2} + \frac{2}{r}\frac{\partial V}{\partial r} + \frac{1}{r^2 \tan\theta}\frac{\partial V}{\partial \theta} + \frac{1}{r^2}\frac{\partial^2 V}{\partial \theta^2} = -\frac{\rho}{\varepsilon_o} , \qquad (10)$$

can be cast into a five-point difference form, which up to the fourth order term of a Taylor series, is given as follows (for notations see Figure 2):

$$V_o = B_1 \left(1 + \frac{h_\theta}{2\tan\theta}\right) V_E + B_1\left(1 - \frac{h_\theta}{2\tan\theta}\right) V_W + B_2 V_N$$

$$+ B_3 V + B_4 \frac{h_s^2 \rho}{\varepsilon_o} \qquad \text{(off-axis)}$$

$$V_o = B_{10} V_E + B_{20} V_N + B_{30} V_S + B_{40} \frac{h_s^2 \rho}{\varepsilon_o} \qquad (11)$$

(on-axis)

where ρ is the space charge density (negative value for electrons, positive for ions) in coulomb/m^3, $\varepsilon_0 = 8.854 \times 10^{-12}$ F/m is the permittivity of free space, B_1, B_2, B_3, B_4 and B_{10}, B_{20}, B_{30}, B_{40} are functions of h_θ only.

That is to say, after h_θ is chosen, the B coefficients are constants, a factor which saves considerable computing time in electric potential iterations.

Virtual Source Size

Two experimental methods can be used to obtain values of d_v. Obviously, from Eq.(5) experimental measurements of d vs α can be used to evaluate d_v provided C_s and C_c are known or α is sufficiently small that the chromatic and spherical aberrations can be neglected. This is generally difficult since small values of d must be measured accurately. Another experimental method of measuring d_v is from the contrast patterns of a two beam interference behind an electrostatic biprism of the Mollenstedt type. This method was used to obtain a value of d_v = 31 nm for a room temperature W(310) FE source where α = 0.1 mr and I' = 7 mA/sr [11].

In addition to the experimental methods of obtaining d_v, computer trajectory calculations of the type first carried out by Weisner and Everhart can be employed [16]. Whereas the latter authors utilized an emitter shape that lends itself to an analytical solution for the electric potential, the SCWIM method can be applied to an emitter and extractor electrode of arbitrary shape.

The structure of field emission guns may vary a great deal. Among them a triode-type gun is most common. Figure 4 shows an example of a gun structure used with the ZrO/W(100) TFE source and, after some minor size modifications, with the LMIS. The suppressor electrode V_s can be used as a current control electrode and, in the case of the low work function ZrO/W(100) TFE, as a means of reducing thermal emission from the emitter shank.

ZrO/W(100) TFE Source

Following the approach of reference 16 we have used the SCWIM computer method to calculate

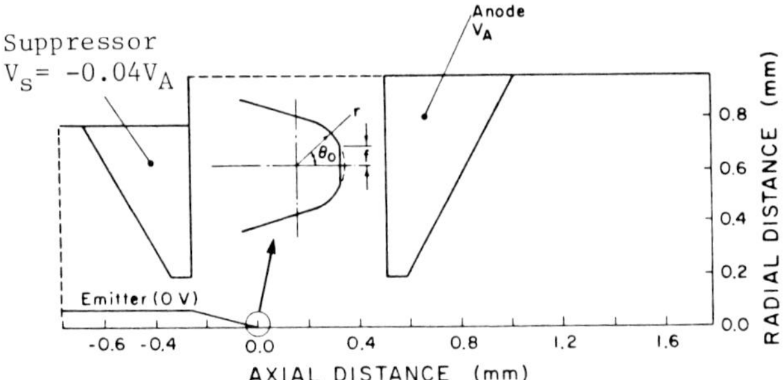

Figure 4.

Diagram of electrode configuration used in the gun structure for the ZrO/W TFE source. Insert shows an expanded view of the emitter of apex radius r for the rounded and faceted case. The launch angle for trajectories is θ and the flat radius is f.

d_V for the ZrO/W(100) TFE emitter for the realistic electrode configuration of Figure 4. As indicated in the inset of Figure 4 the ZrO/W(100) forms a large facet on the low work function (100) plane where the ratio of flat radius f to apex emitter radius r (when a rounded end form is assumed) is found experimentally to be f/r = 0.3. The reference plane from which extrapolation of trajectory tangents is made (see Figure 1) is R = 2.5 mm.

Values of d_V were calculated as a function of aperture angle α and emitter radius r for the faceted ZrO/W(100) TFE. The value of V_A was adjusted so that a constant value of $I' \cong 1$ mA/sr was maintained. An energy spread ΔV = 0.8 eV, T = 1800 K and work function ϕ = 2.8 eV were assumed for these calculations. The results given in Figure 5 and Table 2 show that d_V exhibits a minimum at small values of α. This is due to a balance between the diffraction aberration disk which varies as 1/α and the Gaussian, spherical and chromatic aberration terms all of which increase as various powers of α. Interestingly, the minimum value of d_V decreases with decreasing emitter size r; this is primarily due to a decrease in the Gaussian contribution since the spherical and chromatic terms are negligible. For emitter sizes typically used, e.g. r = 500 to 1000 nm, the minimum values of d_V are 9 to 12 nm. This corresponds to a cathode axial current density of $J_c = 4 \times 10^3$ to 1.5×10^4 A/cm^2 and an axial brightness $B = 0.9 \times 10^9$ to 1.6×10^9 A/cm^2 sr.

The unusually low values of J_c associated with I' = 1 mA/sr and the large values of B are due to the fact that the angular magnification $m \cong 0.22$ is unusually low. This follows from Eq.(2) where J_c is shown to vary inversely with m. For rounded emitter shapes typically $m \cong 0.5$ so that values of J_c are correspondingly small for a specific value of I'.

Using the same emitter parameters, except for a rounded end form as shown in Figure 4, the Figure 6 results were obtained. Again, the calculated values d_V show a minimum as α is varied. Plotting the minimum values of d_V vs r, we obtain the results shown in Figure 7 for both the faceted and rounded end form. From the latter

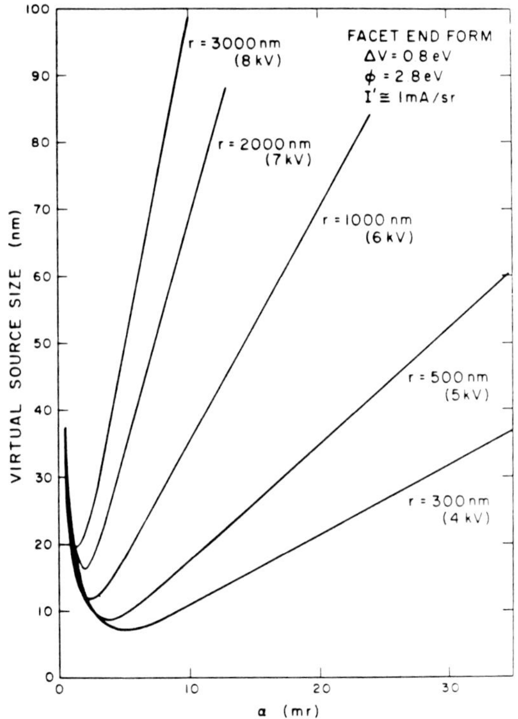

Figure 5.

Plot of virtual source size vs beam angle for the facet end form emitter for the indicated emitter radii and anode voltages.

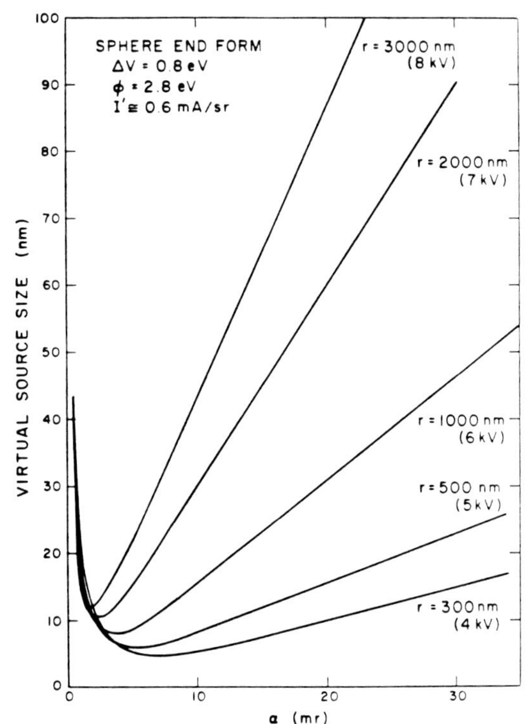

Figure 6.

Plot of virtual source size vs beam angle for the round end form emitter for the indicated emitter radii and anode voltages.

TABLE 3

Effect of Facet Formation on Angular Intensity for the ZrO/W(100) Emitter Where ϕ = 2.8 eV

Emitter Radius (μm)	Anode Voltage (kV)	Apex Field (V/Å)	J (A/cm^2)	m	I'(mA/sr)
0.3 (round)	4	0.14	1.40×10^5	0.45	0.61
0.3 (facet)	4	0.11	3.58×10^4	0.20	0.72
1.0 (round)	6	0.089	1.36×10^4	0.50	0.54
1.0 (facet)	6	0.070	5.29×10^3	0.22	0.97
3.0 (round)	8	0.053	2.05×10^3	0.55	0.62
3.0 (facet)	8	0.041	1.04×10^3	0.24	1.46

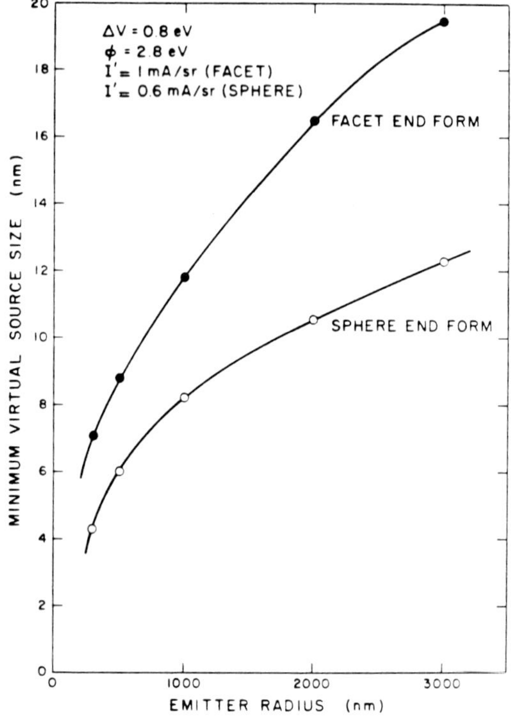

Figure 7.

Plot of the minimum virtual source size vs emitter radius for the facet and round end forms.

results we may conclude that the effect of the facet is to increase d_v by ~ 75% throughout the range of r values investigated.

At present no direct experimental measurement of d_v for the ZrO/W(100) TFE has been carried out. Indirect measurements involving the use of Eq.(5) to compare predicted and experimental values of d in focused beam studies using the ZrO/W(100) TFE source generally support a value of $d_v \leq 20$ nm for r = 1 μm, I' = 1 mA/sr and α = 2 to 6 mr [14,15].

LMIS

The value of d_v for the LMIS can be determined by the SCWIM computer program in the same manner as described in the previous section for the ZrO/W(100) TFE source. The primary modification of the input parameters is to the values of F and r which become F ≅ 2 V/Å and r ≅ 10 nm. Unfortunately, the detailed shape of the emitting portion of the LMIS is not well known so a model calculation using the SCWIM computer program would be premature. Originally the shape was believed to be a 45° half angle cone truncated by a hemispherical emitter of radius < 10 nm. More recent computer modeling of the emission characteristics and stability criteria of the liquid cone has suggested that the apex of the 45° half angle cone has an elongation of the order of 0 to 100 nm depending on the current [5,6]. This has been confirmed recently by TEM of an operating LMIS [2].

Using an elongated cone shaped emitter and the SCWIM program a preliminary estimate of d_v = 0.2 nm was obtained for a Ga^+ source [5]. The very small value of d_v for the LMIS is due to the elimination of diffraction aberration and the small transverse velocity associated with field evaporated ions. Nevertheless experimental measurements of d_v using results of focused ion beam measurements at small values of α suggest the d_v ≅ 50 nm [9,12]. This large discrepancy between prediction and experiment is not completely clear at this time, however if further results confirm this discrepancy its origin may be due to the strong Coulomb interaction known to occur in the high charge density region within a few emitter radii from the emitter surface. This Coulomb interaction comes about due to the random fluctuations in the charge density and manifests its effect in terms of an increase in the energy spread and d_v. A considerable discussion of this effect has taken place in connection with electrons and is known as the "Boersch effect." More recently this important effect has been considered in connection with high J point sources of electrons and ions [7,17] and at present is not included in the SCWIM program.

Angular Intensity

According to Eq.(2) the angular intensity I' of a point emitter is determined by the emitter current density, which is intrinsic to the emission process, and the angular magnification m, which is determined by the emitter and surrounding extractor electrode shape and potentials. The value of m can be determined by SCWIM via Eq.(1); the dependence of J on F, T and ϕ requires a theoretical model of the emission mechanism.

For electron emission the J(ϕ,F,T) relationship can be determined either numerically or from analytical expressions which cover the range of F, ϕ and T extending from pure field emission at high F and low T to pure Schottky emission at low F and high T. The general emission equation based on the Sommerfeld model of metals is given by

$$J(T,\phi,F) = \int_0^\infty J(\varepsilon)d\varepsilon \qquad (12)$$

where ε is the electron energy relative to the Fermi level E_F and

$$J(\varepsilon) = \frac{4\pi me}{h^3} f(\varepsilon) \int_0^{E_F+\varepsilon} D(W)dW \qquad (13)$$

where $f(\varepsilon)$ if the Fermi-Dirac function, W is the kinetic energy associated with the normal component of electron momentum and D(W) is the one-dimensional transmission function [10]. The numerical integration of Eq.(12) is performed in the SCWIM program for electrons thereby providing a value of J as a function of the local field.

For the LMIS an analytical relationship for J is not presently available and is not likely to be developed due to the uncertainty of the emitter shape. Nevertheless, using a dynamic model for stabilization of an elongated cone, a field evaporation model for ion formation and the SCWIM program to calculate trajectories an empirical relationship between I' and J has been developed [6].

ZrO/W(100) TFE Source

Trajectories were calculated for the faceted and rounded end form using the SCWIM program and the Figure 4 electrode structure. The results listed in Table 3 show that the value of m for the faceted end form is a factor of 2 lower than the rounded end form.
Thus, the value of J_C for a specified I' is reduced by a factor of 4 due to the facet and the current density to achieve ~ 1 mA/sr emission intensity remains low (< 10^4 A/cm^2) throughout the range of practical emitter radii. Empirically it has been found that Eq.(2) is of the form

$$I' = 20.42\, J_C\, r^{1.87}$$

for the faceted end form emitter. The low values of J_C required to obtain a value of I' ≅ 1 mA/sr for r > 0.3 μm mean that the continuum space charge effect on trajectories and surface field is negligible as shown earlier [4].

Typically, the ZrO/W(100) cathode can be operated stably for values of I' and r in the range of 0.1 to 2 mA/sr and 0.3 to 1.5 μm respectively. The apparent source brightness, given by

$$B_A = \frac{4I'}{\pi d_V^2},$$

can be calculated for I' = 1 mA/sr and d_V ≅ 10 nm to be $B_A = 1.3 \times 10^9$ A/cm^2 sr.

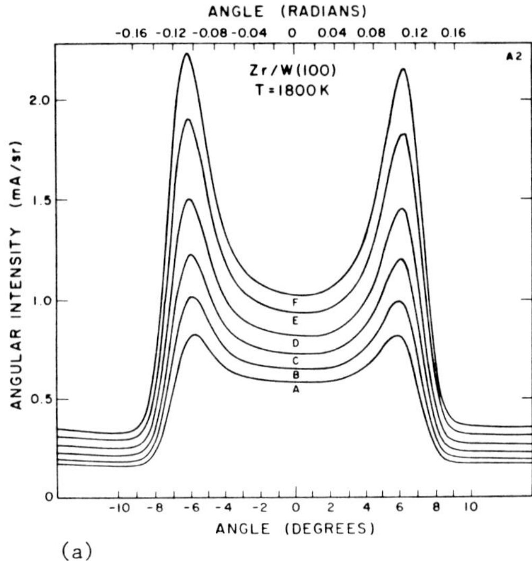

(a)

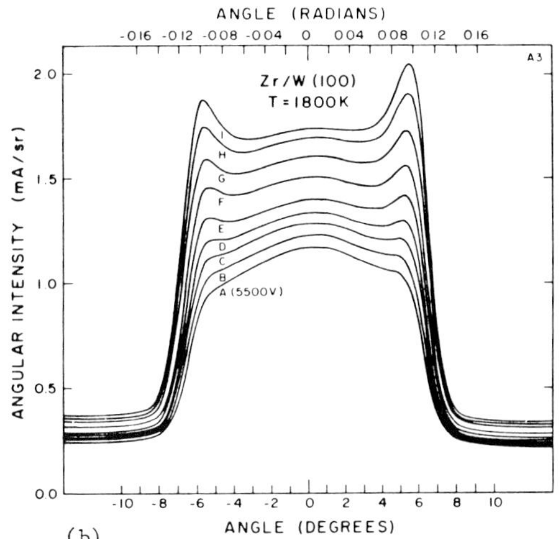

(b)

Figure 8.

Experimental values of the angular current distribution for a ZrO/W(100) TFE source at 100 V anode voltage increments: (a) Emitter radius was 0.8 μm and anode voltage for curve A was 4300 V; (b) Emitter radius was 2.0 μm and anode voltage for curve A was 5500 V.

The experimental angular intensity distribution for two emitter radii is given in Figure 8. In Figure 9 are the predicted angular intensity distribution curves using the SCWIM program for a 0.3 and 3.0 μm faceted emitter when the axial value of I' ≅ 1 mA/sr. The predicted and experimental curves show remarkably good agreement. The high value of I' at the edge of the facet is due to the locally high field strengths. However, as r increases or V decreases the emission mode shifts from TF to pure Schottky emission thus reducing the effect of local field on the emission distribution.

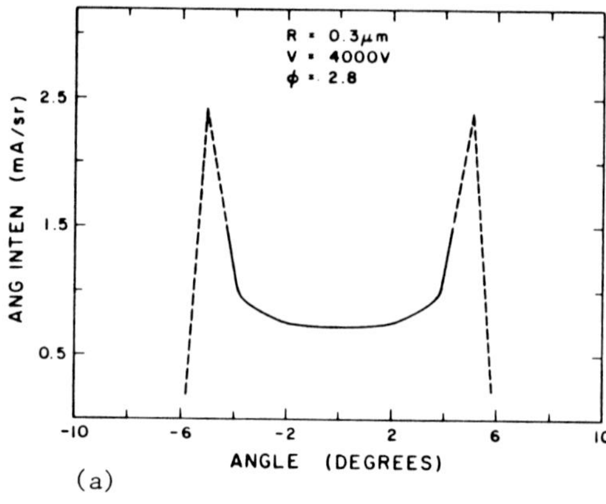

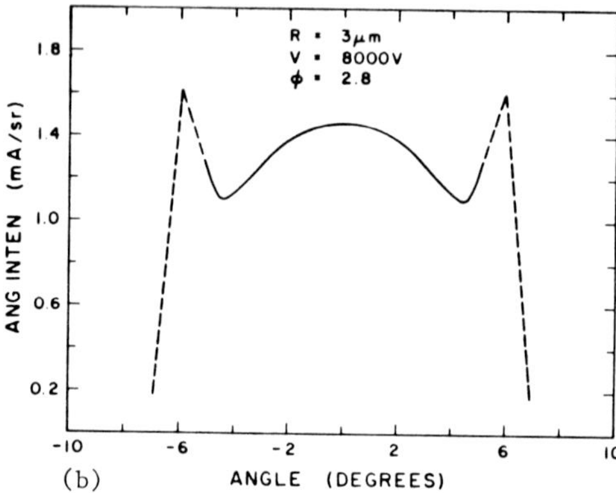

Figure 9.
Calculated angular distribution curves for a faceted emitter using the SCWIM computer program: (a) Emitter radius was 0.3 μm; (b) Emitter radius was 3.0 μm.

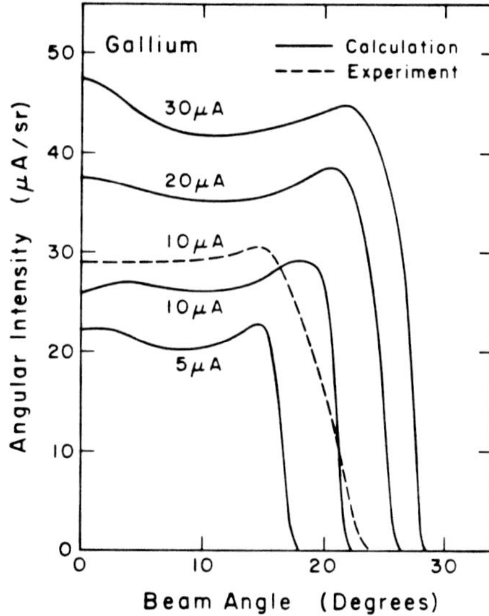

Figure 10. Calculated angular intensity in the exit plane as a function of beam angle for a Ga source at currents between 5 and 30 μA. The experimental results for a Ga LMIS at 10 μA are also shown.

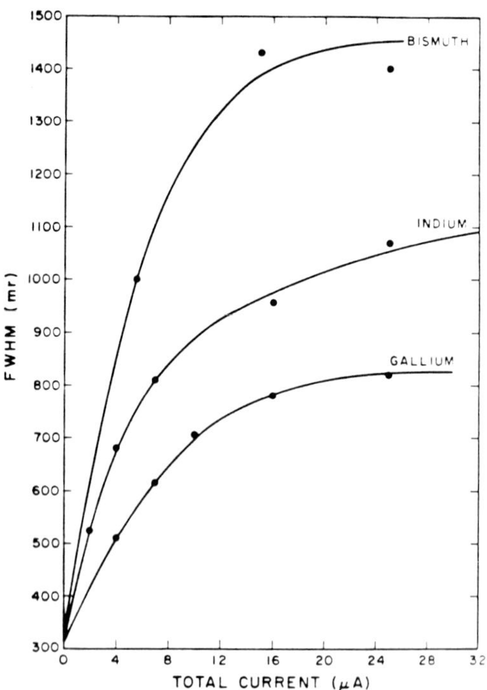

Figure 11. Curves show the experimental values of full width at half maximum (FWHM) of the Figure 10 angular distribution curves vs current for the various LMIS.

LMIS

The angular intensity distribution for the LMIS is similar for all sources and typically shows a uniform distribution whose angular divergence increases with current and mass. Figure 10 shows the current dependence of the angular distribution for a Ga LMIS. The dashed line is a theoretical prediction based on an elongated emitter shape using the SCWIM program for trajectory calculation [6]. Figure 11 shows the mass dependence of the full width at half maximum of the angular divergence. For a low mass LMIS such as Al the value of I' at the onset of current is ~ 40 μA/sr. In contrast for a Bi LMIS the onset value is I' ≅ 10 μA/sr. Because of the small size of the emitting area of the LMIS, the current density is remarkably large--believed to be in excess of 5×10^8 A/cm^2. Thus, space charge effects in the beam are important. Nevertheless the analysis shows that the increase in beam divergence with current and mass is due to both space charge effects and trajectory modification due to elongation of the cone shaped liquid metal emitter [6].

The substantial decrease in experimental axial values of I' with mass for I = 10 μA is shown in Figure 12. This result is in agreement with the Figure 11 results and is due to the mass dependent space charge effects and trajectory modification as mentioned earlier. Using the elongated, cone shaped model of the LMIS and the SCWIM program, the calculated curve in Figure 12 for I' vs mass at I = 10 μA shows good agreement

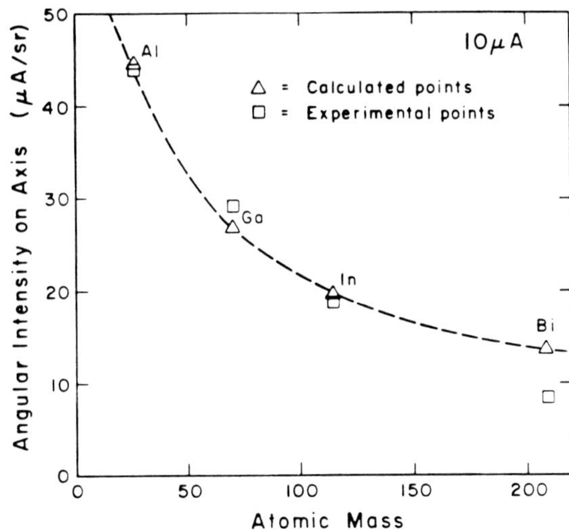

Figure 12. Points show the experimental and calculated (using the SCWIM program) values of the axial angular intensity vs atomic mass for a beam current of 10 μA.

with experimental results [6].

If one assumes the previously mentioned experimental values of d_v = 50 nm and I' = 20 μA/sr, then one obtains a source brightness for a Ga LMIS of B = 1 × 10^6 A/cm^2 sr. Although higher values of B can be obtained at larger values of I', as will be shown below, the increase in beam energy spread negates any optical advantage of larger B.

Energy Distribution

According to Eq.(6) the beam current in a focused beam dominated by chromatic aberration is inversely proportional to ΔE^2. In addition to an intrinsic energy spread ΔE_i, it has been found that an additional, current and emitter radius dependent contribution $\Delta E_c(I,r)$ occurs for all high field sources. The origin of ΔE_c for high field, point sources has been a matter of some discussion and is believed to be due to relaxation of initial Coulomb potential energy caused by random density fluctuations of the charged particles [7,17].

ZrO/W(100) TFE Sources

From the numerical integration of Eq.(13) the theoretical energy distribution and, hence, $\Delta E_i(\phi,F,T)$ can be obtained for a FE source. The results of such a calculation are given in Figure 13 as a function of J (obtained from Eq.(12)) for specified ϕ and J values. For emitters with ϕ < 3.0 eV and J $\lesssim 10^5$ A/cm^2 and operating temperature T $\gtrsim$ 1500 K the value of ΔE_i decreases with increasing T. Thus for typical operating parameters of the ZrO/W TFE ($\phi \cong 2.8$ eV, J = 10^3 to 10^5 A/cm^2, T = 1800 K) the value of ΔE_i < 0.6 eV. On the low temperature side of the maxima in the Figure 13 curves the emission mode is primarily field emission, whereas on the high temperature side of the maxima the emission mode is primarily Schottky emission and ΔE_i becomes small and approaches kT. The unique

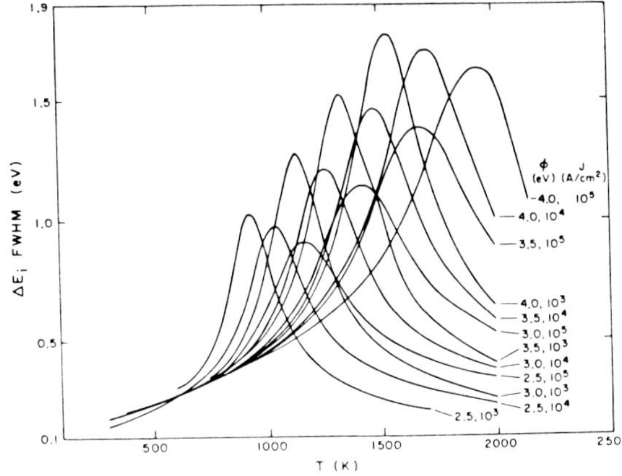

Figure 13. Calculated values of the full width at half maximum of the total energy distribution curves as a function of temperature. The plots are given for the indicated values of work function ϕ and current densities J.

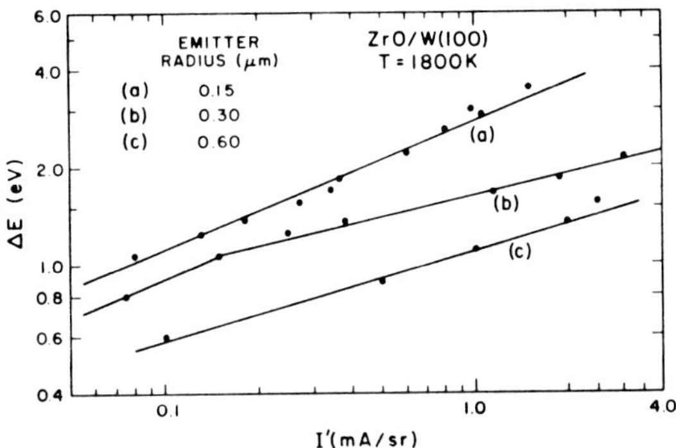

Figure 14. Experimental values of the full width at half maximum of the energy spread ΔE vs angular intensity I'. The results are given for three values of emitter radius.

operating parameters of the ZrO/W(100) emitter allow it to function as a pure FE at low T and high F or as a pure Schottky emitter at high T and low F.

Figure 14 shows experimental values of ΔE vs I' for ZrO/W(100) emitters of differing emitter radius. For values of I' $\lesssim$ 0.1 mA/sr ΔE approaches the theoretical value of $\Delta E_i \lesssim 0.6$ eV; however as I' increases or as r decreases ΔE increases due to the aforementioned stochastic Coulomb interactions. Thus, in order to reduce the chromatic aberration contribution at high values of I' the ZrO/W(100) emitter radius should be large. In practice typical values of r are in the range 0.5 to 1.0 μm and at I' = 1 mA/sr the relatively low values of J = 1.5 × 10^3 to 5.4 × 10^3 A/cm^2 are obtained.

LMIS

As in the case for electron emission from the ZrO/W(100) TFE it has been observed that the LMIS also shows an increase in the FWHM of the total energy distribution (TED) with current. Figure 15

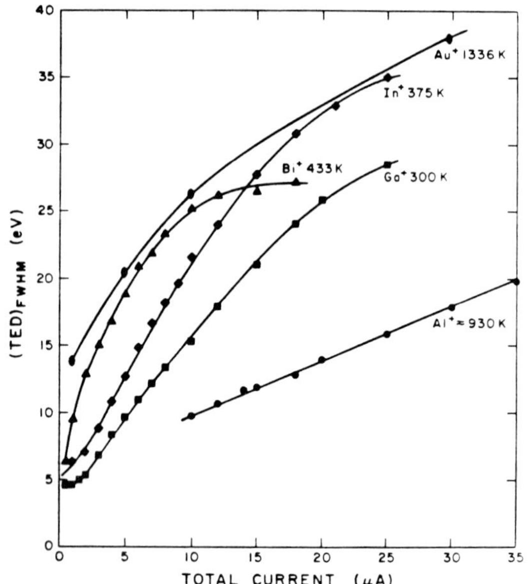

Figure 15. Experimental values of the full width at half maximum of the total energy distribution vs current for the indicated ions of a single component LMIS.

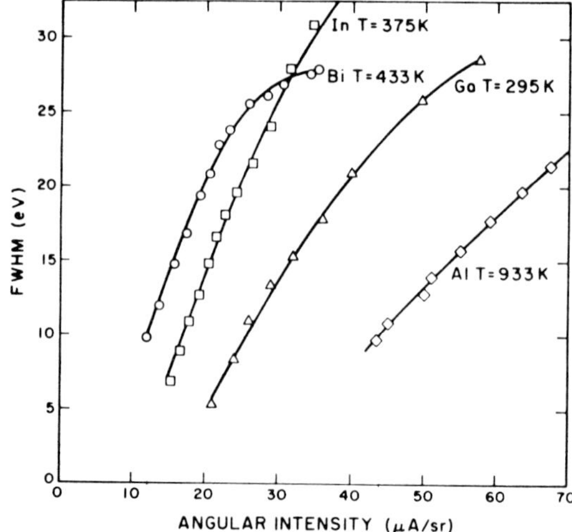

Figure 16. Experimental values of the full width at half maximum of the total energy distribution vs angular intensity for the singly charged ions of the indicated single component LMIS.

shows the experimentally observed variation of ΔE with total source current for several pure metal LMIS [12]. As $I \rightarrow 0$ the value of $\Delta E \rightarrow 5$ eV for nearly all the LMIS investigated whereas at high values of I a strong mass dependence of the ΔE values is observed.

The mass dependence of the ΔE values can be seen more dramatically in Figure 16 where its variation with the axial value of I' is shown. According to Eq.(6) the source figure of merit for chromatically limited micro-focus applications is given by the value of $I'/\Delta E^2$ when $d/m > d_v$. Clearly, from Figure 16 one can see that $I'/\Delta E^2$ decreases rapidly with increasing mass at a fixed value of ΔE. For example, at $\Delta E = 10$ eV the figure of merit values for Al and Bi LMIS are 0.45 and 0.1 μA/sr $(eV)^2$ respectively.

Although few theoretical analyses of the energy broadening of LMIS with increasing I and mass have been carried out to date, it is generally believed that, as in the case of the TFE source, a stochastic Coulomb interaction in the beam is the primary mechanism responsible for the Figures 15 and 16 results.

Conclusions

The analysis of two promising electron and ion field sources shows that high angular intensities and small values of virtual source size can be expected. Typically source brightness values in excess of 1×10^6 and 1×10^9 A/cm^2 sr can be obtained from the LMIS and ZrO/W TFE electron source. Because of the increase in energy spread with angular intensity for these sources, one must carefully optimize the source operating conditions for a specific optical column. In addition, for LMIS an increase in ion mass reduces angular intensity and increases energy spread for a given source current.

Acknowledgments

This work was supported in part by NSF grant ECS-8206796. The author is grateful to A. Bell for helpful discussions and to G. Schwind for obtaining many of the experimental results.

References

[1] Bell A, Schwind G, and Swanson L. (1982). The Emission Characteristics of an Al Liquid Metal Ion Source. J. Appl. Phys. 53, 4602-4605.

[2] Gaubi H, Sudraud P, Tence M, and van de Walle J. (1982). Some New Results About In Situ TEM Observations of the Emission Region in LMIS, in: Proc. 29th Int. Field Emission Symposium, Goteborg, Sweden, H. O. Andren and N. Norden (ed), Almqvist and Wiksell Int., Stockholm, 357.

[3] Kang N, Orloff J, Swanson L, and Tuggle D. (1981). An Improved Method for Numerical Analysis of Point Electron and Ion Source Optics. J. Vac Sci. Technol. 19, 1077-1081.

[4] Kang N, Tuggle D, and Swanson L. (1983). A Numerical Analysis of the Electric Field and Trajectories With and Without the Effect of Space Charge for a Field Electron Source. Optik 63, 313-331.

[5] Kang N, Swanson L. (1983). Computer Simulation of Liquid Metal Ion Source Optics. Appl. Phys. A30, 95-104.

[6] Kingham D, Swanson L. (in press). Shape of a Liquid Metal Ion Source; A Dynamic Model Including Fluid Flow and Space-Charge Effects. Appl. Physics A.

[7] Knauer W. (1981). Energy Broadening in Field Emitted Electron and Ion Beams. Optik 59, 335-354.

[8] Komuro M. (1983). Liquid Metal Ion Sources--Mass Spectrometry Study of Ga, In, Sn, Au, Pb, and Bi. Proc. Int. Ion Eng. Congress, Kyoto, 337-348.

[9] Komuro M, Hiroshima H, Tanoue H, and Kanayama T. (1983). Maskless Etching of a Nanometer Structure by Focused Ion Beams. J. Vac. Sci. Technol. B1, 985-989.

[10] Murphy E, Good R. (1956). Thermionic Emission, Field Emission, and the Transition Region. Phys. Rev. 102, 1464-1473.

[11] Speidel R, Kurz D. (1977). Richtstrahlwertmessungen an einem Strahlerzeugungssystem mit Feldemissionskathode. (Brightness measurements at an electron gun with field emission cathode.) Optik 49, 173-185.

[12] Swanson L. (1983). Liquid Metal Ion Sources: Mechanisms and Applications. Nucl. Inst. Meth. Phys. Res. 218, 347-353.

[13] Swanson L, Bell A. (1973). Recent Advances in Field Electron Microscopy of Metals. Adv. Electronics & Electron Phys. 32, 193-309.

[14] Swanson L, Tuggle D, and Li J. (1983). The Role of Field Emission in Submicron Electron Beam Testing. Thin Solid Films 106, 241-255.

[15] Tuggle D, Swanson L, and Orloff J. (1979). Application of a Thermal Field Emission Source for High Resolution, High Current e-Beam Microprobes. J. Vac. Sci. Technol. 16, 1699-1703.

[16] Wiesner J, Everhart T. (1973). Point-Cathode Electron Sources--Electron Optics of the Initial Diode Region. J. Appl. Phys. 44, 2140-2148.

[17] Yau Y, Groves T, and Pease R. (1983). Space Charge Effects in Focused Ion Beams," J. Vac. Sci. Technol. B1, 1141-1144.

Discussion with Reviewers

P. B. Sewell: For the computations of Fig. 13, work function values of 3.5 and 4.0 are used, rather than values close to 2.8 eV as generally used for the ZrO/W(100) emitter. Is there any reason for this? Also, with reference to the results of Fig. 13, could you comment further on the physical origin of the increase in ΔE_i between the extreme values for the field emission and Schottky modes?

Author: The values of work function used in Fig. 13 vary from 2.5 to 4.0 eV. The value typically measured for the ZrO/W(100) emitter at room temperature is 2.7 ± .2 eV. The maximum in the Fig. 13 curves is due to broadening with increasing temperature of the occupied electron states above the Fermi level from which electrons tunnel. At low temperatures and high electric field ΔE_i is small due to the fact that few states above the Fermi are occupied and at high temperatures ΔE_i is small because the field is low and all electrons are forced surmount the work function barrier.

E. Munro: Table I - it was not very clear to me what the angle α represents in this table. Is it the angle at which electrons are emitted from the cathode, or the angle at which they emerge from the gun? Also the meaning of V was not clear to me, I presume it means the beam voltage at emergence from the gun?

Author: The angle α is the beam angle, i.e. the angle when the electrons emerge from the gun. The symbol V refers to the beam voltage as the particles emerge from the gun region.

Electron Optical Systems (pp. 149-162)
SEM Inc., AMF O'Hare (Chicago), IL 60666-0507, U.S.A.

0-931288-34-7/84$1.00+.05

A COMPARISON OF LANTHANUM HEXABORIDE, COLD FIELD EMISSION AND THERMAL FIELD EMISSION ELECTRON GUNS FOR LOW VOLTAGE SCANNING ELECTRON MICROSCOPY

J. Orloff

Oregon Graduate Center
Department of Applied Physics and Electrical Engineering
19600 N.W. Walker Road
Beaverton, Oregon 97006

Phone no: (503) 645-1121

Abstract

A comparison of lanthanum hexaboride, cold W(310) field emission and Zr/W thermal field emission cathodes was made by calculating the current-spot size relationship for each, using comparable lenses, to determine which would be suitable for high current operation at 1 keV beam energy, with a focused beam diameter $\lesssim$ 0.05 µm. On the criteria of highest current, reasonable operating conditions for the gun for low noise operation and long term cathode stability it was found that the lanthanum hexaboride and cold field emission cathodes are inadequate or marginal and that the best performance is obtainable from the thermal field emission cathode.

Key Words: Field Emission, Thermal Field Emission, Cold Field Emission, LaB_6, Electron Gun, Scanning Electron Microscopy, e-Beam Inspection

Introduction

Electron beam testing of semiconductor devices is a subject of rapidly increasing importance. In fact, e-beam testing may well be the most important use of low voltage scanning electron microscopy. It is of interest to make a comparison of the cathodes which can be used for low voltage SEM, because the performance of an electron gun is quite different at the ≈ 1 keV beam energy appropriate for e-beam testing than at the more usual energies for SEM, E ≈ 20 keV. High energy beams can only be used for testing robust devices which will not be damaged by the penetration of the beam through the passivation layer [59].

The problem of e-beam testing of semiconductor devices is a difficult one because, for a variety of reasons one would often like to work at high speed, which requires high beam current, yet a reasonably high resolution, 0.05 µm or better is usually necessary. The exception to this would be voltage contrast microscopy at a point, but for inspection or line width measurement--perhaps the most important applications in terms of the volume of work--high current and good resolution will both be necessary at low beam energy. This is a difficult requirement.

In this paper we compare three electron guns: lanthanum hexaboride (LaB_6); cold field emission (CFE); and thermal field emission (TFE), for low voltage, high current operation and to indicate which would be best suited for high current, moderately high resolution low voltage SEM. This is done by first comparing the current-spot size relations for the three guns using realistic optics in the voltage, current and spot size regime necessary for e-beam testing; such a comparison allows one to determine which cathode will give useful performance. Next, we consider the noise current, stability and lifetimes of the three kinds of cathodes, including design and vacuum constraints imposed by each.

The criteria for the best cathode are: (1) maximum current into a ≈ 0.05 µm beam spot at low (1 keV) energy; (2) sufficient long-term stability and reliability to be usable in a semiconductor fabrication line.

Symbols

Q charge measured in coulombs

C_s spherical aberration coefficient on lens object side (mm)

C_{si} spherical aberration coefficient on lens image side (mm)

C_{co} chromatic aberration coefficient on lens object side (mm)

C_{ci} chromatic aberration coefficient on lens image side (mm)

δ electron source optical size (μm)

M linear magnification

m angular magnification

α_o divergence angle of beam entering optical system (mrad)

α_i convergence angle of beam on target (mrad)

E beam energy (eV)

ΔE beam energy spread (eV)

V_o voltage of beam on object side of lens (volts)

V_i voltage of beam on image side of lens

V_B beam voltage

h Planck's constant

e electronic charge

d focused beam diameter (Å or μm)

I electron current (A)

I_t total cathode current (A)

I_b focused beam current (A)

J_c cathode current density (A/cm^2)

$\frac{dI}{d\Omega}$ angular intensity (A/sr)

β source brightness (A/cm^2 sr)

F electric field (volts cm^{-1})

ΔZ_i shift of image position due to lens aberrations (μm)

f frequency (Hz)

ΔI noise current (A)

T temperature (K)

θ cone angle of field emitter

γ surface tension (joule/m^2 or dyne/cm)

P pressure (torr)

Properties of LaB_6, CFE and TFE Cathodes

LaB_6

The best thermionic cathode for SEM, in terms of brightness and lifetime, is LaB_6. Properties of this material and its applications as a cathode have been thoroughly treated in the literature [17,18,27,33,39,43,45,46,47,53,54,58,66,70]. The most commonly used thermionic cathode in electron microscopy is made from tungsten (W). While rugged, W cathodes are not able to provide high cathode current density (J_c) and long life simultaneously. The reason is that the work function of W is high--about 4.5 eV--so that it has to be operated at a very high temperature in order to achieve large values of J_c. For example, at a temperature of 2700 K J_c = 1.63 A cm^{-2} and the evaporation rate is 3.2×10^{-8} gm cm^{-2} sec^{-1}. At this rate the cathode life is only ~ 50 hours. Since the evaporation rate depends exponentially on the temperature, the lifetime is greatly shortened by further heating. Thus, while J_c is roughly doubled by raising the cathode temperature from 2700 K to 2800 K, the evaporation rate is increased by a factor of 3.5, from 3.2 to 11×10^{-8} gm cm^{-2} sec^{-1}, and the lifetime is reduced correspondingly.

LaB_6 is a rather unusual material in that its volatility is low when J_c is high by comparison with W, because it has a much lower work function. The work function of the (100) crystal plane of LaB_6 is approximately 2.5 eV [54], and J_c of 1.5 A cm^{-2} can be drawn from it at a temperature of ≈ 1500 K. At this temperature the evaporation rate is ≈ 10^{-13} gm cm^{-2} sec^{-1} [54]. If the temperature is raised to 1700 K, J_c = 13 A cm^{-2} and the evaporation rate is ≈ 10^{-10} gm cm^{-2} sec, implying a cathode life more than an order of magnitude greater than W. Clearly, LaB_6 can be a far superior cathode to W and, indeed, it is successfully used in many commercial SEMs and e-beam lithography systems. Because of the high J_c, LaB_6 cathodes can be made with a small emitting area and in single-crystal form, so that the emission is essentially drawn from only a few or even one crystal plane, and is uniform and stable [17,18,39,47,54,58].

About the only disadvantage of the LaB_6 cathode vis-a-vis the W cathode is the requirement for high vacuum. LaB_6 forms volatile oxides of La and B in the presence of water vapor or oxygen [47], consequently the vacuum must be better than that generally acceptable for W in order to achieve long life. Unfortunately, this means it is not simple to retrofit an SEM designed for a W cathode with a LaB_6 cathode. In our laboratory we have observed that different crystal planes oxidize at different rates, so that emission patterns from the cathode will change with time in poor vacuum ($P > 10^{-7}$ torr) [46]. We have successfully operated cathodes for 3000 hours at 1×10^{-9} torr, while at 10^{-7} torr significant degradation of the cathode is seen in ~ 500 hours.

The brightness β of LaB_6 cathodes has been measured by a number of workers [18,27,47,58] to lie in the range 5×10^5 - 2×10^6 A cm^{-2} sr^{-1} at a cathode temperature of 1800 K and at 20 kV. Variations occur depending on the precise gun geometry and on the crystallographic orientation and shape of the cathode. Hohn et al [27] found the relative brightness of several orientations of conical cathodes having apex radii of 2 μm to be $\beta(100) = \beta(321) > \beta(210) > \beta(311)$. Takigawa et al [58] measured the brightness of <100>, <110> and <111> oriented LaB_6 cathodes with 15 μm radii and found $\beta(100) > \beta(110) > \beta(111)$. Of equal significance are the emission characteristics of cathodes with different end radii. If a conical shaped cathode has a small end radius, both the tip and part of the cone will contribute current to the crossover. If the end radius is large or the end of the cone has a flat ground on it, then it can be arranged that only the end or the flat will contribute significant current to the crossover. Furukawa et al [17] found that a cathode with a 100 μm end radius was capable of a uniform angular distribution at a total current I_t = 0.7 mA, whereas the angular distribution became nonuniform for smaller radii cathodes at lower currents, due to emission from the side of the cone. Although smaller crossover diameters could be achieved at high currents with large radii cathodes, the emittance, i.e. product of crossover size and angular spread of the beam, was also larger.

For use in SEM, the cathode radius seems not to be critical. The orientation is important, to maximize emission. The cone angle is also important, because of the anisotropy of volatilization rates for different crystal planes [46,54], unless the gun vacuum is very good ($\lesssim 10^{-7}$ torr).

At 1 kV, β will be reduced from its value at 20 kV by a factor of 20. This cannot be avoided, whether the electron gun is operated at 1 kV or at higher voltage with the beam decelerated by an electrostatic lens following the gun, however it may be useful to operate in the latter mode. It is well known that a high current electron beam can be spread spatially due to beam interaction effects, and this phenomenon is a function of beam energy [30]. If the beam were extracted from the gun at high voltage and decelerated after much of it had been removed by an aperture, beam interaction effects would be reduced.

The noise current ΔI associated with a beam current I_b will determine over what ranges of I_b and bandwidth an instrument can be employed. All cathodes will exhibit shot noise (statistical fluctuations proportional to $I_b^{1/2}$), and they may also suffer from additional (flicker) noise due to thermal motion of atoms, adsorption and desorption of gas molecules which affect the work function, current spikes from microgeometric changes due to ion bombardment etc. A discussion of the effect of noise as it relates to the above example of e-beam testing will be given below. In our laboratory we have measured the spectral density function and the fractional noise current $\frac{\Delta I}{I_b}$ for a LaB_6 cathode in a commercial SEM, and found that the flicker noise decreased as 1/f, as expected, where f is the frequency, with the spectral density function decreasing to the shot noise level at f ≈ 400 Hz. Shot noise current is given by

$$\Delta I_{shot} = \sqrt{2eI_b f} \quad (1)$$

and for the LaB_6 cathode we found

$$\frac{\Delta I}{I_b} = \left[9.07 \times 10^{-9}\, \ell n\left(\frac{400}{0.1}\right) + \frac{2e(f-400)}{I_b}\right]^{1/2} \quad (2)$$

For a bandwidth of f = 10^6 Hz, and I_b = 1 nA, ΔI is essentially all due to shot noise.

Measurements by Pfeiffer [40] indicate that the energy spread ΔE for an electron beam drawn from a pointed LaB_6 cathode is proportional to $\sqrt{\beta\delta}$, where δ is the optical source size, i.e., the crossover diameter. For $\beta = 1.5 \times 10^5$ A cm^{-2} sr^{-1}, Pfeiffer measured ΔE = 1 eV, when δ ≈ 10 μm.

CFE

CFE has been successfully exploited in commercial and laboratory SEMs, conventional transmission electron microscopes and scanning transmission electron microscopes and has been treated extensively in the literature [4,6-11,13,15,16,20,22,23,32,35,36,41,42, 49,52,60,61,63,64,69]. Field emission is a process whereby electrons are extracted from a conductor, usually a refractory metal, by deforming the potential barrier at the vacuum-metal interface to such an extent that electrons can tunnel through it [23]. This is in contrast with thermionic emission, where thermal energy has to be imparted to electrons to enable them to surmount the potential barrier. The barrier is deformed by applying an electric field F of the order of 10^7 V cm^{-1}. Such a high field can be produced with a reasonable voltage only if the field emitter has a very small radius of curvature, typically 0.01 to 0.3 μm; thus field emitters are made in the form of extremely sharp needles. For a field emitter, J_c is given by [23]

$$J_c = \frac{1.54 \times 10^{-6}\, F^2}{\phi\, t} \exp\left[-6.83 \times 10^7 \frac{\phi^{3/2}}{F} v\right] \text{ A cm}^{-2} \quad (3)$$

where ϕ is the work function and t and v are slowly varying functions of F and ϕ and which are of the order of unity.

It is well known from classical electrodynamics that the stress due to an electric field is proportional to the square of the field. At the high fields necessary for a field emission cathode the emitter is highly stressed [15] and so the emitter must usually be fabricated from a refractory material. The most commonly used material for electron microscope cathodes is W, although there may be more suitable materials, as will be discussed below.

The current distribution from a CFE cathode is usually contained within a cone of half-angle ~ 20° [6]. If one assumes an approximately uniform distribution, then at 20 μA total current the angular intensity $\frac{dI}{d\Omega} \approx 5 \times 10^{-5}$ A sr^{-1} and the corresponding brightness, measured at the emitter is $\beta \approx 2 \times 10^8$ A sr^{-1} cm^{-2}, some 2-3 orders of magnitude greater than that of thermonic cathodes. In addition, the best thermionic cathodes achieve $\beta \sim 10^6$ at relatively high voltages, ≈ 20 kV, whereas the CFE cathode can achieve its high brightness at much lower voltages, with typical operating voltages lying in the range of 3-6 kV.

CFE requires very good vacuum for long-term, stable operation [10,11,35,36,42,60]. There are several reasons for this. Residual gas molecules which adsorb on the field emitter will cause a change in the work function with a consequent change in the field emission current. From Eq.(3), we see that $\frac{dI}{I} \sim 1.5 \frac{d\phi}{\phi}$. A 1% change in ϕ can cause a ≈ 15% change in I [35]. If the adsorbed molecule then diffuses about the emitter surface, the current will fluctuate; this is the source of flicker noise. Another source of noise is sputter induced damage to the emitter caused by ion bombardment of gas molecules ionized by the electron beam. Such damage results in a local change in the radius of curvature of the emitter and therefore of the electric field. The current change is $\frac{dI}{I} \sim \frac{dF}{F}$ and a 1% change in F can cause ≈ 10% change in I. This usually manifests itself in the form of random current spikes. Also, adsorbed gas molecules will be sputtered off the cathode surface, which will result in a random work function and current change. In addition to noise, ion bombardment which causes local changes in radius can lead to emitter failure by the initiation of a vacuum arc. The arc is caused by the increase of field emission current which heats the emitter near the sputtered asperity. The heated region deforms and becomes sharper under the stress of the field. Runaway emission follows, which destroys the emitter [4,6,35]. Since instabilities are seen even at $P \sim 10^{-9}$ torr, it has been found necessary to periodically "flash" the cathode to a high temperature, ≈ 2000 K, to both anneal the emitter and desorb gases [41,50]. The high voltage usually must be shut off during this procedure, necessitating shutting off the SEM. If in the electron gun $P < 10^{-9}$ torr, such tip conditioning is necessary on a time scale ~ 50 hours [10]. Long term operation without flashing requires $P \sim 10^{-12}$ torr [35,36].

Todokoro et al [60] and Saitou [42] have shown that virtually all of the ions which impinge on the emitting region of the cathode are formed very close to the cathode, generally within a few tip diameters, if the pressure is greater than about 2×10^{-11} torr. The percentage fluctuation $\frac{\Delta I}{I}$ due to residual gas pressure was found [42] to be proportional to $\log (P \times I/9 \times 10^{-15})$, where P is in torr and I in amperes, with $\frac{\Delta I}{I} \approx 1\%$ at $P \times I = 9 \times 10^{-15}$ A torr. Thus at $P = 5 \times 10^{-9}$ torr and $I = 40\ \mu A$, $\frac{\Delta I}{I} \approx 3\%$ due to ion bombardment. At $I = 20\ \mu A$, $\frac{\Delta I}{I} \approx 2.5\%$. This underscores the need for high vacuum in the electron gun if high currents are to be produced. An additional source of noise in the CFE cathode is the migration of atoms across the crystal planes. It has been found [52] that for W emitters, this source of noise can be significant at room temperature on the (310) plane, which is commonly the plane oriented on the optical axis of a CFE cathode. The threshold for onset of noise on the (310) plane due to migration of W atoms, is 300 K, while the threshold temperatures on the (112) and (100) planes are 650 K and 1000 K respectively; unfortunately the work functions of the (112) and (100) planes are 4.65 eV and 4.52 eV, respectively, compared with $\phi = 4.35$ eV for the (310) plane. Because of the exponential dependence on $\phi^{3/2}$, much higher electric fields would be required to obtain useful emission from the (110) or (100) planes than from the (310) plane, and it is not practical to orient CFE cathodes along those directons. Based on the results for the measurements of $\frac{\Delta I}{I_b}$ on the (310) plane of W at 900 K [34], it is estimated that at 300 K $\frac{\Delta I}{I_b}$ would be about 0.5%, over a bandwidth 50 Hz < f < 10^5 Hz [20].

In our laboratory we have modified a commercial CFE SEM (CWIKSCAN Model 100) for use with both CFE and TFE cathodes [56]. At $P \approx 7 \times 10^{-9}$ torr, measurements of the noise current of a CFE W(310) cathode, at room temperature, gave $\frac{\Delta I}{I_b} \approx 5\%$ at $I_b = 0.7$ nA and bandwidth 0.1 Hz < f < 25 kHz. The spectral density function fell to the shot noise level at f ≈ 20 kHz. The shot noise current in the bandwidth 0.1 Hz < f < 25 kHz at I = 0.78 nA is $\frac{\Delta I}{I_b} = 0.3\%$ [48], so the noise current is mainly due to flicker noise. Zaima et al [73] found similar results, although with a smaller bandwidth, at $P \approx 8 \times 10^{-10}$ torr.

From these results we see that shot noise and flicker noise due to the thermal motion of atoms on the emitter surface are negligible at room temperature but that there will be significant flicker noise and noise from sputter induced damage and desorption of adsorbed gas unless the pressure is very low. The magnitude of these effects is highly instrument dependent, and will be a strong function of the quality of the design of the electron gun and its vacuum system.

Energy spread measurements indicate that ΔE ≈ 0.2 eV at $\frac{dI}{d\Omega} = 1 \times 10^{-4}$ A sr^{-1}, increasing to ΔE = 1 ± 0.2 eV at $\frac{dI}{d\Omega} = 5 \times 10^{-4}$ A sr^{-1} [3].

Because of the substantial noise current seen with W CFE cathodes, even when $P \approx 10^{-9}$ torr, interest has revived in development of cathodes which are less affected by residual gas, i.e., have a lower probability for adsorption. Martin and coworkers reported [36] field emission from ZrC, which is quite refractory but, perhaps because of the ease of fabrication of W cathodes in comparison with the carbides, it has been little used. Zaima et al [73] recently investigated emission from TaC and found that at $P \lesssim 3 \times 10^{-10}$ torr the flicker noise was absent, although there were still current spikes. These were attributed to sputtering events, as their number was proportional to P × I. The absence of flicker noise was believed to be due to a very low probability for O_2 and N_2 to adsorb on TaC, compared with W. When the pressure was increased to 2×10^{-9} torr, flicker noise was again seen. It could be made to disappear by flashing to 1500 K, when the pressure had been lowered to 3×10^{-10} torr again.

Similar results obtain with TiC emitters. This is a more difficult material to work with because it is hard to produce stoichiometric TiC:

HfC might be a better choice. However, Futamoto et al [19] found quite different results with TiC, noting little improvement over W(310). This was attributed to the reactivity of Ti with O_2 and N_2, when the concentration at the surface increased after heating. This seems to indicate the difficulty of producing, or maintaining the correct stoichiometry at the surface.

TFE

TFE developed out of attempts to overcome the stability problems of CFE by operating the field emitter at high temperature, $T \gtrsim 1500$ K, to anneal sputter-caused damage and to remove adsorbed gas molecules which would cause current fluctuations [14]. This was thought to be essential if field emission cathodes were to be employed on a practical basis, as it was clear that the level of vacuum required for long life without frequent flashing of the cathodes, $P \lesssim 10^{-12}$ torr, was impossible to achieve in any but the most highly specialized instrumentation. The TFE cathodes which have been developed have proven to be unique in their capabilities; they have been carefully studied [3,12,14,34,48,50,51,55,62] and a number of technologically important applications have been reported [28,29,31,56,52,65,67,71,72].

When a field emitter is heated in the absence of an electric field, the atoms migrate from the emitter apex towards the emitter shank [2,14], with the rate of increase of radius (or dulling) of the apex given for W, by

$$\frac{dr}{dt} = 2.6 \times 10^{-11}\ \theta\ \exp\left[\frac{-\ 36300}{T}\right]\ (Tr^3)^{-1}\ \left(\frac{cm}{sec}\right), \quad (4)$$

where θ is the cone angle of the emitter, and T the temperature [41,42]. If an electric field is applied, the rate of dulling is modified by the factor $(1 - \frac{rF^2}{8\pi\gamma})$, where γ is the surface tension [14,48]. If $F = F_c = (\frac{8\pi\gamma}{r})^{1/2}$, the emitter should be stable; if F is less than or greater than F_c, the emitter will either become duller or sharper with time, respectively. Of course, the local radius is different at different locations near the apex of the emitter, so the value of F can vary and may exceed F_c at some points and equal F_c at others. If $F > F_c$, the emitter behavior is rather complex, because different crystal planes have different energies for the nucleation of new atomic layers and some will facet more quickly than others. The result is the emitter assumes a polyhedral shape ("build-up") [5]. Emission current can be very high at the intersections of the crystal planes, leading to further heating and eventual destruction of the emitter; consequently, there are few stable shapes.

The surface tension of W is 2.9 joule m^{-2} = 2900 dyne cm^{-1}, so $\frac{dr}{dt}$ nominally vanishes when $F_c = 8.1 \times 10^4\ r^{-1/2}$ V cm^{-1}. For clean W, the work function of 4.5 eV requires F ~ 4 to 8 × 10^7 V cm^{-1} in order to obtain useful currents, that is I_t ~ 1 to 1000 μA, according to Eq.(3). If F = 6 × 10^7 V cm^{-1}, corresponding to I ~ 10 to 100 μA, r < 0.02 μm in order to avoid buildup. This is an extremely small value for the radius.

There are two practical, tungsten TFE cathodes. One is clean W, oriented in the <100> crystalline direction (W(100)), the other is zirconiated, <100> oriented W (ZrO/W(100)) [51]. No other practical TFE cathodes have been reported, although it is possible to fabricate TFE cathodes from other refractory materials. We limit our discussion to these two.

The W(100) TFE cathode is formed by operating a slightly oxidized emitter at ≈ 1800 K and allowing build-up to occur. The apex of the emitter changes shape as the (110), (112) and (310) planes facet at the expense of the (100) plane [48]. After a fairly short time, the (100) plane is reduced to a very small area at the end of a pyramidal shape. This area is $\lesssim$ 100 Å in diameter and consequently the local radius of the emitter is quite small, so electron emission is very intense and localized to within ≈ 6° of the axis of the emitter [65] and $\frac{dI}{d\Omega}$ = 1 mA sr^{-1} is easily attained. Long lifetimes have been measured for this cathode, and it can be operated reliably at pressures up to 1 × 10^{-8} torr [51].

There are two difficulties with the W(100) cathode. Because the area of emission is extremely small J_c is extremely high, ~ 10^7-10^8 A cm^{-2} at $\frac{dI}{d\Omega} = 10^{-3}$ A sr^{-1}. Consequently, the energy spread in the beam is quite large [3,51], ΔE ≈ 2-3 eV. This severely tests the electron optics of any system. A second problem stems from the very small area of emission; since only a rather small number of atoms are included in this area, any change in the number or position of these atoms will cause a significant fluctuation in the beam current [22]. Noise studies on W(100) typically show $\frac{\Delta I}{I}$ ≈ 3-10% in the frequency interval 1 Hz < f < 10^4 Hz at currents ranging from 30 nA to 220 nA. These two characteristics make it difficult to apply the cathode for electron beam testing.

The ZrO/W(100) TFE cathode takes advantage of the fact that ZrO selectively lowers the work function of the (100) plane of W to ≈ 2.6 eV [23]. From Eq.(3) we see that a reduction of ϕ from 4.5 eV to 2.6 eV would permit a reduction of F by a factor of ~ 2 while maintaining J constant. Consequently, it is possible to operate the ZrO/W cathode with a radius ~ 0.1 μm at high angular intensities while remaining below the field strengths that would cause build up [48,50,51]. It is actually possible to use cathodes with even a larger emitting area, ~ 1 μm in diameter, because the low work function (100) plane forms a relatively stable facet after which the emission current is unchanged.

Many more emitting sites are included than in the case of W(100), and noise studies confirm this, with $\frac{\Delta I}{I_b}$ typically < 1% in the interval 1 Hz < f < 10 kHz and I_b ranging from 25 nA to 250 nA [51,62]. At frequencies up to ~ 25 kHz the main component of noise in the current is flicker noise due to thermal motion of the atoms in the emitting area. The spectral density function falls off slowly in

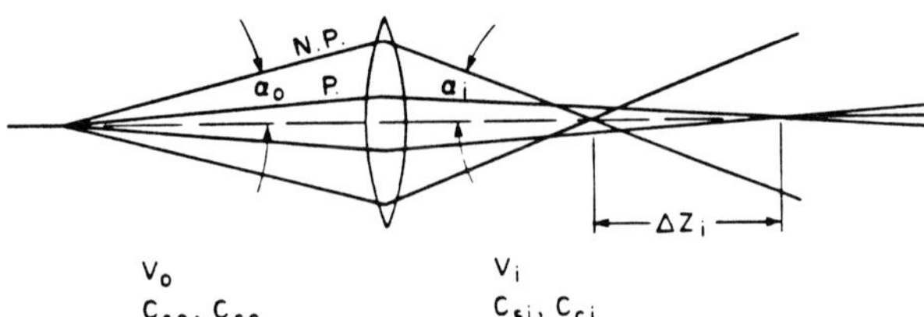

Figure 1. A schematic representation of a lens with spherical and chromatic aberration coefficients C_{so} and C_{co} referred to object space, C_{si} and C_{ci} referred to image space. The spherical aberration causes a shift Δz_i in the image position for non-paraxial (N.P.) rays subtending angle α_0, compared with paraxial (P.) rays. V_0 and V_i are the potentials of object and image space, respectively, and $\alpha_i = m\alpha_0$.

the range 1 Hz-10 kHz and then decreases as ~ $\frac{1}{f}$ [62], reaching the shot noise level at roughly 25 kHz [48]. Empirically, it has been found [34] that

$$\frac{\Delta I}{I_b} = \left[\frac{2.53 \times 10^{-9}}{\alpha_0} + \frac{2e\,(f - 25{,}000)}{I_b}\right]^{1/2} \tag{5}$$

At a temperature $T \approx 1800$ K and $P < 1 \times 10^{-8}$ torr, current spikes are not seen in either the W(100) or the ZrO/W cathode.

Method of Calculation

We now make a comparison of the current-spot size relations for the three electron guns. This is done by calculating the contribution to the final beam spot size of: (1) the optical size of the electron source; (2) the spherical and chromatic aberrations of the optical system; and (3) the effect of diffraction at the beam limiting aperture. For the relatively small viewing areas involved, the off-axis aberrations such as coma are not important and are ignored. This topic has been thoroughly developed and notation standardized in numerous articles on electron optics; the reader is referred to the standard textbooks for complete treatments, e.g., Klemperer and Barnett [30], Grivet [24], Septier [44], Hawkes [26], Glaser [21], and Zworykin et al. [74]. We briefly review the concept.

Spherical aberration is the result of a lens focusing rays which are farther from the axis more strongly than those which are close to the axis, as shown in Figure 1. The resultant minimum beam diameter for a point object is

$$d_s = \frac{1}{2} C_{si}\, \alpha_i^3 \tag{6}$$

where C_{si} is the spherical aberration coefficient (units-length) referred to the image side of the lens. The aberration coefficient when referred to the object side of the lens is $C_{so} = \left(\frac{V_0}{V_i}\right)^{3/2} M^{-4} C_{si}$, in which case $d_s = \frac{1}{2} M C_{so} \alpha_0^3$. α_i and α_0, the angles of the trajectories with respect to the lens axis are defined in Figure 1. The linear magnification of the lens is M and the corresponding angular magnification is $m = M^{-1} \left(\frac{V_0}{V_i}\right)^{1/2}$. Here, V_0 and V_i refer to the energy, or voltage of the electron beam on the object and image sides of the lens, respectively. For a magnetic lens $V_0 = V_i$; V_0 is often different from V_i for an electrostatic lens.

Chromatic aberration is a lens defect caused by the inability of a lens to focus particles of different energies initially following identical trajectories, to the same point. This effect is proportional to the spread of energy in the beam, thus if $E = E_0 \pm \Delta E$,

$$d_c = C_{ci} \frac{\Delta E}{E_i} \alpha_i = M\, C_{co} \frac{\Delta E}{E_0} \alpha_0 \tag{7}$$

where C_{ci} and C_{co} are the image and object side aberration coefficients, respectively, $E_0 = eV_0$ and $E_i = eV_i$ are the nominal beam energies on the object and image sides of the lens, respectively and ΔE the spread in the energy of the beam, usually taken to be the full width at half maximum of the current vs energy distribution. The units of C_c are length and $C_{co} = (V_0/V_i)^{3/2} M^{-2} C_{ci}$.

The wave nature of matter is expressed by the deBroglie relation $\lambda = \frac{h}{p} = \frac{h}{\sqrt{2mE}}$. For an electron, $\lambda \approx \frac{12}{\sqrt{V}}$ Å, where V is the voltage through which the electron has been accelerated from rest. This manifests itself in the diffraction of a beam of electrons when it passes through a small aperture. If the aperture is on the image side of the lens, the effect of diffraction is to contribute to the final beam size an amount

$$d_d \approx \frac{15}{\sqrt{V_i}\, \alpha_i} \text{ Å} \tag{8}$$

If the aperture is on the object side of the lens, $d_d \approx \frac{M\,15}{\sqrt{V_0}\, \alpha_0}$ Å.

Finally, there is the contribution of the optical size of the source. This is $d_g = M\delta$, where δ is the optical diameter of the crossover and M is the total linear magnification of the optical system: if the system consists of several lenses with magnification $M_1, M_2 \ldots M_k$, then $M = M_1 \times M_2 \ldots \times M_k$. In a field emission gun there is no actual, physical crossover; δ is the "virtual" crossover diameter, determined by tracing the tangents to the trajectories far from the field emitter, back inside the emitter [16,69]. The waist of these tangents gives δ (see Fig. 2). The crossover in a gun with a thermionic

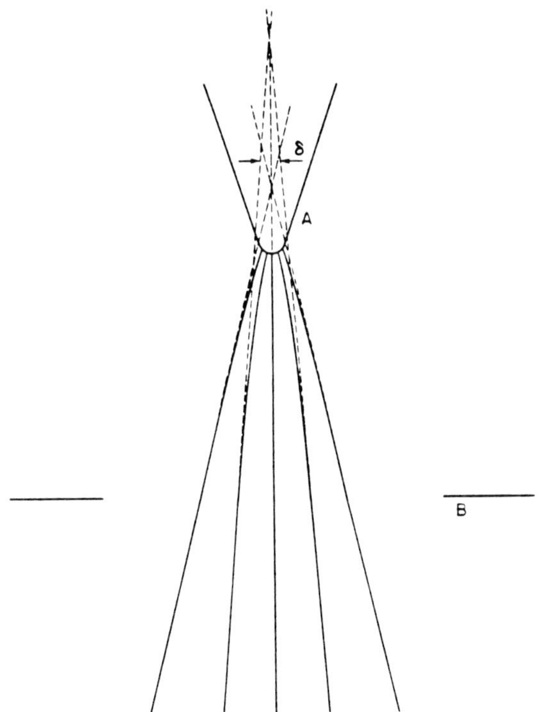

Figure 2. Schematic diagram indicating the origin of the virtual source size δ for a field emitter A. δ represents the minimum diameter subtended by the tangents to the trajectories when extended back from the aperture B to their intersection inside the emitter. Drawing is not to scale.

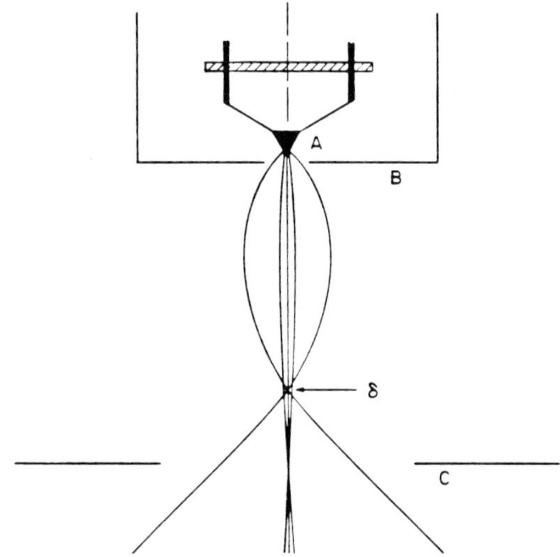

Figure 3. Schematic diagram of the crossover with diameter δ in a thermionic gun. Cathode is A, Wehnelt is B, anode is C. Trajectories are greatly exaggerated. In actuality, the angles would be very small and an image of the cathode would be formed below the anode.

cathode is the waist of the current distribution produced by an optical system consisting of the cathode, an anode and a control electrode (wehnelt), as shown in Figure 3. By proper location of the physical elements with respect to one another and by application of appropriate voltages, the crossover can be made small and uniform in cross section [1,25,57]. The final beam spot size is estimated by [68]

$$d^2 = d_s^2 + d_c^2 + d_d^2 + M^2\delta^2 \tag{9}$$

A useful measure of the current, crossover size and angular confinement of the electron beam from a gun is the brightness, β, which has units of amperes per square centimeter per steradian:

$$\beta = \frac{I}{\frac{\pi}{4}\delta^2 2\pi[1 - \cos\alpha_0]}.$$ For small angles,

$\alpha_0 << 1$,

$$\beta = \frac{I}{\frac{\pi}{4}\delta^2 \pi \alpha_0^2} \quad . \tag{10}$$

The solid angle containing the beam current I is determined by the angle α_0. In the limit $\delta \rightarrow 0$, $\alpha_0 \rightarrow 0$, $I \rightarrow 0$, $\frac{\beta}{V}$ is a conserved quantity for the optical system, where V is the beam voltage. For a finite α_0 and δ, $\frac{\beta}{V}$ is degraded by the lens aberrations. It can be shown [37] that $\beta \approx \frac{J_c}{\pi}\frac{eV}{kT}$ for a thermionic cathode, where J_c is the current density at the cathode surface. The larger β, the more current can be delivered to a given spot size within a given solid angle on the target. Typically, at V = 20 kV, $\beta \approx 10^4$-10^5 for W and $\beta \approx 10^6$ for LaB_6.

It is sometimes convenient to characterize an electron gun in terms of the angular intensity, dI/dΩ. This is true if the source dimensions are small compared with the desired focused beam size, as in a field emission gun; in that case the beam diameter will generally be determined by the aberrations of the optical system.

β is then given by $\frac{4dI/d\Omega}{\pi\delta^2}$ and, for $\alpha_0 < 10^{-1}$ rad

$$I_b = \pi\alpha_0^2 \frac{dI}{d\Omega} \tag{11}$$

Since J_c depends exponentially on F for a field emitter the brightness is exponentially dependent on the applied voltage, which determines the field. Because of the strong dependence of the current on the voltage, one usually chooses a fixed operating or extraction voltage, V_E and varies the beam energy by means of an electrostatic lens.

In an SEM with a thermionic cathode there are usually two or three lenses and the beam is demagnified at each lens. δ is typically 10-50 µm, depending on the geometry of the gun and the type of cathode employed. If the final beam spot

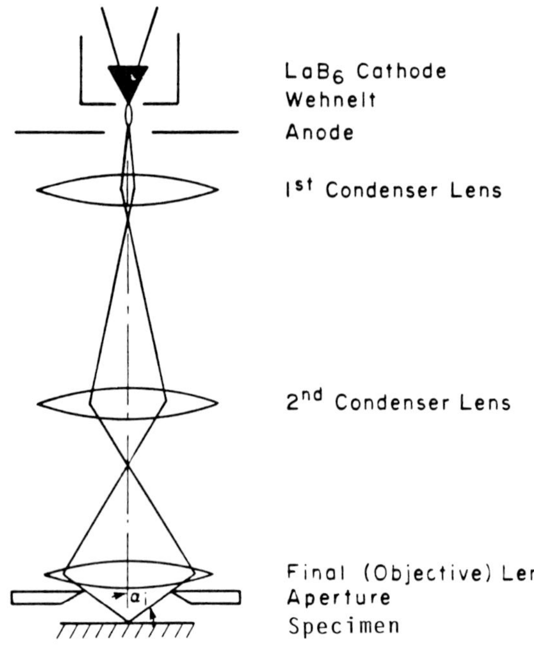

Figure 4. Schematic diagram of the optical system for an SEM employing a LaB_6 thermionic cathode. Scan coils, astigmatism correction coils and spray apertures are not shown, for simplicity.

is to be, say 100 Å in diameter, then if δ = 25 µm, M $\lesssim 4 \times 10^{-4}$, with the equality holding only if the aberrations are negligible compared to 100 Å. The situation is very different for field emission. Here, $\delta \approx$ 50 Å for a CFE source and $\approx$ 150 Å for a TFE source [16,69], hence M $\approx$ 1, and fewer lenses may be needed; the very different system magnifications result in very different designs for thermionic and field emission optical systems.

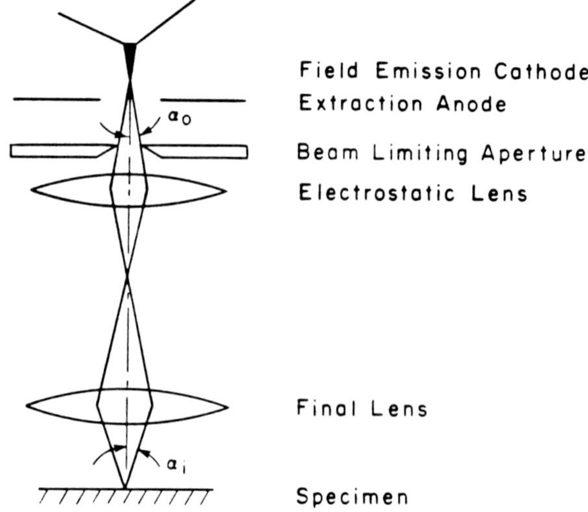

Figure 5. Schematic diagram of the optical system for an SEM with a field emission cathode. Scan and astigmatism correction coils are not shown. The beam limiting aperture is placed above the electrostatic lens so that I_b remains constant as the beam energy is changed, since α_i is a function of e_{beam}.

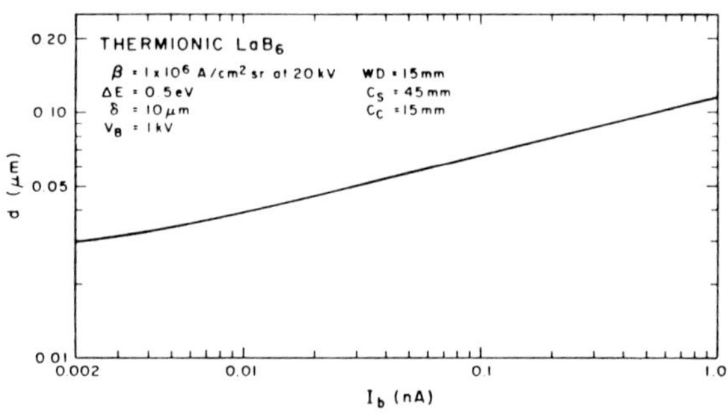

Figure 6. Focused spot size d vs beam current I_b for a thermionic LaB_6 electron gun, with optical parameters and objective lens parameters as shown.

Parameters for Calculations

For the purpose of comparing the current-spot size relations of these three cathodes, we assume that only the final, objective lens is important for the LaB_6 system, and for the calculations, employ a magnetic lens with good aberration coefficients. We may do this because, the contributions to the final spot size of the lens aberrations are important only for reasonably large values of the angle α. Since the magnification of the final lens will be $\lesssim$ 0.1, the angular magnification will be $\gtrsim$ 10 and so the angle subtended in lenses preceding the final lens will be negligible. For the field emission cathodes, we assume a two-lens system consisting of an electrostatic gun lens and a magnetic final lens, with overall magnification $\approx$ 1. We assume the same final lens as for the LaB_6 system, although operated with different magnification. Consequently, the aberration coefficients of the final lens are different in the field emission cases than in the LaB_6 case.

In all cases we assume a working distance of 15 mm from the polepiece of the final lens. For the gun lens we chose a particular design of a three-element, asymmetrical electrostatic lens with good chromatic aberration properties [38], which is used to decelerate the beam. The optical systems used for the calculations are shown schematically in Figures 4 and 5, while the details of the system parameters are shown in Tables 1 and 2.

From the brightness relation we find $\frac{dI}{d\Omega}$ for the LaB_6 cathode to be $\frac{dI}{d\Omega} = \beta \frac{\pi\delta^2}{4}$. With δ = 10 µm, $\frac{dI}{d\Omega} = 3.9 \times 10^{-2}$ A sr^{-1} at 1 kV. If we now replace α_0 by $\left(I_b/\pi \frac{dI}{d\Omega}\right)^{1/2}$ in equation (9), we find the optimum value for M, M_{opt} as a function of I_b by differentiation. M_{opt} was calculated for each value of I_b, and then used to find d. The values ranged from $M_{opt} = 1.53 \times 10^{-3}$ at $I_b = 10^{-12}$ A, to $M_{opt} = 1.35 \times 10^{-2}$ at $I_b = 10^{-9}$ A. A β of 1×10^6 A cm^{-2} sr^{-1} at 20 kV was chosen, which is reduced to 0.5×10^5 A cm^{-2} sr^{-1} at 1 kV, and with δ = 10 μm, Pfeiffer's results [40] were used to assign a value of ΔE = 0.5 eV.

For the field emission guns, d and I_b were calculated using Eqs.(9) and (10), with α_o the variable and the other parameters taken from Table 2.

Table 1

Parameters for the LaB_6 Gun

β	1×10^6 A cm^{-2} sr^{-1} at 20 kV
ΔE	0.5 eV
δ	10 μm
V_B	1 kV
C_{si}	45 mm
C_{ci}	15 mm

Results of Calculations and Discussion

The results of calculations of d vs I_b are shown in Figures 6 and 7. In addition, the convergence half-angles of the beam on the target, α_i, are shown in Figure 8.

For the case of LaB_6, I_b = 1 nA at d ≈ 0.1 μm and I_b ≈ 0.03 nA at d = 0.05 μm. In addition, at I_b = 1 nA, $\alpha_i \approx 10^{-2}$ rad, implying a depth of field of approximately 10 μm.

The case of CFE is quite different. We have plotted d vs I_b for four cases: $\frac{dI}{d\Omega} = 1 \times 10^{-4}$ and 5×10^{-4} A sr^{-1} with V_E = 3 kV and 6 kV. The two extraction voltages were chosen to show the effects of chromatic aberration, which are also evident from the difference in spot size for the two values of $\frac{dI}{d\Omega}$ for a given value of I_b. There could be an uncertainty of ± 0.2 eV in ΔE and so curves C and D could be lower or higher by 20%, since the minimum in the d vs I_b curve is determined by the value of d_c. Higher values of angular intensity were not used because of the rapid increase of ΔE.

The highest resolution, d = 0.03 μm, was achieved with the CFE cathode at $\frac{dI}{d\Omega} = 1 \times 10^{-4}$ A sr^{-1} and I_b = 0.3 nA. This is a factor of 2.5 better than LaB_6 at the same I_b. However, CFE operation at the higher angular intensity results in much larger values of d.

The TFE cathode, Zr/W(100), provides good resolution, with d having a minimum of ≈ 0.057 μm and I_b = 2.5 nA: five times as much current as

Table 2

Parameters for the Field Emission Guns

	CFE				TFE	
	A	B	C	D	E	
$\frac{dI}{d\Omega}$	1×10^{-4}	1×10^{-4}	5×10^{-4}	5×10^{-4}	7.5×10^{-4}	A sr^{-1}
ΔE	0.2	0.2	1.0	1.0	1.0	eV
δ	50	50	50	50	150	Å
I_{tot}	40	40	200	200	300	μA
J_c	4×10^5	4×10^5	2×10^6	2×10^6	4.5×10^4	A cm^{-2}
V_E	6.0	3.0	6.0	3.0	6.0	kV
V_B	1.0	1.0	1.0	1.0	1.0	kV

C_{so} (electrostatic lens)	1.49×10^4 mm
C_{co} (electrostatic lens)	265 mm
C_{si} (magnetic lens)	10 mm
C_{ci} (magnetic lens)	5 mm
M	1.28

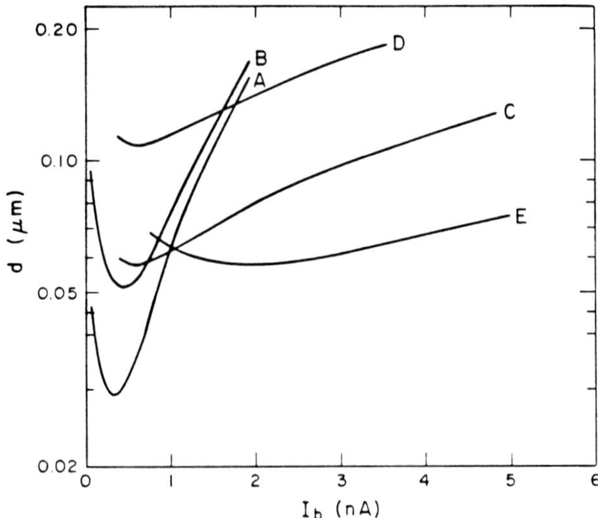

Figure 7. Focused spot size d vs beam current I_b for field emission electron guns. Curves A and B are CFE at 6 kV and 3 kV, respectively and $\frac{dI}{d\Omega} = 10^{-4}$ A sr^{-1}. Curves C and D are for CFE with V_E = 6 kV and 3 kV, respectively and $\frac{dI}{d\Omega} = 5 \times 10^{-4}$ A sr^{-1}. Curve E is for TFE with V_E = 6 kV and $\frac{dI}{d\Omega} = 7.5 \times 10^{-4}$ A sr^{-1}. See Table 1 for other parameters.

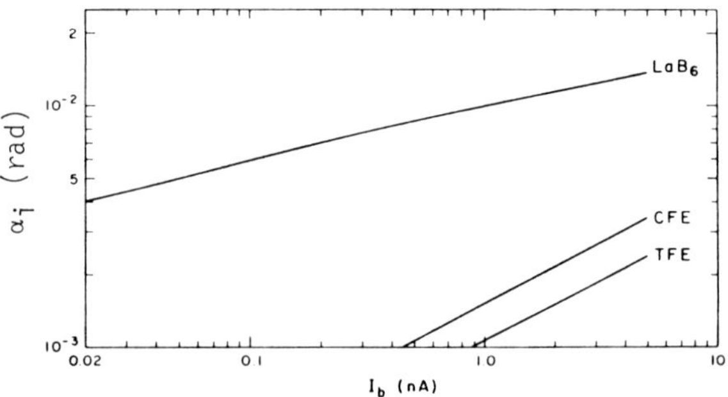

Figure 8. The convergence angle α_i of the focused beam on the target for LaB_6, CFE and TFE electron guns, as a function of beam current I_b. The depth of focus is inversely proportional to α_i.

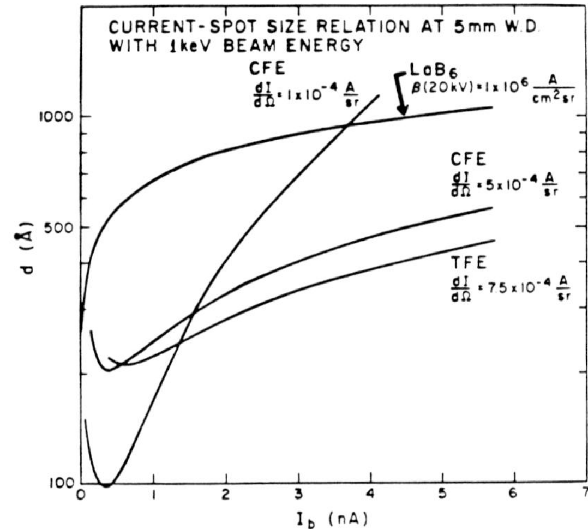

Figure 9. Focused beam spot size vs. beam current at 5 mm working distance for LaB_6, CFE and TFE guns, with the same optical systems as were used for 15 mm working distance. The gun voltage was 6 kV for the field emission cases and ΔE was the same as for the 15 mm working distance case.

the CFE cathode operated at either $\frac{dI}{d\Omega} = 1 \times 10^{-4}$ or 5×10^{-4} A sr^{-1}, and 25 times as much current as the LaB_6 cathode, at the same spot size. This is because at ΔE = 1 eV, $\frac{dI}{d\Omega} = 7.5 \times 10^{-4}$ A sr^{-1} [65] and at M = 1.28, the effect of δ is unimportant. An additional benefit of the field emission optical columns is that at I_b = 1 nA, $\alpha_i \approx 10^{-3}$, rad, implying a depth of field of ≈ 50 μm.

Based upon the d vs I_b calculations, LaB_6 is quite inadequate for high throughput applications where low beam voltage (1 kV) and moderately high resolution (≈ 0.05 μm) are required. CFE is best suited for high resolution, d ≈ 0.03 μm, with fairly high current, I_b ≈ 0.3 nA. TFE provides lower resolution, d ≈ 0.05-0.06 μm, but also an order of magnitude higher current, ≈ 4 nA at d = 0.06 μm, and shows much better performance than CFE over most of the current range.

These statements must be carefully qualified by noting that we have chosen optical systems with a working distance of 15 mm. A shorter working distance would result in improved values of d for a given I_b or improved values of I_b for a given d, as shown in Figure 9. However it seems highly unlikely that sufficient improvement could be made for LaB_6 to be useful for high throughput applications, and LaB_6 certainly would not be competitive with field emission. We add a parenthetical comment that workers in our laboratory have measured d vs I_b at low voltages with commercial SEMs employing LaB_6 cathodes, and have found that at E = 1 keV, when d ≈ 0.03-0.06 μm I_b ~ 0.001-0.01 nA, at working distances in the range 15-30 mm.

As reported by Swanson et al [56], a CWIKSCAN Model 100 SEM, modified to operate with a Zr/W TFE cathode has achieved I_b = 9 nA at E = 1.33 keV and d = 0.22 μm, at 30 mm working distance. The optical system employs a two-element immersion lens, and hence has little flexibility in the beam voltage. Nonetheless, the performance was noteworthy for the high current.

Besides the d vs I_b relations, it is important to consider the noise in the beam, since a cathode may not be usable if $\frac{\Delta I}{I_b}$ is too large. If we use the criteria of Wells [68], the signal to noise $\frac{I_b}{\Delta I} \geq 20$, for a system with 8 gray levels. For example, if we require a bandwidth $f = 10^6$ Hz, a reasonable figure for high speed inspection, we note that $\frac{I_b}{\Delta I_{shot}} = 20$ at $I_b = 0.13$ nA. From Figure 6, the LaB_6 results give $I_b = 0.05$ nA at d = 0.055 μm, and from Eq.(3) $\frac{I_b}{\Delta I} = 13$. Therefore, from the point of view of signal to noise, LaB_6 is quite marginal. From Figure 7, the TFE results give $I_b = 2.3$ nA when d = 0.057 μm, and from Eq.(5), with $\alpha_0 = 10^{-4}$ rad $\frac{I_b}{\Delta I} = 79$. The main component of the noise current is ΔI_{shot}, because of the bandwidth of 10^6 Hz. The flicker noise component of ΔI equals the shot noise component only when f is reduced to 43 kHz.

The situation for CFE is more complicated. If the vacuum were extremely high the noise would be entirely shot noise. At $P = 7 \times 10^{-9}$ torr, $\frac{I}{\Delta I}$ for a W(310) cathode was measured [48] to be ≈ 5% for f = 25 kHz and I = 0.7 nA. Above this frequency the noise is dominated by shot noise, which is negligible. In this case $\frac{I_b}{\Delta I} \approx 20$ and the cathode would be usable in terms of its signal to noise ratio. Based on results of Zaima et al [73], the signal to noise ratio of a TaC CFE cathode is at least an order of magnitude superior to that of the W(310) CFE cathode, at $P = 3 \times 10^{-10}$ torr. Therefore, from the point of view of noise current, such a cathode would be an improvement over W. Results for TiC are uncertain and seem to be sensitive to surface stoichiometry. However, a vacuum in the electron gun of $P < 10^{-9}$ torr is still necessary.

It is clear from the discussion so far that the TFE cathode is best suited for high throughput e-beam testing, taking into consideration current and beam spot size at E = 1 kV, noise and vacuum requirements of the electron gun necessary for reliable, long term cathode life. CFE is marginal in terms of noise if the W(310) cathode is used; carbide cathodes may offer an improvement in this regard. CFE is substantially better than LaB_6 in terms of current and beam spot size at E = 1 kV, but is inferior to TFE. Its vacuum requirements are an order of magnitude higher.

The issue of stability is usually brought up when field emission cathodes are considered, and this is an important issue from the point of view of practical applications. By stability is meant random current spikes or increasing current fluctuations with time, which lead to cathode failure. In the case of TFE, current spikes are not seen; in the case of CFE, even at $P < 10^{-9}$ torr and with carbide cathodes they are present.

The increases in current fluctuations with time seen in CFE, which are totally absent in the case of TFE, are due to increasing surface roughness due to ion bombardment. The rate at which this happens is proportional to P × I and occurs with all CFE cathodes. The problem cannot be made to go away at any practically attainable vacuum, but it can be controlled by periodic flashing of the cathode to high temperature. The frequency of flashing is determined by P, but it is impractical to operate an electron gun at $P \sim 10^{-10}$ torr in conjunction with frequent specimen changes. For this reason, CFE is not suitable for long term work requiring minimal operator attendance.

Taking all of the issues into account, it appears that the TFE cathode is best suited in terms of beam current with acceptable noise at E = 1 kV, or below.

References

[1] Andersen WHJ, Mol A. (1970). "Simultaneous Measurements of the Brightness and the Energy Distribution of Electrons Emitted from a Triode Gun," J. Phys. D 3,965-979.

[2] Barbour JP, Charbonnier FM, Dolan WW, Dyke WP, Martin EE, Trolan JK. (1960). "Determination of the Surface Tension and Surface Migration Constants for Tungsten," Phys. Rev. 117, 1452-1459.

[3] Bell AE, Swanson LW. (1979). "Total Energy Distribution of Field Emitted Electrons at High Current Density," Phys. Rev. B 19, 3353-3363.

[4] Bennette CJ, Swanson LW, Charbonnier F. (1967). "Electrical Breakdown between Metal Electrodes in High Vacuum. II. Experimental," J. Appl. Phys. 38, 634-640.

[5] Bettler PC, Charbonnier FM. (1960). "Activation Energy for the Surface Migration of Tungsten in the Presence of a High Electric Field," Phys. Rev. 119, 85-93.

[6] Charbonnier FM, Bennette CJ, Swanson LW. (1967). "Electrical Breakdown Between Metal Electrodes in High Vacuum. I. Theory," J. Appl. Phys. 38, 627-633.

[7] Cleaver JRA. (1975). "Field Emission Guns for Electron Probe Instruments," Int. J. Electronics 38, 513-529.

[8] Cleaver JRA. (1978). "Field Emission Electron Gun Systems Incorporating Single Pole Magnetic Lenses," Optik 52, 293-303.

[9] Crewe AV, Wall J, Welter LM. (1968). "A High-Resolution Scanning Transmission Electron Microscope," J. Appl. Phys. 39, 5861-5868.

[10] Crewe AV, Eggenberger DN, Wall J, Welter LM. (1968). "Electron Gun Using a Field Emission Source," Rev. Sci. Inst. 39, 576-583.

[11] Crewe AV, Isaacson M, Johnson D. (1968). "A Simple Scanning Electron Microscope," Rev. Sci. Inst. 40, 241-246.

[12] Danielson LR, Swanson LW. (1979). "High Temperature Coadsorption Study of Zirconium and Oxygen on the W(100) Crystal Face," Surf. Sci. 88, 14-30.

[13] Denizart M. (1981). "Triode and Tetrode Guns for Field Emission Electron Microscopes," Ultramicroscopy 7, 65-80.

[14] Dyke WP, Charbonnier FM, Strayer RW, Floyd RL, Barbour JP, Trolan JK. (1960). "Electrical Stability and Life of the Heated Field Emission Cathode," J. Appl. Phys. 31, 790-805.

[15] Eaton HC, Bayuzick RJ. (1978). "Field-Induced Stresses in Field Emitters," Surf. Sci. 70, 408-426.

[16] Everhart TE. (1967). "Simplified Analysis of Point-Cathode Electron Sources," J. Appl. Phys. 38, 4944-4957.

[17] Furukawa Y. et al. (1982). "Emission Characteristics of Single-Crystal LaB_6 Cathodes with <100> and <110> Orientations," J. Vac. Sci. Technol. 20, 199-203.

[18] Furukawa Y, Yamabe M, Inagaki T. (1983). "Emission Characteristics of Single-Crystal LaB_6 Cathodes with Large Tip Radius," J. Vac. Sci. Technol. A1, 1518-1521.

[19] Futamoto M, Yuito I, Kawabe U. (1982). "Study on Titanium Carbide Field Emitters by Field Ion Microscopy, Field-Electron Emission Microscopy, Auger Electron Spectroscopy, and Atom-Probe Field-Ion Microscopy," Surf. Sci. 120, 90-102.

[20] Gesley M, Swanson LW. (1984). "Spectral Analysis of Adsorbate Induced Field Emisson Flicker (1/f) Noise - Canonical Ensemble," Phys. Rev. B (in press).

[21] Glaser W. (1952). "Grundlagen Der Elektronenoptik," Springer-Verlag, Vienna, Ch. 16-23.

[22] Gomer R. (1973). "Current Fluctuations from Small Regions of Adsorbate Covered Field Emitters," Surf. Sci. 38, 373-393.

[23] Good RH, Jr, Muller EW. (1956). "Field Emission," Handbuchder Physik, S. Flugge, Ed., Springer-Verlag, 1956, Berlin, 176-234.

[24] Grivet P. (1972). "Electron Optics," Vol. 1., 2nd Ed., Pergamon Press, Oxford, 1972, Ch. 4, 5, 7.

[25] Haine ME, Einstein PA. (1952). "Characteristics of the Hot Cathode Electron Microscope Gun," Brit. J. Appl. Phys. 3, 40-46.

[26] Hawkes PW. (1972). "Electron Optics and Electron Microscopy," Taylor and Francis, Ltd., London, Ch. 2.

[27] Hohn FJ, Chang THP, Broers AN, Frankel GS, Peters ET, Lee DW. (1982). "Fabrication and Testing of Single-Crystal Lanthanum Hexaboride Rod Cathodes," J. Appl. Phys. 53, 1283-1296.

[28] Kang NK, Tuggle D and Swanson LW. (1983). "A Numerical Analysis of the Electric Field and Trajectories With and Without the Effect of Space Charge for a Field Electron Source," Optik 63, 313-331.

[29] Kelly J, Groves T, Kuo HP. (1981). "A High-Current, High Speed Electron Beam Lithography Column," J. Vac. Sci. Technol. 19, 936-940.

[30] Klemperer O, Barnett ME. (1971). "Electron Optics," 3rd Ed., Cambridge University Press, Ch. 1-2, 6-7.

[31] Kuo HP, Siegel B. (1978). "A High Brightness, High Current Field Emission Electron Probe," Proc. Symp. on Electron and Ion Beam Sci. Tech., R. Bakish, Ed., The Electrochemical Soc., Pennington, NJ, 3-10.

[32] Kuroda K, Suzuki T. (1972). "Electron Trajectory and Virtual Source of Field Emission Gun of SEM," Jpn. J. Appl. Phys. 11, 1390-1391.

[33] Lafferty JM. (1951). "Boride Cathodes," J. Appl. Phys. 22, 299-309.

[34] Li JZ. (1984). "The Prospects of Field Emission for e-Beam Inspection," J. Vac. Sci. Technol. (in press).

[35] Martin EE, Trolan JK, Dyke WP. (1960). "Stable, High Density Field Emission Cold Cathode," J. Appl. Phys. 31, 782-789.

[36] Martin EE, Charbonnier FM, Dolan WW, Dyke WP, Pitman HW, Trolan JK. (1960). "Research on Field Emission Cathodes," WADD Technical Report 59-20, Ch. 5, Office of Tech. Services, U.S. Dept. of Commerce, Washington, D.C.

[37] Moss H. (1968). "Narrow Angle Electron Guns and Cathode Ray Tubes," Academic Press, New York, 205-215.

[38] Orloff J, Swanson LW. (1979). "An Asymmetric Electrostatic Lens for Field Emission Microprobe Applications," J. Appl. Phys. 50, 2494-2501.

[39] Oshima C, Aono M, Tanaka T, Kawai S, Shimizu R. (1980). "Thermionic Emission from Single-Crystal LaB_6 Tips With [100], [111] and [210] Orientation," J. Appl. Phys. 51, 1201-1206.

[40] Pfeiffer HC. (1971). "Experimental Investigation of Energy Broadening in Electron Optical Instruments," Proc. 11th Symp. on Electron, Ion and Photon Beam Technology, R. F. M. Thornley, Ed., San Francisco Press, Inc., 239-246.

[41] Roques S, Denizart M, Sonier F. (1983). "Optical Characteristics of the Electron Source Given by a Tetrode Field Emission Gun," Optik 64, 299-312.

[42] Saitou N. (1977). "Trajectory Analysis of Ions Formed in the Field Emitter Inter-Electrode Region," Surf. Sci. 66, 346-356.

[43] Schmidt PH, Joy DC. (1978). "Low Work Function Electron Emitter Hexaborides," J. Vac. Sci. Technol. 15, 1809-1810.

[44] Septier A. (1967). (Editor), "Focusing of Charged Particles," Academic Press, New York, Ch. 2.2, 2.3, 2.5.

[45] Sewell PB, Ramachandran KN. (1978). "Grid Aperture Contamination in Electron Guns Using Directly Heated Lanthanum Hexaboride Sources," Scanning Electron Micros. 1978; I: 221-232.

[46] Sewell PB. (1980). "High Brightness Thermionic Electron Guns for Electron Microscopes," Scanning Electron Micros. 1980; I: 11-24.

[47] Shimizu R, Shinike T, Tanaka T, Oshima C, Kawai S, Hiraoka H, Hagiwara H. (1979). "Brightness of Single Crystal LaB_6 Cathodes of <100> and <110> Orientations," Scanning Electron Microsc. 1979; I: 11-18.

[48] Swanson LW, Crouser LC. (1969). "Angular Confinement of Field Electron and Ion Emission," J. Appl. Phys. 40, 4741-4749.

[49] Swanson LW, Bell AE. (1973). "Recent Advances in Field Electron Microscopy of Metals," Adv. in Elect. and Elect. Phys. 32, Academic Press, pp 193-309.

[50] Swanson LW, Martin NA. (1975). "Field Electron Cathode Stability Studies: Zirconium/Tungsten Thermal-Field Cathode," J. Appl. Phys. 46, 2029-2050.

[51] Swanson LW. (1975). "Comparative Study of the Zirconiated and Built-Up W Thermal Field Cathode," J. Vac. Sci. Technol. 12, 1228-1233.

[52] Swanson LW. (1978). "Current Fluctuations from Various Crystal Faces of a Clean Tungsten Field Emitter," Surf. Sci. 70, 165-180.

[53] Swanson LW, McNeely DR. (1979). "Work Functions of the (001) Face of the Hexaborides of Ba, La, Ce and Sm," Surf. Sci. 83, 11-28.

[54] Swanson LW, Gesley MA, Davis PR. (1981). "Crystallographic Dependence of the Work Function and Volatility of LaB_6," Surf. Sci. 107, 263-289.

[55] Swanson LW, Tuggle D. (1981). "Recent Progress in Thermal Field Electron Source Performance," Appl. Surf. Sci. 8, 185-196.

[56] Swanson LW, Tuggle D, Li JZ. (1983). "The Role of Field Emission in Submicron Electron Beam Testing," Thin solid Films 106, 241-255.

[57] Swift DW, Nixon WC. (1962). "Characteristics of a Point Cathode Triode Electron Gun," Brit. J. Appl. Phys. 13, 288-293.

[58] Takigawa T, Yoshii S, Sasaki I, Motoyama K, Meguro T. (1980). "Emission Characteristics for <100>, <110> and <111> LaB_6 Cathodes," Jpn. J. Appl. Phys. 19, L537-L540.

[59] Taylor DM. (1978). "The Effect of Passivation on the Observation of Voltage Contrast in the Scanning Electron Microsope," J. Phys. D 11, 2443-2454.

[60] Todokoro H, Saitov N, Yamamoto S. (1982). "Role of Ion Bombardment in Field Emission Current Instability," Jap. J. Appl. Phys. 10, 1513-1516.

[61] Troyon M. (1980). "High Current Efficiency Field Emission Gun System Incorporating a Preaccelerator Magnetic Lens. Its Use in CTEM," Optik 57, 401-419.

[62] Tuggle D, Swanson LW, Orloff J. (1979). "Application of a Thermal Field Emission Source for High Resolution, High Current e-Beam Microprobes," J. Vac. Sci. Technol. 16, 1699-1703.

[63] Veneklasen LH. (1972). "Some General Considerations Concerning the Optics of the Field Emission Illumination System," Optik 36, 410-433.

[64] Veneklasen LH, Siegel BM. (1972). "A Field Emission Illumination System Using a New Optical Configuration," J. Appl. Phys. 43, 4989-4996.

[65] Veneklasen L, Yew N, Wiesner J. (1978). "Application of Field Emission in High Current Electron Beam Lithography Optics," Proc. Symp. on Electron and Ion Beam Sci. Tech., R. Bakish, Ed., The Electrochemical Soc., 11-16.

[66] Verhoeven JD, Gibson ED. (1976). "Evaluation of a LaB_6 Cathode Electron Gun," J. Phys. E 9, 65-69.

[67] Wardley GA. (1973). "Potential of Field Emission Cathodes for Microfabrication," J. Vac. Sci. Technol. 10, 975-978.

[68] Wells OC. (1974). "Scanning Electron Microscopy," McGraw-Hill, New York, Ch. 2 and 4.

[69] Wiesner JC, Everhart TE. (1973). "Point-Cathode Electron Sources - Electron Optics of the Initial Diode Region," J. Appl. Phys. 44, 2140-2148.

[70] Windsor EE. (1969). "Construction and Performance of Practical Field Emitters from Lanthanum Hexaboride," Proc. IEEE 111, 348-350.

[71] Wolfe JE. (1975). "Abstract: Electron Gun for Data Storage Micromachining," J. Vac. Sci. Technol. 12, 1169.

[72] Wolfe JE. (1979). "Operational Experience with Zirconiated T-F Emitters," J. Vac. Sci. Technol. 16, 1704-1708.

[73] Zaima S, Saito K, Adachi H, Shibata Y, Hojo H, Ono M. (1980). "Field Emission from TaC," Proc. 27th Int. Field Emission Symp., Dept. of Metallurgy and Materials Science, University of Tokyo, 348-357.

[74] Zworykin VK, Morton GK, Ramberg EG, Hillier J, Vance AW. (1945). "Electron Optics and the Electron Microscope," John Wiley and Sons, New York, Ch. 10-17.

Additional Discussion

P. B. Sewell: The superior performance characteristics of the TFE source have been well established by many excellent publications such as the above. However, the general acceptance of a 'new' source seems to be influenced by many practical operating factors. For example, the slow turnaround time for the vacuum processing of LaB_6 cathodes as compared with tungsten, is considered by some to be a disadvantage in routine SEM applications. Could the author comment on the cycle time from loading of a TFE source to its stable operation in the electron optical column?

Author: To assure reliable operation of the ZrO/W cathode a vacuum level $< 1 \times 10^{-8}$ torr is necessary. To achieve such a vacuum in an electron gun connected to a specimen chamber which operates at the usual SEM vacuum level ~ 10^{-6} torr, it will be necessary to differentially pump the gun through a small (~ 0.5 mm) aperture which should be located some distance from the cathode. This means the gun chamber will have its own independent vacuum pumps and should be capable of sustaining a mild bake to speed pumping of water vapor. For example, in the rebuilt CWIKSCAN 100 SEM at the Oregon Graduate Center, the gun chamber is pumped by two 20 1/s ion pumps connected by high conductance lines. The turnaround time for a cathode change is typically 12 hours, including the bake cycle. This is probably a typical number for a system which will operate at 5×10^{-9} torr and, while it may seem long compared with the 1/4 hour cycle time typical for a thermionic W cathode, it should be kept in mind that the cathode life will be 1000 - 2000 hours, barring accident. This is 15 - 30 times the life of a thermionic W cathode (assuming a 60 hour life) and when the cycle is amortized over the life, it is only 1 - 2%. The cycle time for the thermionic W cathode is 0.5%, which is not much different.

P. B. Sewell: As the TFE emitter and single crystal LaB_6 emitters have been under development for about the same time, could the author comment on possible reasons for the slow acceptance of this type of source in electron optical instruments. Does any currently manufactured commercial instrument employ the TFE type source?

Author: Undoubtedly the main reason that LaB_6 cathodes caught on quickly was that they are a greatly improved version of the type of cathode that was in wide use already. This means that only minor, if any changes in the electron optics were required in the instrumentation. The main change required was in the gun vacuum system, which has not been difficult for the manufacturers to put into place. Use of the TFE cathode requires a completely new electron optical design, which is a much more serious proposition.

The commercial applications of the TFE cathode have been primarily in e-beam lithography machines. To date A.T.& T. Bell Labs, E-Beam Corporation, Hewlett-Packard Corporation have built such instruments. The latter two efforts have not been commercially successful, but the reasons for this are not clear. A number of CWIKSCAN 100 SEM's have been converted to use the ZrO/W cathode and electron guns with this cathode are being sold by FEI Co. At this time there are no complete systems employing the TFE cathode which are commercially available.

Electron Optical Systems (pp. 163-170)
SEM Inc., AMF O'Hare (Chicago), IL 60666-0507, U.S.A.

THERMIONIC EMISSION STUDIES OF MICRO-FLAT SINGLE CRYSTAL LANTHANUM HEXABORIDE CATHODES

P.B. Sewell* and A. Delâge

Division of Electrical Engineering
National Research Council of Canada
Ottawa, Ontario K1A 0R6
Canada

Abstract

A new type of high brightness thermionic cathode has been developed. The cathode utilises emission from a small flat surface (generally less than 50 μm diameter) prepared parallel to a specific crystal plane on a single crystal of Lanthanum Hexaboride. Emission from the rest of the single crystal is suppressed by the use of a high work function coating of pyrolytic graphite. When used in a conventional triode gun, the maximum total electron emission is controlled by the area of the micro-flat, the work function and the temperature of the emitter. The Wehnelt potential serves a minor role in controlling the divergence of the beam. At certain emitter height settings, the gun produces the maximum axial brightness at zero bias. The field at the surface of the micro-flat is higher than that for pointed emitters in a conventional configuration and no longer limits the gun brightness. As the emitting region is now parallel to a specific crystallographic surface, the emission anisotropy of LaB_6 can now be utilized in developing emitters of optimum brightness. The new sources reduce Wehnelt aperture contamination and offer long lifetimes under favourable vacuum conditions.

Key Words: Thermionic Emission, High Brightness Triode Guns, Lanthanum Hexaboride, Single Crystals, Micro-flat Cathodes.

*Address for correspondence:
Division of Electrical Engineering
National Research Council of Canada
Bldg. M-50, Montreal Rd.
Ottawa, Ontario
Canada K1A 0R6 Phone No.: (613) 993-2304.

Introduction

Lanthanum hexaboride electron sources are now commonly used in electron optical instruments such as scanning electron microscopes and electron beam writers. When correctly employed they provide both a higher gun brightness and longer filament life than that normally achieved with tungsten emitters. The application of LaB_6 cathodes has closely followed the general practice of tungsten cathodes in that the improved performance has been sought through changes in operating conditions in the conventional high voltage triode electron gun. In order to make use of higher cathode loadings at the emitting tip of the cathode, the axial electric field in this region is generally enhanced by the use of sharply pointed single crystal LaB_6 sources. This pointed emitter is the most commonly used LaB_6 electron source and is available from numerous commercial sources.

Unfortunately, this form of conical cathode structure presents several problems when attempting to optimise its performance in a triode gun. One limitation results from the operating characteristics of the triode gun itself. A second is the complex crystallographic nature of the emitting region of the pointed cathode. A third is the evaporation and oxidation of the large area of LaB_6 exposed to the internal surface of the Wehnelt aperture. These points warrant further discussion before describing a new cathode structure which overcomes some of these limitations.

The triode gun controls electron emission by changing the electric field at the surface of the cathode. At the "cut-off" potential, the zero equipotential of the gun is adjusted to be just in front of the cathode surface so that no electron emission occurs. As the Wehnelt potential is made less negative the zero equipotential begins to intercept the cathode surface so that emission is possible. When emission just begins, the field at the surface is very low and the cathode emission at normal operating temperatures (LaB_6 e.g. 1800-1900K) is strongly space charge limited. As the Wehnelt becomes less negative the total emission from the cathode increases as the conical area of the tip within the accelerating field region increases.

The field strength over the surface of this cone varies in a complex manner, being maximum at the apex and dropping to zero at the zero equipotential intercept. To increase further the field at the tip, the Wehnelt potential is reduced further and the total emission increases rapidly as the area of the emitting surface increases. For optimum performance of the gun, a balance is developed between a practical value of total emission and a maximization of the field at the apex. These conditions result in a complex beam structure which may contain an intense central spot surrounded by a diffuse halo or multiple lobes as observed for single crystal cathodes.[1 2 3] At high temperatures, the full theoretical brightness of a LaB_6 cathode is rarely obtained, even on the axial beam.

The area of electron emission around the tip of a conical single crystal emitter cannot be defined simply in terms of any particular single crystal face and hence complex emission characteristics come into play as the tip is allowed to emit by control of the Wehnelt potential. Tip structures developed by thermal evaporation and associated lobe patterns of emission have been described by several authors.[4 5] These emission patterns depend critically on the alignment of the tip within the Wehnelt aperture. To utilise the thermionic emission anisotropy that exists between specific crystal planes of LaB_6 [6 7] a simpler form of emitting surface is desirable. A flat cathode cut parallel to a specific crystal plane is preferred for fundamental studies of emission anisotropy.

Considerable data on the evaporation of LaB_6 at typical operating temperatures have been published. [7 8] In addition to direct evaporation, the evaporation of oxidation products is also of importance if operating pressures in the gun are above about 7 x 10^{-6}Pa (5 x 10^{-8} Torr). [9] As a result of these processes a material deposit accumulates on the surface of the Wehnelt facing the cathode. In time this deposit may result in electrostatic charging effects or produce physical spalling of material. The development of whiskers or spalled layers of LaB_6 on the Wehnelt aperture can cause gun instabilities even in ultra-high vacuum conditions. A reduction in the area of exposed LaB_6 would reduce potential problems of this type.

For the above reasons efforts have been made to develop a new type of single crystal LaB_6 cathode. The initial interest was to explore thermionic emission anisotropy at high cathode loadings from small surfaces cut parallel to specific crystal planes. The resulting cathodes show promise as practical long life emitters operating at the full theoretical brightness anticipated from work function measurements.

Experimental

Filament Structure

In the new structure, electron emission from the major portion of the cathode has been suppressed by the use of a thermally stable, high work function coating of pyrolytic graphite. Two types of cathode configuration have been tested. One is the "mesa" or pedestal type structure formed on the arc bonded single crystal of LaB_6 by electric discharge machining (EDM). Steps in the formation of the cathode are shown in Fig. 1. After bonding of the single crystal to the rhenium heater[10], the "mesa" is formed by trepanning with a small aperture using very low energy electric discharge machining. The filament after this stage is shown in Figs. 1a and 1b. The height of the "mesa" is controlled to be about equal to the diameter. The next step requires heating the filament to normal operating temperatures in about 800 Pa (60 torr) of a hydrocarbon such as propane or ethylene to form a coating of pyrolytic graphite 10 to 20 µm thick[11]. Fig. 1c shows the filament of Fig. 1a after coating with graphite. The final stage is to support the filament and polish off the carbon coating on the "mesa" with a fine diamond abrasive until a polished LaB_6 surface is produced. This surface is surrounded by a shell of inert graphite as shown in Fig. 1d. A similar treatment can produce micro-flats on conventional pointed emitters, although in this case the diameter of the final flat varies as the cone tip is ground away.

The desired diameter of the micro-flat can be determined from a consideration of the cathode loading (A/cm^2) and the total emission current that can be tolerated in the column, as indicated in Table 1. Here, source diameters of 10 to 70 µm are listed with their respective area in µm^2. The total emission current at selected cathode loadings from 2 to 50 A/cm^2 are indicated. It is seen that in order to limit total emissions to reasonable values (e.g. $\leq$ 400µA) cathode diameters of some 30 to 50 µm are desirable. Most cathodes tested to date have been between 30 and 70 µm diameter and the smallest "mesa" structure formed by electric disharge machining is about 20 µm diameter.

Cathodes are individually calibrated for heater current versus temperature in a vacuum comparable to the operating conditions of the gun. Due to increased radiation losses, coated

TABLE 1

Emission Current µA from Circular Micro-Flat Cathodes

Source Dia. µm	Area µm^2	Cathode Current Density A cm^{-2} 2	5	10	25	50
10	78	1.6	3.9	7.9	19.6	39.3
20	491	6.3	15.7	31.4	78.5	157.1
30	761	14.1	35.5	70.7	176.7	353.4
40	1257	25.1	62.8	125.7	314.2	
50	1963	39.3	98.2	196.3		
60	2827	56.5	141.4	282.7		
70	3848	77	192.4	384.8		

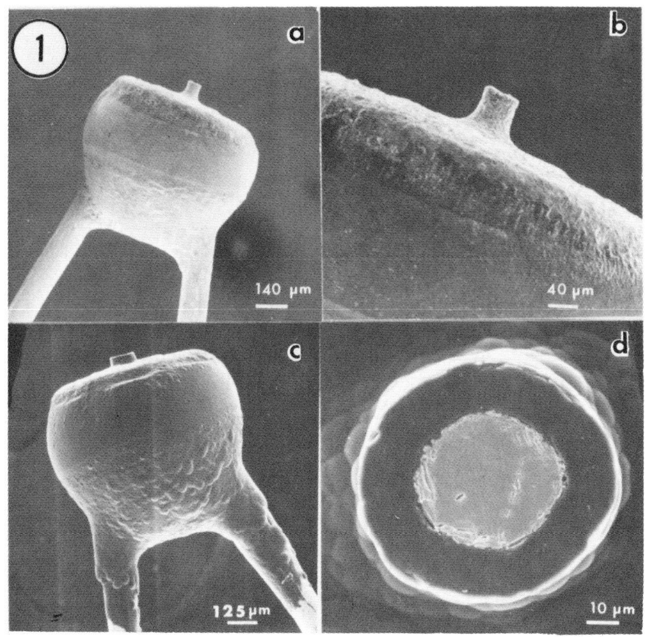

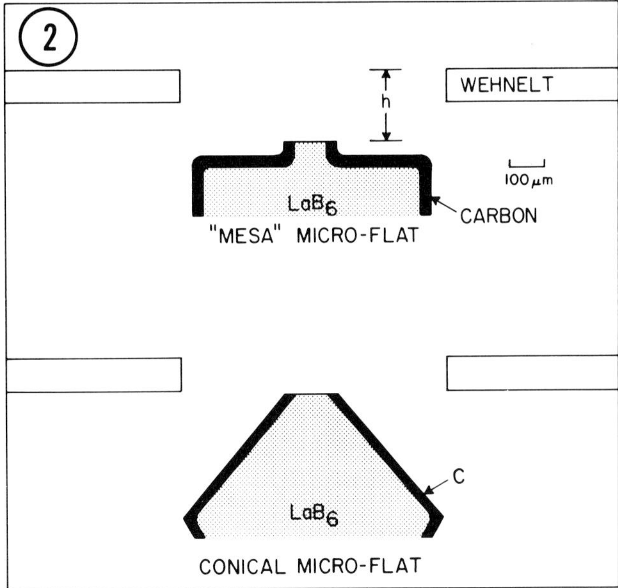

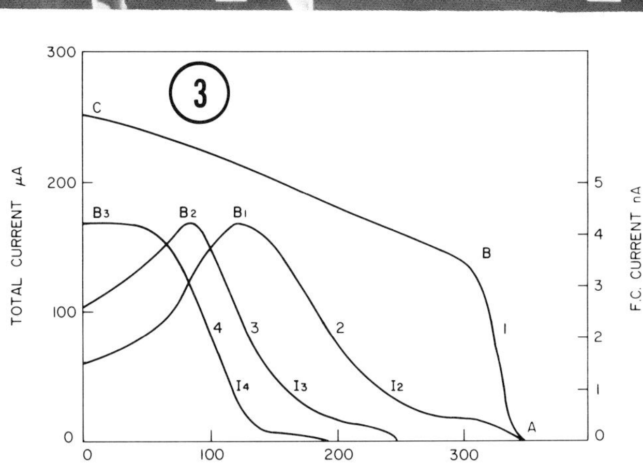

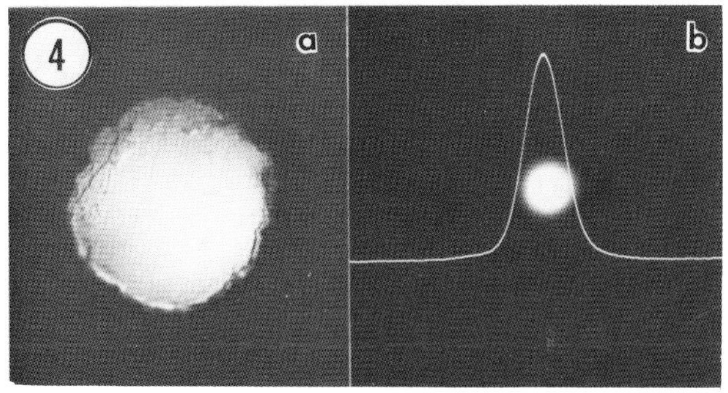

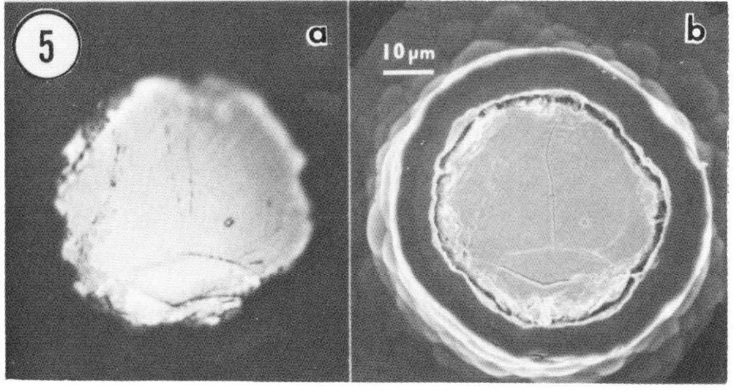

Fig. 1 Stages in the preparation of "mesa" type micro-flat cathodes from a single crystal of LaB_6. (a) Crystal arc-bonded to rhenium heater wire with 50 micron diameter "mesa" formed by EDM methods. (b) Higher magnification of "mesa" in a. (c) Specimen a after coating with 20 microns of pyrolytic graphite. (d) (001) emitting surface of LaB_6 surrounded by 20 microns graphite after removal of graphite cap.

Fig. 2 Schematic showing location of two types of micro-flat cathodes within the Wehnelt aperture.

Fig. 3 Electron emission characteristics of micro-flat cathode as described in text.

Fig. 4 a) Beam profile image of a 50 μm diameter micro-flat cathode at I_2 of curve 2, Figure 3. Original magnification as a thermionic emission image x 550. b) Beam profile of same source at the maximum brightness position B_1 of curve 2.

Fig. 5 Comparison of thermionic emission image and SEM image of (001) micro-flat cathode after 150 hours of operation at 1800K. (a) image from the beam profile mode under conditions similar to Fig. 4a, original image magnification x 500. (b) secondary electron SEM image after removal of cathode from gun.

cathodes require slightly more current for operation than uncoated cathodes. A typical "mesa" cathode required about 2.4 A at 1700K and 2.9 A at 1900K. Temperatures of the cathode during electron optical testing were estimated from the operating current and the calibration curves.

Gun Vacuum Conditions

To date, the new filaments have been evaluated in a Nanolab 7 SEM equipped for LaB_6 operation having independent bias and both beam profiling and specimen current monitoring facilities. The gun pumping conditions were such that even during operation at 30kV and 400μA emission, a base pressure in the gun was maintained at 1.3 x 10^{-5} Pa (10^{-7} torr) or better. A small quadrupole mass spectrometer fitted to the gun housing was used to monitor both total and partial pressures. At the base pressure, the major gas component was found to be water vapor with smaller contributions from nitrogen and oxygen resulting from gas permeation through the viton seals of the vacuum system. To reduce outgassing from the copper liner tube in the double condenser portion of the column, this unit was outgassed in a vacuum furnace at 900K for several hours prior to insertion in the microscope. This pretreatment markedly reduced liner tube outgassing during high power operation of the gun. Modifications to the gun pumping system permitted a starting vacuum of 3 x 10^{-4} Pa (2 x 10^{-6} torr) to be reached within about 20 minutes. This was sufficient to check newly installed filaments for basic operating performance. All long term tests were conducted close to the base pressure of 1.3 x 10^{-5} Pa (1 x 10^{-7} torr).

The gun was the conventional triode used in The Nanolab 7 SEM. A removable tantalum Wehnelt aperture was used, having a thickness of 100 μm and an aperture diameter of 750 μm. The Wehnelt to anode separation was 4 mm. The configuration of the coated filaments in the region of the Wehnelt aperture is shown in Fig. 2, where the normal convention for the filament height setting is indicated. Due to the diameter of the "mesa" type cathodes, these could not be set with a filament height of zero with the 750 μm Wehnelt aperture. However for optimum performance this was not necessary.

Results

Electron Optical Characteristics

The behaviour of the gun can best be compared with conventional sources by measuring the total emission current as a function of Wehnelt potential for a fixed accelerating potential and filament temperature. In addition an estimate of the peak axial brightness is obtained by measuring the current in a Faraday cup at the specimen level with a high excitation of the double condenser lens and a small beam limiting aperture. This Faraday cup current is directly proportional to the axial brightness and serves for comparison of the relative brightnesses of other filaments operating under the same conditions of column excitation.

Fig. 3 illustrates the general behaviour of all the limited emission micro-flats tested. These results are from a "mesa" type cathode set at an initial height of 0.2 mm in a Wehnelt aperture of 750 μm diameter. These results were obtained at 15kV and are similar to those obtained at 5 and 30kV. The behaviour of the total emission current and the specimen current for this initial setting is shown in curves 1 and 2 respectively. As the Wehnelt potential is reduced below the "cut-off" value of -350 V, the total emission current rises rapidly over the region A-B as an increasing area of the micro-flat is allowed to emit. Over the regio B-C the current rises less rapidly as the field at the surface of the micro-flat increases, but no further area is available for emission. At zero bias, the total current is limited by the temperature of the micro-flat, the work function of the surface and the surface area. This behaviour is markedly different from the normal pointed LaB_6 source where the total emission at zero bias would rise continuously to many thousands of microamperes as an increasing area of the cone is allowed to emit.

For the same setting of the filament, the Faraday cup current rises to a maximum B_1 at about -130 V and then decreases as the bias is further reduced. As the filament height setting is increased to 0.25 and 0.3 mm the behaviour of the current is shown in curves 3 and 4. For curve 4 the maximum Faraday cup current and hence the maximum axial brightness of the gun is obtained at zero bias. This behaviour would seem to be a unique feature of this type of cathode and offers simplification in the design of the high voltage gun.

An additional observation that can be obtained during the course of the above measurements is that of the beam profile. This type of measurement has been described in detail [12 13] and is now a regular feature of most commercial SEM's. At the Wehnelt potential I_2 on curve 2 the beam profile appears as a thermionic emission image of the micro-flat cathode. A similar image appears at the potential I_3 and I_4 of curves 3 and 4 respectively. At the potentials B_1, B_2 and B_3 of curves 2, 3, and 4 the beam profile decreases in diameter to uniform circular beam of near gaussian profile. Images typical of these two conditions are illustrated in Figs. 4a and 4b.

The direct association of the beam profile with the thermionic emission image of the source is shown in Fig. 5. Here, as shown in Fig. 5a another cathode is imaged in the beam profile mode at a potential about 100 V below "cut-off". Shortly after this observation, this cathode was removed from the gun for structural examination in the SEM. A regular SEM image of the micro-flat is shown in Fig. 5b, where the correlation of surface features with the beam profile image Fig. 5a is obvious. It is clear from this comparison that the carbon coating and any reaction products between the carbon and the LaB_6 are not contributing to the thermionic image. Hence estimates of the current density from the cathode can be made from a consideration of the measured surface area of LaB_6 and the total emission current.

Current Density Measurements

To date, seven micro-flat cathodes of near (001) orientation have been studied in some detail. The longest period of operation for any one cathode has been about 300 hours and the total accumulated time for all cathodes about 800 hours. All seven cathodes have shown the same general emission characteristics as described in the previous section.

For comparison of performance, the zero field cathode current density has been estimated from the measured cathode surface area and the total emission at zero bias. The results from four different cathodes at various temperatures are shown in Fig. 6. These results are compared with calculated values of current density for various values of work function shown as the solid curves A, B and C. The results are in good agreement with a work function value of 2.69 eV, close to the value suggested by Swanson [14 21] for the (001) surface of LaB_6.

Comparison with other Cathodes

Although direct axial brightness measurements have not been made to date, the new cathodes have been compared with conventional cathodes in the same experimental arrangement. The Faraday cup current at the specimen level has been compared for emitters operating at the same accelerating potential, with the same beam limiting aperture and over the same range of condenser lens excitation. All gun parameters were adjusted for optimum performance of the cathodes. The results of such comparisons are shown in Fig. 7. For the Nanolab 7 SEM, the condenser excitation of spot size setting range from 2 to 3 corresponds to normal high resolution operation.

For a tungsten hair-pin cathode operating at about 2900°K a typical specimen current at spot size 2-0 is 3-4 x 10^{-12} A and the change in current over the range from 2-0 to 3-0 is shown. For pointed (001) LaB_6 cathodes (conical cathodes) operating at 1850 to 1900K specimen currents are higher and a typical current at 2-0 is about 10^{-11} A. The performance of the micro-flat cathodes has been consistently better than the regular pointed cathodes and higher specimen currents are recorded even with cathodes operating at lower temperatures. In the extreme

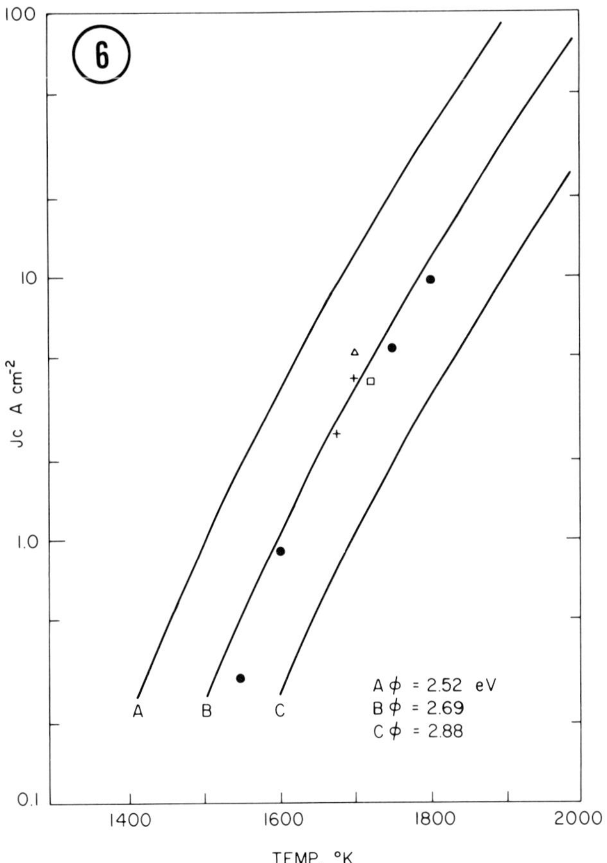

Fig. 6 Plot of measured zero field cathode current density as a function of temperature for a series of LaB_6 micro-flat cathodes, ●, □ , +, Δ. Curves A, B, C theoretical values for work function 2.52, 2.69 and 2.88 respectively.

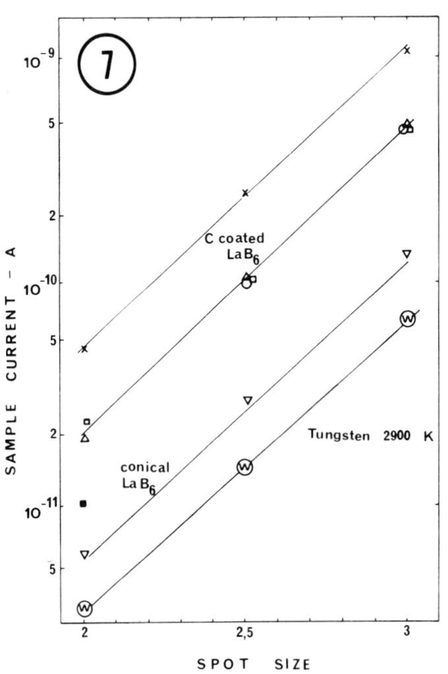

Fig. 7 Comparison of specimen currents (and hence relative brightness) for cathodes in the same gun at 15 kV and conditions of optimum brightness. Tungsten hair-pin at 2900K (w); pointed [00] single crystal LaB_6 1850K (▼); 1900K (■); micro-flat cathodes near (001) orientation at various temperatures (Δ, □ , o); microflat cathode J-4 at 1750K (x)

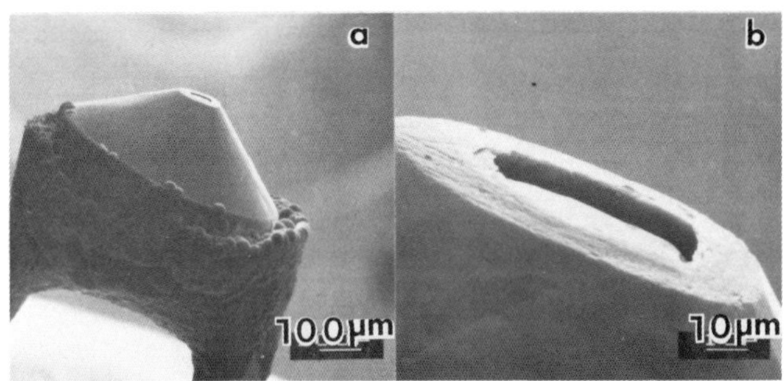

Fig. 8 Conical micro-flat after 265 hours of operation between 1750 and 1800K. Set-back of LaB_6 about 6 µm due to evaporation and oxidation.

case the conical micro-flat cathode J-4 gave a specimen current of 5 x 10^{-11} A at 2-0 when operating at 1750K. This cathode was found to be 14° off the [001] zone towards the [120] zone.

Material Loss from Cathode

For some cathodes operated in poor vacuum for extended periods of time, it was noticed that for a fixed cathode temperature and gun operating parameters, that the Faraday cup current at the brightness maxima began to decrease. This decrease in brightness is associated with the loss of material from the cathode surface by evaporation and oxidation. Since the pyrolytic carbon coating has proven to be extremely stable, the level of the LaB_6 emitting surface recedes below the level of the outer carbon coating. For a micro-flat surface some 50 µm diameter, it has been found that this reduction in brightness becomes measurable after about 10 µm of material has been lost from the surface of the emitter.

A set-back of some 6 µm is shown for a conical micro-flat in Figs. 8a and 8b. This emitter was examined after 265 hours of operation between 1750K and 1800K. At this point no loss of brightness had been detected. From published data[7], the evaporation loss at 1800K for 265 hours would be about 3 µm. The remaining loss is attributed to evaporation of the oxidation product formed during operation at 1.3 x 10^{-5} Pa (1 x 10^{-7} torr). From observation of the mass spectrum at this pressure the oxidizing species (H_2O and O_2) contribute about 50% of the gaseous species. From estimates of oxidation rates at 7 x 10^{-6} Pa (5 x 10^{-7} torr)[9] the material loss after 265 hours is about 6 µm. The estimated loss of 3 µm corresponds to a partial pressure of the oxidizing species of 3 x 10^{-6} Pa (2.5 x 10^{-8} torr). The discrepancy is within the limits of error of the estimations for this process. However, the importance of maintaining low operating pressures for the cathodes is clearly estalished. Another effect associated with the background pressures is the sensitivity of the total emission from these cathodes to small changes in the gas pressure. This was noticed as a reduction in total emission current as a function of time after the beam was allowed to impinge on the liner tube. No rise in pressure at the gauge was recorded during this behaviour. However, the effect was attributed to gas from local outgassing of the liner tube impinging on the single crystal surface of the cathode which would be in direct line of sight of such a gas source. Similar effects have not been observed when using pointed single cathodes which contain many different crystal faces. The problem has been eliminated by high temperature outgassing of the liner tube.

Grid Aperture Contamination

The physical location micro-flat cathodes within the Wehnelt aperture is shown in Fig. 2. The protective coating of pyrolitic graphite inhibits the evaporation of LaB_6 and thus reduces the amount of material depositing of the inner surfaces of the aperture. The area (a) a micro-flat is typically some 2 x 10^{-5} cm^2, about 200 times less than the surface area of the exposed conical surface of regular pointed emitters. Furthermore, due to the location of the micro-flat most evaporation and oxidation products are directed towards the anode rather than the Wehnelt aperture. Examination of top-hat apertures removed from the Wehnelt structure after 200 hours of filament operation has not detected aperture contamination by weight gain measurements or in optical or electron micrographs. This is in contrast to the results reported previously[9] for regular LaB_6 filaments.

Material Reactions

Pyrolytic graphite is widely used as an electrical contact for LaB_6 emitters in the compression type structure developed by Vogel[15]. The reaction of graphite with LaB_6 in such emitters appears to have no deletarious effects after several hundreds of hours of operation. However, in these structures the graphite is not in close proximity to the emitting area at the tip of the emitter.

For the present emitters a graphite-LaB_6 interface is in the immediate vicinity of the emitting surface and problems associated with the diffusion of carbon into or onto the surface of the LaB_6 micro-flat might be anticipated. Evidence of interdiffusion is seen in micrographs of the interface regions, with the diffusion of components from the LaB_6 into the carbon being most readily observed. However, to date no deleterious effects due to the long term reactions between these materials have been evident in the electron emission experiments. Thermionic emission images of the cathode never show any detectable emission from the graphite shell surrounding the LaB_6. Conversely, no reduction in the emission from micro-flats has yet been attributed to the diffusion of carbon into the emitting region although such an effect might be expected [4 16].

With increasing time of operation, a gap develops between the graphite and the LaB_6 as seen in Fig. 5b. This material loss, presumably due to evaporation, results in a reduction in the diameter of the micro-flat and this in turn may well set the limit to the useful life of these emitters.

Discussion

The electron optical behaviour of the restricted emission micro-flat LaB_6 cathode, resembles that of the oxide-cored cathode developed by Uyeda[17]. This is not surprising as the physical configuration of the two cathodes is very similar. For the oxide-cored cathode, a central emitting core of barium-strontium oxide about 70 μm diameter is encased by a high work function tube of platinum. The electron optical characteristics of such cathodes have been studied in some detail[18 19]. In the article by Ando et al[18], their Fig. 8 resembles closely the results presented here in Fig. 3. The similarity in performance is particularly interesting as the results of Ohno[19] for a cored cathode suggest that electron energy spread from such cathodes is less than from conventional cathodes. It was concluded that anomalous energy broadening is caused by the effect of the biased Wehnelt on the electrons emitted from off central areas of conventional cathodes.

Loeffler[20] developed a complex cathode structure which limited the area of emission from a polycrystalline LaB_6 cathode and established that this type of operation resulted in experimental electron energy spread of about 0.6V in good agreement with the theoretically estimated value at 1850K. A thin rhenium foil containing a 7.2 μm diameter aperture was located within 100 μm of the cathode surface and operated at a potential of +100V with respect to the cathode. Only 2.7μA of the 40mA of total cathode emission to the rhenium electrode passed through the aperture and was used for beam formation. This corresponded to about 6.6 A/cm^2 from the cathode operating at 1850K and is in reasonable agreement with the current density estimated from curve C of Fig. 6 for a work function of 2.88 eV, a value commonly quoted for polycrystalline LaB_6.

A more practical type of micro-flat structure on a single crystal LaB_6 cathode has been described by Swanson et. al.[21]. This emitter, referred to as a truncated cathode, has a small flat about 40 μm diameter ground normal to the cone axis of a regular pointed emitter. Such a cathode still emits from the conical surface and can give rise to complex emission patterns as the electric field at the tip increases. This structure serves to increase the area of the (001) plane that is involved with emission, a factor which has been shown to be important from the correlation of electron emission characteristics and the thermal faceting of pointed cathodes[4].

The present cathode combines features of the truncated cathode and the cored oxide cathode and was developed specifically for the study of thermionic emission anisotropy of LaB_6 in high brightness triode guns. To date studies have been conducted only on (001) emitters. It is anticipated that studies on lower work function surfaces may lead to improve brightness at lower temperatures. Such increases in the cathode loading can be estimated from Fig. 6 where curve A for a work function of 2.52 eV indicates cathode loadings of 10 A/cm^2 at 1700K and about 50 A/cm^2 at 1900K. Recent work[20] suggest a work function of 2.41 eV for the (346) surface of LaB_6.

For cathodes operating at 1700K the evaporation losses would be about 1 μm in 380 hours[7]. If the base pressure in the gun is further reduced to about 4×10^{-6} Pa (3×10^{-8} torr), the oxidation losses would also be about 1 μm. Under such conditions, a cathode would loose about 6 μm of material over 1000 hours of operation. Hence it is not unreasonable to anticipate that practical changes in the conditions of operation and the selection of cathode surfaces of low work functions will result in a micro-flat cathode operating for 1000 hours at 1700K with a higher brightness than that obtained from a conventional pointed (001) emitter operating between 1850 and 1900K.

Conclusions

A new type of single crystal LaB_6 cathode structure has been developed. Preliminary observations on the electron optical performance of this cathode have been presented and show that in the present electron gun, the cathode provides a higher brightness than pointed cathodes operating under similar conditions.

Some advantages to the new cathode are as follows:

1. The crystallographic orientation and the area of the emitting surface is known.
2. At specific height settings of the cathode, the gun can be operated at zero bias, permitting high fields at the cathode surface.
3. The beam profile (and hence the cross-over) is circular and near gaussian containing no complex lobes from the walls of the cathode.
4. The total loss of material from the cathode by evaporation and oxidation is greatly reduced as only a small area of LaB_6 is directly exposed to the vacuum. Hence, troublesome contamination of the Wehnelt apertures is eliminated.
5. The carbon coating contributes to the mechanical strength and stability of the bond between the LaB_6 and its support.

The main problems that have been experienced with these cathodes are:

1. The sensitivity of the single crystal surface to electron beam induced outgassing from the walls of the column.
2. Material loss at the interface between the LaB_6 and the carbon coating resulting in a reduction in diameter of the emitting area.

Neither of these problems would appear to limit the practical usefulness of the micro-flat cathodes. Further detailed electron optical studies of these new emitters are in progress.

Acknowledgements

The authors wish to thank J. LeGeyt and R. Côté of the NRC Electron Physics laboratory for assistance in exploratory studies and fabrication of cathodes by electric discharge machining. We are also indebted to Dr. C. Bouchard of Bausch and Lomb Canada for assistance with carbon coating, polishing of micro-flats and the provision of arc-bonded LaB_6 single crystal emitters.

References

1. Takigawa T, Yoshii S, Sasaki I, Motoyama K. and Meguro T. (1980). Emission characteristics for (100), (110), and (111) LaB_6 cathodes. Jap. J. Appl. Phys. 19, L537-L540.

2. Takigawa T, Sasaki I, Meguro T, and Motoyama K. (1982). Emission characteristics of single-crystal LaB_6 gun. J. Appl. Phys. 53, 5891-5897.

3. Furukawa Y, Yamabe M, Itoh A, and Inagaki T. (1982). Emission characteristics of single-crystal LaB_6 cathodes with (100) and (110) orientations. J. Vac. Sci. Technol. 20, 199-203.

4. Hagiwara H, Hiraoka H, Terasaki R, Ishii M, and Shimizu R. (1982). Crystallographical and geometrical effects on thermionic emission change of single crystal lanthanum hexaboride cathodes. Scanning Electron Microsc. 1982; II: 473-483.

5. Kato T, Shigetomi A, Watakabe Y, Hagiwara H and Hiraoka H. (1983). Evaluation of single crystal LaB_6 cathodes. J. Vac. Sci. Technol. B1, 1, 100-106.

6. Hafner P and Bas EB. (1976). Investigation on bolt-cathodes with floating zone melted polcrystalline and monocrystalline LaB_6 emitters. 7th Int. Conf. Electron and Ion Beam Tech., R Bakish (ed), Electrochem. Soc. Inc., 3-17.

7. Swanson LW, Gesley MA and Davis PR. (1981). Crystallographic dependence of the work function and volatility of LaB_6. Surface Sci. 107, 263-289.

8. Ames LL and McGrath L. (1975). Vaporisation studies of the rare earth hexaborides. High Temp. Science 7, 44-54.

9. Sewell PB and Ramachandran KN. (1978). Grid aperture contamination in electron guns using directly heated LaB_6 cathodes. Scanning Electron Microsc. 1978; I: 17-23.

10. Ramachandran KN. (1975). Rhenium bonded LaB_6 electron source. Rev. Sci. Instrum. 46, 1662-1663.

11. Bokros JC. (1969). Deposition, structure and properties of pyrolytic carbon. in:Chemistry and physics of carbon. P.L. Walker Jr. (ed), Marcel Dekker Inc., New York.

12. Broers AN. (1973). A new high resolution electron probe. J. Vac. Sci. Technol. 10, 979-982.

13. Sewell PB. (1980). High brightness thermionic guns for electron microscopes. Scanning Electron Microsc. 1980; I: 11-24.

14. Swanson LW and McNelly DR. (1979). Work functions of the (001) face of the hexaborides of Ba, La, Ce and Sm. Surface Sci. 83, 11-28.

15. Vogel SF. (1970). Pyrolytic graphite in the design of a compact inert heater of a lanthanum hexaboride cathode. Rev. Sci. Instrum. 41, 585-587.

16. Oshima C, Bannai E, Tanaka T and Kawai S. (1977). Carbon layer on lanthanum hexaboride (100) surface. Jap. J. Appl. Phys. 16, 965-969.

17. Uyeda R. (1956). Discussion on cored-oxide cathode in: Proc. 1st Reg. Conf. Electron Microsc. in Asia and Oceania. Organising Committee (ed.), Electrotechnical Laboratory, Tokyo, 146.

18. Ando K, Kamigaito O, Kamiya Y, Takahashi S and Uyeda R. (1959). Oxide-cored cathode. J. Phys. Soc. Japan 14, 180-185.

19. Ohno T. (1974). Effect of emitting area on the energy distribution of thermionic emission. J. Electron Microsc. 23, 1-7.

20. Loeffler KH. (1970). A new cathode. Septième Congrès international de microscopie électronique, Grenoble. (Pub) Société française de microscopie électronique, Paris, 77-78.

21. Swanson LW, Davis PR and Gesley MA. (1982). Rare earth electron emitter materials fabrication and evaluation. Oregon Graduate Centre, Report No. RADC-TR-82-12, 1-119. (Available from L. Swanson, see his paper this volume).

Discussion with Reviewers

L. Swanson: In order that the beam current be proportional to brightness as the Wehnelt bias or emitter structure is varied, the beam size and angle at the specimen plane must be changed. Can the authors comment on this?

Authors: In the SEM the overall demagnification of the source (cross-over) is about 10,000 times. At the specimen plane, or entrance to the Faraday cup, the collection angle at the source as seen by the specimen is only about 10^{-6} radians, a very small percentage of the total beam divergence angle (typically 0.01 - 0.02 radians). Small changes in the total beam divergence from the gun or small changes in the cross-over position have little effect on the apparent axial brightness as estimated from the specimen current. In this particular case, the interest is to compare sources in the gun set to operate under best performance conditions of smallest cross-over consistent with maximum axial brightness. The measurements are relative and not absolute.

Electron Optical Systems (pp. 171-178)
SEM Inc., AMF O'Hare (Chicago), IL 60666-0507, U.S.A.

STUDY OF SPACE CHARGE LIMITATION OF THERMIONIC CATHODES IN TRIODE GUNS

A. Delâge and P.B. Sewell

Division of Electrical Engineering
National Research Council Canada
Ottawa, Ontario K1A 0R6
Canada

Abstract

A computer program has been developed to compute the emission of cathodes from the fully space charge limited diode equation. The emission of four different gun arrangements with thermionic cathodes of the type used in electron microscopes is investigated. Microflat cathodes with a small (≤ 50μm in diameter) fixed emitting area show several advantages for high brightness possibilities whereas other types of cathodes for which the emitting area increases with the brightness are usually limited by too great a total emission.

KEY WORDS: Thermionic emission computation, Space-charge simulation, Thermionic cathodes, Microflat cathodes, Triode electron gun, Diode electron gun, Low voltage gun

Address for correspondence:
A. Delâge, Division of Electrical Engineering
National Research Council Canada, Ottawa, Ontario
Canada, K1A 0R6 Phone No.: (613) 993-2304.

Introduction

In electron microscopes equipped with thermionic sources, the filament is often used in the regime of space charge limitation in order to provide the gun with a higher stability. In this regime the electron emission is in principle determined entirely by the cloud of electrons produced in front of the cathode and is independent of the characteristic emission of the cathode itself, provided it can emit a sufficient number of electrons. The emission is then controlled by the field in front of the cathode and can be adjusted by varying the Wehnelt potential.

Since the life of the cathode is very dependent on its operating temperature, this is generally chosen such that the filament is on the border of the space charge limitation. This regime makes the emission independent of the filament temperature and the nature of the filament itself. The direct replacement of the tungsten filament by a better emitter such as LaB_6 would be almost imperceptible if the gun operating parameters remain the same.

In order to improve the brightness obtained from triode guns a detailed investigation of space-charge limitation is necessary.

Thermionic Emission

A computer program has been developed to calculate the emission from a cathode taking into account the space charge and the possible temperature limitation of the cathode. This program follows the technique described by Kirstein et al[3]. It is based on the theoretical maximum current that can be drawn from a one-dimensional diode. Before discussion of this technique, some basic concepts of thermionic emission are considered.

The simple model of the free electron gas in a metal[7] describes successfully the thermionic emission of solids in terms of only two parameters: the temperature T and the work function ϕ associated with a particular metal. The model predicts a current density given by Richardson's equation:

List of Symbols

A	Constant in the Richardson's equation $= 120\ A.cm^{-2}.K^{-2}$
D	Diameter (μm) of the emitting area in equation (14)
E	Electric field (Vm^{-1} if not specified)
G	$(4/9)\varepsilon_o\ (2e/m)^{\frac{1}{2}} = 2.34 \times 10^{-6}\ A.V^{-1.5}$
G'	G/z_o
I(r)	Current emitted by an annulus on the cathode, of mean radius r and a width Δr
I(r,n)	I(r) after the nth iteration
$I_{tot}^{(n)}$	Total current emitted after the nth iteration
J	Current density in A/cm^2
J_o	J given by Richardson's equation, also called the temperature limitation
J(T,E)	J given by Schottky's equation
J_{sc}	Space charge limited J, also called the space charge limitation
$J_D(E,D)$	J_{sc} for the diode gun with a limited emitting area
N	Number of trajectories issued from each annulus on the cathode
N_o	Density of electrons in the model of the electron gas in metal
Q(R,Z)	The space charge density term for each mesh of the grid used in the network method
R	Radial matrix coordinate (integer) in the network method
S	Surface on the emitting cathode
T	Absolute temperature (K)
V_m	Potential minimum (V)
V_A	Anode potential corresponding to the energy of the electrons at the anode (V)
V(z)	Axial potential in the gun (V)
$V_{el}(R,Z)$	Solution of the potentials exclusively due to the electrode arrangement
$V_{sc}(R,Z)$	Solution of the potentials exclusively due to the space charge in the gun
$V_{sc}^{(n)}$	V_{sc} at the nth iteration
Z	Axial matrix coordinate (integer) in the network method
d	Separation cathode-anode in the diode model
dt	Integration step of the trajectory
e	Electronic charge (1.602×10^{-19}C)
h	Mesh size of the grid in the network method
k	Boltzmann's constant $1.38 \times 10^{-23}\ J.K^{-1}$
m	Electronic mass 9.1×10^{-31} kg
$n(v_z)dv_z$	Density of electrons having a speed between v_z and $v_z + dv_z$
r	The radial coordinate
r_d	Radius of the small area in the center of the cathode
v_x,v_y,v_z	Electron speeds in the model of electron gas in metal
z	The axial coordinate
z_m	Position of the potential minimum
z_o	Axial position where the Child-Langmuir equation is applied
β	Brightness in A $cm^{-2}sr^{-1}$
γ	Under-relaxation factor ($0 \leqslant \gamma \leqslant 1$; ideal value 0.667)
Δr	width of an emitting annulus on the cathode
Δvol	Volume element associated with each cell of the mesh in the network method
ε_o	Vacuum permittivity 8.85×10^{-12} C/V.m
∅	Work function of the cathode material (V)

$$J_o(T,\emptyset) = A\ T^2 \exp(-\ e\emptyset/kT) \qquad (1)$$

This equation gives the thermal limitation of the cathode. In fact the presence of the electron cloud or the presence of a purely attractive field E (without the electron cloud) will influence the height of the barrier that the electrons have to cross at the surface of the solid. In the first case the space charge gives a supplementary barrier (V_m on Fig. 1b) that will be added to ∅ while in the second case the reduction of the height of the barrier will lead to Schottky emission[8]:

$$J(T,E) = J_o \exp\{0,4403\ \sqrt{E}/T\} \qquad (2)$$

(J and J_o in A/cm^2, E in V/m and T in degree K).

In the mode of space charge limitation the barrier V_m produces an emission density simply given by[4]:

$$J_{sc} = J_o(T,\emptyset)\ \exp\ (-eV_m/kT) \qquad (3)$$

As mentioned previously the local field in front of the cathode will determine the height of V_m. Unfortunately the relation between this field (prior to emission) and V_m is not known except for the case of a one-dimension diode.

In a general approach it is possible to compute the emission from this model. It consists of evaluating numerically the height of V_m by tracing the trajectories of the electrons considering a gaussian velocity distribution as stated in the original gas model:

$$n(v_z)dv_z = N_o \exp\{-(v_x^2 + v_y^2 + v_z^2)m/2kT\} \quad (4)$$

This approach is very tedious since the number of trajectories to be calculated is very large and only an insignificant portion succeed in going through the barrier to form the beam[9]. It also contains another difficulty since it assumes that we have to resolve the distance z_m between the barrier and the cathode, a distance very small compared to the length of the beam.

However this procedure can be overcome by using a slightly different approach that includes the evaluation of the emission for the case of the one-dimension diode. The potential on the axis of such a cathode is presented in Fig. 1a with some enlargement on Fig. 1b for the region near the cathode. The curve A represents the potential prior to emission or for a highly temperature limited diode where the beam is so small that it does not affect the potential. The curve B represents the diode fully space charge limited. It is possible to find an analytical solution for this problem if we neglect the size of V_m and z_m by solving the Poisson equation in one dimension. This leads to a current density given by:

$$J[A/cm^2] = G\ V_A^{1.5} / d[cm]^2 \quad (5)$$

with $G = (4\ \varepsilon_o/9)\ \sqrt{(2e/m)} = 2.34 \times 10^{-6}\ A/V^{1.5}$.

Since the potential on the axis is given by $V(z)=V_A(z/d)^{4/3}$ an interesting conclusion from this approach is the following relation for the current density:

$$J = G\ V(z)^{1.5}/z^2 = G\ V_A^{1.5}/d^2 \quad (6)$$

This means that the current density can be evaluated at any location where the gun is fully space charge limited.

Another supposition is implicit in equation (6); the initial velocity of the electrons is zero so that both V_m and z_m are neglected. Some refinement of this solution consists of using the Child-Langmuir expression that takes into account the axial velocities in the beam. In the present case since the computations are done on a micro-computer and are not aimed at very large accuracy, this refinement is neglected.

However the conclusion for this diode may be generalised to the case of the triode gun (or any other gun geometry). For the triode gun the presence of the negatively charged Wehnelt modifies the axis potential in a way very similar to the space charge in the diode except for its external control. This control modifies continuously the potential in the gun from curve A to B in Fig. 1a without changing the temperature of the filament.

On the other hand two main differences exist. (1) The electron beam is now concentrated near the axis instead of filling all the space and consequently only the potential near the axis

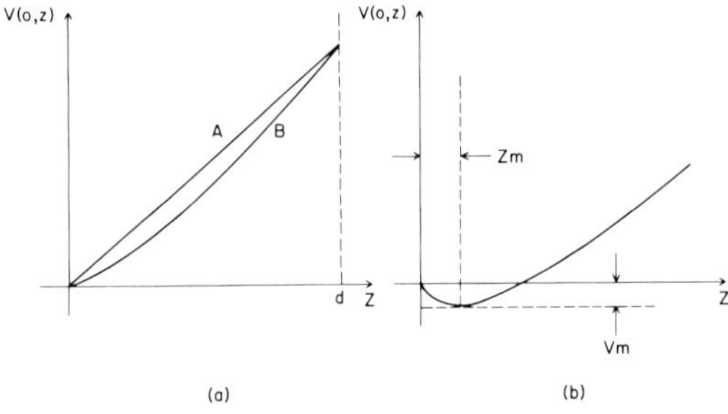

Fig. 1a Axial potential distribution in a diode (A) prior to emission and (B) at the full space-charge limitation.

Fig. 1b Enlargement of curve B in the vicinity of the origin.

will be affected by the beam. (2) The effect of the charge distribution on the Wehnelt has a much longer range than the space charge of the beam so that in practice only the region near the cathode where the slow electrons produce a large space charge will be affected by the space charge itself.

It is concluded from these remarks that to be valid the Langmuir equation (6) must be applied at a distance z_o near the cathode much larger than z_m.

Numerical Evaluation of Cathode Emission

The technique used here to compute the emission of the gun has been adapted from the procedure described by Kirstein et al[3]. It consists of the following iterative procedure:

1. Solution of Laplace Equation; this solution is called here $V_{el}(R,Z)$.
2. Division of the cathode in small areas according to the symmetry; in this case a disk and concentric annuli of width Δr; a number (N) of trajectories issuing from the areas at different angles and/or energies.
3. Computation of I(r) the total current associated to each beamlet or I(r)/N for each trajectory.

   $$I(r) = 2\ \pi\ r\ \Delta r\ \ G\ \ V(r,z_o)^{1.5}/z_o^2.$$

4. Computation of the space charge from the trajectories. At each step of integration in the trajectories the following matrix element is computed:
   $Q(R,Z) = Q(R,Z) + I(r)dt\ h^2/(\varepsilon_o\Delta Vol)$, h being the mesh size and ΔVol the volume element associated to each mesh.
5. Solution of Poisson Equation with 0 volt on the electrodes such that $V(R,Z) = V_{el}(R,Z) + V_{sc}(R,Z)$.
6. Repeat from 3 until I(r) converges.

Technically it appears that some time can be saved by introducing near the cathode a smaller boundary than the entire gun. The space charge is then computed numerically only in this region. The boundary itself is affected by the space charge but only in a small amount especially if it has been chosen properly.

Usually this boundary "converges" much more rapidly than I(r) and very often only one iteration is necessary.

In the present approach it has been found very efficient to compute this boundary from the charge density method and to apply the finite difference method inside as suggested by Birtles[1].

The choice of z_o may be somewhat controversial but since the mesh size (h) used inside the boundary is large, z_o = h seems adequate.

It has been proposed[3] to increase the convergence of this problem by under-relaxing the choice of I(r) so that the nth iteration is as follows:

$$I(r,n) = I(r,n-1)(1-\gamma) + \gamma I(r) \qquad (7)$$

$$\text{with } 0 < \gamma \leqslant 1 .$$

This technique can be improved by assuming that the space charge is directly proportional to the total current I_{tot} in the beam. This current is given by the following integral:

$$I_{tot}^{(n+1)} = G^1 \int_S (V_{el} + V_{sc}^{(n)})^{1.5} \, dS. \qquad (8)$$

Ideally when the value of the current converges, the solutions of the nth and (n+1)th iterations are equal for large n so that we can write:

$$I_{tot}^{(n+1)} = G^1 \int_S (V_{el} + V_{sc}^{(n+1)})^{1.5} \, dS. \qquad (9)$$

If now we use this expression as very good approximation for $V_{sc}^{(n+1)}$:

$$V_{sc}^{(n+1)} = I_{tot}^{(n+1)} \; V_{sc}^{(n)} \; / I_{tot}^{(n)} \qquad (10)$$

a much more accurate total beam current for the next iteration is given by solving equation (9). This later relation includes I_{tot} on both sides of the equation but it pays to solve it numerically since the time required for one full iteration cycle represents five to ten minutes on a small computer (HP 9836) compared to a few seconds ($\leqslant$ 5s) to solve equation (9).

Once $I_{tot}^{(n+1)}$ is found, the much more accurate $V_{sc}^{(n+1)}$ as defined by equation (10) is used instead of $V_{sc}^{(n)}$ to compute I(r). In this procedure the convergence applies only to the shape of the beam and the distribution of electrons in it, never to the total current. This improves the convergence because the value of current has a tendency to be highly oscillatory. Results show that 10% of the final value is reached at the first iteration and usually less than 1% at the second one.

Results and Discussion

This technique has been applied to the four different guns described in Table 1. The first is highly space-charge limited due to its flat cathode and large Wehnelt-anode distance of 9mm. The emission of this gun is presented in Figure 2 as a curve of brightness vs total current; the theoretical model is compared to experimental measurements obtained in our laboratory. The theoretical brightness is obtained by applying Langmuir's equation[5] for $V_A >> kT$:

$$\beta \; (A/cm^2 sr) = I(0) e V_A / \; \pi^2 \; r_d^2 \; k \; T \qquad (11)$$

where $I(0)/\pi r_d^2$ is the current density of the central beamlet leaving the cathode on the axis from a small circular area of radius r_d. The agreement between the model and the experimental results is remarkably good. The prime interest of this gun was to study the effect of reducing the field on the performance of the overall source. The results show a large space-charge limitation in the brightnesses and also some deterioration in the size of the cross-over as the distance anode-Wehnelt is increased, due to large interaction between electrons.

The second gun with a conical cathode on a shorter Wehnelt-anode separation is typical of a triode gun used in electron microscopes. The current density in the apex region and the total current as a function of the logarithm of the excess of the Wehnelt voltage (V_w) over the cut-off voltage (V_{co}) are presented in Fig. 3. The emitting area of the cone is also presented. The current density can be converted into brightness using the following equivalent to equation (11):

$$\beta \; (A/cm^2 sr) = J(A/cm^2) \; e \; V/\pi \; k \; T \qquad (12)$$

For the example presented here the current density is converted into brightness by multiplying $J(A/cm^2)$ by $\sim 3 \times 10^4 \; sr^{-1}$.

This graph shows how the emitting area increases with the field which is directly proportional to $V_W - V_{co}$. This increase is responsible for the difference in slope between the current density and the total current curves. From Figure 3 it is seen that a cathode operating at 10 A/cm^2 would emit about 150 µA. However at higher cathode loadings such as 30 or 50 A/cm^2, the total current would be 0.9 and 2 mA respectively. Such currents are beyond the range of commercial electron microscope power supplies.

This engineering limitation associated with large power dissipation is not the only argument against the conical cathode. Since the current reaching the sample is many orders of magnitude less than the emitted current, only the electrons emitted near the axis contribute finally to the signal while all others interfere with the quality of the beam optics.

In the first place, there is an energy dispersion caused in the gun by the interaction between electrons originating from the cone and the axial electrons[5] and must be minimized when chromatic aberration of the lenses is the limiting factor encountered in the column.

TABLE 1
Gun Geometries

Gun	I	II	III	IV
Cathode Type	Flat Disk	Cone on Cylinder	Microflat on Cone	Flat Diode
Cathode Size[a] (μm)	Ø: 500 t: 750	Cone Angle: 90° Base Ø: 400	Cone Angle: 90° μFlat Ø: 90 Emitting Ø: 60	Emitting Circle of Diameter D
Cathode Height (μm)	250	200	100	-
Wehnelt Diameter (mm)	1	1	1	-
Wehnelt-Anode (mm)	9	4 in Fig.3 10 in Fig.4	10	4[b]
High-Voltage (kV)	15	15	15	Variable

a) t = thickness b) Distance cathode-anode

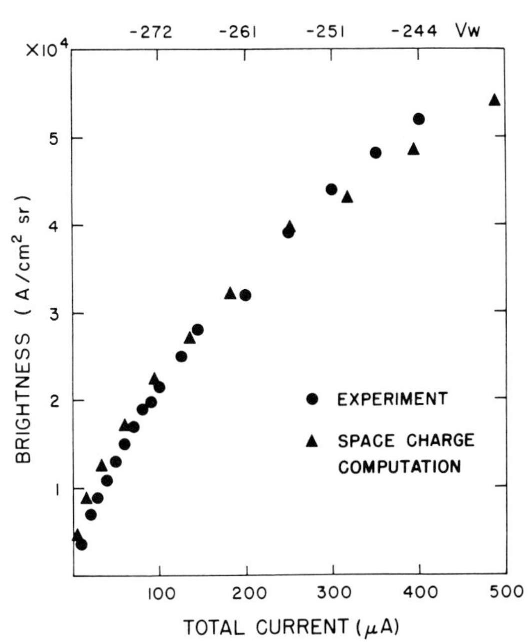

Fig. 2 Brightness vs total current characteristic for a flat cathode (gun I in Table 1).

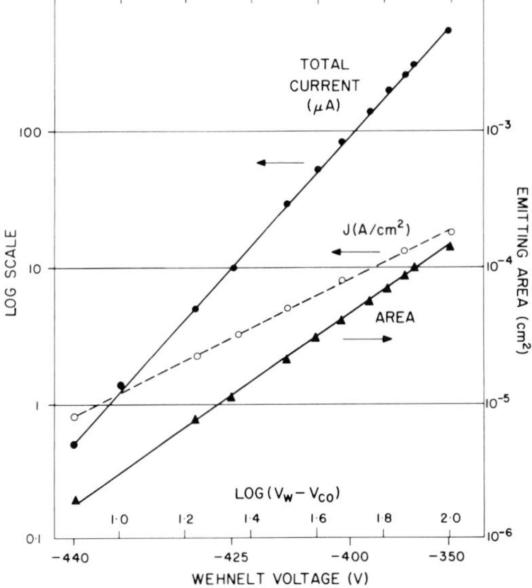

Fig. 3 Emission of a conical cathode (gun II in Table 1) as a function of the drive voltage ($V_W - V_{co}$). The left hand log scale gives the values for the total current in μA and the current density at the apex in A/cm^2.

Secondly the elimination of the undesirable electrons is not always hundred percent effective and stray electrons may contribute to the deterioration of the ideal performance of lenses and the sharpness of the central beam.

Finally an emitter tip with a finite radius of curvature can be represented by a portion of a sphere on the top of a truncated cone. This geometry contains two different electron optical systems with their own characteristics. At high negative bias, electron emission is confined to the tip region which results in good gun performance with high brightness[10] and near-gaussian beam profile. At lower negative bias, emission from the walls of the conical surface leads to a complex beam form. For sintered LaB_6 cathodes this results in an additional complex ring structure around the central beam and for single-crystal LaB_6 an additional lobe pattern with crystallographic symmetry[2]. In such a case only a small portion of the beam is due to the desired paraxial electron emission. For the example discussed in Figure 3, if a conical cathode emitting at 10 A/cm^2 had a tip radius of curvature of 5 μm, only 3% of the total current would come from the surface of the spherical tip.

The coated microflat cathode[11] has been designed to solve this fundamental difficulty. Some preliminary results on the emission of such cathode is presented in Figure 4. This conical cathode is coated and ground in such a way that the flat is 90 μm in diameter and exposes an LaB_6 area some 60 μm in diameter. The anode-Wehnelt distance is 10 mm and its performance can be directly compared to previous results obtained on flat cathodes (curve D) redrawn from Figure 2. The results for a pointed cathode in a 10 mm anode-Wehnelt spacing gun are also presented for comparison (curve C). The emission of the microflat is bounded in this figure by two curves: (1) the full space-charge limitation curve B and (2) the temperature limitation (curve A) given by the Richardson and Schottky equation:

$$\beta = J(T,E) \ / \ \pi \ k \ T \qquad (13)$$
$$I_{tot} = J(T,E) \times \text{Microflat area}$$

The space charge limitation is computed from the model presented in this paper. If in this model we limit the cathode loading, the emission characteristic of this cathode goes from curve B to A following the paths shown in Figure 4. The numerical values on these paths correspond to the cathode loading in A/cm^2. The curve C for the pointed cathode indicates much better performance near the cut-off but the effect of space-charge limits the current density to about 6 A/cm^2 at 400 μA of total emission while the fully space-charge limited microflat (curve B) is much less limited.

For the microflat cathode, this limitation is to a first approximation proportional to the total current and will be highly dependent on the area of emission. The fourth gun of Table 1 has been studied to investigate the influence of this area on the space-charge limitation. Microflat cathodes with small emitting areas can be used at zero bias potential (V_W = 0 V) and give good performance, the current density being improved by Schottky emission[11]. This suggests that this kind of cathode can be used in a diode gun. Considering that the diode gun is a limiting case for the triode gun as far as the field is concerned, some conclusions on the maximum brightness of a triode gun can be computed from this study.

The model of the diode gun consists simply of two planes, one of which forms the anode at potential V and a second one, the cathode, is allowed to emit only from a small circular region. The emission properties are studied as a function of the potential V and the diameter (D) of the emitting area. If the distance anode-cathode (here 4 mm) is much larger than D, the current density emitted depends mainly on the mean field E in the diode and the size of the emitting area. For computation of E in the range $0.01 \leqslant E \leqslant 0.5$ kV/mm and $20 \leqslant D \leqslant 80$ μm the axial current density is very well represented by the empirical formula:

$$J_D \ (A/cm^2) \cong 260 \ E(kV/mm)^{1.5}/D(\mu m)^{0.5} \qquad (14)$$

where practical units have been used and are given in parenthesis.

Figure 5 gives a representation of the space charge limitation in terms of brightness reduction as a function of the voltage (representing the beam energy). The space-charge limitation effects occur only at low voltage. The two sets of curves give the percentage brightness loss for two different cathode temperatures 1800 and 1900 K for which J(T,E=0) is respectively 10 and 30 A/cm^2. Each set of curves has been computed for D = 5, 10, 25, 50 and 100 μm assuming that relation (14) can be extrapolated when necessary. Those results are computed from the following relation:

$$\beta \text{ Loss } (\%) = 100\% \ (1 - J_D/J(T,E)) \qquad (15)$$

where J_D is given by equation (14) and J(T,E) by equation (2).

For the lower temperature case (J = 10 A/cm^2), the gun starts to be space charge limited for beam energies around 2.2 keV when the emitting area is 100 μm. This pin-points the difficulty of using the standard triode gun for scanning electron microscopy application at low voltages. If a 10 μm microflat can be made, one can expect to use this kind of cathode without space charge limitation at 1 keV. Such low voltage operation is desirable for inspection of microelectronic devices without causing serious damage[6].

Severe limitations are placed on the size of the microflat if full advantage is to be made of the emission of LaB_6 at low voltages. The set of curves at higher temperature gives a threshold of 2 keV for a 10 μm diameter microflat before space-charge limitation occurs.

Conclusion

The agreement between experimental and theoretical results presented in this paper indicates that the procedure described by Kirstein et al.[3] does apply very well to triode guns. This technique has been used to study the space-charge effects of different types of cathodes, namely conical and microflat cathodes. The conical uncoated cathodes give very good performance near the cut-off where only the very tip is involved in the emission. On the other hand the emission from the cone usually gives too much current before the field at the apex increases enough to reach the full theoretical axial brightness. The coated microflat cathode which suppresses emission from the cone is much less prone to space-charge limitation and is more suitable for application where high brightness is required, especially at low voltages.

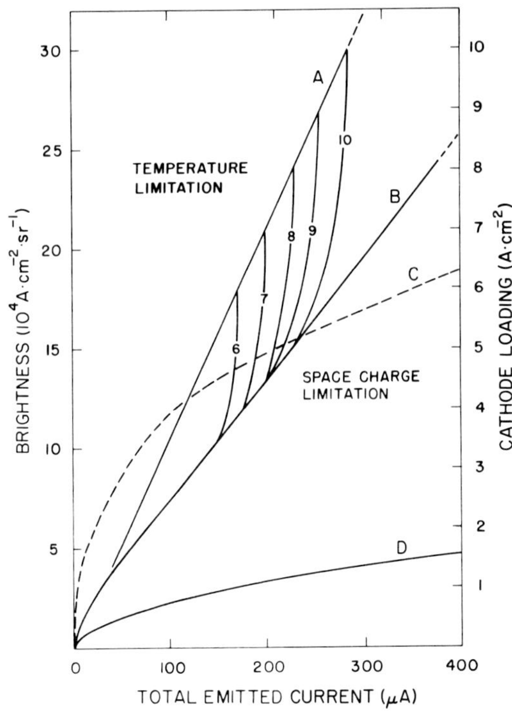

Fig. 4 Brightness vs total current characteristic for a microflat cathode (gun III in Table 1) for different cathode loadings (index in A/cm^2). Curves A and B are respectively the temperature and the space-charge limitation curves. The microflat is compared to a flat cathode (curve D) and a pointed cathode in a similar gun (curve C).

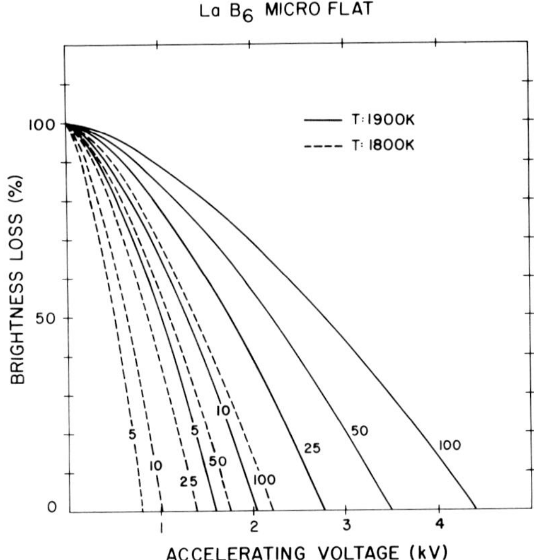

Fig. 5 Brightness loss at low energy for a microflat cathode in a diode gun (gun IV in Table 1) for two temperatures; curve indices are the microflat diameters in μm.

References

1. Birtles A.B., (1972) "An efficient technique for electrostatic field computation in some axially symmetric electron optics system", Int. J. Electronics 33, 649-657.

2. Furukawa Y., Yamabe M. and Inagaki T., (1983) "Emission characteristics of single-crystal LaB_6 cathodes with large tip radius", J. Vac. Sci. Technol. A1, pp 1518-1521.

3. Kirstein P.T., Kino G.S., Waters W.E., (1967) Space-Charge Flow. McGraw Hill, New York, Chapter VIII, pp377-399.

4. Kirstein P.T., Kino G.S., Waters W.E., (1967), ibid., Chapter VI, pp263-276.

5. Lauer R., (1982) "Characteristics of triode electron guns", in Advances in optical and electron microscopy, Vol. 8, Barer R. and Cosslett V.E. Eds., Academic Press, London, pp137-202.

6. Lukianoff G.V. and Langner G.O., (1983) "Electron beam induced voltage and injected charge modes of testing", Scanning 5, pp 53-70.

7. Nottingham W.B., (1956) Thermionic Emission, in Handbuck der Physik, Volume XXI, Springer-Verlag Berlin. Section 17-18, pp14-17.

8. Nottingham W.B., ((1956), ibid., Section 27, 28-33.

9. Renau A., Read F.H. and Brunt J.N.H., (1982) "The charge-density method of solving electrostatic problems with and without the inclusion of space-charge", J. Phys. E: Sci. Instrum. 15, pp347-354.

10. Sewell P.B., (1980) "High brightness thermionic electron guns for electron microscopes", Scanning Electron Microsc. 1980; I: pp11- 24.

11. Sewell P.B. and Delâge A., (1984) "Thermionic emission studies of Micro-flat single crystal LaB_6 cathodes", in: Electron Optical Systems, 3rd Pfefferkorn Conf. Proceedings, SEM, Inc., AMF O'Hare, IL 163-170.

Discussion with Reviewers

J. Orloff: The coating of carbon on the microflat cathode will have an evaporation rate which is much less than that of the LaB_6. The rate of evaporation or thickness loss, of LaB_6 is about 0.01 micrometers per hour at 1800 K and about 0.1 micrometers per hour at 1900 K. This means that the electron-emitting surface of the cathode will retreat behind a shell of carbon at some rate, depending on the cathode temperature. Can you comment on the effect this will have on the performance of the cathode?

Authors: Essentially the recess of LaB_6 with time will cause a non-uniform decrease in the field in front of the cathode, the drop being more important on the outermost part of the emitting area near the carbon shell.

When the cathode is used in a standard gun with a field larger than 1 or 2 kV/mm the resulting electrical field will not affect the axial brightness which is usually temperature limited for this type of cathode.

On the other hand the effects of the recess will appear at low energy below the threshold where the emission is space-charge limited. These effects are (1) a drop in the emission current, (2) a decrease in the axial brightness and finally (3) an increase in the convergence of the beam due to the non-uniformity in the change in the field.

The total emission will be reduced significantly in those conditions because the part of the cathode that emits the most of the current is the most affected; consequently, the brightness will not decrease as much since the space-charge due to the total current will be less; this effect will partially compensate for the drop in the field.

The following table gives some figures for those effects; the computations have been carried out on the gun IV for a field of 0.1 kV/mm on a microflat of 50 μm. Results show that after 1000 hours of operation (at a loss of .01 μm/h) the brightness will be 54% of its original value for a microflat 50 μm in diameter.

RECESS (μm)	TOTAL CURRENT (%)	BRIGHTNESS (%)
0.0	100	100
2.5	67	92
5.0	45	80
7.5	32	67
10.0	23	54
12.5	17	43

Electron Optical Systems (pp. 179-186)
SEM Inc., AMF O'Hare (Chicago), IL 60666-0507, U.S.A.

0-931288-34-7/84$1.00+.05

DETERMINATION OF OPTICAL PROPERTIES OF A FIELD EMISSION GUN COUPLED WITH A LINEAR ACCELERATOR FOR HIGH VOLTAGE MICROSCOPY

M. Denizart*, S. Roques, F. Sonier and B. Jouffrey

Laboratoire d'Optique Electronique du C.N.R.S.
29, rue Jeanne Marvig, 31055 Toulouse Cedex, France

Abstract

The electron optical properties of a field emission gun plus a twenty-stage linear accelerator system have been determined in the case where the entrance pupil of the system is defined by one of the electrode apertures.

The first part of the calculation concerns the electrical and geometrical parameters of triode and tetrode field emission guns which can give a fixed location of the source and a small spherical aberration coefficient.

The second part is related to the possibility of placing a tetrode field emission gun at the top of the linear accelerator of the scanning high voltage electron microscope which is being constructed in the Toulouse Laboratory. Interesting conditions can be obtained when the electrical parameters are fixed as follows : either $V_i/V_e = 0$ or $V_i/V_e = 20$ (V_e is the extracting voltage, V_i is the potential of the second anode of the gun) whatever the accelerating voltage may be.

KEY WORDS : Field emission, Field emission guns, High voltage electron microscopy.

*Address for correspondence:
For reprints and other information please contact M. Denizart at above address.
Phone no.: (61) 52-65-96.

Introduction

The electron optical properties of field emission sources and their advantages in conventional transmission electron microscopy as well as in scanning electron microscopy are now well known (Kasper, 1982). The Toulouse Laboratory being concerned with the construction of a high voltage scanning transmission electron microscope (named M.E.B.A.H.T. : Microscope Electronique à Balayage à Haute Tension) operating at 1.6 MV, there has been some interest in the problems of placing a field emission gun (F.E.G.) at the top of a linear accelerator.

Two approaches (Denizart et al. 1981 ; Garg et al. 1984) have been used to find an optimal configuration compatible with the requirements of field emission and high voltage (1.6 MV). We report here the results of our approach.

The study has been concerned with two questions : 1) Is it possible to design a gun-accelerator system such that the source position is unchanged when the beam current or the electron energy is varied ? 2) Can the diameter of the source obtained with such a system be kept sufficiently small to provide a good resolution in the M.E.B.A.H.T. ?

In a first stage, the electron optical properties of the F.E.G. have been studied, and the electrical and geometrical parameters giving a stable position and a small diameter of the effective source have been determined. In the following stage, the best gun geometry was used to determine the geometrical parameters of the accelerator and the electrical parameters of the gun-accelerator system.

Throughout this study, we have striven to find solutions that are simple from a technological point of view and flexible from the practical point of view. A triode F.E.G. was thus designed and tested in an electron microscope and the lessons which have emerged from its use induced us to study the optical properties of a tetrode F.E.G.

The results of a systematic study of the electron optical properties of triode and tetrode guns and of a tetrode F.E.G. + a twenty-stage linear accelerator system, are presented in this paper.

List of symbols

C_{so}	spherical aberration coefficient referred to the object space.
C_{co}	chromatic aberration coefficient referred to the object space.
M_L	linear magnification.
Z_i	source position referred to the origin defined on fig. 1.
V_a	acceleration voltage for the guns.
V_e	extracting voltage.
k	ratio of V_a to V_e for a triode gun.
V_i	potential of the intermediate anode of the tetrode guns.
k_1	ratio of V_i to V_e.
k_2	ratio of V_a to V_e for a tetrode gun.
$V_1, \ldots V_{20}$	potential of the first,... twentieth electrode of the accelerator.
$E_1,\ldots,E_{20}$	first,..., twentieth electrode of the accelerator.
A_1	first anode of a gun.
A_2	second anode of a gun.
A_3	third anode of a tetrode gun.
$D_{A_1A_2}$	distance between the anodes A_1 and A_2.
e	thickness of the first anode in the triode guns.
α	cone angle of the aperture in the anode A_1 of a triode gun.
D_{CA}	distance between the cathode and the first anode A_1.
e_1,e_2,e_3	thickness of A_1, A_2, A_3 in a tetrode gun.
ϕ_1,ϕ_2,ϕ_3	diameters of the aperture of A_1, A_2, A_3 in a tetrode gun.
β_1,β_3	cone angle of the apertures in the anodes A_1 and A_3 of a tetrode gun.
e_E	thickness of the accelerator electrodes.
R	radius of the electrode apertures.
f_o	object focal length.
f_i	image focal length.

Generalities

The three types of systems studied are shown in fig. 1. We have only considered the case in which the entrance pupil of the system under study is defined by one of the electrode apertures : the best geometrical configuration will then be the one which has the smallest spherical and chromatic aberration coefficients C_{so} and C_{co} (referred to the object space) for any linear magnification M_L (Denizart, 1981 ; Munro, 1971).

We have therefore examined the influence of the different geometrical parameters (the electrode thickness, the radius and the shape of the electrode aperture, the distance between the electrodes and between the tip and the first anode) on the position of the source (denoted by $Z_i = OS_i$ (fig. 1) and on the values of C_{so}, C_{co} and M_L.

The purpose of the study was not to establish the exact properties of a small number of systems but to define the general characteristics of systems capable of giving a source stable in position and small in radius whatever the working conditions of the M.E.B.A.H.T. may be. A systematic study of many systems was therefore necessary.

The determination of the electron optical properties of an electrostatic system requires a knowledge of the potential distribution throughout the system. In view of the large number of geometries to be studied, it was not possible to take into account the real potential distribution between the tip and the first anode, although this can be calculated using the methods proposed by the different authors such as Kern et al. (1978), Hoch et al. (1978) or more recently by Kang et al. (1983) or Uchikawa et al. (1983). For this study, the space between the tip and the first anode has been considered to be equipotential. The finite element method has been employed to determine the potential using the program of Munro (1971), which has been adapted to obtain the electron optical properties of each system.

We note that by using different methods for calculating the properties of F.E.G., the discrepancies for the same gun between the finite element method and that of Kern can be shown to be quite acceptable, compared to the experimental precision (Denizart, 1981).

For the triode F.E.G., the properties are commonly studied as functions of a single electrical parameter $k = V_a/V_e$ (V_e is the extracting voltage and V_a the accelerating voltage). It has been shown (Denizart et al. 1981) that it is possible to characterize the properties of a tetrode F.E.G. as well as those of a tetrode F.E.G. + a twenty stage linear accelerator system in terms of two independent electrical parameters : $k_1 = V_i/V_e$ and $k_2 = V_a/V_e$ (V_i is the potential of the intermediate anode A_2 and V_a the accelerating potential of the tetrode F.E.G. (fig. 1b), furthermore $V_a = V_1 = V_{20}/20$ in the case of the complete system, V_{20} being the accelerating voltage of the system (fig. 1c).

The principal results of the study are presented for each system as functions of the geometrical and electrical parameters that have been previously defined.

Triode Field Emission Guns

Two types of triode guns have been considered : symmetrical (A_1 and A_2 are identical) and unsymmetrical (A_1 is the same as in the preceding case but A_2 is plane and 1 mm thick) (fig. 1a).

The thickness has been varied between 1 mm and 8 mm, the distance $D_{A_1A_2}$ between 3 and 12 mm, the cone angle of the anode aperture between 20° and 90° and the cathode-anode distance D_{CA} between 4 and 12 mm.

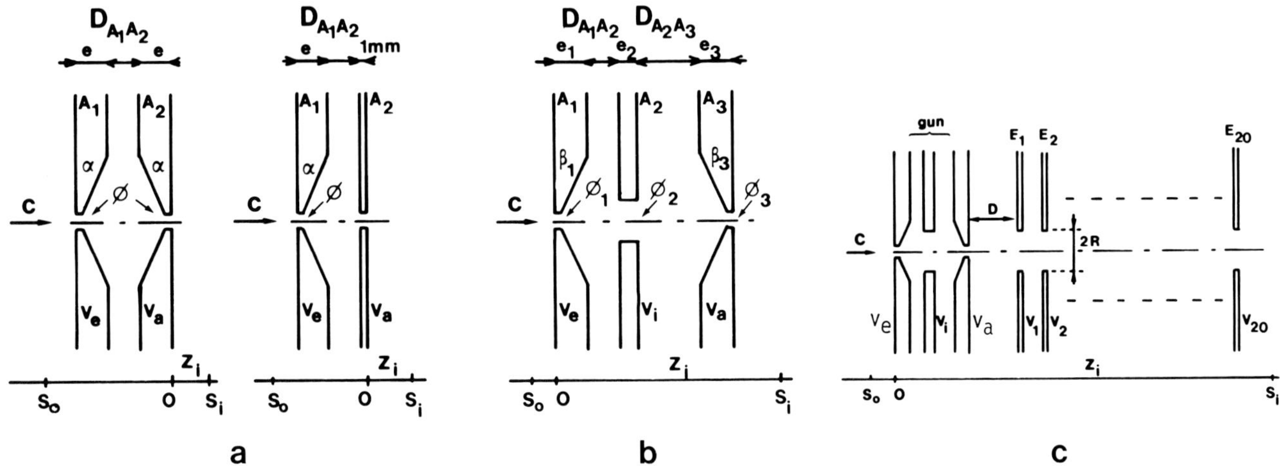

Fig. 1. Electrical and geometrical parameters of a triode gun (a), of a tetrode gun (b) and of a tetrode plus accelerator system (c).

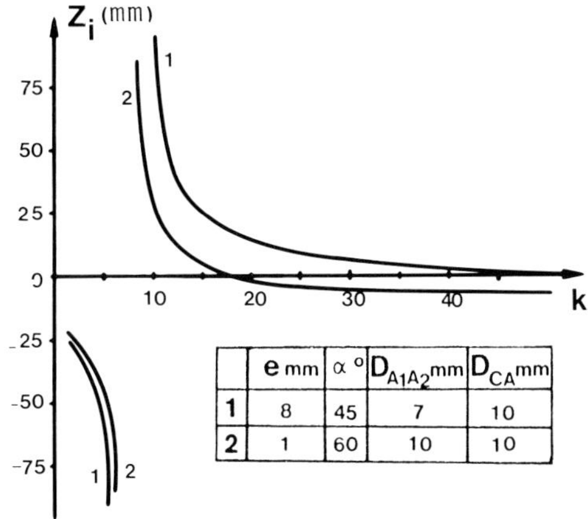

	e mm	α °	$D_{A_1A_2}$ mm	D_{CA} mm
1	8	45	7	10
2	1	60	10	10

Fig. 2. Typical variation $Z_i = Z_i(k)$.

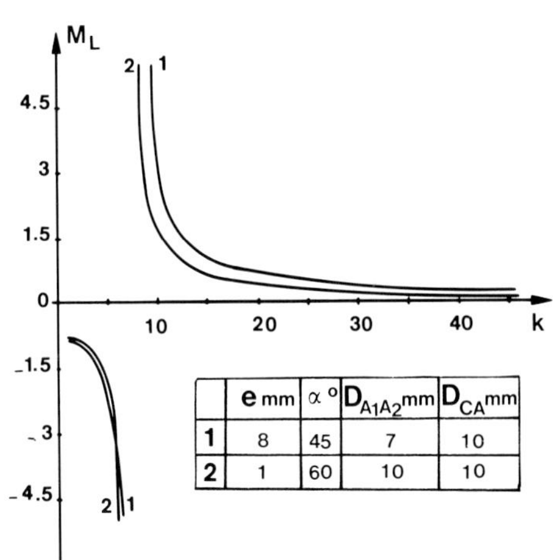

	e mm	α °	$D_{A_1A_2}$ mm	D_{CA} mm
1	8	45	7	10
2	1	60	10	10

Fig. 3. Typical variation $M_L = M_L(k)$.

From a general point of view, it can be said that the best results are obtained with the unsymmetrical geometries and that the values of C_{co} are 100 to 1000 times smaller than those of C_{so}, so only the results concerning the unsymmetrical systems and C_{so} are given here.

We first consider the problem of the source position stability ; the curves plotted in fig. 2 give an example of the variation of Z_i as a function of k. They show a discontinuity, the position of which depends on the geometrical parameters under consideration ($e, \alpha, D_{A_1A_2}, D_{CA}$), as do the M_L curves shown in fig. 3. Thus flexible operating conditions of the gun and therefore of the microscope will not be obtained for k values lower than about 15. It will thus be necessary to choose working conditions such that $k > 15$ and $30\ kV < V_a < 80\ kV$, which corresponds to the condition $V_e < 2kV$. These can be satisfied with short D_{CA} distances or with good tips (Denizart, 1981).

Furthermore, the study shows that Z_i is less sensitive to k if α and D_{CA} are large - but if D_{CA} is large V_e cannot be kept sufficiently low - and e and $D_{A_1A_2}$ small. So if $D_{A_1A_2}$ has to be as small as possible for each value of V_a, it will be necessary to modify the $D_{A_1A_2}$ distance in vacuum and under tension, to avoid breakdown problems between the anodes. Even, if this can be done on an experimental low voltage gun, it is not realistic on the M.E.B.A.H.T.

Thus, if only the problem of the source position stability is considered a geometry such as $\alpha \simeq 60°$, $e \simeq 1$ mm, $D_{A_1A_2} = 1$ mm would give good results for $V_a > 50$ kV that is to say for $k > 20$.

Furthermore, the choice of the best geometry has to take into account the values of C_{so} and the variations of Z_i which have to be, at the same time, as small as possible. As we are considering the case where the entrance pupil is defined by the conjugate of one of the anodes, the study shows that, whatever the linear magnification M_L may be, the geometry that gives the smallest values of C_{so} is such that $\alpha \simeq 45°$, $e \simeq 8$ mm, $D_{A_1A_2} \simeq 7$ mm and $D_{CA} = 10$ mm (fig. 4).

Furthermore, as can be seen in fig. 2, the source position stability is quite acceptable for this last configuration if k is chosen greater than 20.

This gun geometry has been built and adapted on a microscope (Denizart, 1981 and Denizart et al. 1981) and it has been possible to obtain a source for which the transverse coherence length determined by the Young's holes method (Munch, 1975 ; Denizart, 1981) is 0.6 µm.

Nevertheless, such a gun is not suitable for an adaptation to the M.E.B.A.H.T. accelerator especially considering the life-time problem created by the poor protection of the tip against breakdown. For this reason, a similar study of the behaviour of the tetrode F.E.G. has been made.

Tetrode Field Emission Guns

Since the optical properties of the tetrode gun can be defined in terms of two independent electrical parameters $k_1 = V_i/V_e$ and $k_2 = V_a/V_e$, the influence of each geometrical parameter on the source position and on the aberration coefficients has been studied as a function of k_1 and k_2.

The study has covered the following ranges of the electrical and geometrical parameters : $0 < k_1 < 25$, $5 < k_2 < 50$ and $1\text{ mm} < e_2 < 6$ mm, $1\text{ mm} < \phi_2 < 6$ mm, $35° < \beta_1 = \beta_3 < 60°$, $1\text{ mm} < \phi_1 = \phi_3 < 6$ mm, $3\text{ mm} < e_1 = e_2 < 10$ mm, $3\text{ mm} < D_{A_1A_2} < 10$ mm and $5\text{ mm} < D_{CA} < 10$ mm. We set out from an initial geometry derived from the results obtained for the triode guns and from the preliminary studies. This geometry is such that : $e_1 = e_3 = 4$ mm, $\phi_1 = \phi_3 = 1$ mm, $e_2 = 2$ mm, $\phi_2 = 3$ mm and $\beta_1 = 45°$.

The values of k_1 that give a stable position of the source, whatever k_2 may be, are $k_1 = 0$ and $k_1 > 15$. If $0 \lesssim k_1 \lesssim 7$, the curves $Z_i(k_2)$ may exhibit discontinuities (fig. 5a). It can be seen that for $k_1 > 15$, C_{so} takes constant and small values but for $k_1 = 0$, C_{so} is greater than 10^4 mm and has not been indicated on fig. 5b. Furthermore, the corresponding values of the linear magnification M_L cannot compensate the values of C_{so} (fig. 5c).

In spite of the poor C_{so} value obtained for $k_1 = 0$, we have considered this case in the following ; it is of practical interest due to the tip protection it can provide against the breakdown phenomena (Engel and Sauer, 1979).

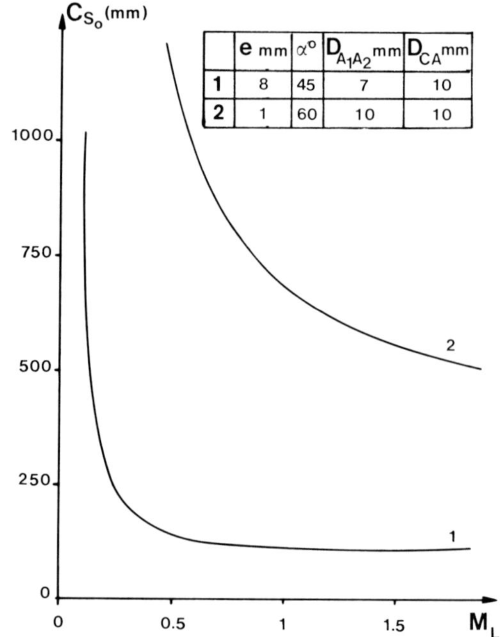

Fig. 4. Typical variation $C_{so} = C_{so}(M_L)$.

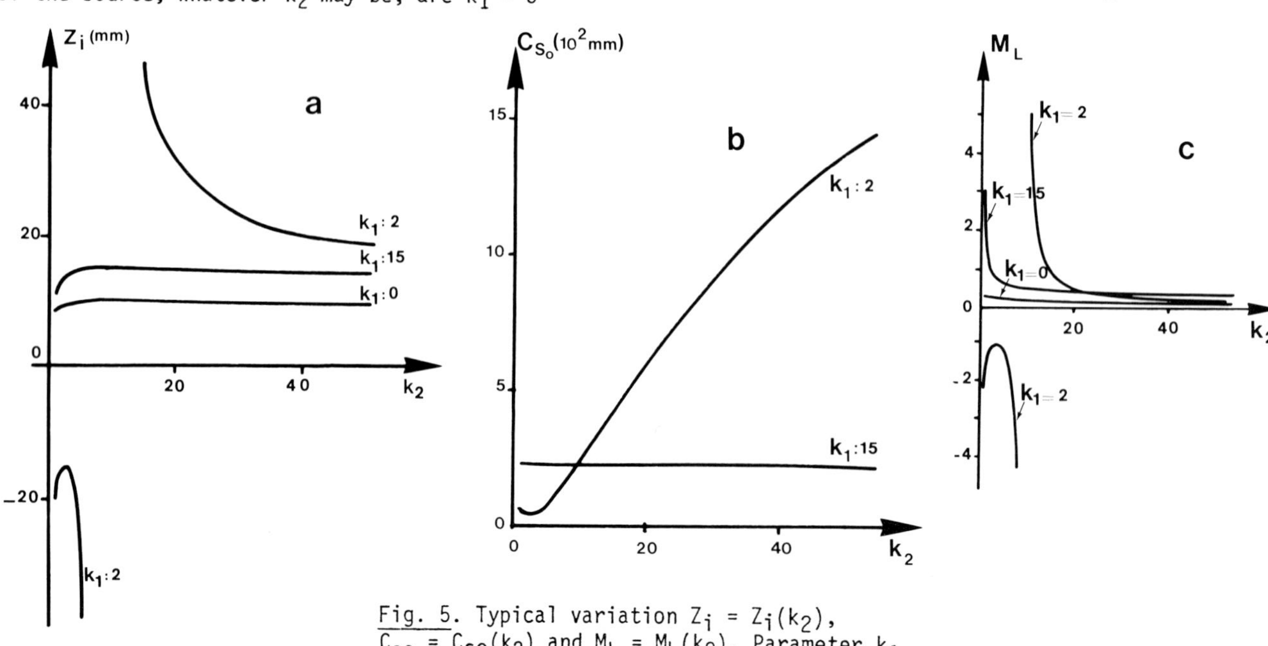

Fig. 5. Typical variation $Z_i = Z_i(k_2)$, $C_{so} = C_{so}(k_2)$ and $M_L = M_L(k_2)$. Parameter k_1.

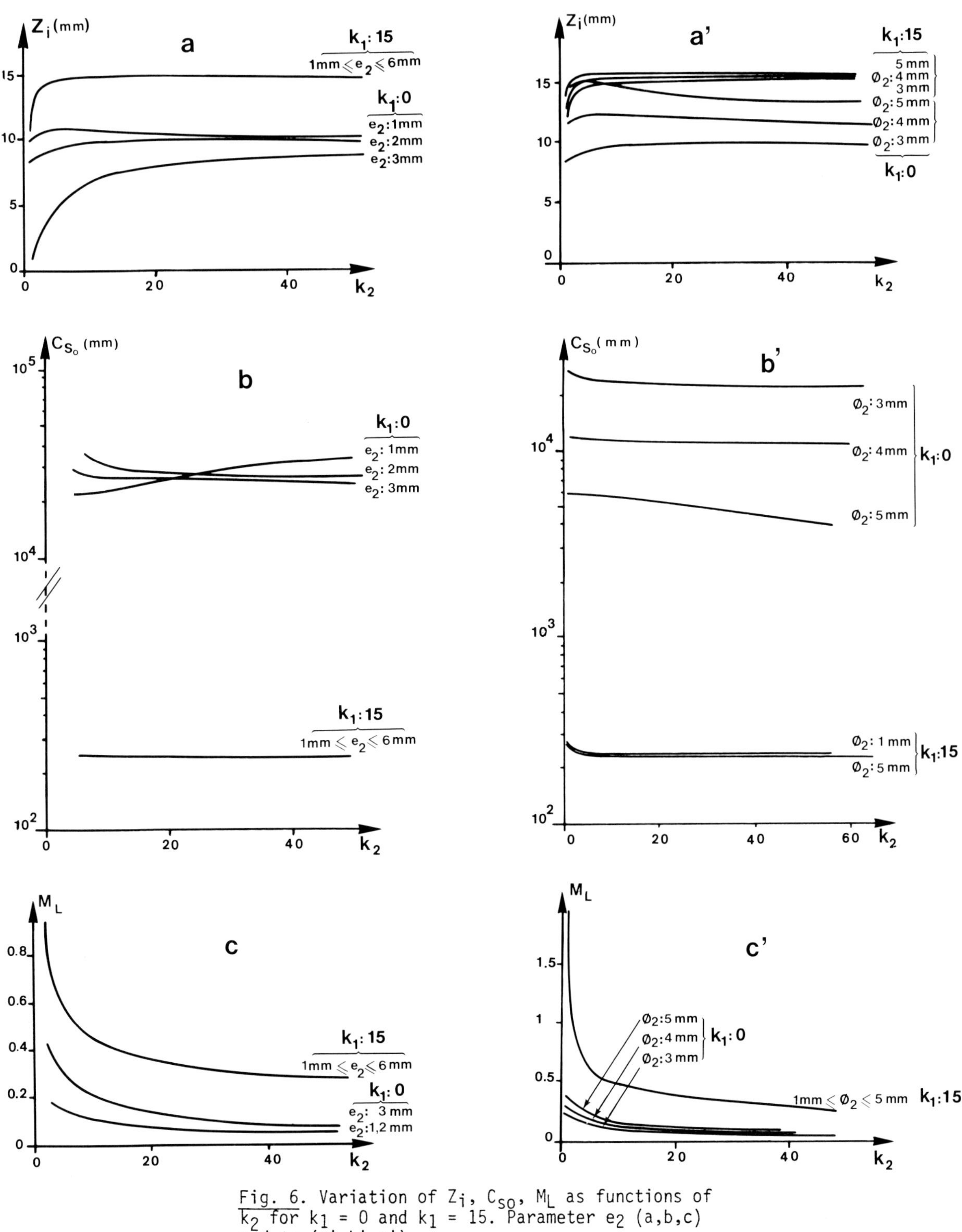

Fig. 6. Variation of Z_i, C_{so}, M_L as functions of k_2 for $k_1 = 0$ and $k_1 = 15$. Parameter e_2 (a,b,c) and ϕ_2 (a',b',c').

In a first stage, the geometrical parameters of the intermediate anode A_2 have been studied. The best results correspond to 3 mm < ϕ_2 < 5 mm and 1 mm < e_2 < 3 mm for $k_1 = 0$ and $k_1 > 15$, but the values of C_{so} are always very large for $k_1 = 0$ and not compensated by M_L. For $k_1 > 15$, Z_i, C_{so} and M_L are less sensitive to the variations of e_2 and ϕ_2 than for $k_1 = 0$ (fig. 6).

In a second stage, taking $e_2 = 2$ mm and $\phi_2 = 4$ mm, the influence of the geometrical parameters of A_1 and A_3 have been studied. A preliminary study of some geometries where A_3 was plane showed that the stability of the source position was poor, particularly for $k_1 = 0$, and we have therefore considered the case where A_1 and A_3 are the same (fig. 1b) and we put : $\phi_1 = \phi_3 = \phi$, $e_1 = e_3 = e$ and $\beta_1 = \beta_3 = \beta$. From the study, we draw the following conclusions :

- the angle β has practically no influence on the optical properties of the tetrode gun particularly for $k_1 = 0$, while C_{so} is always about 3.10^4 mm. For $k_1 > 15$, Z_i, C_{so}, M_L vary slowly and the smallest C_{so} is obtained for $\beta = 48°$.
- the diameter ϕ does not affect Z_i and M_L and C_{so} increases with ϕ for $k_1 = 0$. The same is true for $k_1 = 15$ if $\phi < 3$ mm.
- the thickness e has a greater influence on the optical properties than β or ϕ but the corresponding variations do not strongly modify these properties. The best values are obtained for $e \simeq 4$ mm.
- the distance $D_{A_1A_2}$ has the same type of influence as e and taking into account the breakdown problems between the anodes, it seems reasonable to choose 4 mm $< D_{A_1A_2} <$ 5 mm.

A "best geometry" would thus be such that : $\phi = 1$ mm, $e = 4$ mm, $\beta = 48°$, $\phi_2 = 4$ mm, $e_2 = 2$ mm, $D_{A_1A_2} = 4$ mm and $D_{A_2A_3} = 10$ mm.

Concerning the influence of D_{CA}, it can be said that if the geometry is the "best" one, the variations of D_{CA} have no influence on the properties of the gun ; this is not true if the geometrical parameters do not have the optimal values and discontinuities appear in the $Z_i(k_2)$ curves.

All the properties which have been calculated for the "best geometry" are summarized on fig. 7a for $k_1 = 0$ and on fig. 7b for $k_1 = 15$. It is this geometry which has been chosen for our study of the optical properties of an F.E.G. + a linear accelerator.

Tetrode Field Emission Gun Plus Twenty-Stage Linear Accelerator System

Taking into account the technological constraints related to the construction of the electrodes of the accelerator, we have considered plane electrodes.

The study covers electrical parameters in the ranges $0 < k_1 < 30$ ($k_1 = V_i/V_e$) and $0 < k_2 < 100$ ($k_2 = V_a/V_e = (V_{20}/V_e)/20$) and geometrical parameters in the ranges 1 mm $< e_E <$ 10 mm, 2.5 mm $< R <$ 60 mm and 5 mm $< D_{CA} <$ 10 mm. The distance d between electrodes is taken to 50.5 mm, which is the lower limit for a maximum potential difference of 80 kV and the distance D (fig. 1c) between the third anode (A_3) of the gun and the first electrode (E_1) of the accelerator has been determined in such a way that the accelerator field does not perturb the field in the gun and hence the optical properties of this gun. The optimal value of D is 40 mm.

Concerning the source stability problem, an example of the results obtained for all the configurations which have been studied is shown in fig. 8 for two values of D_{CA}. A stable position of the source is obtained for $k_1 = 0$ and $k_1 > 20$ whatever k_2 may be. For $k_1 = 15$ and $D_{CA} = 10$ mm, the source stability is good but is not satisfactory for $D_{CA} = 5$ mm.

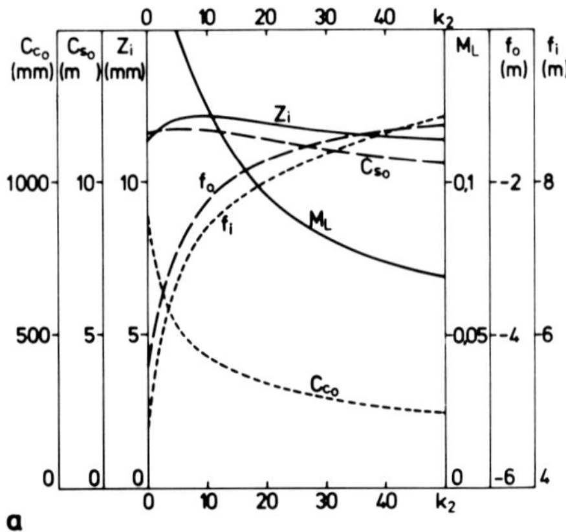

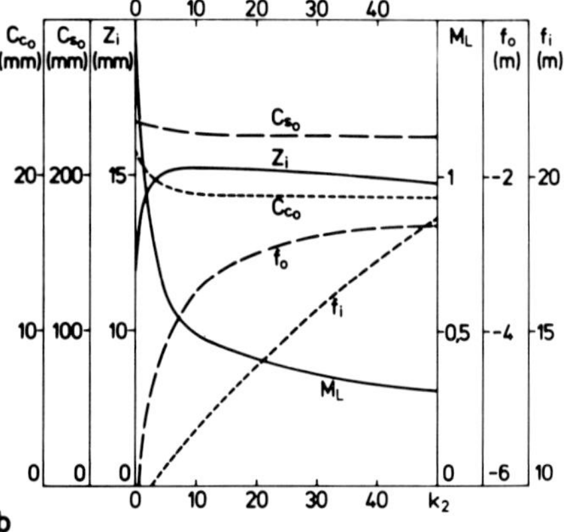

Fig. 7. Variation of Z_i, C_{so}, C_{co}, M_L, f_o and f_i as a function of k_2 ; for $k_1 = 0$ (a) and $k_1 = 15$ (b).

Turning to the variations of C_{so}, it can be seen (fig. 9) that for $k_1 = 0$, the very large values taken by C_{so} are not compensated by the linear magnification. On the other hand for $k_1 = 20$, $D_{CA} = 5$ mm and $R > 7.5$ mm the values of C_{so} lie between 20 and 50 mm. Giving priority to the source position stability, k_1 must be taken between 15 and 20 and $D_{CA} = 5$ mm.

The electrical parameters now being determined, the influence of e_E and R on Z_i and C_{so} has been studied and the smallest variations ΔZ_i of Z_i are obtained for $k_1 = 20$, $D_{CA} = 10$ mm, R = 10 mm and $e_E < 3$ mm. For these values, $\Delta Z_i \simeq 1.4$ mm if $10 < k_2 < 90$ and $\Delta Z_i \simeq 0.4$ mm if $10 < k_2 < 50$ (figs. 10a and 10b).

An example of the C_{so} variations as a function of R is shown in fig. 9 for $k_2 = 50$, $k_1 = 0,20$ and $D_{CA} = 5,10$ mm, the thickness being 1 mm. The smallest C_{so} (and the greatest M_L) correspond to $k_1 = 20$ and $D_{CA} = 5$ mm whatever R may be. However, as shown in Table 1, M_L cannot compensate the great values taken by C_{so} for $k_1 = 0$.

Note also that for $k_1 = 20$ and $D_{CA} = 5$ mm, C_{so} is constant for $R > 7.5$ mm if $e_E = 1$ mm, $R > 15$ mm if $e_E = 3$ mm and $R > 30$ mm if $e_E = 10$ mm. Furthermore C_{so} is not very sensitive to k_2 for $R > 5$ mm.

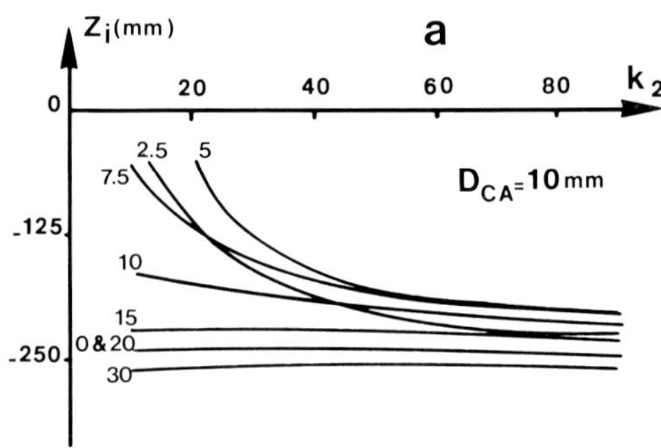

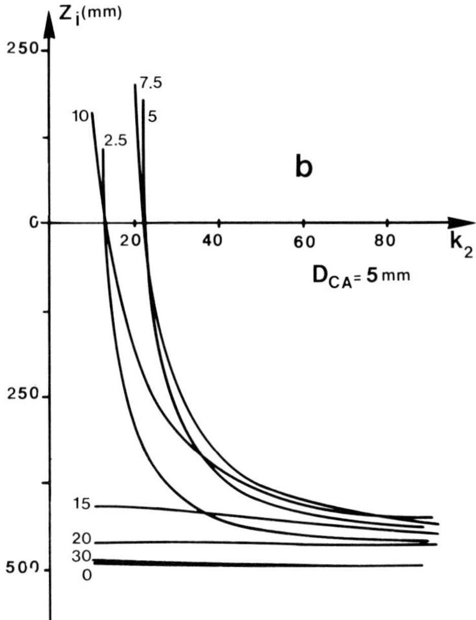

Fig. 8. Typical variation $Z_i = Zi(k_2)$ for D_{CA} = 10 mm (a) and D_{CA} = 5 mm (b). Parameter R (mm).

k_1	D_{CA} mm	C_{so} mm	M_L
0	5	$9.5\ 10^3$	0.09
	10	$3.4\ 10^4$	0.06
20	5	36	0.8
	10	80	0.25

Table 1. Typical values of C_{so} and M_L for k_1 = 0,20 and D_{CA} = 5,10 mm.

We recall finally that C_{co} is always 100 to 1000 times smaller than C_{so}. It thus seems reasonable to regard following geometrical parameters of the accelerator electrodes as the best that can be achieved : 10 mm < R < 15 mm and $e_E \simeq 1$ mm.

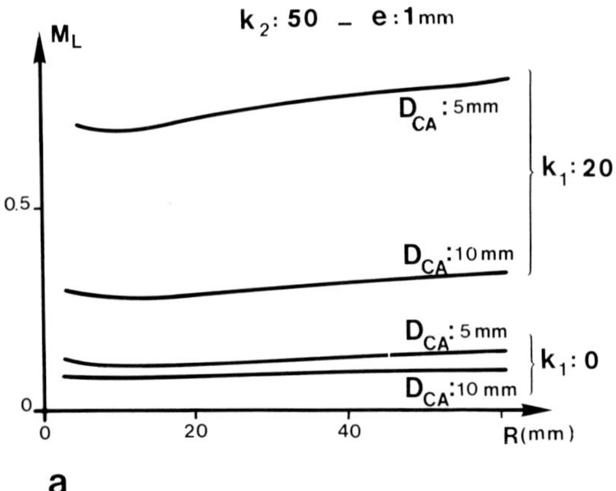

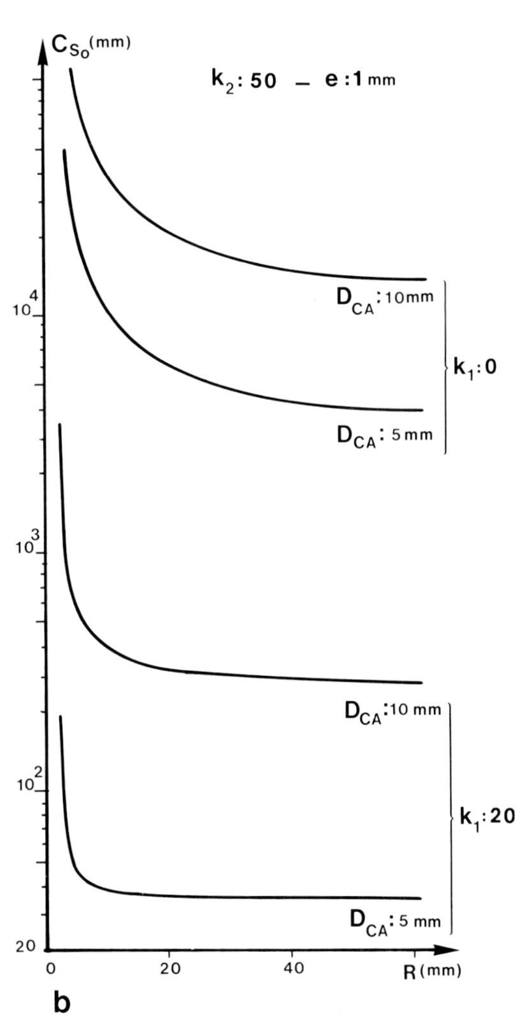

Fig. 9. Variation of M_L (a) and C_{so} (b) as function of R for e = 1 mm, k_2 = 50, k_1 = 0.20 and D_{CA} = 5.10 mm.

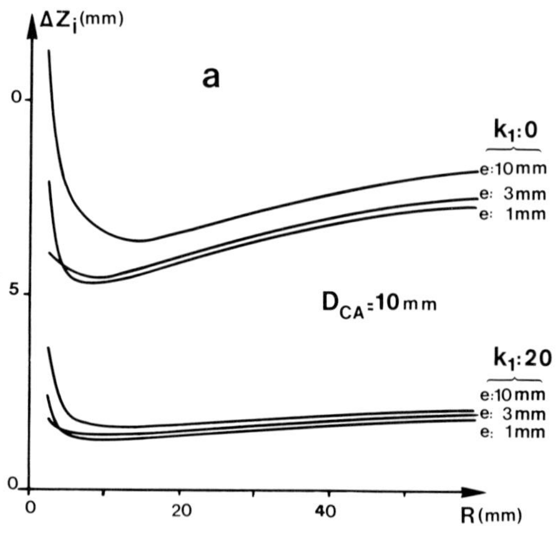

	1	2	3	4	5	6
D_{CA}(mm)	5	7.5	10	5	7.5	10
k_1	0			20		

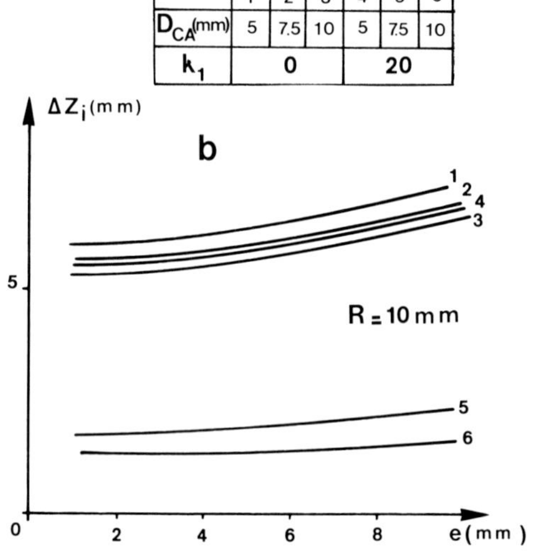

Fig. 10. Variation of the source position as a function of R (a) and as a function e (b) for k_1 = 0.20.

Conclusion

The results given above form part of a study of the gun plus linear accelerator system for the Toulouse 1.6 MV (M.E.B.A.H.T.). They show that it should be possible to obtain a perfectly stable position of the source.

The radius of the source will lie between 200 and 800 nm for V_e = 1 kV, k_1 = 15, D_{CA} = 5 mm and between 150 and 300 nm for V_e = 3 kV, k_1 = 15, D_{CA} = 5 mm. These values could be reduced by introducing a diaphragm between the gun and the accelerator. The determination of the characteristics of a system with such an aperture is the main object of our present work.

References

Denizart M, (1981). Emission de champ en microscopie électronique (Field emission for electron microscopy), Thèse d'Etat, 997, Université Paul Sabatier, Toulouse, France.

Denizart M, Roques S, Sonier F, Jouffrey B, Trinquier J. (1981). Triode and tetrode guns for field emission electron microscopes, Ultramicroscopy 7, 65-80.

Engel W, Sauer J. (1979). A field emission gun with a three-electrodes accelerating lens, Ultramicroscopy 5, 262.

Garg RK, Séguêla A, Jouffrey B. (1984). Proc. of the 3rd Asia-Pacific Conf., Singapore, August 1984, in press (Copy available from B. Jouffrey).

Hoch H, Kasper E, Kern D. (1978). Darstellung stationärer rotationssymmetrischer elektromagnetischer Felder durch Superposition von Lochblenden-und Kreisringfeldern (Representation of stationary rotationally symmetric electromagnetic fields by superimposing the fields of coaxial circular apertures and charged rings) Optik 50, 413-425.

Kang NK, Tuggle D, Swanson LW.(1983). A numerical analysis of the electric fields and trajectories with and without the effect of space charge for a field electron source, Optik 63, 313-331.

Kasper E. (1982). Field Electron Emission Systems. Advances Optical and Electron Microscopy. Barer E., Cosslett V.E. (Eds) Academic Press (London) 8, 207-260.

Kern D, Kurz D, Speidel R. (1978). Electronenoptische Eigenschaften eines Strahlerzeugungssystems mit Feldemissionskathode (The electron optical characteristics of a field emission gun), Optik 52, 61-70.

Munch J. (1975). Experimental Electron Holography, Optik 43, 79-99.

Munro E. (1971). Computer-aided-design methods in electron optics. Dissertation, Trinity College, Cambridge, U.K.

Uchikawa Y, Ozaki K, Ohye T. (1983). Electron optics of point cathode electron gun, J. Electr. Microsc. 32, 85-98.

Discussion with Reviewers

J. Orloff: The readers may be interested in the following references which are relevant to the topic of this paper:

Kuroda K, Ebisui H, Suzuki T. (1974). Three-anode accelerating lens system for the field emission scanning electron microscope, J. Appl. Phys. 45, 2336-2342.

Orloff J, Swanson LW. (1979). An asymmetric einzel lens for field emission microprobes, J. Appl. Phys. 50, 2494-2501.

Riddle GHN. (1978). Electrostatic einzel lenses with reduced spherical aberration for use in field emission guns, J. Vac. Sci. Tech. 15, 857-860.

Authors: Thank you.

Electron Optical Systems (pp. 187-195)
SEM Inc., AMF O'Hare (Chicago), IL 60666-0507, U.S.A.

0-931288-34-7/84$1.00+.05

ELECTRON DETECTORS USED FOR IMAGING IN THE SCANNING ELECTRON MICROSCOPE

V.N.E. Robinson

Faculty of Applied Science
The University of New South Wales
P.O. Box 1
KENSINGTON N.S.W. 2033
AUSTRALIA
Phone: (61-2) 662-2335

Abstract

Electron detectors used for imaging in the scanning electron microscope include those which detect secondary electrons, various portions of the backscattered electron signal, and the residual specimen current. The use of a different detector will often produce a different image of the same specimen. The information contained in these images depends upon the signal detected and the properties of the detector used. The choice of detector to be used depends upon the information desired.

KEY WORDS: Electron signals, secondary electron detectors, backscattered electron detectors, topography contrast, atomic number contrast, imaging.

Introduction

The image obtained from a scanning electron microscope (SEM) is not just a magnified presentation of the specimen being studied. Like all images, it is a transformed representation of the specimen. For example, at a magnification of x100, the optical microscope (OM) and SEM images of the same object may look very different. Both images are magnified representations of the specimen. Both are correct presentations and yet they may be very different. Such images are transformed representations of the specimen, where the transformation process is a function of the physics of the imaging technique.

In the SEM, this transform function depends upon the electron beam conditions, the scan circuitry, the specimen itself, the signal detected, the properties of the detector employed, signal processing and the linearity of the imaging system. For example, different accelerating voltages will produce different images of the same specimen. Essential to the production of the SEM image is the detector used to produce the image. Its properties, plus those of the signal detected, greatly influence the type of information that can be gained from the SEM image.

This paper discusses the electron detectors that are used, for imaging purposes, in SEMs. These are the secondary electron (SE) detectors, the backscattered electron (BSE) detectors, the low loss backscattered electron (LLBSE) detector and the specimen current (SC) detector. The information presented relates to the properties of the detectors as seen by the user, not just the properties of these detectors. This has been done to enable users to better understand the consequences of changing SEM operating parameters, when different detectors are employed.

The Electron Signals

Figure 1 illustrates the electron beam-specimen interaction, showing the various signals that are generated. Electrons from the beam enter the specimen, are scattered approximately as indicated and, within this scattering volume,

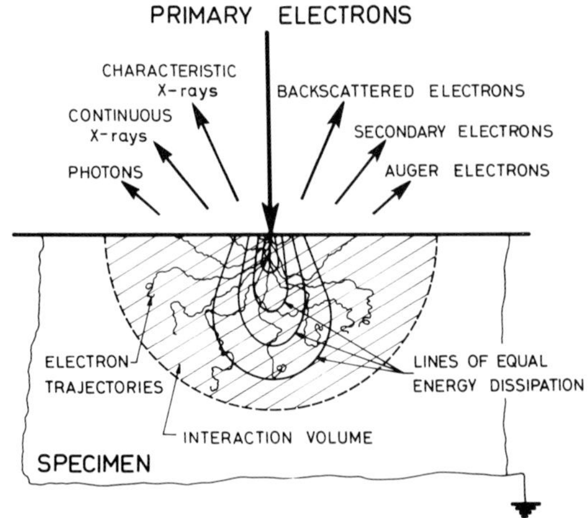

Figure 1. Schematic illustration of the electron-specimen interaction which gives rise to the signals detectable in SEM.

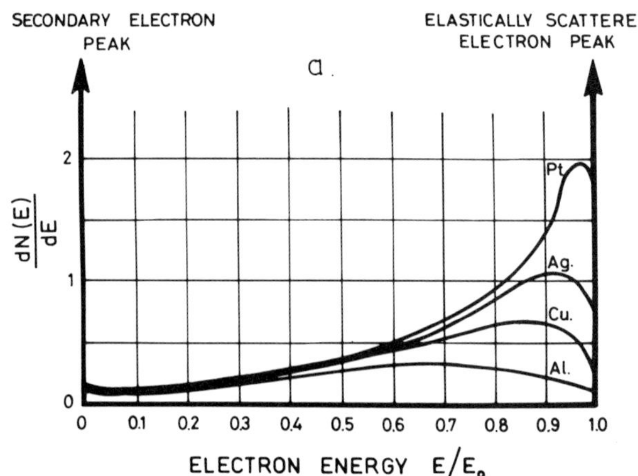

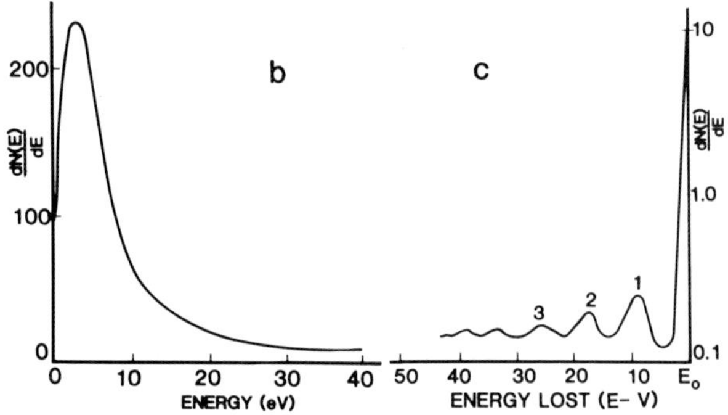

Figure 2. Electron energy emission spectra for an electron beam of energy E (> 5 keV) vertically incident upon a flat surface. (a) Total emission spectrum (after Kulenkampf and Spyra, 1954); (b) details at the low energy end of the spectrum (secondary electron emission after Kollath, 1947); (c) details at the high energy end (elastic scattering and plasmon loss, schematic).

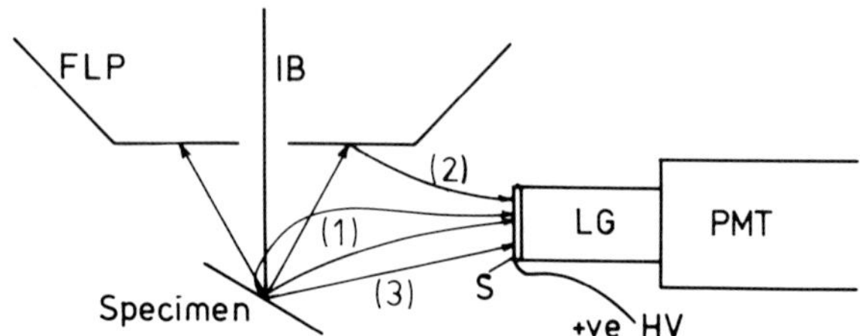

Figure 3. Schematic illustration of the construction of, and the electron contributions to, the signal output from an Everhart-Thornley detector.

they generate their signals. The signals of interest in this paper are only the backscattered electron (BSE), secondary electron (SE) and the specimen current (SC) signals.

Figure 2 shows the complete electron emission spectrum of all electrons emitted from a sample struck by high energy, say 20 keV, electrons vertically incident upon the various flat surfaces, as indicated. Figure 2a shows the complete spectrum. Figure 2b shows the detail at the low energy end of the spectrum, that is secondary electron emission. Figure 2c shows details at the high energy end, that is the elastically scattered end of the spectrum. Note that the vertical and horizontal scales are different in the separate figures. Although the height of the BSE peak is much lower than either the SE or elastically scattered BSE peaks, its much greater width makes it the largest signal emanating from the specimen (Robinson, 1973).

The three categories of electron detectors to be mentioned, each make use of one of these three signals, shown in Figure 2, namely SE detectors, Figure 2b, BSE detectors, Figure 2a and LLBSE detector, Figure 2c.

Secondary Electron Detection

The most common form of secondary electron detector employed in SEMs is that due to Everhart and Thornley (1960). Its essential features are displayed in Figure 3. These consist of a scintillation surface (S), connected via a light guide (LG) to a photomultiplier tube (PMT). A positive high voltage (HV), usually of about +10 kV is applied to the scintillation surface and some form of electrostatic shielding is employed to ensure that the secondary electrons from the specimen impinge upon the scintillator surface, and not just the conductor supplying the positive high voltage. There are several variations of this shielding, involving the presence or absence of a positive low voltage, approximately +250V, grid between the scintillator and the specimen. Provided that the grid works properly, there is no noticeable performance difference between these different types of electrostatic attraction-shielding combinations.

These detectors work by providing an

electrostatic field strength of approximately +50 to +100V/cm at the surface of the specimen. This is sufficient to attract the SEs, typical energy +2V to +5V, towards the detector, without greatly influencing the primary beam. As they travel towards the scintillator, they are accelerated into it where they impinge with a high energy. They generate a number of photons, which are channelled through a light guide to the photomultiplier tube. Here they are converted to electrons and amplified. This complexity is used because it still gives the most noise free signal. This principle also ensures that every SE detected, gives the same signal output from an Everhart-Thornley (E-T) detector. As such the signal output of the detector, as seen by the operator, will depend very much upon the number of emitted secondary electrons.

Figure 3 also illustrates the method by which electrons contribute to the signal from the E-T detector. These are the signals from:-

(1) SEs released from the specimen as it is struck by the incident beam (Type I and Type II SEs)

(2) SEs released as BSEs strike the polepiece (Type III SEs).

(3) BSEs impinging directly upon the surface of the E-T detector.

Type III SEs account for approximately 15% to 20% of the E-T detector signal output (Moll et al., 1978, 1979), whilst the direct impingement of BEs into the E-T detector can contribute up to 2 - 5% of the total signal output. Of the SEs released from the specimen, approximately 80% are type II, i.e. those released as the scattered primary electrons pass out of the specimen surface (Robinson, 1974a; Wooldridge, 1939). It is easy to see from this that the signal output by the Everhart-Thornley detector contains a great deal of BSE information, as well as SE information. Figure 4 is a brief summary of properties of secondary electrons. Figure 4a shows variation of the secondary electron yield (δ) with atomic number (Z). Figure 4b shows the variation of the SE yield δ with accelerating voltage (kV). Figure 4c shows the variation of SE yield δ with sample tilt.

The uniformity of response of the E-T detector to each low energy electron, means that, to the user, the detector response with variations of atomic number, accelerating voltage and sample tilt are approximately the same as the variation in SE yield. That is, to increase secondary electron yield, use a higher atomic number sample, a lower accelerating voltage or a higher specimen tilt. Many SEM users will recognise the concept of gold or gold-palladium coating and operating with a 30 or 45 degree tilted specimen, to get a higher signal to noise ratio image. The additional contributions due to backscattered electrons are only such as to vary the absolute magnitude of the response of the detector away from the signal curves. They do not vary the intent or direction of any of these curves. As such, the variation in the signal output of an E-T detector with atomic number, accelerating voltage or specimen tilt will be similar to the SE yield curves shown in Figure 4.

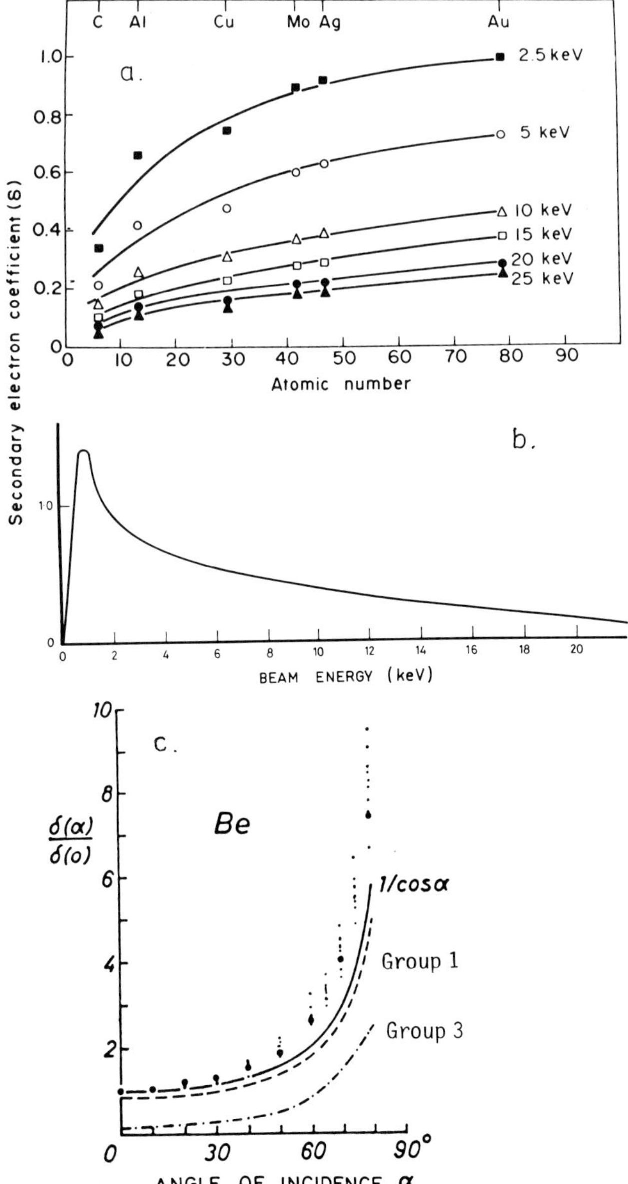

Figure 4. Variations of the secondary electron yield with (a) atomic number, (Moncrieff and Barker, 1978); (b) accelerating voltage (schematic); and (c) specimen tilt (after Drescher et al, 1970).

Another effect which has not been well studied, which makes quantitative results with E-T detectors difficult, is that due to the electrostatic attraction of secondary electrons reducing with increasing distance from the detector. This effect is easily seen on micrographs taken on flat, untilted surfaces at low magnification. It becomes more apparent when a side positioned E-T detector is used at short working distances.

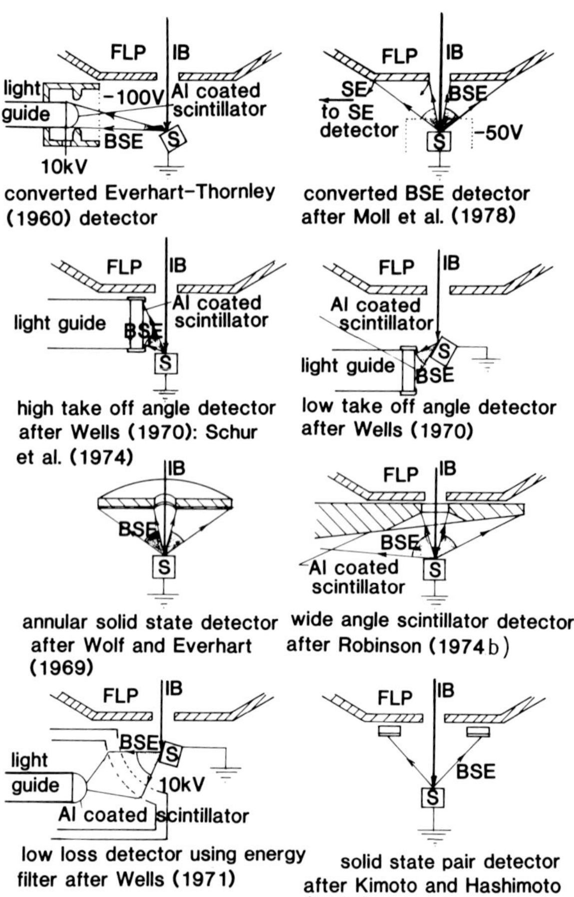

Figure 5. Illustrations of the many different types of backscattered electron detectors. FLP = final lens pole-piece, IB = incident beam, S = specimen.

As the distance from the detector to the region of the sample where the secondary electrons are being emitted, is increased, the signal received diminishes. This effect can be quite large, encompassing some 10% to 20% change at low magnifications, i.e., the variation can be as much as 1% or 2% of the SE signal per mm. It can often be seen on micrographs taken of flat polished surface untilted at low magnification. Similarly, as the working distance is reduced, the secondary electron signal decreases. This is probably due to a drop off in the electrostatic potential at the surface of the specimen. These effects almost disappear when SEs are detected through the lens.

Until such time as there is a full investigation of these properties of the Everhart-Thornley detector, the obtaining of quantitative results using this detector will be difficult. It becomes even more difficult when additional effects are introduced, e.g. those due to charging artifacts, and edge brightness, i.e., strong SE emission from very small particles or from very close to the edge of heavily sloping surfaces. All of these effects have yet to be properly understood before quantitative results can be achieved, using the E-T secondary electron detector, over a wide range of surfaces.

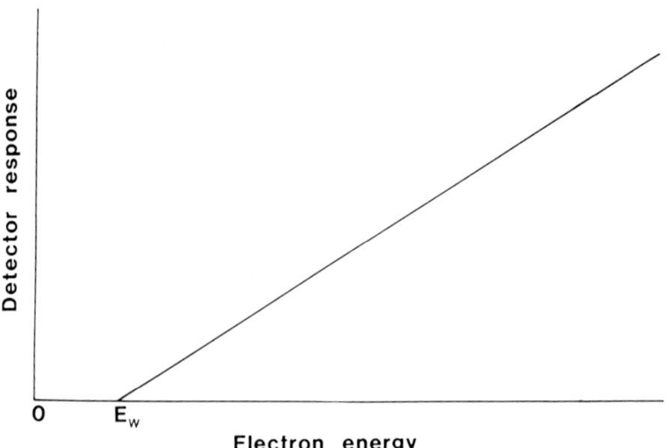

Figure 6. Variation of BSE detector response with energy of an incident electron.

Backscattered Electron Detectors

The second group of electrons that is widely used for imaging purposes in the SEM, is the backscattered electrons (BSEs), as shown in Figure 2a. These are the electrons that have suffered considerable energy loss, from a few eV to a few thousand eV. This group of electrons does not have a very high intensity at a given energy, compared to either the secondary electrons, or the elastically scattered electrons. However, it is extremely wide, spreading over greater than 99% of the energy of the electron beam, for beam energies greater than 10kV (Robinson, 1973). Its width makes it the most abundant type of electrons emitted from a specimen when struck by an electron beam, and it has by far the largest amount of energy associated with any electron emission group.

Over the years many fundamentally different types of backscattered electron detectors have been constructed to detect these electrons. Some have been based on the Everhart-Thornley detector, such as the unbiased detector, and the converted BSE detector due to Moll et al., (1978), Reimer and Volbert (1979). These are illustrated in Figure 5. Another group of detectors which has been employed is those which are specifically designed to detect only a certain fraction of the emitted BSE electrons. These include the solid state pair detector, after Kimoto and Hashimoto (1966), the high take-off angle detector, after Schur et al., (1974) and Wells (1970), the low take-off angle detector after Wells (1970), as well as some detectors made by SEM manufacturers specifically for their SEM. Several variations of these have been made, usually by extending the light guide of an E-T detector and placing an unbiased scintillator on the end of it (Zeldes and Tassa, 1979, Fitch et al., 1984).

The third type of detector is the non-directional BSE detector, that is the wide angle detector which is designed to detect as many BSEs as possible. The two different types are the silicon solid state, either surface barrier or the shallow diffused p-n junction (Wolf and Everhart, 1969; Stephen et al, 1975) and the wide angle scintillator type (Robinson, 1974b) detector.

All of these different types of BSE detectors have been summarised by Robinson (1980) and are illustrated in Figure 5. With the exception of the converted BSE detector, (Moll et al, 1978; Reimer and Volbert 1979), all other BSE detectors shown are passive detectors. That is, they rely entirely upon the energy of the BSEs themselves to give rise to the signal from the detector. They do not have an applied voltage as does the E-T detector. This means that, unlike the E-T detector, most BSE detectors give a different signal output response for electrons of different energies, as illustrated in Figure 6 (Robinson, 1975; Baumann and Reimer, 1981).

Another type of electron detector which is finding increasing use is the channel plate detector (Griffiths et al., 1972). It is useful for detecting both SEs and BSEs. The channel plate detector is not as sensitive to SEs as is the E-T detector, nor to high energy BSEs as are some scintillator type BSE detectors. However, it is a detector which is sensitive to low energy, less than 5 kV, BSEs. These detectors have been discussed by Russell (1984). They are of interest for the study of integrated circuits at low accelerating voltages. Figure 7 shows the variation in the backscattered electron yield (η) with changes in SEM parameters. Figure 7a shows the variation of η with atomic number (Z). Figure 7b shows the variation of η with accelerating voltage. Figure 7c shows the variation with specimen tilt. The response of a BSE detector is a combination of that fraction of the signal emitted, (see Figure 2a) that is detected by the detector, in conjunction with the response of the detector to the energy of the electrons, as shown in Figure 6, plus the properties of the backscattered electrons themselves as shown in Figure 7, plus variations in the shape of the curves shown in Figure 2a with variations of tilt. All in all a complex situation. However, all of these properties are either known or can be easily determined and quantitative calculations with BSEs can be performed. One example of this is the new technique of composition analysis in which the signal output of a BSE detector is matched to the composition or chemical formula of a specimen (Robinson et al., 1984).

To the user, BSE detectors have the following properties:-

(i) Greater signal output with increasing atomic number.
(ii) Greater signal output with increasing accelerating voltage. At voltages above approximately 18kV to 20kV, a good high collection efficiency scintillator type BSE detector will have a higher signal to noise (S/N) ratio output than the E-T detector, for the beam vertically incident upon a flat surface (Robinson, 1975; Baumann and Reimer, 1981). Below 15kV accelerating voltage, the E-T detector has higher S/N

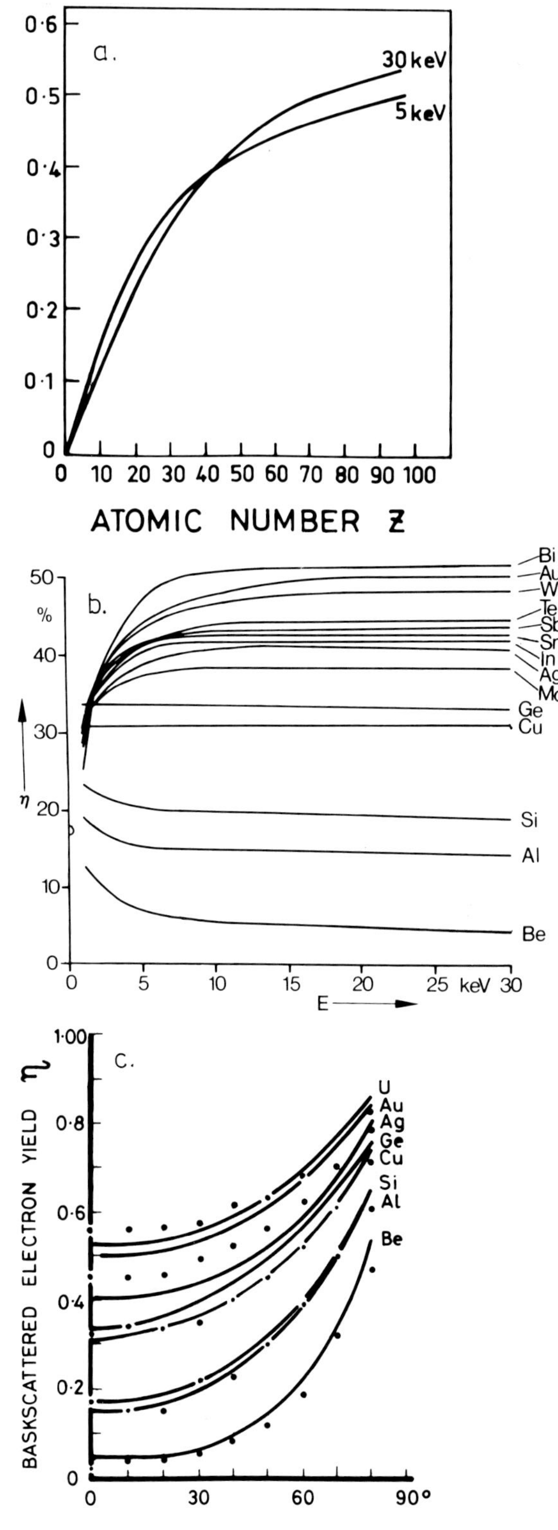

Figure 7. Variations of the backscattered electron yield with (a) atomic number (after Bishop, 1966, Colby, 1969); (b) accelerating voltage (after Reimer and Tolkamp, 1980) and; (c) sample tilt (after Drescher et al, 1970).

than a BSE detector. Below 10kV accelerating voltage, the performance of a scintillator type BSE detectors is seriously impaired. Below 5kV, it is essentially unworkable and at about 2kV it no longer functions. The relative merits of the different types of scintillator detectors have been summarised by Baumann and Reimer (1981). The solid state (silicon surface barrier or p-n junction) BSE detector show the same trends as the scintillator types but have a lower S/N ratio at the same operating conditions. The converted BSE detectors (Moll et al., 1978) have a high S/N at a low kV (Baumann and Reimer, 1981).
(iii) The variation of signal output with specimen tilt will depend upon the type of detector employed. Using a low take-off angle BSE detector and tilting the specimen towards the detector will increase the signal output. Tilting the specimen when a high take-off angle or wide angle detector is used will result in a decrease in signal output with increasing tilt because the BSEs are directed away from the detector.

Low Loss Backscattered Electron Detectors (LLBED)

These detectors are designed to detect the electrons shown in the spectrum in Figure 2c, i.e. the electrons which have gone into the sample and then scattered with little or no loss of energy. These electrons have undergone only one or two scattering events and have not penetrated very far into the surface. They constitute about 10^{-3}% of the emitted electron signal, making it a very weak signal. The only way which these electrons have been successfully detected has been by electrostatic suppression, i.e., allowing all of the electrons to enter an electrostatic field which is nearly as strong as the voltage of the incident beam (Wells 1971). The other electrons are then suppressed by the electrostatic filter and only those few electrons which have suffered little or no loss of energy are allowed through to a detector similar to the Everhart Thornley detector, as shown in figure 5. The work of Munro (1974) has shown that these electrons cannot be separated from the rest of the backscattered electrons, i.e. those which have lost a lot of energy, by the magnetic field of a pre or post specimen lens. These electrons are extremely sensitive to surface effects such as contamination and do not suffer from penetration effects as do the secondary electrons and other backscattered electrons. They do not show any subsurface features and are strictly surface imaging detectors (Wells, 1974). Despite these advantages, they have a low S/N ratio and are not widely used.

Specimen Current Imaging

The charge balance existing when incident beam electrons impinge upon a conducting specimen can be expressed mathematically by the equation:-

$$I_B = I_{SE} + I_{BSE} + I_{SC}$$

where I_B is the incident beam current, I_{SE} is the emitted secondary electron current, I_{BSE} is the emitted backscattered electron current and I_{SC} is the specimen current. Under most usual SEM operating conditions, accelerating voltage greater than 5 to 10 keV, $I_{SE} + I_{BSE} \neq I_B$ leaving a residual charge in the specimen. This can be conducted away to give a specimen current which has a reverse contrast to the SE + BSE contrast images, see for example Newbury (1977). This signal can be measured using a specimen current meter and amplified to form an image.

This mode of imaging can only be successfully employed with conducting specimens. It has the disadvantage of a low bandwidth. This is due to the inherently large capacitance of a specimen stub and holder. It requires long scan times for imaging purposes. It has been successfully used for channelling contrast studies (Coates, 1969) and the occasional atomic number and topography study. Specimen current imaging is not widely used.

Summary

Having described a number of different types of detectors and described their properties, as seen by the user, it is only fitting to describe the situations of what type of detector to use in particular applications. In commencing this, I should point out, that when you purchase a scanning electron microscope, the manufacturers have usually predetermined that you shall use an Everhart-Thornley detector for most situations. In the past few years, most manufacturers modified this a little by offering the option of one or two types of backscattered electron detectors as well. There are a number of laboratories where people have removed their Everhart-Thornley detector and do all imaging with a backscattered electron detector. When choosing a detector to use, there are two principal criteria to be invoked. (1) Available signal to noise; (2) the type of information you wish to detect. Baumann and Reimer (1981) have summarised the relative merits of the various scintillator type detectors used in scanning electron microscopes. For the sake of simplicity, I will categorise the types of information desired into three types; (1) atomic number contrast and topography contrast subdivided into (2) flat surface contrast and (3) edge contrast. Edge contrast is the type of contrast that you get when you have a surface such as pollen grains or micro-organisms which have a large number of very small, heavily curved surfaces.

For general topographic imaging purposes, the E-T detector is still the most widely used general purpose detector. It is so widely used, that there is a tendency to regard the SE image as the correct representation of the sample. Many users tend to forget that it is only one representation of the specimen. Figure 8 shows the wide angle BSE and SE images of an aluminium fracture surface, imaged simultaneously using a Robinson detector and an Everhart-Thornley detector respectively. These images have a very different appearance as each detector provides

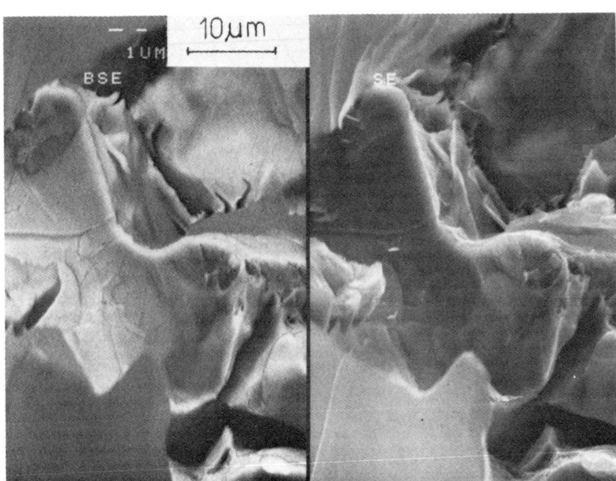

Figure 8. Wide angle BSE (Robinson) and SE (Everhart-Thornley) images of a fractured aluminium sample, 0 tilt, 25kV.

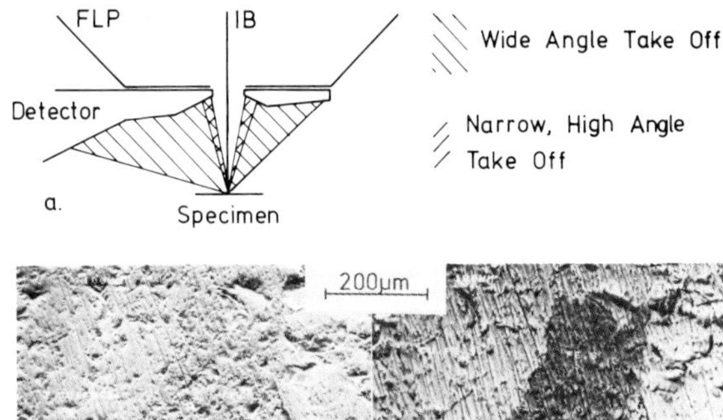

Figure 9. Schematic illustration of wide angle and narrow high take-off angle BSE detection (a) and micrographs of wide angle (b) and narrow, high, take-off (c) of a sample of a Cu, Fe and S containing ore, with a rough surface, 25kV, 0 tilt.

its own appropriate transformation to the image formation process. Both images are magnified transformations of the specimen. The SE image obtained with the E-T detector cannot be regarded as the sole accurate magnified representation of the specimen. A second transformation using a different detector to present a separate image is essential to avoid the situations where users consider properties of the transformation process as accurate representations of the specimen.

In many situations, the image obtained with the E-T detector will be an adequate representation of the specimen. This is particularly true of situations where the strong SE signal from edges is a desirable feature. This includes many high resolution topography imaging situations and studies of curved surfaces, such as cellular structures. In situations where the edge signal is too strong, a wide angle BSE detector placed above the specimen will not produce this strong edge signal and will produce images showing improved flat surface contrast. These detectors are also desirable when the E-T SE image displays charging artifacts.

The low take-off angle BSE detector with the specimen placed in the high field region of a condenser objective lens has shown itself to be useful for high resolution imaging of highly tilted specimens (Wells et al, 1973). Appropriately positioned and variable position high take-off angle BSE detectors have been successfully employed in studies of topography by providing the appropriate illumination conditions (Reimer et al, 1978; Schur et al, 1974).

If it is desired to observe atomic number contrast and the sample can withstand high voltages, > 15kV, then the best detector to use is a wide angle BSE detector. Wide angle BSE detectors give better S/N ratio images and will show smaller signal differences - smaller Z variations, than narrow angle detectors. The Everhart-Thornley detectors, detecting SEs, do not show as good a variation with Z as do BSE detectors. BSE detectors mounted to one side of the specimen do not show good signal variation with Z.

The one exception to this is if you have topography and atomic number contrast variations on the same sample and you wish to search for the smallest variation in atomic number that can be seen above the topography contrast. The best detector to use is a high take-off angle BSE detector which surrounds the beam and detects only those BSEs which have been scattered through close to 180 degrees, that is those electrons that have come down and been scattered back up towards the direction from which they just came. Electrons so scattered show very little response to changes in topography and as such the signal obtained from topography variations is quite small whilst the signal variations from variations in atomic number is still relatively large. This is illustrated in Fig. 9 which shows the wide angle BSE detector image and the narrow, high take-off angle BSE image of the same region of the same sample. Within these limitations, scintillator type BSE detectors tend to have a higher S/N ratio than the solid state type.

Electron channelling contrast has been best studied using either high take-off wide angle BSE detector or imaging in the specimen current mode. Crystallographic orientation effects have been best studied using a high take-off, narrow angle BSE detector. Magnetic and voltage contrast effects are beyond the scope of this paper.

Conclusions

Many different types of electron detectors are either available commercially or have been built for experimental purposes. The Everhart-Thornley detector is by far the most widely used detector for imaging purposes in the scanning electron microscope. There is a lot of additional information to be gained from using a second type of detector. Over the past few years, there has been an increase in the use of wide angle backscattered electron detectors, both the scintillator and solid state types.

References

Baumann W, Reimer L. (1981) Comparisons of the noise of different electron detection systems using a scintillator-photomultiplier combination, Scanning, 4, 141-151.

Bishop HE. (1966) Some electron backscattering measurements for solid targets. Fourth Cong. Int. X-ray Opt. Mic., Orsay, Hermann, Paris, 153-158.

Coates DG. (1969) Pseudo-Kikuchi orientation analysis in the scanning electron microscope. Scanning Electron Microsc., 1969: 27-40.

Colby JW. (1969) Backscattered and secondary electron emission as ancillary techniques in electron probe analysis. Adv. Electron. Electron Phys. Suppl. 6. Tonsimis AJ, Martin L, (Eds), Academic Press, New York, 177-196.

Drescher H, Reimer L, Seidel H. (1970) Backscattered coefficient and secondary electron yield of 10-100 keV electrons (In German) Z. Angew. Phys., 29, 331-336.

Everhart TE, Thornley RFM. (1960) Wide-band detector for micro- microampere low energy electron currents, J. Sci. Instrum., 37, 246-248.

Fitch RK, Johnson JD, Walker AR. (1984) A close approach high-energy electron detector for examination of insulating materials in a scanning electron microscope, J. Phys. E: Sci. Instrum., 17, 25-27.

Griffiths BW, Pollard P, Venables JA (1972) A channel plate detector for the scanning electron microscope, Proc. 5th Europe Cong. Electron Microsc. Manchester, Inst. of Physics, Bristol and London, 176-177.

Kimoto S, Hashimoto H. (1966) Stereoscopic observation in scanning microscopy using multiple detector, in McKinley TD, Heinrich KFJ, Wittry DB. (Eds) The Electron Microscope (Proc. Symp., Washington 1964) John Wiley and Sons, New York, 480-489.

Kollath R. (1947) On the energy distribution of secondary electrons (In German) Ann. Phys. (Leipzig), 1, 357-380.

Kulenkampf H, Spyra W. (1954) The energy distribution of backscattered electrons (In German), Z. Physik 137, 416-425

Moll SH, Healey F, Sullivan B, Johnson W. (1978) A high efficiency, non-directional backscattered electron detection mode for SEM, Scanning Electron Microsc. 1978; I: 303-310.

Moll SH, Healey F, Sullivan B, Johnson W. (1979) Further development of the converted backscattered electron detector, Scanning Electron Microsc. 1979; I: 149-154.

Moncrieff DA, Barker PR. (1978) Secondary electron emission in a scanning electron microscope, Scanning, 1, 195-197.

Munro E. (1974) The mechanisms of surface-image formation in a condenser-objective lens using low-loss scattered electrons, Electron Microscopy, 1974, Proc. 8th Int. Cong., Ed. JV Sanders and DJ Goodchild, Aust. Acad. Sci., Canberra, 218-219.

Newbury DE. (1977) Fundamentals of scanning electron microscopy for physicist: contrast mechanisms, Scanning Electron Microsc. 1977; I: 553-568.

Reimer L, Popper W., Brocker W. (1978) Experiments with a small solid angle detector for BSE, Scanning Electron Microsc. 1978; I: 705-710.

Reimer L, Tollkamp C. (1980) Measuring the backscattering coefficient and secondary electron yield inside a scanning electron microscope, Scanning, 3, 35-39.

Reimer L and Volbert B. (1979) Detector system for backscattered electrons by conversion to secondary electrons, Scanning, 2, 238-248.

Robinson VNE. (1973) A reappraisal of the complete electron emission spectrum in scanning electron microscopy, J. Phys. D: Appl. Phys. 6, L105-107.

Robinson VNE. (1974a) The origins of the secondary electron signal in scanning electron microscopy, J. Phys. D: Appl. Phys., 7, 2169-2173.

Robinson, VNE. (1974b) The construction and uses of an efficient backscattered electron detector for scanning electron microscopy, J. Phys. E: Sci. Instrum., 7, 650-652.

Robinson VNE. (1975) Backscattered electron imaging, Scanning Electron Microsc. 1975: 51-60.

Robinson VNE. (1980) Imaging with backscattered electrons in a scanning electron microscope, Scanning, 3, 15-26.

Robinson VNE , Cutmore N G , Burdon R G. (1984) Quantitative composition analysis in a scanning electron microscope, Scanning Electron Microsc. 1984; II: 483-492.

Russell PE (1984) Microchannel plates as specialized scanning electron microscope detectors, in Hren J, Lenz F, Munro E, Sewell PB (Eds) The Electron Optical Systems (Proc. 3rd Pfefferkorn Conference, Ocean City, MD, 1984), SEM Inc., AMF O'Hare, IL pp. 197-200.

Schur K, Blaschke R, Pfefferkorn G. (1974) Improved conditions for backscattered electron SEM micrographs on polished sections using a modified scintillator detector, Scanning Electron Microsc. 1974: 1003-1010.

Stephen J, Smith BJ, Marshall DC, Wittam EM. (1975) Applications of a semiconductor backscattered electron detector in a scanning electron microscope, J. Phys. E: Sci. Instrum., 8, 607-610.

Wells OC. (1970) New contrast mechanism for scanning electron microscope, Appl. Phys. Lett., 16, 151-153.

Wells OC. (1971) Low-loss image for surface scanning electron microscope, Appl. Phys. Lett., 19, 232-235.

Wells OC, Broers AN, Bremer CG. (1973) Method of examining solid specimens with improved resolution in the scanning electron microscope (SEM), Appl. Phys. Lett, 35, 353-355.

Wells OC. (1974) Scanning Electron Microscopy, McGraw-Hill, New York, Ch. 6.

Wolf ED, Everhart TE. (1969) Annular diode detector for high angular resolution pseudo-Kikuchi patterns, Scanning Electron Microsc. 1969: 41-44.

Wooldridge DE. (1939) Theory of secondary emission, Phys. Rev., 56, 562-578.

Zeldes N, Tassa M. (1979) Conversion of existing SEM components to form an efficient backscattered electron detector and its forensic applications, Scanning Electron Microsc. 1979; II: 155-158.

Electron Optical Systems (pp. 197-200)
SEM Inc., AMF O'Hare (Chicago), IL 60666-0507, U.S.A.

0-931288-34-7/84$1.00+.05

MICROCHANNEL PLATES AS SPECIALIZED SCANNING ELECTRON MICROSCOPY DETECTORS

P. E. RUSSELL

JEOL U.S.A., INC.,
11 Dearborn Road, Peabody, MA 01960

Phone no.: (617) 535-5900

Abstract

The need for a specialized detector for low beam voltage and low beam current applications has led to the investigation of a microchannel plate detector for SEM. The application requirements are described in detail, with the case of integrated circuit metrology used as an example. The microchannel plate (MCP) detector has proven to meet almost all of the design objectives of a low voltage metrology SEM detector. The symmetry of the detector and the ability to mount it directly to the final pole piece are among the most important features.

KEY WORDS: Detectors, metrology linewidth measurement, microchannel plate, low voltage microscopy

Introduction

While the conventional Everhart-Thornley secondary electron detector has proven to be extremely versatile in scanning electron microscopy (SEM) applications[1,3], there are some applications where development of a specialized detector is warranted. Two such applications are currently receiving active attention due to the needs of the semiconductor industry in the development of VLSI (very large scale integrated circuits). These applications are integrated circuit (IC) metrology[5] or linewidth measurement and quantitive voltage contrast microscopy.[2] Both of these techniques involve low signal levels and place special requirements on the symmetry of the detector and associated electric fields. Thus, at least one of the potential disadvantages of a microchannel plate (MCP) detector is not of concern; that is the inability to handle very high signal levels. The major problem of concern in these applications is the sensitivity of the MCP to contamination.

The IC metrology need is growing as the feature size in IC's begins to reach the one micron or smaller level. At this feature size, traditional optical microscopy based metrology systems cannot provide the required measurement accuracy and precision, primarily because of the wavelength of visible light involved. An SEM based metrology system must retain the non-destructive aspects of the optical counterparts, as well as the ease of operation and reliability. The non-destructive requirement and the fact that most samples (typically photoresist lines on insulating layers) are non-conducting require the use of low beam voltage (500-2000 volts) and low beam currents (0.1 - 5 pA). The topographical nature of features to be measured, and the need for high precision (~100Å) measurement capability in all directions, requires the use of an electron detector which is symmetric around the measurement point and which is very sensitive to low level signals. A microchannel plate detector, such as the one described by Griffiths, et al in 1972[4] easily satisfies these requirements. The detector requirements of the IC metrology application and the results of a MCP detector for this application will now be described in detail.

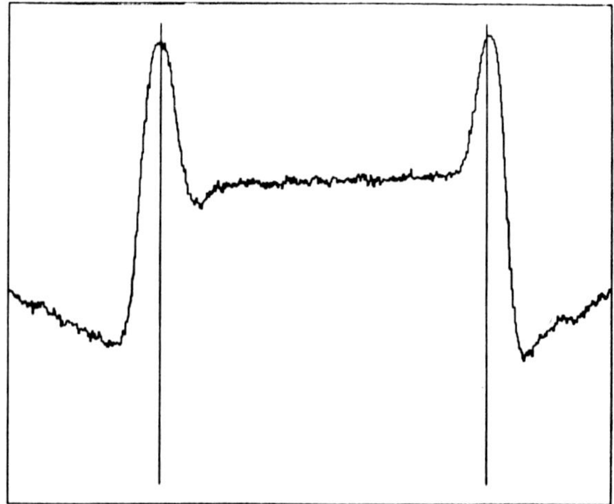

Figure 1. Secondary electron waveform obtained by scanning a 1 keV electron beam across a 1 micron photoresist line; with the measured line pointing directly along the Everhart-Thornley detector axis, and the beam scan direction perpendicular to this axis.

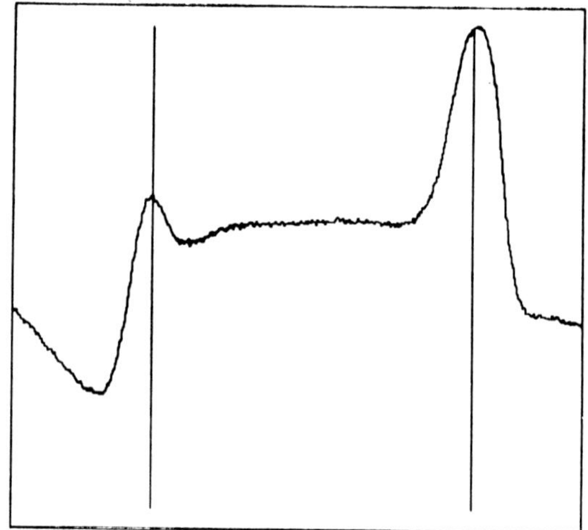

Figure 2. Secondary electron waveform obtained by rotation of the line and scan direction used for Figure 1 by 90°.

Application Requirements

Line width measurement in an SEM is accomplished by scanning the electron beam across the feature of interest and measuring electron emission as a function of beam position. Since the electron beam scan distance can be accurately calibrated, the measured electron emission can be related to features along the measurement direction; in particular to line edges. Secondary electron imaging is known to provide excellent edge detection capabilities and hence the detection of secondary electrons is more desirable than backscattered electron detection alone. (See the paper in this volume by Robinson[6] for a review of electron detectors used in SEM). In using a conventional Everhart-Thornley secondary electron detector, it is found that edge detection capability is strongly affected by the geometry of the detector. If the line to be measured runs directly along the axis of the detector, such that the beam can be scanned perpendicular to the detector axis, both edges of the line can be detected easily. A typical waveform illustrating this is shown in Figure 1. However, if the orientation of the line is rotated by 90°, such that the beam must be scanned along the detector axis, the two line edges are detected very differently, with one edge having substantially reduced signal to noise. A typical waveform illustrating this geometry is shown in Figure 2. Notice the loss of symmetry in going from one orientation to another. This is because the electric field from the detector is not symmetric at the sample; it is actually one-dimensional. There are two major problems with waveforms of the type shown in Figure 2. First, the signal to noise ratio of one edge signal is severely degraded, such that under poor signal conditions edge detection may not be possible. Second, for automated edge detection by simple algorithms, a symmetric waveform (such as that in Figure 1) is less complex and can be handled faster and more reproducibly. Thus, a new detector is required which will provide symmetric waveforms from a line of any orientation in the X-Y sample plane.

The SEM operating conditions required for linewidth measurement included low keV (~ 1 keV) and low beam current (~ 0.1 to 1 pA). The low keV requirement is imposed by two conditions; the desire to avoid charging by working near the condition of unity electron emission and by the requirement to avoid possible radiation damage which higher keV beams induce. The low beam current requirement is imposed by the sensitivity of specimens to total accumulated dose and by the desire to eliminate charging. Both the low voltage and low current requirements present problems when using a conventional ET detector. The electric field of the ET detector, typically in the range of 100V/cm at the measurement site, causes one-dimensional defocusing of the beam, which is increased in effect as the beam voltage is lowered. The low beam current requirements present signal collection problems. The problem is increased by the need to use very short working distance to obtain small spot size at low keV. This poses geometrical constraints on the ability of the ET collector fields to reach the region between the sample and the pole piece.

Detector Requirements

For the above reasons, a new detector was sought with the following basic specifications:

1) Symmetric collection field so as not to distort a low voltage beam,
2) Short working distance configuration,
3) No requirement for tilting specimen to achieve optimum signal collection,
4) High sensitivity,

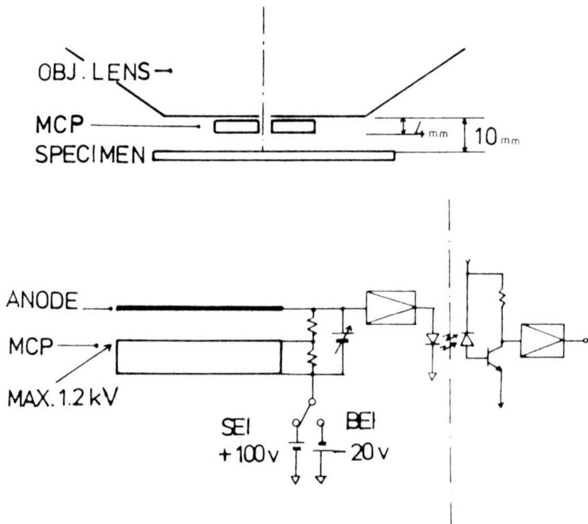

Figure 3. Microchannel plate detector geometry and electrical configuration.

5) Capability to determine angular properties of electron emission, and
6) Ability to detect backscattered electrons.

A microchannel plate detector, similar to that described by Griffiths et al[4], has been successfully implemented to meet these objectives. The geometry of the microchannel plate detector is shown in Figure 3.

The detector assembly is mounted directly on the final lens assembly and requires only 4mm of space below the pole piece. The working distance of 10mm has been chosen in this example, due to other design considerations. The detector itself is disk shaped with a shielded center hole for the primary electron beam. A high voltage across the MCP provides the electron multiplication; this voltage is 1000-1200 volts. The front surface of the detector is biased at +100V for enhanced collection of secondary electrons and at -20V for suppression of secondaries; i.e., for backscattered electron imaging. The entire assembly is electrically isolated from the SEM by use of optical decoupling. The electrical configuration is shown schematically in the lower portion of Figure 3. The actual signal measured is the current collected by the MCP anode plate.

Results

The low beam voltage performance of the MCP system is illustrated in Figures 4 and 5.

Figure 4 shows the detector output current as a function of extraction (or bias) voltage applied. The extraction voltage is seen to strongly increase the overall gain as the voltage is increased from 0 to 100 volts. Above 100 volts the increase is small. Data is shown for 0.5, 1.0 and 3.0 keV beams, all showing similar effects. Based on this data 100 volts was chosen as the optimum voltage for secondary electron collection. The MCP output current

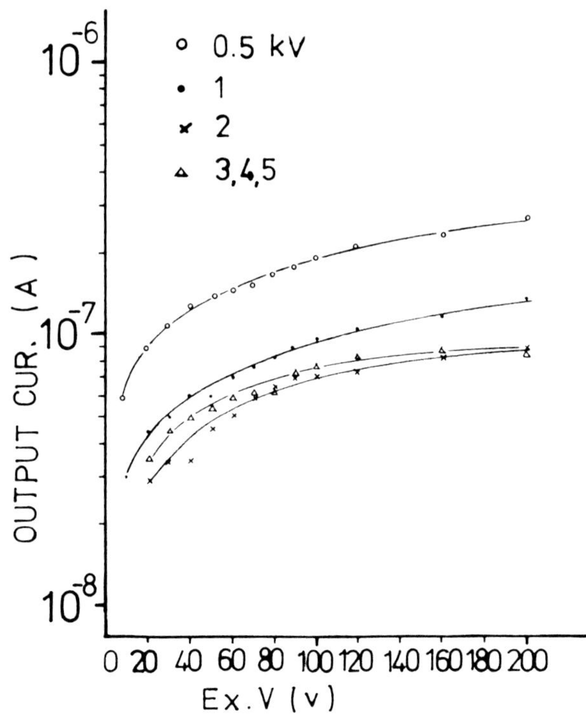

Figure 4. MCP detector output current versus extraction voltage for 0.05, 1, 2 and 3.0 keV primary beam with 1 pA beam current.

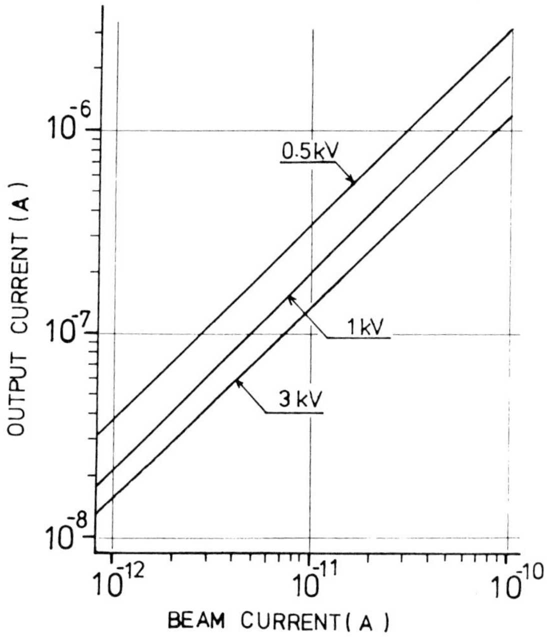

Figure 5. MCP detector output current versus primary beam current for 0.5, 1.0 and 3.0 keV beam voltage. Extraction voltage is 100V.

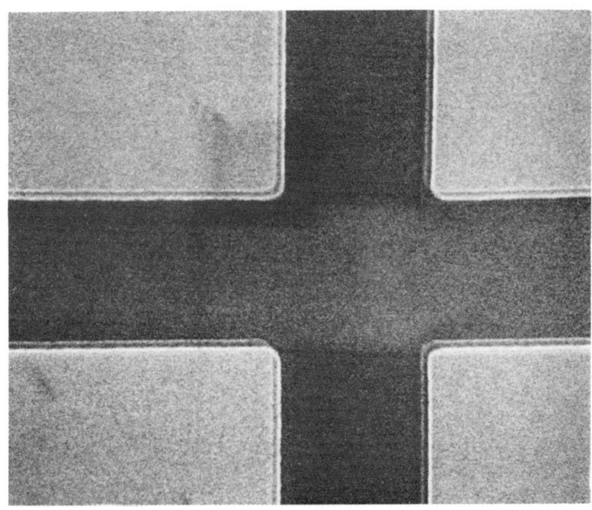

Figure 6. Micrograph of cross pattern (2 micron linewidth) obtained using MCP detector; 1 keV and 1 pA beam conditions.

versus primary beam current is shown in Figure 5, again for 0.5, 1.0 and 3.0 keV beams. The gain is shown to be linear over the beam current range of 10^{-10} to 10^{-12}A. The output current is higher for the low keV beams due to the increase in secondary electron emission. Overall, the MCP detector system is shown to produce acceptable signal levels for the low current, low voltage applications described above.

An image obtained using the MCP detector is shown in figure 6. A 2 micron polysilicon line is shown imaged with 1 keV and 1 pA. All edges are shown to be easily detected. Waveforms obtained from any orientation line result in symmetry and signal to noise such as that shown in Figure 1.

Conclusion

A summary of the attractive features of the microchannel plate as a specialized IC metrology detector is as follows:

- symmetric geometry which allows measurement of features in any orientation
- high sensitivity
- ability to mount directly to lower pole piece, thus allowing very short working distances to be used while maintaining high signal levels
- causes no beam position shift when changing keV (due to symmetry of electric fields involved)

The MCP detector system described above has been implemented on a prototype instrument for approximately 1 year. The system has achieved all design goals and has shown very good stability and reliability. Field installation of several of the detectors in dedicated IC metrology SEM's has been completed. Current investigations include the use of the MCP detector in other applications, the use of a segmented anode plate, and the effects of vacuum level and contamination on detector performance and reliability.

References

1) Everhart T E, Thornley R F M (1960) Wide band detector for micro-microampere low-energy electron currents, J. Sci. Instr. 37, 246-248.

2) Feuerbaum H P (1979) VLSI testing using the electron probe, Scanning Electron Microsc. 1981;I: 285-296.

3) Goldstein J I, Yakowitz H (1975) Practical Scanning Electron Microscopy, Plenum Press, New York, 101-103.

4) Griffiths B W, Pollard P, Venables J A, (1972) A channel plate detector for the Scanning Electron Microscope, Proc. Fifth European Congress on Electron Microscopy, The Inst. of Physics, London & Bristol, 176-177.

5) Nyyssonen D, (1982) Design of an optical linewidth standard reference material for wafers, in: Nyyssonen D (ed.) Proc. SPIE, Vol. 342, SPIE, Bellingham, WA, 27-34.

6) Robinson V, (1984) Electron detectors used for imaging in the scanning electron microscope, in: Electron Optical Systems, J. Hren, F. Lenz, E. Munro, P.B. Sewell (eds.), SEM Inc., AMF O'Hare, IL, 187-195.

Discussion with Reviewers

J.B. Warren: Solid state BSE detectors mounted on the pole piece and specimen current imaging would also seem to meet the criteria listed in the paper for line width measurement. Is the signal-to-noise ratio of the MCP detector superior to these methods for the current and voltage regime described?

Author: It is true that solid state backscattered electron detectors and specimen current imaging in principle provide axially symmetric imaging. However, since edge detection is the major objective of the work described in this paper, and since low beam energies (typically 0.7 keV to 2.0 keV) are required, solid state backscattered electron detectors are not suitable. Also, since the specimens are non-conducting, and the beam voltage is such that absorbed current is essentially zero, absorbed current imaging is not suitable.

Electron Optical Systems (pp. 201-207)
SEM Inc., AMF O'Hare (Chicago), IL 60666-0507, U.S.A.

0-931288-34-7/84$1.00+.05

THE INFLUENCE OF SAMPLE AND DETECTOR ANGLES UPON AUGER ELECTRON SIGNAL

Robert L. Gerlach

Perkin-Elmer, Physical Electronics Division,
6509 Flying Cloud Drive,
Eden Prairie, Minnesota 55344, U.S.A.
Telephone Number (612) 828-6212

Abstract

The effects of the sample to incident electron beam angle and of the detector angle on the Auger electron signal are important for quantitative Auger analysis, particularly for Auger mapping and line scans. A first approximation, single scattering, mean free path model is employed to simulate the Auger signal resulting from a range of sample and detector angles. The model is single scattering in the sense that the excitation path is taken to be a straight line into the solid. A second model approximates multiple scattering by a normal distribution of ionizing flux angles about the incident beam direction.

Since spherical particles exhibit all possible surface angles, they are useful for testing the theoretical models. A coaxial electron gun/cylindrical mirror analyzer (CMA) instrument with an "angle-resolved drum" is employed to analyze 200μm diameter Ti spheres on a Sn substrate. The observations compare favorably with predictions of the model for Ti and O Auger signals, and the multiple scattering approximation is seen to more accurately describe the results.

Key Words: Auger, surface analysis, detector, microprobe

Introduction

The Scanning Auger Microprobe (SAM) has widespread acceptance in pure and applied surface science studies. The effects of surface topography and detector geometry upon the Auger signal are important in the interpretation of Auger line scans and maps. The effects of surface roughness upon Auger electron spectroscopy, where the incident beam is large compared with surface roughness dimensions, was treated by Holloway[4]. It was shown, for example, that using a coaxial electron gun in a cylindrical mirror analyzer (CMA) with electrons impinging normally upon an Au surface having grains of 21° RMS angle from the surface normal gave 10% reduced signal compared with a smooth surface. Also demonstrated was that a smooth Au surface results in less than 15% reduction in Auger signal when tilted up to 80° to the full CMA axis.

Recent SAM instruments employ microbeams of sufficiently small diameter that surface roughness (grain size) dimensions may be large compared with the probe diameter. In this case, we are sampling an individual surface geometric condition and can observe large deviations in signal. Shimizu et al [9] have used Monte Carlo methods to investigate normal and 45° incidence, 10keV electron beams on an Al surface. They predict that a 70% increase in K-ionization results in the near surface region upon tilting from normal to 45°, whereas, only a 40% increase results in the L-ionization. Only these two angles were treated by Shimizu.

In a more recent paper, Shimizu et al [10] examined, theoretically and experimentally, the effects of a sharp edge upon the Auger signal. The Monte Carlo calculation predicts an enhancement of the K and L Auger signals at the edge of an Al sample, which was also observed experimentally. This enhancement results from the fact that Auger electrons are detected from the edge which is parallel to the beam as well as from the perpendicular face.

El Gomati and Prutton [2] performed Monte Carlo calculations of normal and 75° incident electrons upon an Al target. Figures showing trajectories of 100 electrons in the solid are shown, but no quantitative comparison for Auger signal is made for the two angles. An increase in signal for the glancing incidence is apparent from the figures, however. Both the Shimizu and Gomati papers do not account for detector geometry.

Jablonski [5] recently presented various approximations for the Auger backscattering factor as a function of sample angle utilizing Monte Carlo calculations. Results for the full cylindrical mirror analyzer detector are similar to those presented in this paper, but no analysis is given for other detector geometries. Definitive data and theoretical treatment of sample angle and detector geometry effects upon Auger signal have been lacking.

Theoretical Treatment

Single Scattering Approximation

In this section we consider the production of Auger electrons as a function of primary electron incidence angle from a smooth spherical surface. Here we consider the two detector geometries shown in Figure 1 - a full CMA about the Z-axis and a segment of the CMA towards the X-axis direction. The CMA collects electrons from 36° to 48° from the Z-axis.

Referring to Figure 2, assume in our first approximation that the incident beam travels in a straight line into the solid and that the core level excitation (production of Auger electrons) decays exponentially with path length [1,9] governed by the mean free path λ_I. Similarly we assume that the probability for Auger electron escape from the solid decays exponentially with escape length [8]. In a previous paper [3] it was shown that the following equation results for the Auger electron intensity as a function of the sample normal direction (Θ_N, ϕ_N):

$$J_{IA}(\Theta_N,\phi_N) = (J_1\lambda_I/\lambda_e) \int_{\Omega_A} (1+\beta \cos\Theta_{IN}/\cos\Theta_{NA})^{-1}\, d\Omega_A \qquad (1)$$

where λ_e is the core ionization mean free path, λ_A is the mean free path for ejected Auger electrons, $\beta = \lambda_I/\lambda_A$, Ω_A is the solid angle of ejected Auger electrons, Θ_{IN} is the angle between the incident beam and the surface normal, and Θ_{NA} is the angle between the surface normal and the ejected Auger electron.

The integration in Equation (1) is performed only over the detector solid angle. In addition, the inequalities

$$0 \le \Theta_{IN} \le 90^0 \quad \text{and} \quad 0 \le \Theta_{NA} \le 90^0 \qquad (2)$$

must be satisfied.

In the typical case, β =100 such that the integrand in Equation (1) reduces to $\cos\Theta_{NA}\sec\Theta_{IN}$ provided $\Theta_{IN} << 90°$. This $\cos\Theta_{NA}\sec\Theta_{IN}$ relationship was discussed briefly by Janssen and Venables [6].

Approximation of Multiple Scattering

Multiple scattering and reduction of the incident flux energy are included in the Monte Carlo calculations of Shimizu et al [10] and of Gomati and Prutton [2]. If we consider Figure 1 of reference 10 and Figure 3 of reference 2, it is reasonable to assume that the incident ionizing beam I_I may be approximately described by

$$I_I = I_{IO}\, e^{-L_1/\lambda_I} \times \frac{1}{\sqrt{2\pi}\,\phi_{IO}}\, e^{-1/2(\phi_I/\phi_{IO})^2} \qquad (3)$$

where L_1 is the incident beam path length into the solid. The first exponential term is as used in our first approximation (equation 1) and the second term approximates the scattering of the ionizing flux to angles ϕ_I away from the incident beam direction with a normal distribution. Hence, the equation for the Auger signal becomes

$$J_{2A}(\Theta_N, \phi_N) = (J_2\,\lambda_I/\lambda_e\phi_{IO}) \int_{\Omega_A}\int_{\Omega_I} (1+\beta\cos\Theta_{IN}/\cos\Theta_{NA})^{-1} e^{-1/2(\phi_I/\phi_{IO})^2}\, d\Omega_I\, d\Omega_A \qquad (4)$$

where Ω_I is the ionizing flux solid angle. Equation (4), of course, reduces to equation (1) for $\phi_{IO} = 0$.

Numerical calculations of Auger line scans across a spherical sample are shown in Figure 3, where the distance X across the sphere is normalized to unit radius. In addition, the Auger signals are normalized to the center of the sphere for each set of parameters, since we are primarily interested in the shape of the signal response. Figure 3A shows curves for the full CMA with $\beta = \lambda_I/\lambda_A$ =1000, and various ϕ_{IO} values employed in equation 4. The theory predicts relatively flat signal response over the center of the sphere and signal enhancement at the edges. The effect of increasing ϕ_{IO} is to spread the edge enhancement towards the sphere center as is obvious from the form of equation 4.

The effects of the parameter β are shown in Figure 3B for the full CMA and ϕ_{IO} = 18°. Increasing β increases the edge signal enhancement without affecting the signal function shape near the center of the sphere. As the depth of penetration of the incident beam is increased more signal is generated as the beam skims the edge of the sphere where the Auger electrons are very likely to escape and be detected.

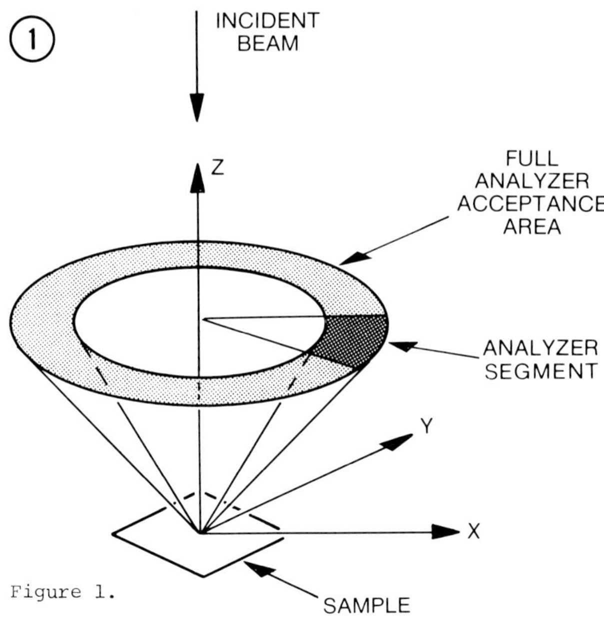

Figure 1.

Isometric drawing of the full and segmented CMA/primary electron beam geometries. In each case the Auger electrons are collected at 36° to 48° from the analyzer axis.

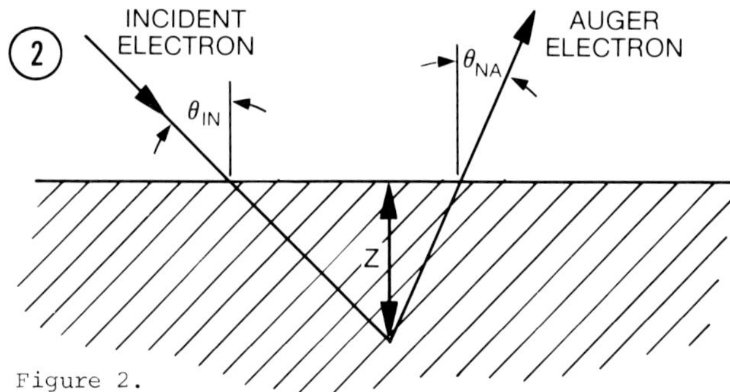

Figure 2.

Diagram of the primary electron beam and of an Auger electron escape from the solid surface.

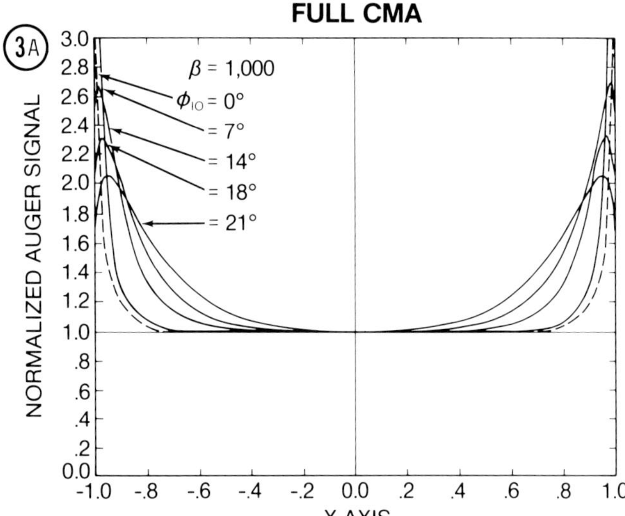

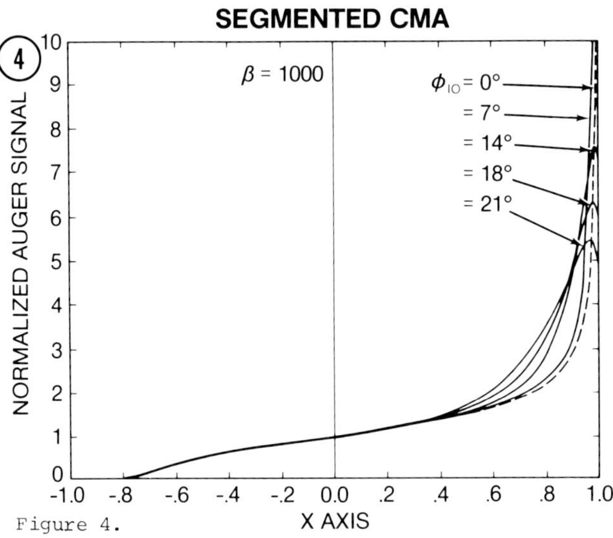

Figure 4.

Normalized Auger signal with the segmented CMA versus distance calculated from equation 4 for β = 1000 and various ϕ_{IO}.

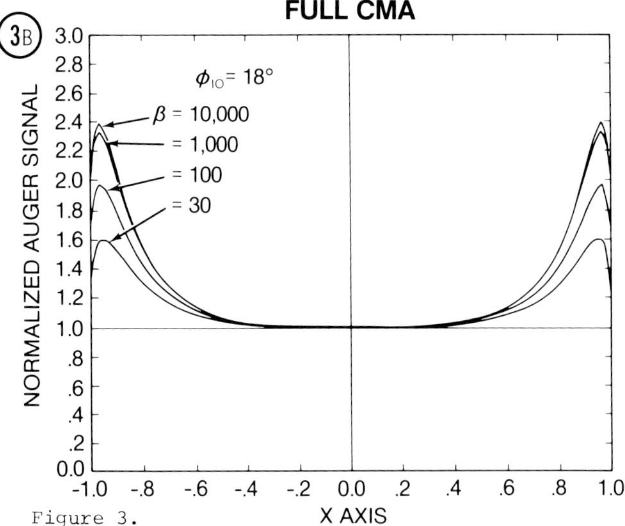

Figure 3.

Normalized Auger signal with the full CMA versus the normalized distance X across the spherical sample calculated from equation 4.

A. For β = 1000 and various incident beam half angles ϕ_{IO}.

B. For ϕ_{IO} = 18° and various β.

Calculations of the Auger signal detected by the segmented CMA (see Figure 1) are shown in Figure 4 for β = 1000 and various ϕ_{IO}. The segmented detector is blind to the left side of the sphere (shadowing effect). The signal rises rapidly with increasing X and reaches a peak near the right-hand edge of the particle. Again, increasing ϕ_{IO} spreads the edge enhancement towards the sphere center.

Comparison of Theoretical and Experimental Auger Intensities

Ti spheres of approximately 200μm diameter were dispersed onto a Sn surface and Auger analyzed with a coaxial CMA/electron gun system equipped with an "angle resolved drum". The electron gun with 4μm beam diameter at 5keV energy was scanned across the centers of the spheres and produced the Auger line scans in Figure 5, where comparisons with the theory are shown.

For comparisons of the theory with experiment we must evaluate β from the known Auger escape depth and excitation mean free path values for Ti. Since the O(KLL) and Ti(LMM) Auger electron emissions are at 503 and 418eV, respectively, their escape depths are estimated [8] to be λ_A = 10A. An estimate of the excitation mean free path from Love et al (see Figure 10 of reference 7) indicates that $\lambda_I \simeq$ 1μm for Ti at 5keV and for a 10:1 ratio for primary beam energy to ionization energy. A ratio β = 10,000 A/10A = 1000 was therefore used in our calculation. The single scattering approximation (equation 1) qualitatively describes the Auger intensities for both the segmented and full CMA. However, the rise in signal at the edges of the sphere is predicted to be too steep and too close to the edges.

Equation 4 with ϕ_{IO} = 18° and β = 1000 fits the data reasonably well as shown in Figure 5. It is physically reasonable that the scatter in the incident beam has an 18° angular divergence [2,10]. The multiple scattering approximation describes the experimental data best but still overestimates the edge enhancement. Surface roughness may account for the lack of agreement at the edges of the sphere.

Surface roughness has two effects upon the measurements. First, the surface normal $\underline{R}_N$ has a distribution about its mean value. Second, the long path lengths at grazing incidence are broken up. To simulate the second effect we limit the grazing incident angle Θ_{IN} according to the following inequality:

$$\cos \Theta_{IN} > 0.03.$$

Physically this limits the grazing incidence to approximately $33\lambda_A \simeq 330$ Å. The resulting curves shown in Figure 5 fit the experimental data reasonably well.

A two dimensional Ti Auger map of the Ti sphere is shown in Figure 6 where the segmented detector was employed. Also shown is a contour map calculated from equation 4 which shows good correspondence to the experimental map.

Other Applications of the Model

Now that we have established that this empirical model describes the Ti sphere data, it is interesting to apply it to other geometries. It should be emphasized that though we have not proven equation 4 to be valid for a range of materials, surface smoothness, and a wide range of ϕ_N, it should give us a qualitative representation. As a first case we examine the effect of placing a segmented (or point) detector at various angles ϕ_D with respect to the primary electron beam axis. Results of this calculation are shown in Figure 7, where the abscissa is now the polar angle ϕ_N of the surface normal (instead of the distance across a sphere). One sees that if the detector could be placed very near the primary beam axis (ϕ_D = 0), then the Auger signal would be relatively insensitive to the sample angle except that the signal falls to zero for ϕ_N = ±90°. As the detector angle ϕ_D is increased, the signal versus sample angle curve becomes increasingly asymmetric and peaked towards the detector side. The signal is cut off for $\phi_N < (-90° + \phi_D)$.

Consider next the effect of the sample holder plane being tilted angle ϕ_S as shown in Figure 8. When the full CMA detector is used, the left-hand side of the detector begins to be cut off to input signal when $\phi_S \simeq \phi_{CMA}$ = 42°. The effects of this are shown in Figure 9 where for computational convenience the CMA is modelled as having entrance angle ϕ_{CMA} =42° with very small Δφ. One sees that for ϕ_S < 42° the signal is an asymmetric function of ϕ_N. When ϕ_S = 60°, the signal for ϕ_N = -60° is less than half that for ϕ_N = +60°.

Typical surfaces are not like Figure 8, where a single sphere lies upon a flat surface at angle ϕ_S, but rather many hills and valleys populate the surface. As a result a hill may tend to shadow another surface feature to angle ϕ_S, producing a loss of signal as represented in Figure 9.

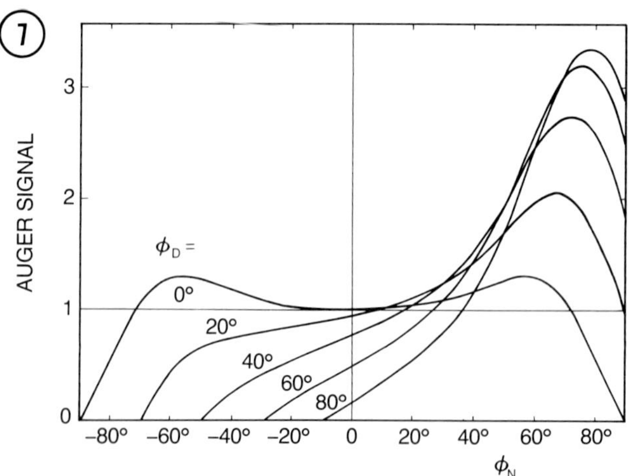

Figure 7.

Computed Auger signal as a function of the sample normal polar angle ϕ_N for various detector polar angles ϕ_D, where ϕ_N, ϕ_D and the Z axis are in the same plane. The curve for ϕ_D = 0 is arbitrarily normalized to unit signal at ϕ_N = 0.

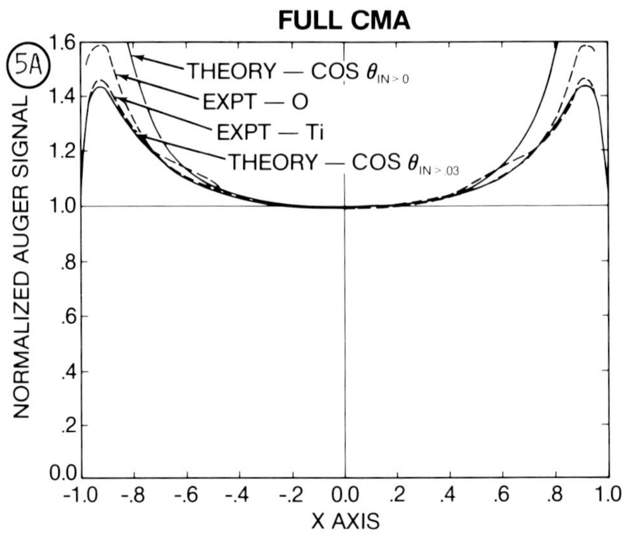

Figure 5. Comparison of theoretical and experimental Auger signal for a line scan across a Ti sphere. The theoretical curves were computed from equation 4 with β = 1000 and ϕ_{IO} = 18°.

A. Full CMA.

B. Segmented CMA.

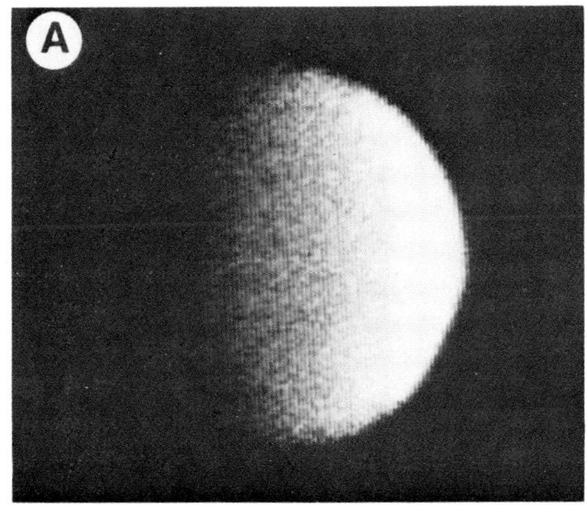

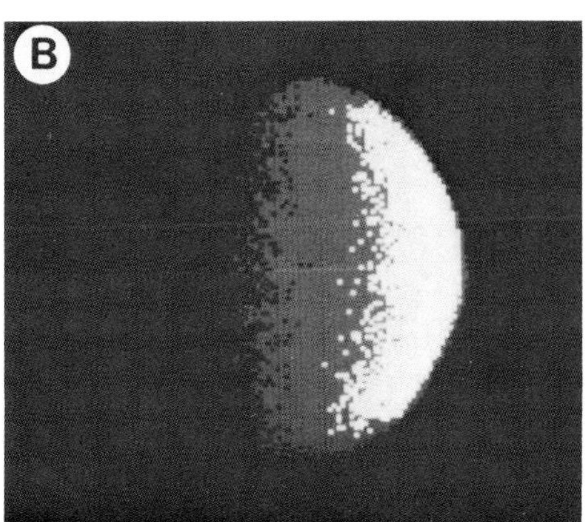

C

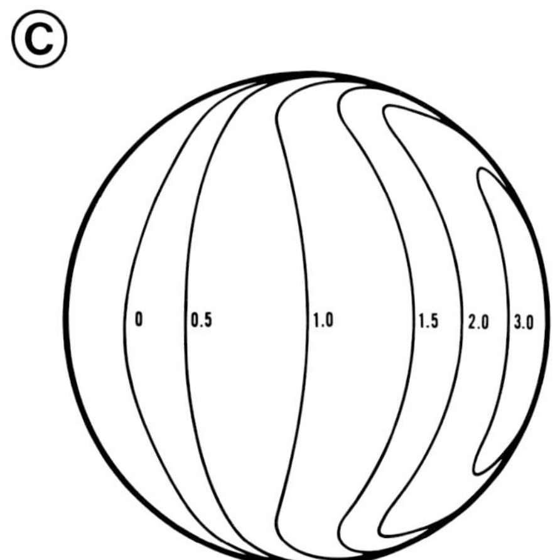

Figure 6. Auger map of the 200μm diameter Ti sphere using a 5 keV primary electron beam and analysis of the 418eV Ti(LMM) peak.

A. Experimental gray scale map.

B. Experimental quadrature map.

C. Theoretical contour map computed from equation 4 with β = 1000, ϕ_{10} =18° and cos Θ_{IN} > 0.03.

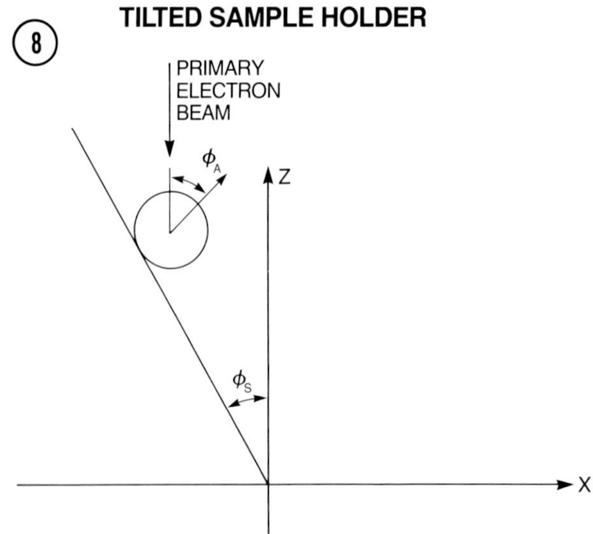

Figure 8.

Diagram of a small sphere on a sample holder tilted at polar angle ϕ_S.

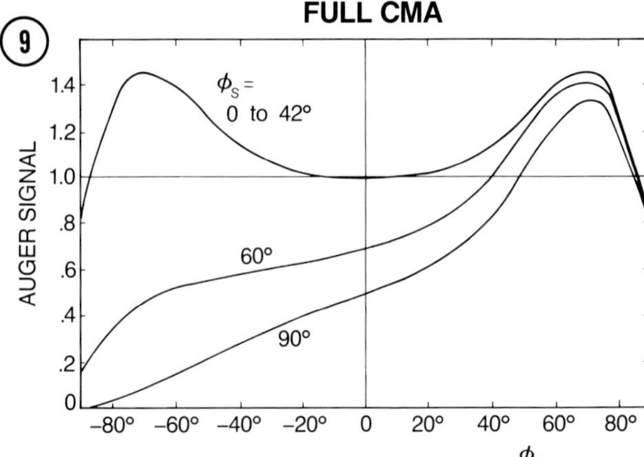

Figure 9.

Computed Auger signal as a function of the sample normal polar angle ϕ_N for various sample holder tilt angles ϕ_S, where ϕ_N, ϕ_S and the Z axis are in the same plane. The curve for ϕ_S = 0 to 42° is arbitrarily normalized to unit signal at ϕ_N = 0.

Conclusions

As a small solid angle detector is brought closer to the primary beam axis, the Auger signal versus sample angle (ϕ_N) curve becomes increasingly symmetric, though the signal at ϕ_N = ±90° goes to zero. However, the detector axis must be within about 20° of the gun axis to give reasonably flat response, which is not a very feasible angle experimentally.

In using the full CMA detector, surface topography and large sample holder tilt angles can block the signal to the CMA. However, there is no signal reduction until an adjacent surface feature is within the ϕ_{CMA} ≈ 42° cone angle of the detector.

Finally, various means are in use to correct for this loss of signal due to surface topography. The most common methods employ the normalization of the Auger signal by the background energy distribution. These methods can be successful provided there is of course a signal from the region of interest and that the corrections are relatively small (≈2x).

References

1. Dudek HJ (1980). Influence of the Electron Beam Angle of Incidence on the Spatial Distribution of the Ionization Density. Proceedings of the Seventh European Congress on Electron Microscopy, The Hague, Netherlands, Aug. 24-29, 1980. Editors P. Brederoo and V.E. Cosslett. Volume 3, 7th European Congress on EM Foundation, Leiden, 22-23.

2. El Gomati MM. and Prutton M. (1978). Monte Carlo Calculations of the Spatial Resolution in a Scanning Auger Electron Microscope. Surface Sci. 72, 485-494.

3. Gerlach RL, Hovland CT, Clough SP, Pinchback TR. (1982). Scanning Auger Electron Images with Orthogonal Versus Coaxial Gun/Analyzer Geometry. 10th International Congress on Electron Microscopy, Hamburg, (ed.) The Congress Organizing Committee, Volume I, Deutsche Gesellschaft fur Electronenmikroskopie, Frankfurt. 719-720.

4. Holloway PH. (1975). The Effect of Surface Roughness on Auger Electron Spectroscopy. J. Electron Spectroscopy and Related Phenomena 7, 215-232.

5. Jablonski A. (1983). Dependence of the Backscattering Factor in AES on the Primary Electron Incidence Angle. Surface Sci. 124, 39-50.

6. Janssen AP, Venables JA. (1978). The Effect of Backscattered Electrons on the Resolution of Scanning Auger Microscopy. Surface Sci. 77, 351-364.

7. Love G, Cox MGC, Scott VD. (1977). A Simple Monte Carlo Method for Simulating Electron-Solid Interactions. J. Phys. D: Appl. Phys. 10, 7-23.

8. Palmberg PW. (1973). Quantitative Analysis of Solid Surfaces by Auger Electron Spectroscopy. Analytical Chemistry 45, 549A-556A.

9. Shimizu R, Aratama M, Ichimura S, Yamazaki Y. (1977). Application of Monte Carlo Calculation to Fundamentals of Scanning Auger Electron Microscopy. Appl. Phys. Lett. 31, 692-694.

10. Shimizu R, Everhart TE, MacDonald NC, Hovland CT. (1978). Edge Effect in High-Resolution Scanning Auger-Electron Microscopy. Appl. Phys. Lett. 33, 549-551.

Discussion with Reviewers

M.P. Seah:
In many practical instances we may take β to be very large. Thus, for instance Figure 3(a) may be taken to represent the practical result expected for various Φ_{10}. Φ_{10} should be a unique value for a given primary electron beam energy, atomic weight and core level ionization energy. It would be interesting in such a three dimensional space, to visualise the contours of surfaces of constant Φ_{10}. Would the author like to comment on the shapes of these surfaces so that the reader can extend the theory presented to other materials, beam energies and transitions?

Author:
Such three dimensional surfaces could be quite valuable. However, without a detailed study of these parameters, it would be premature for me to comment on the surface shape.

Electron Optical Systems (pp. 209-220)
SEM Inc., AMF O'Hare (Chicago), IL 60666-0507, U.S.A.

ELECTRON IMAGE SIMULATION:
A COMPLEMENTARY PROCESSING TECHNIQUE

Michael A. O'Keefe

National Center for Electron Microscopy
Building 72, Lawrence Berkeley Laboratory
University of California
Berkeley, CA 94720

Telephone (415) 486-4610

Abstract

At present it is difficult to use direct image processing techniques to determine the specimen structure from electron micrographs obtained under non-linear imaging conditions, and impossible when the effects of dynamical scattering are strong (as in the case of thicker specimens). However, computing techniques are available to simulate high-resolution transmission electron microscope (HRTEM) images of postulated model structures. With these techniques it is possible to confirm the validity of interpretation of recorded micrographs, to help analyze crystal defects, to characterize microscope parameters, and to determine the ranges of validity of commonly used interpretive approximations. Because processed micrographs can lead to suitable model structures and, in turn, models can indicate optimum directions for processing, the two techniques are excellent complements.

Key Words: Processing, simulation, interpretation, modeling, computing, high-resolution transmission electron microscopy.

Introduction

The ultimate goal of the image processing of electron micrographs is to obtain all possible information about the specimen that may be contained in the micrograph. In fact, since the electron microscope acts somewhat as a spatial frequency filter, more information is available from a "focal series" of micrographs than from any single image. Saxton (1980) shows examples of the application of Schiske's (1973) generalization of the Wiener filter to focal series of micrographs under so-called "linear image" conditions.

Under more general conditions the so-called "second-order" terms can alter the weights of spatial frequencies present in the image and even introduce higher frequencies formed from combinations of the frequencies present in the linear-image contribution to the micrograph. Second-order terms complicate the processing procedure immensely, and no satisfactory method of incorporating them has been published. O'Keefe and Sanders (1976) suggested that it may be possible to remove the second-order component from experimental optimum defocus images by subtracting the minimum contrast (or Gaussian focus) image, since this latter image is formed by selecting the value of focus which produces minimum linear contribution to the image intensity and second-order contributions change only slowly with change in focus. Saxton (1980) has proposed removing the second-order contribution by processing bright-field/dark-field pairs of micrographs, but the method introduces some experimental difficulties. Kirkland (1982) has proposed a computational method, but O'Keefe and Saxton (1983) show that some of the approximations involved are unjustified. At present, therefore, linear transfer can be inverted, but non-linear transfer cannot. Thus in image reconstruction the linear approximation is often chosen even when it is realized that for thicker specimens, the approximation may be poor. After this initial linear step has been used to set up reasonable model structures and yield information on transfer function parameters, simulations can be made both with and without non-linear effects, not only to assist reconstruction, but also in order to assess the importance of the non-linear contribution, and thus the reliability of the initial linear step. In some circumstances the linear approximation can be used to quite high values of crystal thickness (Tanaka and Jouffrey, 1984), whereas in others it fails dramatically by 150Å (O'Keefe and Saxton, 1983).

Present processing methods (and even future ones incorporating the second-order effects) are capable only of determining the complex amplitude (modulus and phase) of the electron wave leaving the exit surface of the specimen. Under some conditions this electron wave is a simple function of the specimen structure projected in the direction of the electron beam, so that the structure is easily derived from the exit surface wave. This procedure is possible when the specimen scatters sufficiently weakly to be considered a "phase-object" or even a "weak phase object". Conditions of strong scattering occur for thicker specimens composed of heavier atoms and viewed down low index planes where atoms superpose exactly. These conditions are often those desirable in studies of periodic structures, especially of defects in such structures. On the other hand, when atom positions do not superpose, as for the case of defects in the thickness direction, interpretation is often impossible even with the aid of image simulation. Fortunately, biological and organic crystals are weak scatterers and hence can be interpreted as phase objects to thicknesses of several hundred Angstrom units; similarly many oxides and silicate minerals have a thickness limit in the 50Å to 100Å range, as do the important semiconductors Si and GaAs. However most metals and alloys have limits of only a few tens of Angstroms. For specimens thicker than the above limits, no simple relationship between the projected specimen structure and the exit-surface wave can be found (e.g. Jap and Glaser 1980) and image simulations must be used.

For the simulation of HRTEM images, a model structure is proposed, assembled in the computer, and images computed incorporating the various microscope parameters. In the decade since simulated high-resolution images of known crystal structures first appeared in the "n-Beam Lattice Images" series of papers (Allpress et al., 1972; Lynch and O'Keefe, 1972, Anstis et al., 1973; O'Keefe, 1973; Lynch et al., 1975, O'Keefe and Sanders, 1975), it has become commonplace to interpret HRTEM images of uncharacterized structures by comparing them with simulated images. Such images are currently produced by a variety of programs written in various laboratories around the world; in fact, commercial packages such as Skarnulis' (1979) interactive program and the more general SHRLI suite of programs (O'Keefe et al., 1978) are now available.

Theory of Simulation

The starting point for image simulation is to model the electron microscope as a simple system of electron beam, specimen and lens system (Fig. 1). Generally the initial electron beam is considered to be a parallel beam of plane wave electrons. The microscope lens system is replaced by one spherically-aberrated lens which can be regarded as representing the objective lens. In a real electron microscope the objective lens has the crucial duty of re-assembling the diffracted beams emerging from the specimen into an image which is merely magnified further by subsequent lenses.

In any simulation three functions representing the electron wave amplitude must be computed at three positions within the model microscope; at the exit surface of the specimen, $f(\underset{\sim}{x})$; at the back-focal plane (diffraction plane) of the objective lens, $f_{\underset{\sim}{k}}$; and at the image plane of the lens, $\psi(\underset{\sim}{x})$. The computation of the exit-surface wave $f(\underset{\sim}{x})$, involves mainly the model specimen structure (the only microscope parameter involved is the energy of the electrons in the incident beam). The wave at the back-focal plane, $f_{\underset{\sim}{k}}$, is obtained via a simple Fourier transform of the exit-surface wave $f(\underset{\sim}{x})$. Calculation of the image from the electron wave at the diffraction plane does not involve the specimen but only microscope parameters, such as objective lens defocus and spherical aberration; together with the objective aperture size and position, these modify $f_{\underset{\sim}{k}}$ before it is transformed into $\psi(\underset{\sim}{x})$, the image amplitude.

Calculation of the Exit-Surface Wave

Self et al., (1983) give details on calculating $f_{\underset{\sim}{k}}$ by various methods and conclude that the multislice method (Goodman and Moodie, 1974) is preferable. In this method, the steps involved in calculating the (structure dependent) values of $f_{\underset{\sim}{k}}$ are:

(i) calculation of $V_{\underset{\sim}{k}}$, the Fourier coefficients of potential (structure factors) at reciprocal lattice points $\underset{\sim}{k}$, lying in the zone perpendicular to the electron beam direction. Summing over all atoms in the unit cell:

$$V_{\underset{\sim}{k}} = \frac{h^2}{2\pi m_e e V_c} \sum_j {}^e f_j \, |\underset{\sim}{k}| \exp\{-2\pi i(\underset{\sim}{k} \cdot \underset{\sim}{x}_j)\} \qquad (1)$$

where V_c is the volume of the unit cell, and ${}^e f_j$ and x_j are the electron scattering factor, and the position of the jth atom respectively. Here the electron scattering factor for each atom is defined as the Fourier transform of the potential distribution for that atom. Inclusion of all V_k within $4Å^{-1}$ of the origin of reciprocal space provides sufficient accuracy for most calculations.

(ii) Fourier transformation of the $V_{\underset{\sim}{k}}$ values produces $\phi_p(x)$ the crystal potential of one unit cell projected in the direction of the electron beam. The effect of such a thin "slice" of crystal on the electron beam is that of a phase object, and the electron "transmission function" for the slice is

$$q(\underset{\sim}{x}) = \exp\{i\sigma\phi_p(\underset{\sim}{x})\Delta z\} \qquad (2)$$

where Δz is the "slice thickness" and σ is the interaction parameter for electrons of the designated energy.

(iii) The exit-surface wave at the desired crystal thickness, $H = m\Delta z$, is found from $q(\underset{\sim}{x})$ by iteration. After m slices the electron wave $f(\underset{\sim}{x})$ is given by

$${}^m f(\underset{\sim}{x}) = [{}^{m-1} f(\underset{\sim}{x}) \star {}^m p(\underset{\sim}{x})]\,{}^m q(\underset{\sim}{x}) \qquad (3)$$

where ${}^m q(\underset{\sim}{x})$ is the transmission function of the mth slice and ${}^m p(\underset{\sim}{x})$ is the small-angle approximation to the free space propagator for the distance between the (m-1)th and mth slices (i.e. the familiar Fresnel propagator); $\star$ represents the convolution operation. The diffraction plane wavefield ${}^m f_{\underset{\sim}{k}}$ is obtained from ${}^m f(\underset{\sim}{x})$ by Fourier transformation. Note that for heavy atoms, or unit cells large in the direction of the electron beam, the slice thickness Δz may need to be chosen smaller than the unit cell height. In such cases a number of projected potentials must be calculated and appropriate transmission functions stored for each slice. Many structures are formed from atoms sufficiently light to allow slice thicknesses from 3Å to 5Å without intracell slicing.

The iteration of (3) is best performed in reciprocal space, where the wavefunction exists only at the discrete points of the reciprocal lattice. Furthermore, as the Fourier components of the wavefunction fall off with increasing order, it is a good approximation to consider the reciprocal space wavefunction as bandwidth limited (i.e. having intensity in only a finite number of beams). In reciprocal space, (3) becomes

$${}^m f_{\underset{\sim}{k}} = [{}^{m-1} f_{\underset{\sim}{k}} \cdot {}^m P_{\underset{\sim}{k}}] \star {}^m Q_{\underset{\sim}{k}} \qquad (4)$$

where P_k and Q_k are the Fourier transforms of $p(\underset{\sim}{x})$ and $q(\underset{\sim}{x})$. This form gives ${}^{m}f_k$ directly and is the form used in the SHRLI programs (O'Keefe et al., 1978) and shown in Figure 1.

Now that large amounts of computer memory are available to an individual user, an alternative to both the real-space (3) and diffraction-space (4) methods has become viable. This method uses the fast Fourier transform (FFT) algorithm of Cooley and Tukey (1965) to replace the convolution by an FFT followed by a simple multiplication and an inverse FFT (Ishizuka and Uyeda, 1977). This procedure is faster than direct convolution for large numbers of beams and extremely fast when programmed on an array processor. One problem is that only 2/3 of the extent of the FFT array can be used in order to avoid the aliasing produced by multiplication of two functions in a pseudo-periodic diffraction space, the pseudo-periodicity resulting from sampling the continuous functions $f(\underset{\sim}{x})$ and $q(\underset{\sim}{x})$. Thus only 4/9 of the two-dimensional array can be used for active beams without the danger of overlap of adjacent diffraction cells over-emphasizing the amplitudes of the outer reflections.

Van Dyck (1980) has used an approximate form of (3) to calculate the scattering in real space. Although the diffraction-space method (4) requires a time proportional to N^2 (where N is the number of diffracted beams), and the FFT method requires a time proportional to N log N, the time for Van Dyck's small-block convolution method is proportional to N. Unfortunately practical trials (Self, 1982; Kilaas and Gronsky, 1983) showed that this method required much larger values of N and smaller values of Δz than both the FFT and diffraction-space methods, resulting in calculations that took 4 to 5 times as long as an FFT multislice to produce an equivalent result.

Calculation of the Image-Plane Wave

In order to compute the image plane intensity from the exit surface wave, we need to include the effects of objective lens defocus and spherical aberration. These parameters act merely to change the phases of the diffracted electron beams passing through the aperture of the lens. Thus:

$$\Psi_{\underset{\sim}{k}} = f_{\underset{\sim}{k}} A_{\underset{\sim}{k}} \exp\{-i\gamma(\underset{\sim}{k})\} \tag{5}$$

where:

$$\gamma(\underset{\sim}{k}) = \pi\lambda \underset{\sim}{k}^2(\Delta f + 1/2\,\lambda^2 C_s \underset{\sim}{k}^2) \tag{6}$$

and $|\underset{\sim}{k}| = 2s = 2\sin\theta/\lambda$, C_s is the spherical aberration coefficient of the lens, and Δf is the amount of defocus of the objective lens from the Gaussian or in-focus position (here a negative value of Δf has been chosen to correspond to the underfocus condition produced by a weakened lens, and a positive value to overfocus). The objective aperture function A_k is unity for beams passing through the aperture and zero for those outside.

The intensity in the image plane can be calculated by Fourier transform of Ψ_k to obtain the amplitude $\psi(\underset{\sim}{x})$ followed by squaring of the amplitude:

$$I(\underset{\sim}{x}) = \psi(\underset{\sim}{x}) \cdot \psi^*(\underset{\sim}{x}) \; . \tag{7}$$

Early programs (Lynch and O'Keefe, 1972) used the above procedure to form the image, relying on the objective aperture function to select the correct number of diffracted beams in order to limit the resolution in the simulated image to that of the electron microscope. However, it was soon found that images calculated with an aperture cutoff corresponding to that used experimentally contained more detail (O'Keefe, 1973) than the experimental micrograph.

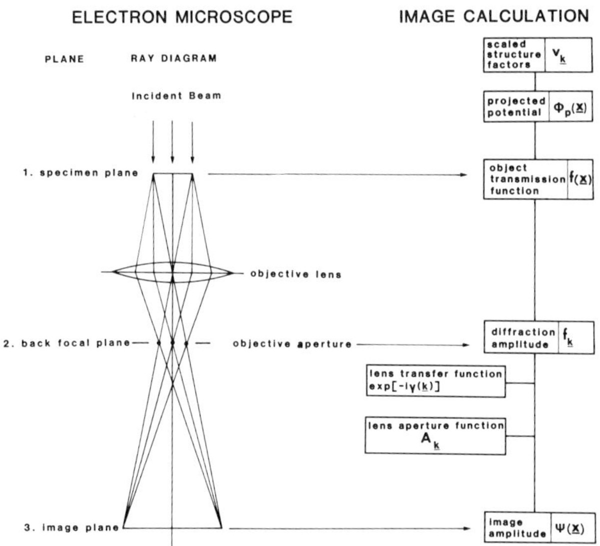

Fig. 1. Model of a simplified electron microscope together with the associated functions for simulating high-resolution images, and showing the three microscope planes at which the electron wavefield is required.

O'Keefe and Sanders (1975) considered two types of "smearing" aberrations which could be responsible for loss of resolution, the convergent character of the incident electron beam and the spread of focus produced by energy spread in the incident electrons. Both these aberrations have the effect of smearing the microscope image by making it a composite of higher-resolution images. Incident beam convergence produces a composite image formed by the summation of many images, each at a different angle within the incident cone. Spread of focus produces a composite formed from images summed over a range of defocus.

O'Keefe and Sanders (1975) measured the experimental value of convergence from a diffraction pattern obtained with the microscope illumination set as for imaging (Fig. 2). Calculation of 49 images at sampling positions within the convergent cone (Fig. 3) produced a set of images (Fig. 4) which could be summed to match the experimental result (Fig. 5). Fejes (1977) included the effect of spread of focus, and also found a degradation in image resolution.

In order to obtain a better understanding of the manner in which resolution is degraded by beam convergence and spread of focus, we can reformulate (7) by taking its Fourier transform to obtain

$$I_{\underset{\sim}{k}} = \Psi_{\underset{\sim}{k}} \star \Psi^*_{-\underset{\sim}{k}} \tag{8}$$

where I_k is the image intensity spectrum, and $\star$ represents the convolution operation. Writing out the convolution gives

$$I_{\underset{\sim}{k}} = \sum_{\underset{\sim}{k}'} \Psi_{\underset{\sim}{k}'} \Psi^*_{\underset{\sim}{k}'-\underset{\sim}{k}} \; . \tag{9}$$

Thus I_k, the image intensity spectrum, the Fourier transform of the image intensity, can be calculated directly from the Ψ_k, the aberrated diffraction-plane wavefield; the image intensity is formed by later Fourier transform of the intensity spectrum.

Figure 6 shows the image intensity spectrum formed by an aberrated diffraction-plane wavefield (or image amplitude spectrum) consisting of five colinear diffracted

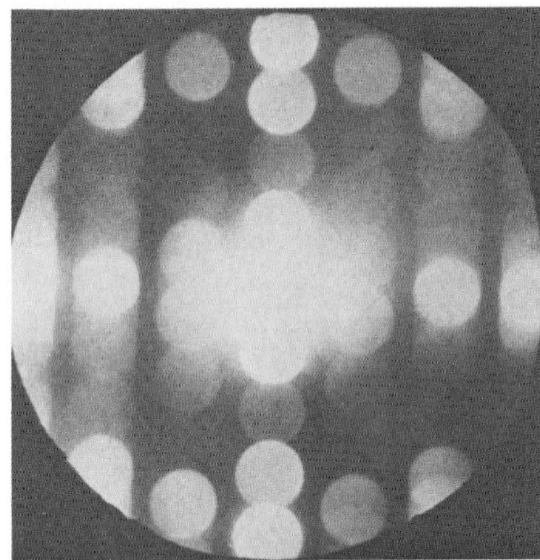

Fig. 2. 001 diffraction pattern of $Nb_{12}O_{29}$ obtained with illumination adjusted for imaging, and showing discs due to electron beam convergence in imaging mode.

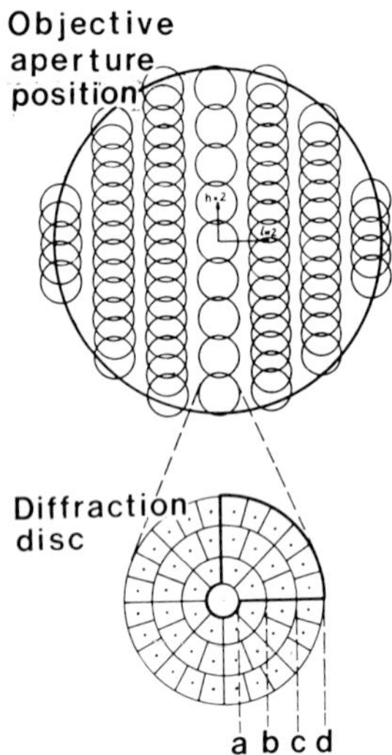

Fig. 3. Method of including beam convergence in the image simulation. Each disc admitted by the objective aperture is sampled at 49 points and images calculated at angles corresponding to each sampling point.

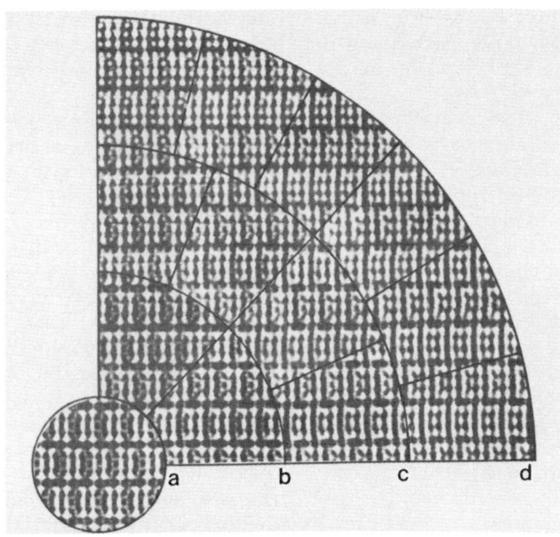

Fig. 4. Tilted-beam images obtained at angles within one quadrant of the convergence cone (the other quadrants are generated from these images by symmetry). The radii (marked a through d) correspond to those so marked in fig. 3. The specimen is 50Å thick $Nb_{12}O_{29}$ imaged at 100keV with an objective aperture size of 0.308 $Å^{-1}$, C_s = 1.8mm, and defocus = -600Å.

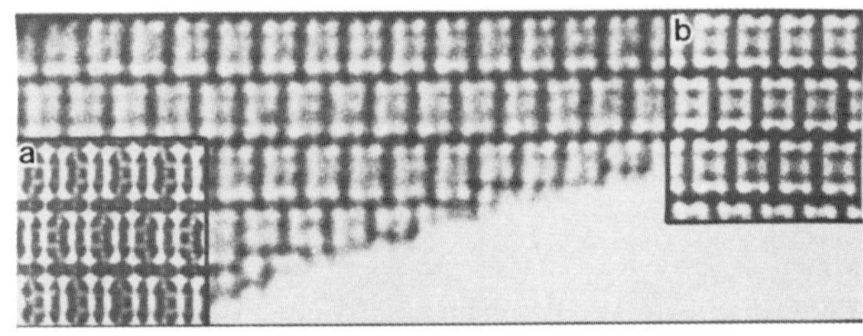

Fig. 5. Experimental image of $Nb_{12}O_{29}$ with inset images simulated without (a) and with (b) inclusion of the convergence effect.

beams. We see that an amplitude spectrum containing terms out to $h = \pm 2$ produces an intensity spectrum with frequencies out to $h = \pm 4$. As shown in the figure, each frequency in the intensity spectrum is made up of a number of terms produced by pairwise multiplication of members of the amplitude spectrum. Terms containing the zeroth order of the amplitude spectrum are called first-order (or linear image) terms (note that they do not occur in frequencies greater than the amplitude spectrum limits of $h = \pm 2$), and those containing no zeroth order are called second-order (or non-linear) terms.

Linear Images

In wavefields (and hence images) from specimens thin enough to be regarded as phase objects, most of the electron intensity resides in the central or zeroth order diffracted beam. In this type of situation the amplitude of the zeroth beam is typically 0.9 of the amplitude of the incident beam, while the stronger diffracted beams may attain values of 10^{-3}. The ratio of the weights of the first order terms to those of the second order is thus approximately one thousand times, meaning that the second-order terms may safely be disregarded in interpreting this type of image.

Linear Image Theory

Dropping second-order terms in (9) we find that only two linear terms remain,

$$^{L}I_{\underset{\sim}{k}} = \Psi_{\underset{\sim}{k}}\Psi_0^* + \Psi_0\Psi_{-\underset{\sim}{k}}^* \tag{10}$$

and this is confirmed by figure 6.

Substituting from (5) and recalling $\Psi_0^* = \Psi_0 \doteq 1$ we get

$$^{L}I_{\underset{\sim}{k}} = f_{\underset{\sim}{k}} \exp\{-i\gamma(\underset{\sim}{k})\} + f_{-\underset{\sim}{k}}^* \exp\{+i\gamma(-\underset{\sim}{k})\}. \tag{11}$$

From (2), the object transmission function for a phase object of thickness H is

$$q(\underset{\sim}{x}) = \exp\{i\sigma\phi_p(\underset{\sim}{x})H\} \ . \tag{12}$$

For a weak phase object (WPO) we can make a kinematic scattering approximation and approximate the transmission function as

$$ {}^{WPO}q(\underset{\sim}{x}) = 1 + i\sigma\phi_p(\underset{\sim}{x})H \qquad (13) $$

so that Fourier transformation gives

$$ {}^{WPO}f_{\underset{\sim}{k}} = \delta(\underset{\sim}{k}) + i\sigma H\, V_{\underset{\sim}{k}} \;. \qquad (14) $$

Substituting for ${}^{WPO}f_k$ in (11) produces the linear image approximation to the image intensity spectrum as,

$$ {}^{L}I_{\underset{\sim}{k}} = \sigma H\, V_{\underset{\sim}{k}} \sin\gamma(\underset{\sim}{k}) + \sigma H\, V^*_{-\underset{\sim}{k}} \sin\gamma(-\underset{\sim}{k}) $$

and since $V^*_{-k} = V_k$, and γ is radially symmetric for a properly-aligned microscope,

$$ {}^{L}I_{\underset{\sim}{k}} = 2\sigma H\, V_{\underset{\sim}{k}} \;\sin\gamma(\underset{\sim}{k}) \qquad \text{for } \underset{\sim}{k} \neq 0 $$

For $k = 0$ the zeroth frequency is approximately $\delta(k)$ so that the full expression is

$$ {}^{L}I_{\underset{\sim}{k}} = \delta(\underset{\sim}{k}) + 2\sigma H\, V_{\underset{\sim}{k}} \sin\gamma(\underset{\sim}{k}) \;. \qquad (15) $$

Thus for a specimen sufficiently thin that (i) the kinematic approximation to electron scattering is adequate, and (ii) the second-order intensity spectrum terms are small, the image intensity may be found by Fourier transformation of (15), or alternatively an experimental micrograph may be interpreted as if the transfer of spatial frequency V_k into the image were controlled by the value of $\sin\gamma$ at that value of k. Since C_s is always positive (for magnetic lenses) and Δf may be chosen to be negative by underfocussing the lens, the shape of the curve of $\sin\gamma(k)$ against k can be controlled to some extent by choosing particular values of underfocus.

Scherzer (1949) searched for the optimum linear image and found that it occurs when a value of defocus near to $-\sqrt{1.5C_s\lambda}$ is chosen. In this case, $\sin\gamma(k)$ is approximately equal to minus one over a band of frequencies extending to

$$ 1.5\, C_s^{-1/4}\lambda^{-3/4} \;. $$

Thus for frequencies within this band

$$ {}^{L}I_{\underset{\sim}{k}} = \delta(\underset{\sim}{k}) - 2\sigma H\, V_{\underset{\sim}{k}} \qquad (16) $$

and the image intensity is

$$ {}^{L}I(\underset{\sim}{x}) = I - 2\sigma H\, {}^{L}\phi(\underset{\sim}{x}) \qquad (17) $$

where ${}^{L}\phi(\underset{\sim}{x})$ is a projected potential with resolution limited to the highest frequency V_k falling within the passband. This type of image is extremely important, because it shows directly the projected potential of the specimen, albeit to a limited resolution, with low intensity (black) in regions of high potential (high concentrations of atoms) and high intensity (white) in low potential regions (tunnels). This type of image is often called a structure image, and the particular value of underfocus is referred to as "Scherzer" defocus. Figure 7 shows a $\sin\gamma(k)$ curve, or linear image "contrast transfer function" for a value of defocus close to Scherzer focus; the passband and cutoff frequency are marked. The cutoff frequency defines the structure image resolution, or Scherzer resolution, for any particular electron microscope.

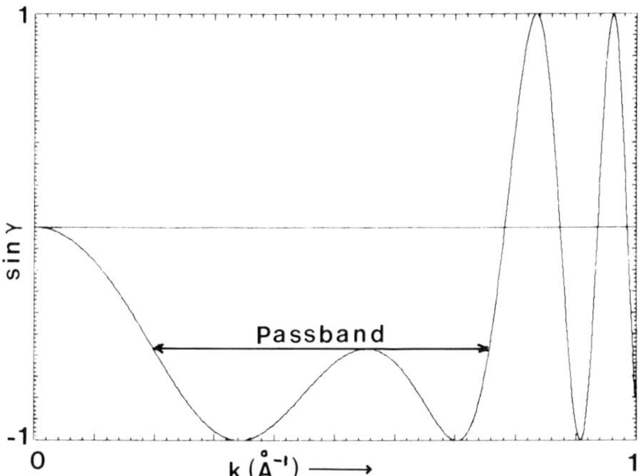

Fig. 7. Linear image contrast transfer function (plot of $\sin\gamma(\underset{\sim}{k})$ against $|\underset{\sim}{k}|$) for an objective lens close to Scherzer defocus. Spatial frequencies are transferred with weights proportional to $\sin\gamma$. Frequencies within the passband (marked) are considered to contribute without significant attenuation. The cutoff frequency that determines resolution in structure images lies at the upper limit of the passband.

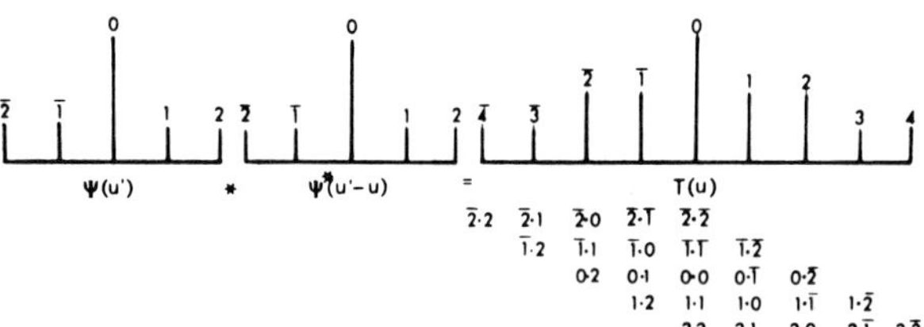

Fig. 6. Representation of the nine spatial frequencies generated in the image intensity spectrum by an amplitude spectrum consisting of five collinear diffracted beams. The terms contributing to each frequency are tabulated in abbreviated form--the term n.m represents multiplication of the pair $\Psi_n.\Psi^*_m$ from the amplitude spectrum.

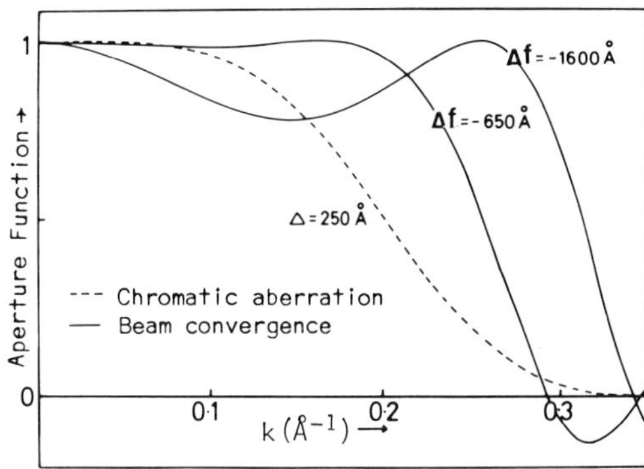

Fig. 8. Plots of the linear damping functions (aperture functions) due to a spread of focus with a gaussian half-width of 250Å (dashed) and a beam convergence with a semi-angle of 1.4 milliradian (solid line) for a 100keV electron microscope with $C_s = 1.8$mm.

Linear Image Resolution

Frank (1973) re-formulated the resolution-limiting effects of both beam convergence and spread of focus into expressions in reciprocal space. He showed that the effect on linear images is to impose envelope functions or soft apertures on the linear image amplitude, limiting transference of the higher spatial frequencies. Aperture functions for a Gaussian spread of focus of half-width 250Å , and for a beam convergence semi-angle of 1.4 milliradian are shown in figure 8. The aperture function due to spread of focus (or chromatic aberration) slopes gradually to zero, transferring higher frequencies only partially. The convergence aperture function is steeper, and its shape and cutoff frequency change with changes in defocus; at higher degrees of underfocus the cutoff frequency is higher, allowing more higher frequencies to contribute to the image, but mid-range frequencies are damped.

Whereas an accurate value of the beam convergence α can be measured directly from a diffraction pattern obtained with focussed illumination (fig. 2), the spread of focus Δ for a particular electron microscope must either be estimated by matching an experimental image with ones computed using different values of Δ, or approximated from known values of the chromatic aberration coefficient C_c and the high voltage and lens current ripple. A good approximation to Δ is obtained from

$$\Delta = C_c\sqrt{(\delta V/V)^2 + 4(\delta I/I)^2 + (\delta E/E)^2} \qquad (18)$$

where $\delta V/V$ and $\delta I/I$ are the high voltage and lens current ripple (usually quoted in ppm) respectively, and $\delta E/E$ is the fractional energy spread of the electron beam.

Figure 9 shows experimental images of a thin crystal of the block oxide $Nb_{12}O_{29}$ obtained on two different electron microscopes. Below each micrograph is the linear image envelope function corresponding to that microscope, and simulated images are inset. Notice that the factor limiting the resolution of the 100keV microscope is the large convergence factor (semi-angle 1.4 milliradian), while the 1MeV microscope is limited by the envelope due to a spread of focus of 500Å. The value of 500Å was obtained by application of equation 18 to a chromatic aberration coefficient of 4.4mm, and high voltage and lens current ripples of 5ppm; the energy spread was taken as 1eV, producing a Δ value of $\Delta = 44\sqrt{25 + 100 + 1} = 494$Å. As it happened, $Nb_{12}O_{29}$ images computed for this 1MeV microscope are not very sensitive to moderate changes in the value of Δ. Images computed for a range of Δ values and compared with experiment show that the correct value of Δ lies between 400Å and 600Å (Fig. 10).

It appears quite common that high voltage electron microscopes have linear-image resolutions limited by the spread of focus effect, presumably because of the difficulty of obtaining low values of ripple. Conversely, lower voltage (100keV) microscopes are usually limited by the effects of the relatively large convergence angles used to obtain sufficient brightness on the fluorescent screen. Of course, spread of focus becomes a significant resolution-limiting factor even in lower-voltage microscopes if Δ is made immoderately large by increasing the energy spread of the electron beam leaving the thermionic filament. Whereas at 1MeV a value of δE of 1eV produces an insignificant contribution to Δ, such a value at 100keV represents 10ppm and contributes a larger proportion than the voltage and current ripple. Krivanek (1975) demonstrated that a high beam current of 30μA (corresponding to an energy spread of 5eV half-width) reduced the linear image resolution of a 125keV electron microscope to worse than 3.4Å.

Non-Linear Images

Images obtained from thicker regions of specimens, as well as dark-field images, cannot be described in terms of linear contrast transfer functions, and hence the concept of the envelope function does not apply.

Non-Linear Resolution

O'Keefe (1979) extended the concept of the linear-image envelope function to include the non-linear terms, and showed that neglect of the "cross-terms" present in the general non-linear (but not the linear) damping functions could lead to the overdamping and loss of higher frequencies in the image intensity spectrum. Figure 11(a) shows how the full damping function due to beam convergence acts on the terms contributing to each frequency in the intensity spectrum, while (b) shows the effect on these terms if linear damping envelope functions are applied. The degree of damping of any intensity spectrum term is read off the plot by considering the two members of the amplitude spectrum making up the term (fig. 6), and reading up or across from the origin of the damping function (located at the center of the plot). The 0.0 term (contributing to the 0 frequency of the intensity spectrum) is located at the origin of the plot and is undamped (white). Similarly, all the n.n terms (falling along the bottom-left to top-right diagonal) are also undamped in the general case (a), but are lost if linear damping is applied (b). Note that those terms which contribute to the linear image (and lie along horizontal and vertical axes through the origin of each plot) are damped equally by both the general and linear damping functions (as indicated by the inset cross-section in (b) showing the familiar linear envelope profile of fig. 8).

A plot of the difference between the general and linear damping functions for convergence (fig. 11c) shows that the main effect of using the linear form to include the effects of beam convergence and spread of focus in a simulated image is to overdamp the n.n terms contributing to the zero frequency. This has the effect of lowering the mean image intensity level, occasionally producing negative values of intensity, but not otherwise seriously affecting the appearance of the simulated image (negative intensities arise because the unphysical nature of the linear damping function violates the principle of particle conservation, as was pointed out by Rose, 1977).

The image changes occasioned by using linear damping for the spread of focus effect are much more serious. Figure 12(a) shows that, as in the convergence case, the general damping function for spread of focus also does not damp the n.n terms. In addition, however, terms of the form $\bar{n}$.n are not damped by the general spread of focus function. Thus the intensity spectrum will always contain frequencies out to twice the order of the highest members of the amplitude spectrum (unless these happen to be damped by a sufficiently large convergence).

We can apply the spread of focus function of figure 12(a) to the intensity spectrum terms listed in figure 6. Consider the $\bar{2}$ frequency in the intensity spectrum; its contributing terms are the linear $\bar{2}$.0 and 0.2 interactions, together with the non-linear $\bar{1}$.1 term. For a sufficiently large value of Δ, the linear terms would be heavily damped, but the non-linear one would not (as marked in figure 12a) so that this frequency would be present in the image. Using the linear damping function (figure 12b) the non-linear term is damped and the frequency will be missing from the image.

Previously we discussed the structure image (or Scherzer) resolution limit determined by the cutoff frequency at which the dominant aperture function blocks linear transfer into the intensity spectrum (as in figure 8).

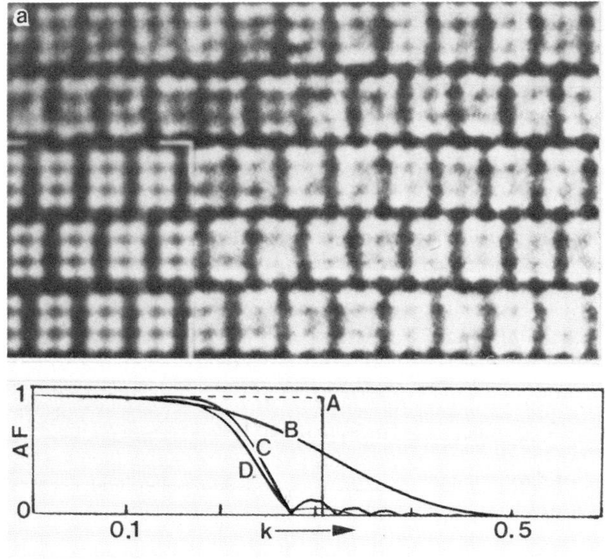

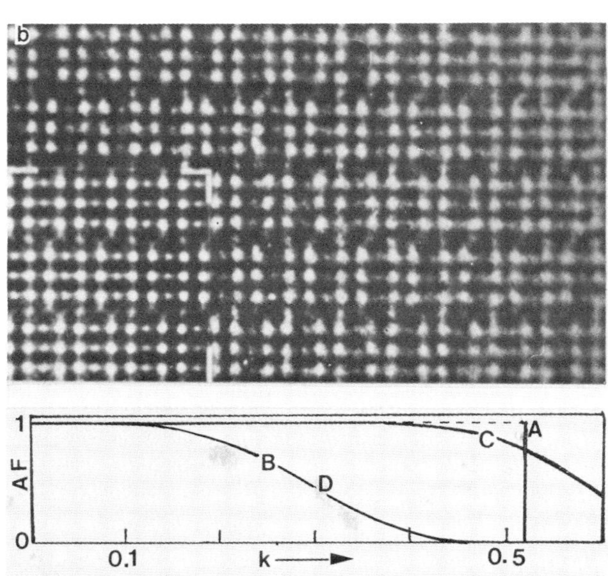

Fig. 9. Structure images of $Nb_{12}O_{29}$ taken at (a) 100kV and (b) 1MV. The inserts at lower left of each micrograph are calculated images for a 38Å thick crystal of $Nb_{12}O_{29}$. The aperture functions (AF, plotted below each image as a function of $\underset{\sim}{k}$ in reciprocal Å) show the resolution conditions under which each calculation was carried out. At 100kV the physical aperture (A) at $\underset{\sim}{k}$ = 0.308Å limits resolution to 3.2Å: the aperture function due to a defocus halfwidth of 100Å (B) limits resolution to 2.4Å; while the aperture function due to an incident beam convergence of 1.4 milliradian (C) restricts it to 3.8Å. The combined effect of these functions (D) results in an image of 3.8Å resolution. At 1MV the physical aperture (A) and convergence aperture function (C) limit resolutions to 1.9Å and 1.5Å respectively. For the calculated image to match the experimental result required a defocus-depth halfwidth of 500Å, resulting in the B curve shown. The combined effect (D) is virtually identical to B and yields an image of 2.5Å resolution.

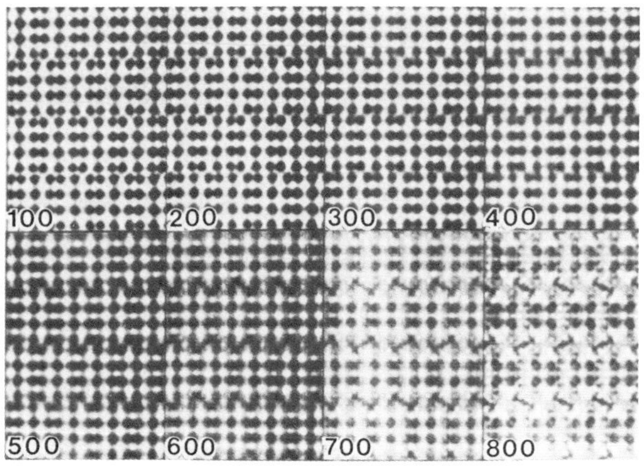

Fig. 10. Series of $Nb_{12}O_{29}$ images simulated for increasing values of Δ, the spread of focus halfwidth (marked in Å). Image resolution is degraded as Δ increases.

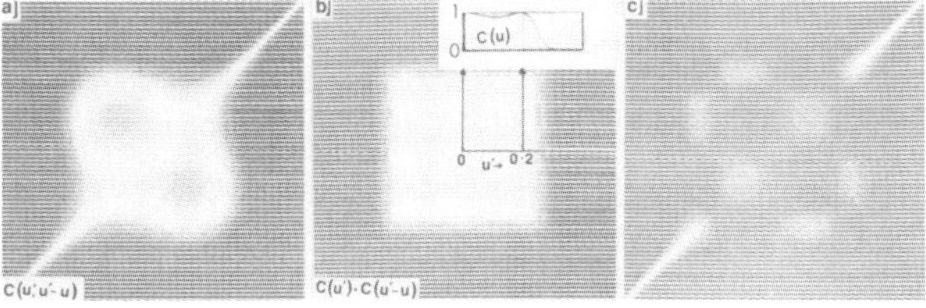

Fig. 11. Intensity spectrum damping functions due to beam convergence: (a) general (linear plus non linear) convergence damping function; (b) linear damping function with inset trace along a linear-image-term direction; (c) the inaccuracy in using linear damping (plot (a) minus plot (b)).

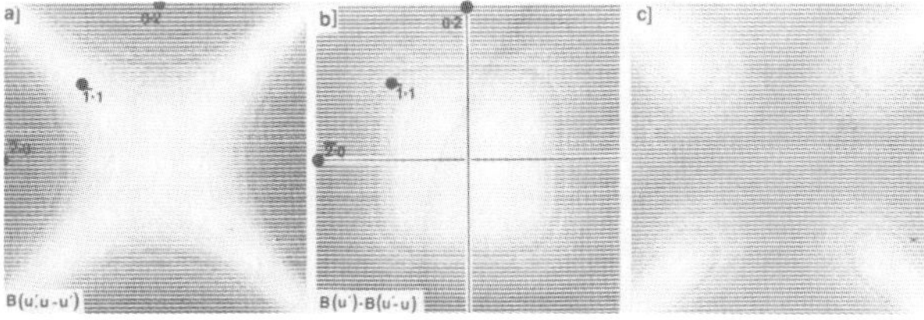

Fig. 12. Intensity spectrum damping functions for spread of focus: (a) general damping function; (b) linear damping function; (c) the difference. The inset spots show interferences contributing to the 2nd order frequency of the intensity spectrum of fig. 6.

Non-linear transfer into the intensity spectrum gives rise to yet another resolution limit. This non-linear resolution limit is sometimes called the "fringe resolution" of the microscope, and is given by the highest frequency transferred into the intensity spectrum. The general damping functions (figs. 11, 12) show that the highest order term transferred will be of the form $\bar{n}.n$ since this type of interference term is undamped by the spread of focus function (fig. 12) and passed by the convergence function (fig. 11) at a maximum value determined only by the value of focus; if the value of focus is chosen to allow the nth order member of the amplitude spectrum to contribute to the image via a linear interference term of

0.n, then the non-linear interference $\bar{n}.n$ will also be passed to appear in the image, even if the linear term 0.n is blocked by the spread of focus function. The fringe resolution of an electron microscope is thus twice as good as the limit set by its linear convergence aperture function (provided, of course, that the microscope is vibration-free and is aligned sufficiently accurately that the $\bar{n}.n$ interference falls within the narrow white arm of the spread of focus function, that is that the $\bar{n}$ and n diffracted beams form equal angles with the optic axis of the microscope).

Real-Space Second-Order Effects

Whereas images from thin crystals show detail to the linear-image resolution of the electron microscope, images from thicker crystals contain linearly-transferred detail to the same resolution plus non-linear detail to the non-linear resolution limit. In structure images the phases of the linear contribution are chosen to be negative (by selecting Scherzer focus) producing black spots at positions corresponding to areas of high potential (atoms or unresolved groups of atoms). In general, strong second-order contributions tend to have positive phase and produce white peaks at positions of high potential. These white peaks are sharper (coming from the high-frequency region of the intensity spectrum) than the linear black peaks, and produce the characteristic 'black donut with white hole' image in thicker structure images. Pirouz (1981) has shown that inclusion of the second-order terms is equivalent to adding a term proportional to the square of the projected crystal potential, and yet another proportional to the projected charge density, to the thin-crystal projected potential image; a negative-going (black) peak in the linear image (due to an atom or group of unresolved atoms) is thus squared to add a sharper positive-going (white) peak into the total (linear plus non-linear) image.

This effect can be seen in both experimental and computed images. Typically an image of a wedge crystal will show periodic black spots at the thin edge and the same black spots with white centers in thicker regions. Similarly, white spots appear in computed images of organic crystals as crystal thickness increases (O'Keefe et al., 1983). A pair of images of an organic crystal computed using the general and linear damping functions is shown in figure 13 together with their difference (formed by subtracting the images using the SEMPER programs of Saxton, 1979). Because the crystal is thin (19Å) the two images appear identical, but subtraction reveals a 3% difference in contrast. This difference appears at both the positions of minima (blacks) and maxima (whites) in the image, since the non-linear contribution arises from a squaring of the linear image contrast (Cowley, 1975). Although the difference (the non-linear contribution) has only 3% of image contrast for a crystal 19Å thick, this figure rises to 20% at 56Å thick. At a thickness of 94Å the contrast difference is 30% and differences in the images are easily visible (fig. 14).

In some structures non-linear contributions can produce images which appear like high-resolution structure images (Smith and O'Keefe, 1983); the "dumbbell" images of CdTe and Si viewed in [110] orientation are good examples of images dominated by non-linear contributions. Fig. 15 shows a series of "weak-phase-object images" of β-SiC in [110] orientation for increasing resolution. These images are the ones expected from very thin crystals imaged at Scherzer defocus; they show that each silicon/carbon atom pair appears as an unresolved black spot at 2.17Å resolution (or alternatively, the tunnels in the structure appear white). As resolution is improved the black spots elongate at 1.54Å resolution then split at 1.09Å resolution. Images computed for the Cambridge University HREM show dumbbell images with this splitting at crystal thicknesses of 150Å, although the linear-image resolution of the microscope is certainly worse than 1.09Å. Smith and O'Keefe (1983) show how the dumbbell images arise from a combination of non-linear and linear terms producing white spots at the atom positions over a limited range of crystal thickness and (non-Scherzer) microscope defocus. Interestingly, images from thick crystals at Scherzer defocus produce a type of dumbbell image where one white spot falls on an atom position, while the other is centered on a tunnel site.

It is considered generally well-known that the use of linear damping functions in image simulations is equivalent to an assumption that the electron beam is perfectly coherent. Rose (private communication - 1984) points out that the entire theory of STEM image formation is based on it--nevertheless it bears repeating to avoid misconceptions (e.g. O'Keefe and Saxton, 1983). Perfect spatial coherence (i.e. a condenser aperture filled with coherent electron waves) is possible in the case of an electron microscope equipped with a field emission gun rather than the more common thermionic kind. In this case the general form for the intensity spectrum (fig. 16, equation 1) may be replaced by one using a linear convergence function (fig. 16, equation 2). However, perfect temporal coherence appears impossible to achieve, so that a linear spread of focus form (fig. 16, equation 3) should not be used for image simulation (except within the unphysical weak phase object domain). Images of β-SiC calculated with linear damping terms (fig. 16, equation 3) do not produce the well-known split white dumbbells at a crystal thickness of 150Å, whereas images calculated with the general form do (fig. 17).

Dark-Field Images

The general form of the intensity spectrum can be used to simulate images in both bright-field and dark-field modes. In dark-field calculations all contributions to the intensity spectrum come from non-linear terms, and incorrect over-damping of these terms must be avoided by using the correct general damping functions. In the case of small-aperture dark field imaging, a large proportion of the contributing beams will have convergence cones that intersect the objective aperture and the simulation program must be capable of modelling this situation. Figure 18 shows diffraction patterns from a crystal of $4Nb_2O_5.9WO_3$ in [001] projection. When a small aperture is centered on the 130 reflection, the defocussed pattern (a) shows 9 diffracted beams passing through the aperture, whereas under imaging conditions portions of 19 discs lie inside the aperture (b). Taking this effect into account Iijima and O'Keefe (1977) successfully matched dark-field images (fig. 19a,b). Centering the aperture on $\bar{1}30$ produced the expected switch in the positions of bright spots in the image (fig. 19c).

Misalignment and Crystal Tilt

One problem that arises in high resolution electron microscopy is accurate alignment of the electron microscope and of the specimen. While determining the structure of takeuchiite by HRTEM, Bovin et al., (1981), found that their best electron micrograph showed lower symmetry than images simulated from proposed models. Examination of the diffraction pattern revealed that the specimen was tilted 23 milliradian off the zone axis so that the center of the Laue circle was at 770. Figure 20 shows the experimental image with inset simulations for crystal thicknesses of 60 Å and 150Å. Note that the lower symmetry is more obvious in the simulation for the thicker crystal.

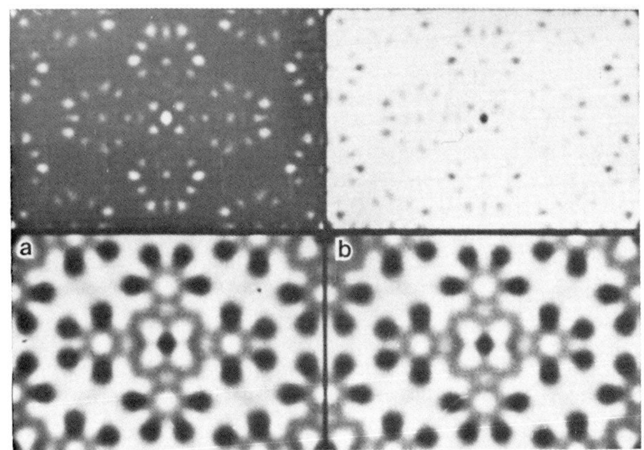

Fig. 13. Images of a thin (19Å) crystal of copper hexadecachlorophthalocyanine using (a) general and (b) linear damping functions. Simulation conditions are as published for the Kyoto HAREM (e.g. Kirkland, 1982). Difference images are shown above; difference contrast is 3% of image contrast.

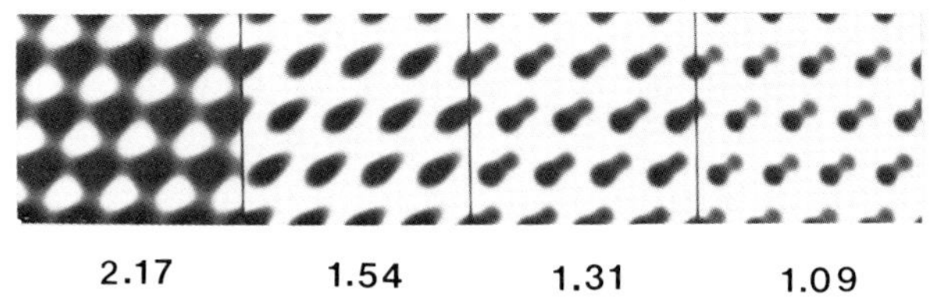

Fig. 15. Resolution series of weak-phase-object images of β-SiC in [110] projection. Resolution (Å) is marked.

Fig. 16. Image intensity spectrum equations for (1) general damping functions, (2) linear convergence and general spread of focus, (3) linear convergence and spread of focus. Here the damping functions due to lack of lateral and longitudinal coherence (i.e. convergence and spread of focus) are designated by $\underset{\sim}{B}$ and E.

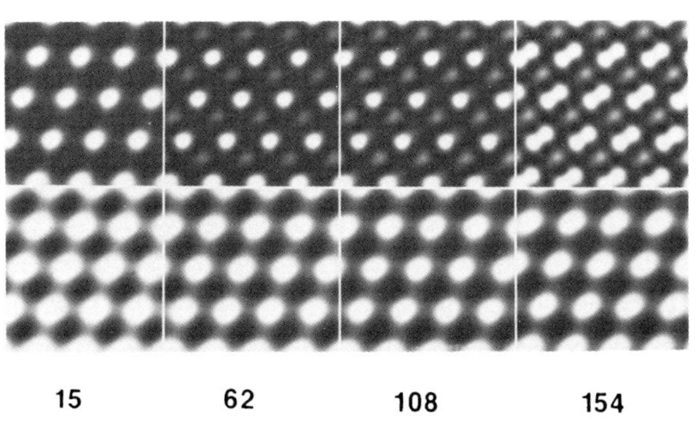

Fig. 17. Images of β-SiC [110] calculated for typical operating conditions (500kV, C_s = 3.5mm, Δf = -1100Å underfocus, convergence half-angle 0.3mrad, rms focus spread 250Å) for crystal thickness (from left) 15, 62, 108 and 154Å ; upper and lower rows calculated using equations (1) and (3) of fig. 16 respectively, and scaled individually to maximum contrast. Note the contrast reversal from the WPO images in fig. 15, due to the non-Scherzer value of -1100Å defocus.

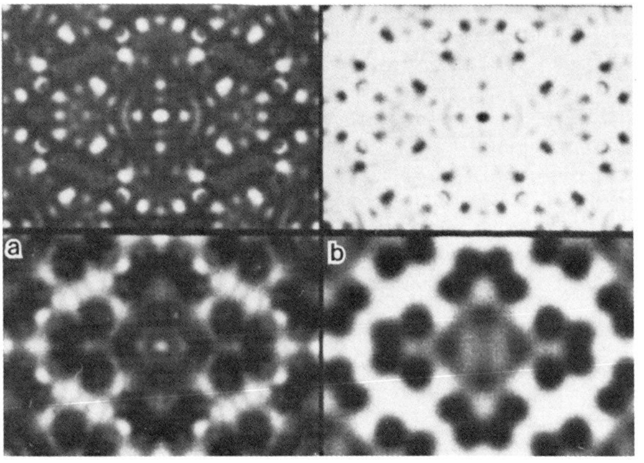

Fig. 14. Images of thick (94Å) copper hexadecachlorophthalocyanine computed with (a) general and (b) linear damping functions. Simulation conditions are for the Cambridge HREM (e.g. Smith and O'Keefe, 1983). Difference images (above) have 30% of image contrast.

(16)

$$I_{\underline{k}} = \sum_{\underline{k}'} f_{\underline{k}'} f^{*}_{\underline{k}'-\underline{k}} \exp\{-i[\gamma(\underline{k}') - \gamma(\underline{k}'-\underline{k})]\}$$
$$\cdot \underline{B}\{[\underline{\nabla}\gamma(\underline{k}') - \underline{\nabla}\gamma(\underline{k}'-\underline{k})]/2\pi\}\ E\{\tfrac{1}{2}(k'^2 - |\underline{k}'-\underline{k}|^2)\} \quad (1)$$

$$I_{\underline{k}} = \sum_{\underline{k}'} f_{\underline{k}'} \exp\{-i\gamma(\underline{k}')\}\ \underline{B}\{\underline{\nabla}\gamma(\underline{k}')/2\pi\}$$
$$\cdot f^{*}_{\underline{k}'-\underline{k}} \exp\{+\gamma(\underline{k}'-\underline{k})\}\ \underline{B}\{\underline{\nabla}\gamma(\underline{k}'-\underline{k})/2\pi\}\ E\{\tfrac{1}{2}(\underline{k}'^2 - |\underline{k}'-\underline{k}|^2)\} \quad (2)$$

$$I_{\underline{k}} = \sum_{\underline{k}'} f_{\underline{k}'} \exp\{-i\gamma(\underline{k}')\}\ \underline{B}\{\underline{\nabla}\gamma(\underline{k}')/2\pi\}\ E\{\tfrac{1}{2}\underline{k}'^2\}$$
$$\cdot f^{*}_{\underline{k}'-\underline{k}} \exp\{+i\gamma(\underline{k}'-\underline{k})\}\ \underline{B}\{\underline{\nabla}\gamma(\underline{k}'-\underline{k})/2\pi\}\ E\{\tfrac{1}{2}|\underline{k}'-\underline{k}|^2\} \quad (3)$$

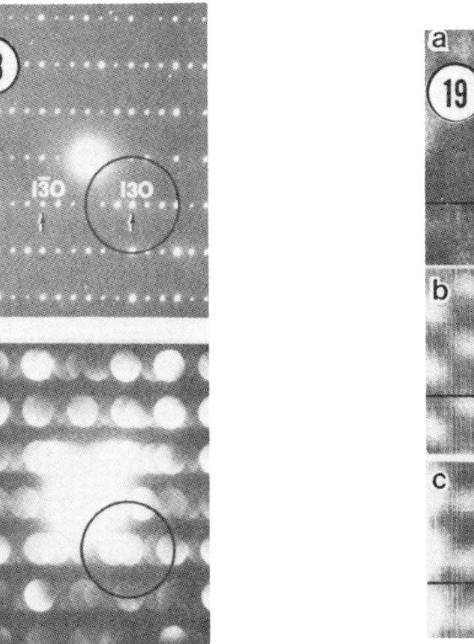

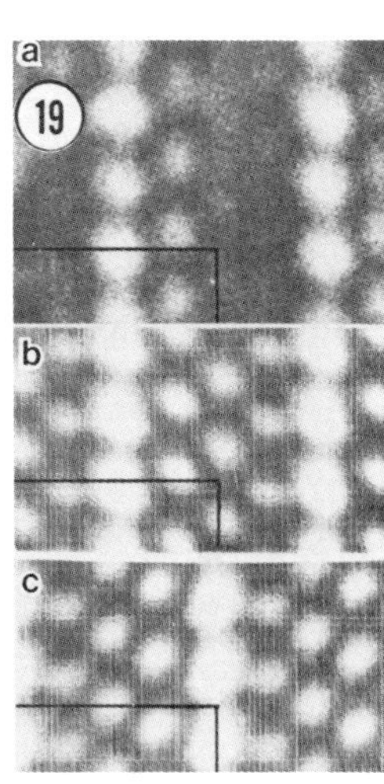

Fig. 18. Diffraction patterns from [001] $4Nb_2O_5.9WO_3$ for parallel (a) and focussed (b) illumination. The objective aperture size and position are shown. Note the many diffraction cones intersecting the aperture in (b).

Fig. 19. Dark-field images of $4Nb_2O_5.9WO_3$ for the conditions of fig. 18. (a) experimental; (b) computed with the aperture centered on 130; (c) computed with the aperture at $\bar{1}30$.

Because the effects of crystal tilts and electron beam misalignment are easily incorporated in simulation programs, the programs are useful in exploring the results of tilts and misalignments on high resolution images. Computations by Smith et al., (1983), revealed that misalignments (the effect of having the electron beam at a slight angle to the optic axis of the objective lens) as small as one half milliradian could change the symmetry in thin (26Å) crystal images (fig. 21), whereas crystal tilts of up to 32 milliradian had little effect on such images. As the crystal thickness was increased the effect of crystal tilt increased; for a crystal of thickness 76Å, significant changes occurred for tilts of 8 milliradian (fig. 22). Thus experimental images in which symmetry is lower merely in the thicker regions (as in fig. 20) indicate specimen tilt (or buckling), while an extension of lower symmetry to the thin edge probably calls for re-alignment of the microscope.

Calculation of Images of Defects

While high-resolution image simulation theory is usually presented for perfect crystals, it is possible to compute images of periodic and non-periodic faults. In programs using a multislice algorithm the method used is to create a large "supercell" with perfect crystal surrounding the fault. Such a calculation was made by O'Keefe and Iijima (1978) for a tetragonal tungsten bronze element embedded in a matrix of ten by ten unit cells of WO_3 (fig. 23). The simulated images matched an experimental focal series of micrographs of such a defect (fig. 24). This "periodic extension" method can be used to extend the multislice calculation to include not only faults, but also crystal edges (e.g. Marks and Smith, 1983).

In simulating images of defects by the periodic extension method outlined above, we are calculating the diffuse scattering from the defect only at certain points in reciprocal space--in effect we are "sampling" what should be a continuous distribution of diffuse scattering at a finite number of points on a grid whose interval is the reciprocal of the real-space supercell. In the calculation of O'Keefe and Iijima (1978), the choice of a ten by ten supercell produced a sampling of the diffuse scattering on a grid ten times finer than the reciprocal lattice containing the Bragg beams from the perfect WO_3 crystal. A finer sampling could have been produced by using a larger supercell, but since calculation size is always restricted (in this case to an array of 128 x 128 points) the calculation would not have extended out far enough in reciprocal space. The trade-off in sampling fineness and extent in reciprocal space restricts 128 x 128 calculations to supercell sizes of less than 40Å and ideally less than 25Å. Larger programs using arrays of 256 x 256 points (Krakow, 1980) and 512 x 512 points could safely handle defect cells to areas of 50Å by 50Å and 100Å by 100Å respectively. These larger areas will enable the extent of the lattice relaxation around the defect to be broadened.

Conclusion

Besides the obvious use of simulated images in testing models of specimen structure, simulation programs are also useful in analysing the effects of microscope parameters like spread-of-focus, in determining the importance of maladjustments such as misalignment and specimen tilt, and in setting limits on how far approximations like the "structure image" concept may be trusted. Modified programs can be used to explore unconventional imaging modes like dark-field hollow-cone (DFHC) and BFHC modes (e.g. O'Keefe and Pitt, 1980).

With the advent of cheaper, larger memory, 32-bit, "super-mini" computers, it should be possible to run large (perhaps 512 x 512) simulation programs on-line in interactive mode. With the addition of an associated array processor, such calculations should produce an image only a few minutes after a new structural model is read in. The image would, of course, be displayed on a video monitor in "split screen" mode for comparison with a digitized experimental micrograph.

Acknowledgement

The author is grateful for helpful comments from F. Lenz and J. Hren and equally helpful discussions with H. Rose. This work is supported by the Director, Office of Energy Research, Office of Basic Energy Sciences, Materials Sciences Division of the U. S. Department of Energy under Contract No. DE-AC03-76SF00098.

References

Allpress JG, Hewat EA, Moodie AF, Sanders JV. (1972). n-beam lattice images. I. Experimental and computed images from $W_4Nb_{26}O_{77}$. Acta Cryst. A28, 528-536.

Anstis GR, Lynch DF, Moodie AF, O'Keefe MA. (1973). n-beam lattice images. III. Upper limits of ionicity in $W_4Nb_{26}O_{77}$. Acta Cryst. A29, 138-147.

Bovin J-O, O'Keeffe M, O'Keefe MA. (1981). Electron microscopy of oxyborates. III. On the structure of takeuchiite. Acta Cryst. A37, 42-46.

Cooley JW, Tukey JW. (1965) An algorithm for the machine calculation of complex Fourier series. Math. Comput. 19, 297-301.

Cowley JM. (1975). Diffraction Physics, North Holland/ American Elsevier, New York, 286-287.

Fejes P. (1977). Approximations for the calculation of high-resolution electron-microscope images of thin films. Acta Cryst. A33, 109-113.

Frank J. (1973) The envelope of electron microscope transfer functions for partially coherent illumination. Optik 38, 519-536.

Goodman P, Moodie AF. (1974) Numerical evaluation of N-beam wave functions in electron scattering by the multi-slice method. Acta Cryst. A30, 280-290.

Iijima S, O'Keefe MA. (1977). Interpretation of dark field images of crystals, in: 35th Ann. Proc. Electron Microscopy Soc. Amer., G.W. Bailey (ed), Claitor's, Baton Rouge, 190-191.

Ishizuka K, Uyeda N. (1977). A new theoretical and practical approach to the multislice method. Acta Cryst. A33, 740-749.

Jap BK, Glaser RM. (1980). The scattering of high-energy electrons. II. Quantitative validity of the single-scattering approximations for organic crystals. Acta Cryst. A36, 57-67.

Kilaas R, Gronsky R. (1983). Real space image simulation in high resolution electron microscopy. Ultramicroscopy 11, 289-298.

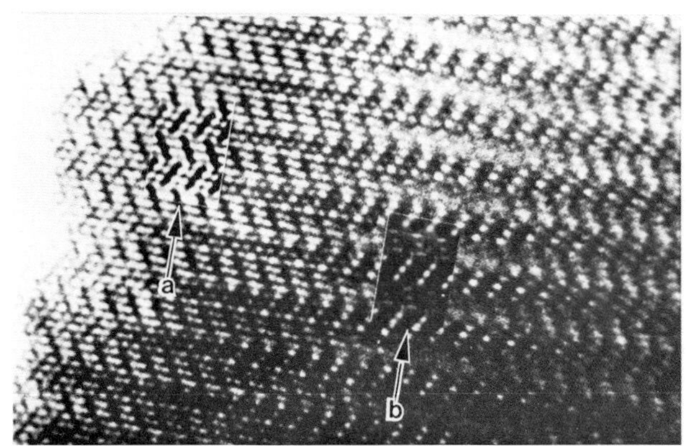

Fig. 20. Experimental image of takeuchiite with inset images computed for the experimentally-determined crystal tilt of 23 milliradians at crystal thicknesses of 60Å (a) and 150Å (b). Note the lower symmetry in the thicker simulated image.

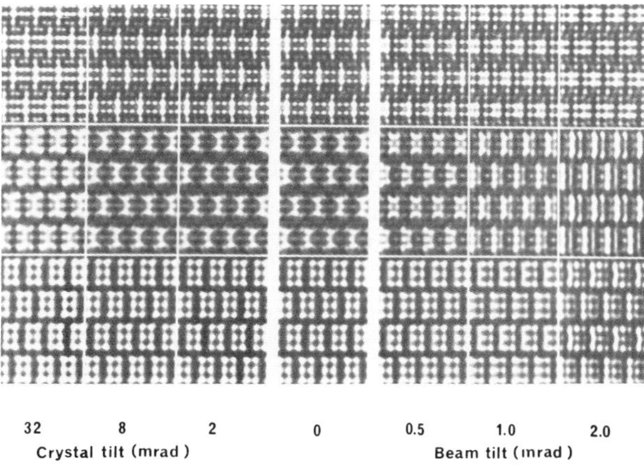

32 8 2 0 0.5 1.0 2.0
Crystal tilt (mrad) Beam tilt (mrad)

Fig. 21. Simulated images of thin (26Å) $Nb_{12}O_{29}$ for increasing degrees of beam and crystal tilt (marked) around the axis lying vertical on the page. Simulated at 100keV for C_s = 1.8mm, convergence semi-angle α_0 = 1.4mrad, Δ = 140Å, and defocus values of -1527Å (upper), -1000Å (center) and -625Å (lower). Small misalignments produce changes in image symmetry. Crystal tilt effects are not as severe.

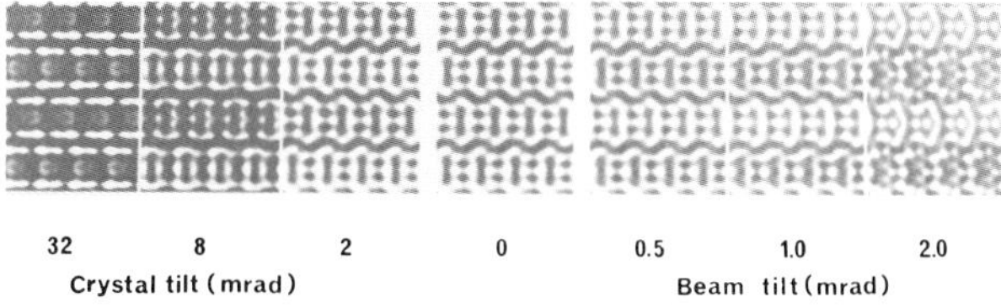

32 8 2 0 0.5 1.0 2.0
Crystal tilt (mrad) Beam tilt (mrad)

Fig. 22. Simulated images of thicker (76Å) $Nb_{12}O_{29}$ for increasing beam and crystal tilt (marked). Conditions as for fig. 21, with a defocus of -600Å. For this thicker crystal, tilt effects are stronger.

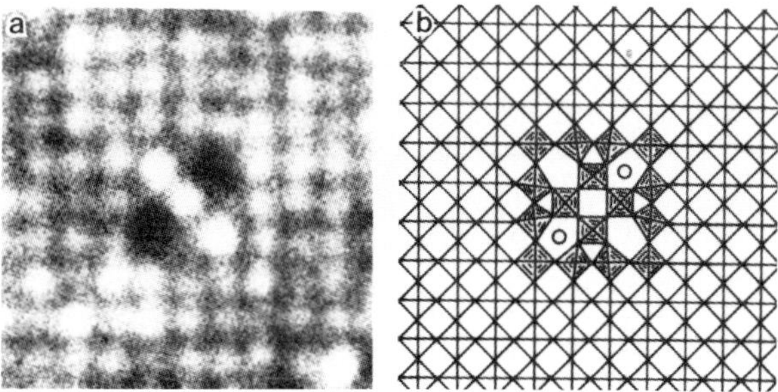

Fig. 23. (a) Structure image of a TTB element in a WO_3 matrix, and (b) the model derived from it.

Experimental

-450 -675 -900 -1350

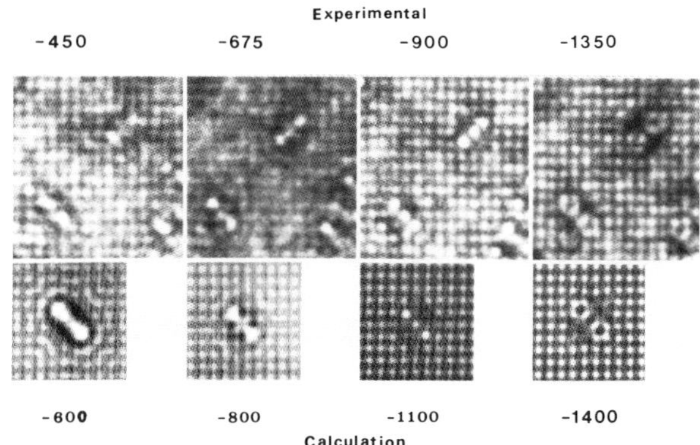

-600 -800 -1100 -1400

Calculation

Fig. 24. Focal series of experimental TTB images with defocus values marked above in Å, together with best-match simulations from the model of fig. 23 computed for values of defocus marked below.

Kirkland EJ. (1982). Nonlinear high resolution image processing of conventional transmission electron micrographs. Ultramicroscopy 9, 45-74.

Krakow W. (1980). Computer modeling of high resolution electron microscope images using 65,536 beams, in: 38th Ann. Proc. Electron Microscopy Soc. Amer., G.W.Bailey (ed), Claitor's, Baton Rouge, 178-179.

Krivanek OL. (1975). The influence of beam intensity on the electron microscope contrast transfer function. Optik 43, 361-372.

Lynch DF, Moodie AF, O'Keefe, MA. (1975). n-beam lattice images. V. The use of the charge-density approximation in the interpretation of lattice images. Acta Cryst. A31, 300-307.

Lynch DF, O'Keefe MA. (1972). n-beam lattice images. II. Methods of calculation. Acta Cryst. A28, 536-548.

Marks LD, Smith DJ. (1983) Direct surface imaging in small metal particles. Nature 303, 316-317.

O'Keefe MA. (1973). n-beam lattice images. IV. Computed two-dimensional images. Acta Cryst. A29, 389-401.

O'Keefe MA. (1979). Resolution-damping functions in non-linear images, in: 37th Ann. Proc. Electron Microscopy Soc. Amer., G.W. Bailey (ed), Claitor's, Baton Rouge, 556-557.

O'Keefe MA, Buseck PR, Iijima S. (1978). Computed crystal structure images for high resolution electron microscopy. Nature 274, 322-324.

O'Keefe MA, Fryer JR, Smith DJ. (1983). High-resolution electron microscopy of molecular crystals. II. Image simulation. Acta Cryst. A39, 838-847.

O'Keefe MA, Iijima S. (1978). Calculation of structure images of crystalline defects, in: Electron Microscopy 1978, J.M. Sturgers (ed), Microscopical Society of Canada, Toronto, 282-283.

O'Keefe MA, Pitt AJ. (1980). The WPO image of $Ti_2Nb_{10}O_{29}$ as a measure of resolution, in: Electron Microscopy 1980, vol. 1, P. Brederos and G. Boom (eds), Seventh European Congress on Electron Microscopy Foundation, Leiden, 122-123.

O'Keefe MA, Sanders JV. (1975). n-beam lattice images. VI. Degradation of image resolution by a combination of incident-beam divergence and spherical aberration. Acta Cryst. A31, 307-310.

O'Keefe MA, Sanders JV. (1976). The phase component of lattice images of a zeolite crystal. Optik 46, 421-430.

O'Keefe MA, Saxton WO. (1983) The 'well known' theory of electron image formation, in: 41st Ann. Proc. Electron Microscopy Soc. Amer., G. W. Bailey (ed.), San Francisco Press, San Francisco, 288-289.

Pirouz P. (1981). Thin-crystal approximations in structure imaging. Acta Cryst. A37, 465-471.

Rose H. (1977) Nonstandard imaging methods in electron microscopy. Ultramicroscopy 2, 251-267.

Saxton WO. (1979). SEMPER--A portable image-processing system applied to electron microscopy, in: Inst. Phys. Conf. Ser. No. 44, Institute of Physics, London, 78-87.

Saxton WO. (1980). Correction of artifacts in linear and nonlinear high resolution electron micrographs. J. Microsc. Spectrosc. Electron. 5, 661-670.

Scherzer O. (1949) The theoretical resolution limit of the electron microscope. J. Appl. Phys. 20, 20-29.

Schiske P. (1973). Image processing using additional statistical information about the object, in: Image processing and computer-aided design in electron optics, P.W. Hawkes (ed), Academic, London, 82-90.

Self PG. (1982). Observations on the real space computation of dynamical electron diffraction intensities. J. Microscopy 127, 293-299.

Self PG, O'Keefe MA, Buseck PR, Spargo AEC. (1983). Practical computation of amplitudes and phases in electron diffraction. Ultramicroscopy 11, 35-52.

Skarnulis AJ. (1979). A system for interactive electron image calculation. J. Appl. Cryst. 12, 636-638.

Smith DJ, O'Keefe MA. (1983) Conditions for direct structure imaging in silicon carbide polytypes. Acta Cryst. A39, 139-148.

Smith DJ, Saxton WO, O'Keefe MA, Wood GJ, Stobbs WM. (1983) The importance of beam alignment and crystal tilt in high resolution electron microscopy. Ultramicroscopy 11, 263-282.

Tanaka M, Jouffrey B. (1984) Lattice-image interpretation of a relatively-small-unit-cell crystal. Acta Cryst. A40, 143-151.

Van Dyck D. (1980) Fast computational procedures for the simulation of structure images in complex or disordered crystals: a new approach. J. Microscopy 119, 141-152.

Additional Discussion

1. Question by J. Hren - Have you (or others) attempted to take into account relaxations associated with the crystal thickness. And, if so, please describe the results.

Answer - Although it is quite possible to include surface relaxations in the simulations (given an appropriate model), I have not done so. The paper by Marks and Smith (1983) describes some results of relaxation at a surface parallel to the beam, but I have never seen any simulations including relaxations at surfaces perpendicular to the incident beam direction.

2. Question by J. Hren - Under what conditions could you expect to image the effects of defects in the thickness direction?

Answer - Again, it is easily done using existing programs, but since an electron micrograph is largely a projection of the specimen structure, defects which produce overlapping structures in the beam direction would lead to confused contrast in both experimental micrograph and simulated image. Given a reasonable thickness of specimen, it might be very difficult to distinguish between simulations produced by different defect models.

3. Comment by F. Lenz - In the earlier literature the "smearing" of "damping" effects due to "convergence" of "spread of focus" were often described as lack of coherence, where "convergence" corresponds to lateral and "spread of focus" to longitudinal coherence.

Answer - Yes, these are certainly effects due to partial coherence. I used the terms "convergence" and "spread of focus" to describe how they were included in the simulation programs.

4. Comment by F. Lenz - The reason why so many authors have neglected second-order terms in the image transfer relations is not only the wish to simplify them. It was well known from the beginning that contrast transfer is a non-linear process, whereas amplitude transfer is linear. For image simulation it is relatively easy to conclude from object properties on image contrast without having to linearize, and it has been done before even for non-periodic objects. The inverse procedure, i.e. the deduction from image contrast of object properties is, however, much more complicated. Linear transfer can be inverted, non-linear transfer cannot. In image reconstruction one has to linearize even if one knows that for thicker specimens this may be a poor approximation. Since our original source of information is usually an electron micrograph, the first step in a reconstruction procedure must be linear. Its results will not only help to set up reasonable model structures but also yield information on the parameters describing the transfer function, such as the coefficients of spherical aberration and axial astigmatism and the defocus at which a micrograph was taken.

Answer - I agree. Image reconstruction and image simulation make excellent complementary techniques for obtaining maximum information from electron micrographs. Of course linear image reconstruction is the ideal method to use with images from thin crystals; however image simulation can also be useful in checking the degree of linearity of the experimental conditions under which the micrograph was obtained, as well as its more usual function of using the structure determined by the reconstruction to simulate images for comparison with experimental images obtained under conditions known to be non-linear, e.g. images from thick crystals, and dark-field images.

Electron Optical Systems (pp. 221-230)
SEM Inc., AMF O'Hare (Chicago), IL 60666-0507, U.S.A.

0-931288-34-7/84$1.00+.05

INTERACTIVE IMAGE PROCESSING FOR ELECTRON MICROSCOPY: MATCHING HARDWARE WITH SOFTWARE

W. O. Saxton+

Department of Metallurgy and Materials Science
University of Cambridge, U.K.

Abstract

The image processing techniques used 'a posteriori' to extract information from electron micrographs are surveyed, including particularly image averaging, selective averaging, 3-D reconstruction, and high resolution focal series restoration; recent developments in on-line image pick up and control have led to fully automatic focussing, stigmating and alignment by a frame store system equipped with a real time correlator board. The diversity of the techniques encountered calls for large integrated program systems with flexible command languages; however, a dilemma exists between providing the user with convenient control of special hardware facilities such as frame stores and array processors, and preventing the programs from becoming so specific that they are extremely short lived. Some of the compromises made in the Semper system are noted.

Key Words: Image processing, image analysis, image averaging, 3-D reconstruction, focussing, stigmating, alignment, high resolution, frame store, array processor

+Address for correspondence:
High Resolution Electron Microscope, Free School Lane, Cambridge, CB2 3RQ, England
Phone No.: (0223) 358381 Ext. 328

Introduction

Anyone concerned with image processing (enhancement or analysis) will be only too well aware of the exceptionally rapid rate of hardware development bearing on the field over the last five to ten years: quite apart from the general move towards 32 bit rather than 16 bit computers, and the widespread introduction of virtual memory systems, which have had a universal impact, the advent of 'frame stores'* and the growing capabilities of 'array' or 'vector' processors† have necessitated a constant re-evaluation of the answers to the common question, "What should I buy as a basic image processing system?".

Frame stores are most important primarily as a means of presenting digitised images in a visual form, being an order of magnitude better for the purpose than any preceding devices, but their former importance as a means of securing reasonably large amounts of inexpensive fast memory in a world dominated by 16 bit computers has already been superseded; they can now be seen as merely one member of the wide range of raster graphics devices which a generation nurtured on home computers now assumes to be present, in a memory-mapped form, in any computer; and it is instead such special capabilities as video-rate filtering and correlation that are now of interest to microscopists. Array processors, formerly hampered seriously for the purposes of image processing by their own limited memory capacities, are now overcoming the problem either through integration into large memory hosts or through massive increases in their own memory sizes.

The sentiment common some years ago that much of what we were doing should be postponed for a few years in anticipation of hardware developments was obviously well founded, and in many respects software is lagging well behind; yet development

* Devices characterised by the ability to store one or several image 'frames', roughly of TV quality, in a form that allows rapid input, output and manipulation of the stored image (from a few microseconds per pixel to real time TV rates) simultaneously with the continuous generation of a video signal from the stored image.

† Devices characterised by the ability to perform various arithmetical or logical operations on an array of operands at an abnormally high rate, through the use of various possible levels of parallelism.

is not slowing down, and the only sensible course for those of us concerned to use such systems is to continue to pursue the hardware with software systems flexible enough to be capable of substantial adaptation. This theme underlies the present review of image processing within electron microscopy: few workers in the field would pretend to be anywhere near the real front of image processing techniques in either hardware or software terms, and the most useful concern is with making the right compromises with relatively conventional approaches.

Survey of applications and problems

After these introductory remarks, an outline follows of the main reasons for the use of image processing in electron microscopy, and the remaining problems associated with some, with a few mathematical details being given in appendices; reviews in book form can be found in [19, 23].

Applications to scanning electron microscopy have largely been of a kind familiar to workers in several other disciplines, including particularly edge enhancement, particle sizing and counting, the related problem of automatic threshold selection, and image texture classification; earlier volumes of the present series contain numerous relevant contributions, to which we might add the UK Electron Microscopy and Analysis series, a recent thorough account of an earlier directional analysis technique due to B M Unitt [41], and an exciting new algorithm for automatic stereometry [3].

A problem common to all forms of microscopy is the reliable achievement of correct instrumental adjustment; the problem is of course most acute in the context of very high resolution (around 0.2nm) fixed beam (conventional) transmission electron microscopy (CTEM), where the image presented on the fluorescent screen is usually of insufficient intensity and/or magnification, and where the best-adjusted condition is in any case difficult to judge from the image appearance. The use of the new technology has recently made dramatic improvements in what is now possible, as described below under the heading of 'on-line processing'.

The sheer bulk of data involved when images are held in digital form has resulted in much effort being devoted in other fields to their compression by various schemes, which commonly reduces bulk for long-term storage by a factor of 8, at the price of rendering data not immediately amenable to display or manipulation. Within EM, R E Burge's group has used the discrete cosine transform as the basis of a very flexible scheme of this type [44]; more importantly, they have also addressed the question of the best image signal, or rather combination of signals, to use given the variety of signals possible in scanning transmission electron microscopy (STEM) by using principal component analysis (App.1) to identify objectively the most strongly differentiated (mutually uncorrelated) linear combinations [5]; this approach allows them at once to record the most informative images possible and to identify which data may be safely discarded, though it does not necessarily provide a simple recipe for image interpretation. Given the variety of signals possible in normal SEM practice, a similar analysis in this context also could well be useful.

My own enthusiasm and experience is centred on the techniques in use with CTEM - characterised perhaps by a relatively high reliance on transforms and large matrix operations (FFTs, special filters, auto- and cross correlation, clustering and 3-D reconstruction) as described below. (FFT = Fast Fourier Transforms)

Biomolecular applications. Micrographs of biological macromolecules remain extremely noisy because of the limited scattering power and radiation tolerance of the specimen, notwithstanding the effort devoted over the last decade to the alleviation of the damage problem, by very low temperature techniques for example [45]; the growing preference for glucose embedded or frozen hydrated preparations over the more artefact-prone use of negative stain will ensure that this remains the case, and as a result the averaging of large numbers of images of identical molecules is indispensable for the recovery of useful images.

Such averaging relied at first on perfect crystallinity in the specimen (usually spatial periodicity, but rotational and even helical periodicities have also been widely exploited), and used Fourier space filters with small 'windows' around the reciprocal lattice sites to suppress image components not sharing the periodicity of the crystal (e.g. [1]). Direct real space superposition was little used, though perfectly viable in most cases (fig.1) and rather simpler, until the more recent interest in averaging molecules in imperfect (i.e. bent) crystals [31,6] and isolated particles [11], the latter in particular due to the stimulus of J Frank. Here, the underlying common technique has been the use of cross correlation with an admittedly noisy reference [10,27] to allow accurate mutual alignment (registration) of individual images for averaging; in automated procedures, spurious correlation peaks due to random structure matching are largely eliminated by a lower threshold imposed on peak heights. The additional problem of unknown orientations in the isolated particle case is dealt with by comparing individual auto-correlation functions (ACFs), which are independent of image translation while preserving orientation and magnification information [11], or by locating individual centres of mass [26], or by simple (subjective) visual alignment [20]. Typically, many hundred images are combined in the case of imperfect crystals, and 20-100 in the case of isolated particles. In the imperfect crystal case at least, such averaging combined with extremely low dose STEM (annular detector mode) of freeze-dried material has allowed direct mass determination of a protein molecule and its major morphological subdivisions [8].

Increasing concern at the possibility that the population of molecules whose images are averaged may in fact be inhomogeneous has stimulated interest in selective forms of averaging. In the context of distorted crystals, restricting the average to the molecules in low strain sites only has been shown to provide a slight improvement in image quality [30]. More excitingly, a different use of principal component analysis has made possible for stained specimens at least, the objective recognition of several distinct groups within a population (clustering of the individual images) [42]: in this approach, the most highly differentiated linear combinations of a set of images are determined and then used to assign individual images to classes on the basis of their degree of similarity to these (measured by the scalar product, sum squared difference or other measure). The direct handling of large numbers of images requires the inversion of large matrices, but averaging might be usefully extended over large populations on the basis of a classification deduced from a limited "learning" set.

The macromolecules studied by these methods are rarely very large relative to the resolution attainable, and a 3-D structure may be fairly easily obtained from a modest number (10-20) of 2-D (projection) images recorded in different directions through the specimen. Most work here (recently reviewed in [2]) has used the transforms of these projections to provide 2-D sections through Fourier space from which the full 3-D transform is built up, various schemes being employed for establishing the precise tilt angles recorded, the appropriate relative normalisation and registration of the projections and reliable sample values of the continuous 3-D transform, given the relatively noisy character that persists in the data, averaged or not. Most work has also been addressed to perfect 2-D crystals, but the averaging methods used for imperfect crystals have also recently been used on the basis for 3-D reconstruction [7,34] and a significant effort has also been devoted to isolated particles, particularly ribosomes, on both sides of the Atlantic [12,18] using other reconstruction methods reviewed in [15], such as filtered back projection, ART, and non-Fourier decompositions. When all the projections used in a reconstruction derive from one single molecule/particle, the final statistical significance is low; but averaging to provide better defined projections requires 'a priori' recognition of the different projections among a random population of molecules and the unambiguous determination of the corresponding projection directions, which continues to present a substantial challenge.

In all cases the interpretation of the 3-D density distribution is less than straightforward, the molecular boundary being invariably anything but sharp, and the problem is compounded for 2-D crystals by the impossibility of recording edge-on views, which leaves the mean density within each layer of the crystal quite indeterminate. The additional information that can be derived about surface profiles from metal-shadowed preparations is likely to prove increasingly important in the future therefore: a thin coating evaporated at an angle to the specimen plane reveals one component of the surface gradient (App.2), from which the actual surface profile can be reconstructed by a suitable Fourier plane filter [38], particularly if the specimen has sufficient symmetry to make recording the other component of the gradient unnecessary.

High resolution studies In studies of inorganic specimens, resolution is commonly of the order of 0.2nm rather than 2nm, and the demands made on image processing techniques quite different. Averaging many unit cells is not indispensable in this context, and it is not therefore universally practised, but it is still frequently helpful (fig.2) and is likely to be employed with increasing frequency. Variations such as averaging over 1-D lattices only are appropriate to materials with planar defects or planar interfaces between different crystalline forms such as are often found in nonstoichiometric materials. Special 'ad hoc' filters have been used to emphasise features difficult to see clearly in the untreated micrographs - e.g., boosting the super lattice reflections only in images of charge density waves (work in progress at J.W. Steed's group) to locate in an image the regions giving rise to particular diffracted beams, and to remove from an image strong crystal periodicities masking fainter local features.

Examination of the power spectra (diffractograms) of disordered image areas has long been used to evaluate instrumental defocus and astigmatism; the 'band-limit' found in these, i.e. the finest periodicity transferred from specimen to image, can also be used to estimate the two main illumination parameters, namely the effective focus spread and beam divergence [22]. A relatively sensitive band-limit estimate can be made by comparing two images recorded in succession, and tabulating in a radial correlation function [31] the level of mutual agreement found for progressively higher spatial frequency bands (fig.3).

The importance of such estimates for high resolution image interpretation lies in the substantial influence such factors have on image detail. Most practical evaluation of high resolution images, except in the very thin limit, has relied on image simulation techniques (reviewed by M A O'Keefe in this volume), in which the correctness of a structural model is tested by the extent to which the images it predicts match those recorded experimentally; even granted general readiness to discard all imperfectly stigmated images, the calculation of images for a reasonable range of specimen thickness, focus, focus spread and divergence is a formidable task, frequently revealing spurious 'matches' with experimental results, and any reliable observed values for some of these parameters makes the structural evaluation much less ambiguous. In this connection, the emerging realisation that small levels of residual beam tilt (around 1mrad) can affect image detail substantially without being detectable in a diffractogram is alarming [36], and this point is taken up again below.

Accurate imaging parameters are of course equally vital for direct image 'deconvolution' - the recovery of artefact free images from the aberrated images directly recorded. This is straightforward even in principle only for sufficiently thin and/or light specimens, which scatter only a small proportion of the incident electrons; in this regime the imaging is at least described by a simple specimen-independent spatial frequency response called a contrast transfer function (CTF), which can be 'divided out' of the image transform. In practice, such a simple procedure is frustrated by substantial bands of low transfer - around the CTF 'zeros' - where division by the CTF results in noise amplification, but a more mature version, drawing on a focal series of images mutually aligned by cross correlation (App.3), and using a 'Wiener' approach to prevent noise amplification [35], provides a highly robust means of recovering faithful images at high resolution [28,16] (fig.4). Computationally, all that is involved is the calculation of a weighted average of the transforms of the individual images (the weighting ranging from point to point), and the only real difficulty of the technique - particularly in the application to images without disordered specimen regions present - is the accurate estimation of the parameters controlling the CTF from the image diffractograms. This technique also figures occasionally in biological applications, and is likely to be crucial to the use of frozen hydrated specimens.

Outside the weak specimen limit, such a variety of schemes was propounded during the '70s to little real avail that it is a matter of some embarrassment to recall much of it! Almost every conceivable combination of images and/or diffraction patterns, bright field, dark field or half-plane apertured has been proposed as the basis of specimen wave recovery schemes, but experimental difficulties, ill conditioning of the data or sheer inability to find the solution to a given set of nonlinear equations variously prevented their providing satisfaction to any but the mathematically inclined! (For full review, see [24].) Few schemes even considered the finite illumination profiles known to affect image quality dramatically, and the most recent [17] relies on an incorrect description of its effects; the most attractive option still open seems to be the subtraction from a bright field image of a corresponding dark-field image recorded with a central beam stop (App.4), for this procedure recovers a difference image exactly described by the (linear) theory applicable to the weak specimen case [25], and deconvolution can accordingly be carried out as described above for that case. At all events, there most certainly remain real problems in interpreting images outside the thin specimen limit; even simulation methods must be suspect beyond 10-20nm thickness, because of absorption effects not yet properly understood.

Real-time processing

The area that has seen the most dramatic impact of the new technology - particularly frame stores - is that of microscope/computer interactions [37,14,39,4]. The more obvious part of this impact has been in image acquisition and presentation: where the direct image is noisy and/or faint, the use of a frame store equipped with TV rate digitisation, the ability to average incoming frames (normally recursively, i.e. taking a weighted average of the incoming frame and that already stored), and present the result continuously on a normal brightness TV monitor overcomes completely the limitations of conventional systems relying on long persistence phosphor, high specimen irradiation levels or scan converter tubes. Microscope interfacing is relatively straightforward for scanning microscopes; but fixed beam instruments currently rely on a TV camera coupled to a fluorescent screen, which is at present the source of considerable loss of resolution, especially in high voltage instruments - a 2:1 or 3:1 tapered fibre optic coupler between the two may provide a simple answer. Given the low contrast of many high resolution images, a real time compensation for the TV camera shading pattern is also normally valuable; this facility depends on the frame store being capable of multiplying one frame by another also at video rates.

A rather less easily anticipated result has been the success of the recently developed procedures for fully automatic instrumental adjustment - particularly in high resolution CTEM applications, where the problems of focussing and stigmating have long been notorious, and those of beam alignment are in fact arguably worse still [36]. In contrast to previous approaches, which for SEM relied mainly on maximising the 1-D signal gradient (being therefore unsatisfactory in high noise situations) and for CTEM on visual assessment of diffractograms or ACF peak profiles, Erasmus and Smith [9] showed that in the (electronic!) hands of a computer the single parameter of image signal variance furnishes an adequate basis for accurate focussing and stigmating in both cases. In the system attached to the Cambridge University 600kV HREM, a minicomputer observes the signal variance at about 40 different values of a supply current in turn, finally selecting the optimum current by locating the required point (usually the centre) of the resulting curve. The variance is determined by using a hardware 'correlator' board to obtain the cross-sum between two successively digitised frames, and several readings are obtained each second. High accuracy is achieved even though the variance is not very sensitive to normal levels of astigmatism because the computer can make use of the higher sensitivity found far from the well stigmated condition.

Given the recent realisation of the importance of perfect beam alignment for high resolution microscopy, and the impossibility mentioned previously of diagnosing residual misalignment unambiguously 'a posteriori', the extension of the variance method to embrace the alignment process too (fig.5), giving a reliable procedure for adjusting all five supply currents automatically within a minute, in spite of their various interactions [33] is potentially equally important.

Any system for digital image acquisition from the microscope provides data in a form immediately suitable for any of the processing techniques described in the previous section, of course; however for CTEM the lower field of view possible with a TV camera (about 1% of the normal recorded field!) means that emulsions and microdensitometry (still surprisingly expensive) will remain with us for some time, even if we assume the widespread introduction of a 1024 square norm for future frame store systems.

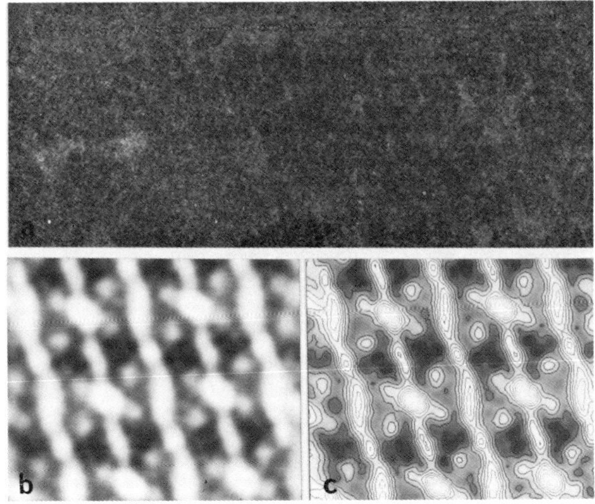

Fig.1 (a) CTEM image of negatively stained Chlamydomonas Reinhardi (due to P J Shaw); (b) average image obtained by superposing corresponding sites in real space, and centro-symmetrising the result by superposing it on a 180-degree rotated copy; (c) the same with density contours marked. (Unit cell dimensions 24 by 13nm.)

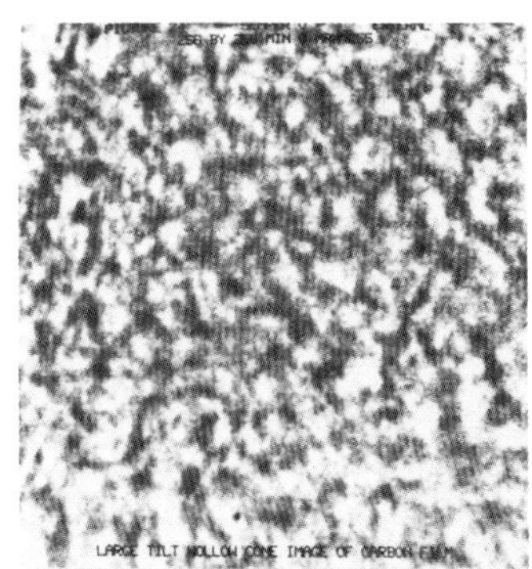

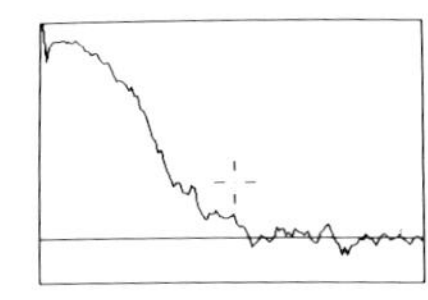

Fig.3 (Left) Photograph of TV monitor display of CTEM image of amorphous carbon film, recorded in the hollow cone illumination mode (due to D J Smith); (right) radial correlation function between the transforms of this image and another recorded immediately subsequently, showing mutual agreement between Fourier components to a period of 0.2nm (marked by the cross).

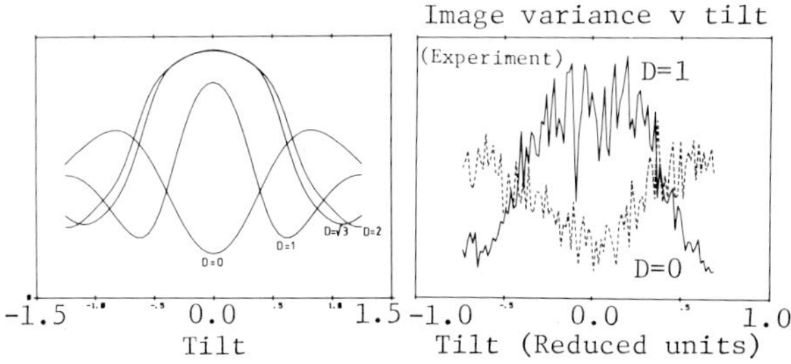

Fig.5 (Left) Theoretical dependence of image variance on tilt over a range of 1.5 reduced units (typically 9mrad) for three focus levels as marked in units of the Scherzer defocus; (right) experimental curves for a slightly lower range of 1 reduced unit, recorded via the 600kV HREM pick up system.

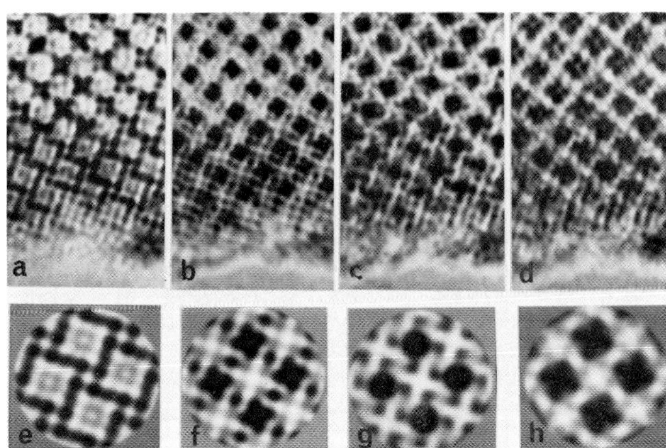

Fig.2 (a-d) CTEM images of a wedge shaped crystal of VNbO, at focus levels of 0.56, 1.67, 2.12 and 2.82 times the Scherzer defocus; (e-h) corresponding clear images obtained by local lattice averaging over the thin crystal areas and P4 symmetrisation.

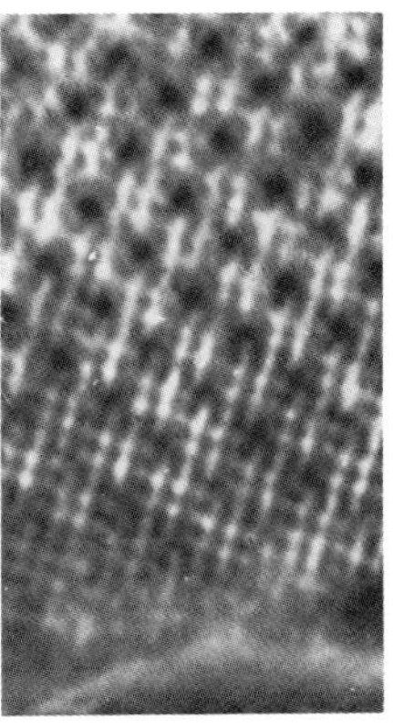

Fig.4 Restoration of focal series in fig.2, using linear method only, so that only the thin edge region is reliable. Tunnels are shown white.

Practical Systems for General Processing

The enormous variety of processing requirements implied by the survey above makes it clear that ultimately the only environment that really ensures adequate flexibility in storage and manipulation of data representing or derived from micrographs is the normal command level of an interactive computer terminal where commands can be entered in turn to apply particular operations to particular pictures, data types and file formats are largely at the user's disposal, and new operations can be defined in the conventional high level languages, such as Fortran*, and

* Is Fortran now in danger of being displaced from its position as the uncontested leader for what the industry calls 'scientific' applications? Perhaps because of the huge burden of Fortran's powerful input/output package more than any real deficiencies of the language, Pascal is increasingly preferred...

compiled to form new programs to be run by subsequent commands.

However, it takes a considerable time to master such an environment for the non enthusiast, and in any case it is in other respects than flexibility rather badly suited to our requirements, involving far too much typing for simple operations, poor inter-program communication, limited looping facilities and error handling, and encouraging the evolution of a plethora of unrelated data formats. It will also be obvious that most of the procedures described above require basic tools at a higher level than Fortran source language statements. For these reasons, almost all groups in the field have developed unified program systems of one kind or another, within which the control facilities approximate more closely to those actually required, and a variety of well established procedures can be applied relatively easily to data in a limited number of formats. The systems are modular, because modularity with well defined interfaces makes subsequent adaptation of the system to changing environments much more feasible. The discipline of writing programs within such a system is repaid directly by a relatively long program lifetime, by a common (and therefore more easily used and documented) user interface, and indirectly - but perhaps most critically - by the convenience of being able to use it in arbitrary (and easily redefinable) combinations with other system commands through the system's command processor.* Interacting with such a system, command by command, preferably with frequent display of intermediate results, is an efficient and satisfactory, if still imperfect, way of working.

J. Frank's account of his group's system SPIDER [13] includes a useful survey of systems suitable for EM image processing, many of which are already ten years old, which indicates reasonable success in keeping pace with technical development. Most frame store manufacturers offer subroutine libraries managing little more than their own peculiar features, though Gould DeAnza supply a somewhat more flexible program system called LIPS, with command processor support, in conjunction with their advanced IP-8500 display, and MicroConsultants offer a system called GPIPS which was developed from our own 'Semper' system. More recently, Logica UK have produced a relatively large (and expensive) image processing system called LUCID (formerly INSIGHT); and a comprehensive subroutine library - also unfortunately called SPIDER! - for memory resident 2-D arrays is available from the Joint System Development Corp., Tokyo (or from Mitsui in the UK at least).

* The usefulness of a program system is only sometimes measurable by how well it does the task it was designed for; in many other cases the relevant question is how easily it can be coerced into doing something different! The considerable progress that has been made was emphasised recently by my finding it possible to implement and test within a few minutes a proposed spatial frequency extrapolation algorithm that would have taken days to explore in the early '70s.

Semper. Our own system, Semper (also available commercially under a British Technology Group license), has been described previously in its Version IV form [29], and illustrates the theme of adapting to hardware changes in having moved easily from the 20KW PDP 8/E computer on which it was first developed to a 32 bit machine (a Systems Engineering Labs 32/27) without appearing grossly inefficient in the new environment [32].

Semper's basic control structure is a cycle of command decoding by an interpreter which relies (apart from a few 'intrinsic' commands) on a command syntax definition array that is updated as fresh commands are added to the system, and calls Fortran subroutines ('extension' routines) appropriately to effect the operations required by the commands encountered. The present extension routines comprise code for almost all the types of processing mentioned above, with the main exceptions of particle counting/sizing and principal component analysis. Within the current interpreter (version V-3), there is provision for command level branches, conditionals and 'FOR' loops, and for dynamically created and edited 'command procedures', as well as for the indirect execution of command sequences prepared outside Semper in ordinary text files. The interpreter manages a table of named numerical variables, which are largely at the user's disposal but which also serve as a means of program intercommunication; it also provides a unified image filing system embracing disc storage, tape and a display (not necessarily of the frame store type), so that commands and extension routines function in a largely device independent manner (e.g., 'Copy 3 to 4' makes a copy of disc picture 3 as a new disc picture numbered 4, while 'Copy 102 to display' causes the file 2 on tape drive 1 to be displayed, without the COPY routine itself being aware of the difference).

Image storage is organised in rows, the extension routines issuing (Semper) system read/write requests as necessary and operating with no more than a few rows available simultaneously; this allows the system to process large pictures on small address space machines - especially 16 bit computers - at some cost in terms of its speed in handling small pictures; we have taken the view that small pictures are handled sufficiently rapidly in any case and that it is the treatment of large ones that matters. Although on small machines the read/write requests result in actual physical transfers, efficiency is achieved in large machines through a 'cache'-like arrangement, Semper's disc input/output routine maintaining a collection of recently used rows in a large memory buffer with physical transfers made less frequently and in larger units. Even on large virtual memory systems such as the DEC VAX 11/780 this arrangement has proved quite successful: an address space of, say, 4MB such as might typically be available to a task under VAX/VMS is still inadequate for large images in floating point representation, and an intermediate buffering or mapping arrangement involving a disc file remains essential. A single large disc file is used with all pictures stored at convenient places inside it, a rudimentary directory being maintained by Semper; this choice - like the decision to link the whole system as a single task rather than as a set of independent tasks run in response to requests from a central command processor task - was made in the interests of portability, a consideration which we have placed above all others, and is by no means essential; however, a very high level of portability has indeed been achieved, with installations on machines made by DEC, IBM, SEL, ICL, Prime, Data General, GEC, Nord and Apollo amongst others, and we are content with the compromise.

Display access presents a particularly difficult problem for a program system that is not to be too closely dependent on any particular display system. Our compromise at least has been to rely within Semper on no more than display erasure, image row output, text and line generation, a user-driven cursor and optional row input too for use with display devices of the frame store type; these functions are provided by 'primitive' routines written in Fortran-callable form to suit any particular display. Currently, all data are scaled suitably on output to the display device, and re-scaled to the original range on recovery, so that the display device can be used as a storage medium (albeit with limited precision), like some kind of visible disc. All other aspects of display manipulation* have been independently provided in our own installation (fig.6), via a keypad and trackerball continuously serviced by a small minicomputer dedicated to frame store management, or via a small host machine program that transmits instructions to the mini in response to single keystroke requests from a terminal† [32]. Different keys request increased or decreased contrast, brightness, upper or lower threshold; grey-scale wrap-round; contrast reversal; increased or decreased zoom; cursor movement or display scroll under trackerball control; alternative display modes for text and line graphics (stored in a separate memory plane in our case); switching between the two frames available; various levels of erasure; and the transfer of an image frame to or from the minicomputer system dedicated to the 600kV microscope. In this way, the display can be manipulated quickly and conveniently irrespective of the host program to which it is currently allocated. Anyone who has used a system where the frame store look-up-table (LUT), in principle always capable of virtually instant alteration, is in practice only alterable through the running of a program which loads a new LUT from a file will see what is meant in this context at least by matching hardware with software!

Likely Future Developments

Several improvements are likely within Semper itself, independent of any special hardware. Work is already in progress replacing the filing system with one supporting a much larger number of

* We have a GEMS Mk I, with a dual 512 by 512 by (8+1) bit configuration.

† It is remarkable how difficult it often is to achieve individual keystroke servicing in large computer environments, an added RETURN being all too frequently necessary.

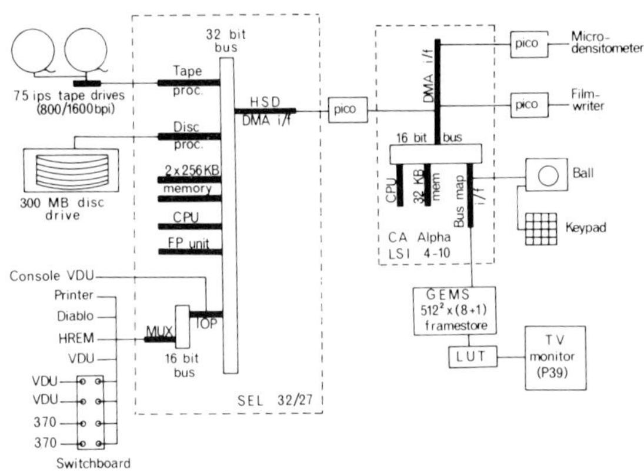

Fig.6 Block diagram of the HREM processing system: SEL computer, and GEMS frame store, with CA minicomputer between the two.

pictures within each physical device, each with a label including a comment or title string and other information (such as a 'protection' flag preventing accidental deletion/overwriting), with provision for multiplanar images in any of four different pixel representations, automatic conversion being performed as necessary by the system row access routines so that (broadly speaking) all routines will operate successfully on all representations. Provision is also being made for access to/generation of tape files in some interchangeable format independent of the Semper filing system, such as the FITS standard [43] - though even this does not propose any truly adequate mechanism for the transport of floating point values, and we are having to propose our own.

Ideas so far less well developed include providing an environment for some extension routines to operate on in-memory 2-D arrays containing the whole of a picture simultaneously (subject to a lower maximum picture size, of course), with automatic assembly/splitting of picture rows by the system as necessary - this seems a sensible move towards growing VM capabilities in future systems; also some of the simpler routines might be reorganised under an intermediate controlling routine that implements something similar to a UNIX 'pipe' (applying several commands in turn to each row of a picture, rather than applying each command to all rows before the next command is begun) - this would improve efficiency again [40], and allow immediate viewing of the cumulative result if the final destination was the display.

But what of more hardware-oriented developments? In spite of the flexibility of the display control mechanism described above, we are still not using all of the capabilities of our display system, nor have we made any provision for the use of an array processor (which we do not have!). The problem is of course that the better any software uses particular hardware, the less easily it is likely to be to transfer it to another environment.

As far as frame store exploitation is concerned, it seems simple and sensible to make use of a column access mode as well - a widely available hardware feature; this could greatly simplify image rotation, transposition and warp correction, for example. By a somewhat less clear route, a mechanism is needed for controlling the use of multiple image frames - e.g., selecting individual planes, or sets of three for three-plane full colour images; the problem here is more one of how the user is to express his intentions in a convenient and memorable form than of actually achieving any particular desired goal, but this frequently occurring problem should not be overlooked!

As far as array processors are concerned, we can perhaps do no more within the context of a standardised portable package than to identify and isolate in subroutine calls vector operations that may reasonably be expected of an array processor; beyond this, adjustments of a very ad hoc character are to be expected, as the present range of architectures and interfaces is already rather wide. In our present context, by far the most useful single operation likely to be encoded in hardware (or at least moderately firmware!) is the 2-D FFT, used both by itself and as part of the process of auto- or cross correlation. Such a capability is already found in some array processors, and at least one frame store, and others will doubtless follow. Depending on precisely how the operation is defined by the hardware, however, we may have to fix or float an image before transformation, move it to a particular address, alter floating point formats, define an address range to be transformed, shift transform origins between corner and centre, pack or unpack real data in complex pairs, insert normalisation factors, and so on. An interesting account of the practical problems of using an existing array processor for FFT calculation appears in [37] (pp85-91). It is all ultimately worthwhile, as the anticipated speeding up is very considerable; but this seems to be the point at which we settle for short-lived - and necessarily home-grown! - software as we explore the new tools.

Acknowledgement

I acknowledge gratefully the support of a Royal Society research fellowship.

Appendix 1: Principal Component Analysis.

Given a set of corresponding pixels $x_1, x_2, x_3 ..$ from different images, the interdependence of the images might be quantified via the covariance matrix

$$v_{ij} = \text{cov}\,(x_i, x_j)$$

(calculated by averaging over all pixels). The matrix v_{ij} is real and symmetric: its eigenvalues λ_i are therefore real, and we may find an orthonormal eigenvector set e_{ij}. If we consider now a set of new images, formed as linear combinations of the original images, with each pixel derived from the original pixels via the equation

$$x'_i = e_{ij}\, x_j$$

(i.e., if we decompose the original pixel vectors in terms of the eigenvectors), we find that the new covariance matrix

$$\begin{aligned} \text{cov}(x'_i, x'_j) &= e_{ip} e_{jq} \,\text{cov}(x_p, x_q) \\ &= e_{ip} e_{jq} v_{pq} \\ &= \lambda_i \delta_{ij} \end{aligned}$$

so that the resulting images $x'_1, x'_2, x'_3 ..$ are <u>uncorrelated</u>, and in this sense at least contain maximally different information. It also follows that λ_i gives the variance of image x'_i, so that the most important 'principal component' images are simply those which correspond to the largest eigenvalues. (The transformation from x to x' is commonly referred to as the 'Karhunen-Loeve' transformation [e.g., 21].)

Appendix 2: Interpreting thin metal coatings.

If a metal coating of thickness d is deposited uniformly at an angle θ to the specimen normal, on a specimen with a height profile h(x), then the coating thickness at x (projected normally) is

$$\begin{aligned} t(x) &= h(x+d\sin\theta) + d\cos\theta - h(x) \\ &= h(x+x_0) - h(x) \end{aligned}$$

if we drop the constant term and write x_0 for $d\sin\theta$, which obviously corresponds to differentiation of h(x) if x is small. On Fourier transformation (denoted by tildes), we have

$$\begin{aligned} \tilde{t}(k) &= \tilde{h}(k).\exp(2\pi ikx\ ) - \tilde{h}(k) \\ &= \tilde{h}(k).[\exp(2\pi ikx\ ) - 1] \\ &= \tilde{h}(k).2\pi ikx \qquad \text{for small k.} \end{aligned}$$

Appendix 3: Linear image deconvolution.

Linear space-invariant imaging is described by a <u>spatial frequency response</u> - the ratio of an image Fourier component to the corresponding object component. In high resolution electron imaging of 'weak-phase' objects [e.g., 23], the 'object' is the projected specimen potential distribution $\phi(x)$, and the 'image' the image plane <u>contrast</u>, or fractional intensity variation c(x); and the spatial frequency response or 'transfer function' p(k) embraces the effects of illumination divergence and focus spread (spatial and temporal coherence), astigmatism, beam tilt, drift and vibration (all largely constants of the series) as well as defocus (deliberately varied from one image to the next). For any one such image, $\tilde{c}(k)/p(k)$ provides an <u>estimate</u> of $\tilde{\phi}(k)$, but the estimate will be poor where p(k) is low or zero; a better estimate is obviously provided by averaging the individual estimates, weighted relatively in some way that reflects their reliability; weighting in proportion to $|p(k)|^2$ gives the result

$$\begin{aligned} \tilde{\phi}'(k) &= \sum (\tilde{c}(k)/p(k)).(|p(k)|^2/\sum |p(k)|^2) \\ &= \sum p^*(k)\tilde{c}(k)/\sum |p(k)|^2 \end{aligned}$$

which is also in fact the estimate minimising the summed squared image residuals; minimising the summed squared object residuals, in expectation, gives the slightly more reliable Wiener solution instead, differing only in the addition to the denominator of an estimate of the spectral power ratio n(k) between image noise and object:

$$\tilde{\phi}'(k) = \sum p^*(k)\tilde{\sigma}(k)/(\sum |p(k)|^2 + n(k)) .$$

Either of these prescriptions yields restorations with essentially flat spatial frequency responses except at very low and very high frequencies where none of the recorded images contains any reliable information. Linear imaging of a weak-phase weak-amplitude specimen producing a Fourier plane scattered wave $\tilde{f}(k)$, yields an image contrast

$$\tilde{\sigma}(k) = \tilde{f}(k)t(k) + \tilde{f}^*(-k)t^*(-k)$$

in terms of a slightly different transfer function t(k); at least two images are now required for any estimate at all to be made of $\tilde{f}(k)$, but similar optimum restoration formulae can be found for exploiting a through-focal series of recorded images.

Appendix 4: Nonlinear image deconvolution.

When nonlinear image intensity components cannot be neglected, the transfer of object information to the image must be described by the equation

$$I(k) = \sum \tilde{f}(q)\tilde{f}^*(q-k)t(q,q-k)$$

in which $\tilde{f}(k)$ is the object transform and $t(k_1,k_2)$ a mutual transfer function, or transmission cross coefficient embracing all the effects noted in App.3 [e.g., 25]. This may be conveniently decomposed into three sets of terms, namely the background term

$$I_0 = |f(k_0)|^2 ,$$

the terms linear in f(k)

$$\tilde{I}_1(k) = \tilde{f}(k_0+k).[\tilde{f}^*(k_0)t(k_0+k,k_0)] + \tilde{f}^*(k_0-k).[\tilde{f}(k_0)t(k_0,k_0-k)] ,$$

and the remaining nonlinear terms

$$I(k) = \sum_{q \neq k, k_0+k} \tilde{f}(q)\tilde{f}^*(q-k)t(q,q-k) .$$

The troublesome terms are $\tilde{I}_2$; but they may be observed independently as the image intensity under central stop dark-field imaging conditions, and subtracted from the full bright-field intensity to leave purely linear terms which can be deconvoluted as in App.3.

References

1 Aebi U, Smith PR, Dubochet J, Henry C, Kellenberger E. (1973). A study of the T-layer of Bacillus brevis. J. Supramol. Struct. 1, 498-522.

2 Amos L, Henderson R, Unwin PNT. (1982). 3-D structure determination by electron microscopy of 2-D crystals. Prog. Biophys. Mol. Biol. 39, 183ff.

3 Atkin P, Smith KCA. (1983). Automatic stereometry and special problems of the SEM, in: Electron Microscopy and Analysis 1983, P.J.Goodhew (ed), Institute of Physics, Bristol. 219-220.

4 Boyes ED, Muggridge BJ, Goringe MJ. (1982). On-line image processing in high resolution electron microscopy. J. Microsc. 127, 321-335.

5 Burge RE, Clark AF. (1981). STEM multiple images: the Karhunen-Loeve transform and data compression, in: Electron Microscopy and Analysis 1981, M.J.Goringe (ed), Institute of Physics, Bristol. 315-320.

6 Crepeau RH, Fram EK. (1981). Reconstruction of imperfectly ordered zinc-induced tubulin sheets using cross-correlation and real space averaging. Ultramicrosc. 6, 7-18.

7 Engel A. (1984) 3-D structure of matrix porin. Ultramicrosc. 13, in press.

8 Engel A, Baumeister W, Saxton WO. (1982). Mass mapping of a protein complex with the scanning transmission electron microscope. Proc. Natl. Acad. Sci. USA 79, 4050-4054.

9 Erasmus SJ, Smith KCA. (1982). An automatic focusing and astigmatism correction system for the SEM and CTEM. J. Microsc. 127, 185-199.

10 Frank J. (1980). The role of correlation techniques in computer image processing, in: Computer Processing of Electron Microscope Images, P.W.Hawkes (ed), Springer-Verlag, Berlin. 187-222.

11 Frank J, Goldfarb W, Eisenberg D, Baker TS. (1978). Reconstruction of glutamine synthetase using computer averaging. Ultramicrosc. 3, 283-290.

12 Frank J, Verschoor A, Boublik M. (1981). Computer averaging of electron micrographs of 40S ribosomal subunits. Science 214, 1353-1355.

13 Frank J, Shimkin B, Dowse H. (1981). SPIDER - a modular software system for electron image processing. Ultramicrosc. 6, 343-357.

14 Herrmann K-H, Krahl D, Rust H-P, Ulrichs O. (1976). Aufbau und Anwendung eines Digitalen Fernsehbild-Halbleiterspeichers fur die Elektronenmikroskopie. (Construction and application of a digital TV-halftone store for the electron microscope.) Optik 44, 393-412.

15 Hoppe W, Hegerl R. (1980). 3-D structure determination by electron microscopy (nonperiodic specimens), in: Computer Processing of Electron Microscope Images, P.W.Hawkes (ed), Springer-Verlag, Berlin. 127-185.

16 Kirkland EJ, Siegel BM, Uyeda N, Fujiyoshi Y. (1980). Digital reconstruction of bright field phase contrast images from high resolution electron micrographs. Ultramicrosc. 5, 479-503.

17 Kirkland EJ. (1982). Nonlinear high resolution image processing of conventional transmission electron micrographs. Ultramicrosc. 9, 45-64.

18 Knauer V, Hegerl R, Hoppe W. (1983). 3-D reconstruction and averaging of 30S ribosomal subunits of Escherichia coli from electron

micrographs. J. Mol. Biol. 163, 409-430; Oettl H, Hegerl R, Hoppe W. (1983). J. Mol. Biol. 163, 431-450.

19 Misell DL. (1978). Image Analysis, Enhancement and Interpretation. North Holland, Amsterdam.

20 Ottensmeyer FP, Bazett-Jones DP, Hewitt J, Price GB. (1978). Structure analysis of small proteins by electron microscopy. Ultramicrosc. 3, 303-313.

21 Rosenfeld A, Kak AC. (1976). Digital Picture Processing. Academic Press, New York. 109-123.

22 Saxton WO. (1977). Coherence in bright field microscopy of weak objects, in: Developments in Electron Microscopy and Analysis 1977, D.L.Misell (ed), Institute of Physics, Bristol. 111-114.

23 Saxton WO. (1978). Computer Techniques for Image Processing in Electron Microscopy. Academic Press, New York.

24 Saxton WO. (1980). Recovery of specimen information for strongly scattering objects, in: Computer Processing of Electron Microscope Images, P.W.Hawkes (ed), Springer-Verlag, Berlin. 35-87.

25 Saxton WO. (1980). Correction of artefacts in linear and nonlinear high resolution electron micrographs. J. Microsc. Spectrosc. Electron. 5, 665-674.

26 Saxton WO. (1980). Digital processing of electron micrographs: a survey of motivations and methods, in: Proc. 7th Eur. Cong. EM, Vol 1, P Brederoo and G Boom (eds) 486-493. 7th Eur. Cong. EM Foundation, Leiden, The Netherlands.

27 Saxton WO, Frank J. (1977). Motif detection in quantum noise-limited electron micrographs by cross correlation. Ultramicrosc. 2, 219-227.

28 Saxton WO, Howie A, Mistry A, Pitt A. (1977). Fact and artefact in high resolution electron microscopy, in: Developments in Electron Microscopy and Analysis 1977, D.L.Misell (ed), Institute of Physics, Bristol. 119-122.

29 Saxton WO, Pitt TJ, Horner M. (1979). Digital image processing: the Semper system. Ultramicrosc. 4, 343-354.

30 Saxton WO, Baumeister W. (1981). Image averaging for biological specimens: the limits imposed by imperfect crystallinity, in: Electron Microscopy and Analysis 1981, M.J.Goringe (ed), Institute of Physics, Bristol. 333-334.

31 Saxton WO, Baumeister W. (1982). The correlation averaging of a regularly arranged bacterial cell envelope protein. J. Microsc. 127, 127-138.

32 Saxton WO, Koch TL. (1982). Interactive image processing with an off-line minicomputer: organisation, performance and applications. J. Microsc. 127, 69-84.

33 Saxton WO, Smith DJ, Erasmus SJ. (1983). Procedures for focusing, stigmating and alignment in high resolution electron microscopy. J. Microsc. 130, 187-201.

34 Saxton WO, Baumeister W. (1984). 3-D reconstruction of imperfect 2-D crystals. Ultramicrosc. in press.

35 Schiske P. (1973). Image processing using additional statistical information about the object, in: Image Processing and Computer-Aided Design in Electron Optics, P.W.Hawkes (ed), Academic Press, London. 82-90.

36 Smith DJ, Saxton WO, O'Keefe MA, Wood GJ, Stobbs WM. (1983). The importance of beam alignment and crystal tilt in high resolution electron microscopy. Ultramicrosc. 11, 263-282.

37 Smith KCA (ed). (1982). Computer Techniques in Electron Microscopy and Analysis. (1982). J. Microsc. 127, 1-126. (Special issue.)

38 Smith PR, Kistler J. (1977). Surface reliefs computed from micrographs of heavy metal-shadowed specimens. J. Ultrastruct. Res. 61, 124-133.

39 Spence JCH, Disko M, Higgs A, Wheatley J, Hashimoto H. (1982). A digital on-line diffractometer and image processor for HREM, in: Proc. 10th Int. Cong. EM (Hamburg), Vol 1, 519-520. Deutsche Gesellschaft fur Elektronenmikroskopie e.V., Frankfurt.

40 Stevens WR, Hunt BR. (1982). Software pipelines in image processing. Comp. Graph. Im. Proc. 20, 90-95.

41 Tovey NK. (1980). A digital computer technique for orientation analysis of micrographs of soil fabric. J. Microsc. 120, 303-315.

42 Van Heel M, Frank J. (1981). Use of multivariate statistics in analysing the images of biological macromolecules. Ultramicrosc. 6, 187-194.

43 Wells DC, Greisen EW, Harten RH. (1981). FITS - a flexible image transport system. Astron. Astrophys. Suppl Ser. 44, 363-370.

44 Wu JK, Burge RE. (1982). Adaptive bit allocation for image compression. Comp. Graph. Im. Proc. 19, 392-400.

45 Zeitler E (ed). (1982). Cryomicroscopy and radiation damage. Ultramicrosc. 10, 1-177. (Special issue.)

Discussion with Reviewers

F. Lenz: In connection with automatized averaging procedures one might ask what the chances are that a given structure of interest is found even in a completely random intensity distribution.

There are cases where structural details are hardly visible in the original micrographs but come out clearly in processed averages. Would this also occur if one searches for such details in an intensity distribution known to be random?

I should like to suggest to use some "figure of credibility" defined as follows: Let us assume that a given structural detail is found N times in a micrograph. If the same structural detail is searched in a random intensity field of the same area, using the same criteria, and it is found n times, then we may define a figure of credibility e.g. by $\log \frac{N}{n+1}$.

Author: Thank you for your comment.

Electron Optical Systems (pp. 231-236)
SEM Inc., AMF O'Hare (Chicago), IL 60666-0507, U.S.A.

0-931288-34-7/84$1.00+.05

DISPLAYS: MARKET AND TECHNOLOGIES

Steve Blazo

Tektronix Inc.
P.O. Box 500
Beaverton, OR 97077
Phone No. (503) 627-4885

Abstract

Displays provide the essential human interface to virtually all electronics instrumentation. The market is large with new applications appearing every year; sometimes with profound impact. Digital watches with liquid crystal displays appeared in the early 1970s and have virtually wiped out the mechanical timepiece industry. Personal computers with cathode ray tube (CRT) displays are proliferating with diverse applications in the industry and in the home. Work stations with high-resolution color displays are changing the way architects, draftsmen, and IC designers perform their job. The CRT is the dominant technology in today's market, and will, no doubt, continue to be for some years to come. High-resolution shadow-mask tubes will capture a larger and larger market share in the coming years. Projection displays will grow rapidly with the introduction of systems based on new technologies and with the advent of high definition TV. The flat panel industry is growing at 30 percent per year, not so much from taking business away from CRTs, as in the creation of new applications. This is an area rich in technologies with many contenders such as plasma, electroluminescence, liquid crystal, vacuum fluorescent, electrochromic, and others.

Key Words: Display, Cathode Ray Tube, Liquid Crystal, Electroluminescence, Plasma, Projection display.

Introduction

Displays are the primary man-machine interface. Information passed through them is input to man's most powerful sensory organ, his eyes. With the addition of peripheral devices such as a digitizing tablet, touch panel, keyboard, and computer, the collection becomes a powerful system for analyzing and controlling information and machines. Therefore it is not surprising that display technology has been a field of active research and development, and that a large number of related technologies exist or are being developed to satisfy this need.

Virtually every physical phenomenon that could be imagined to produce a reversibile visible image has at some time or another been investigated for use as a display medium. These are listed in figure 1 where no attempt has been made to be exhaustive. Most of these methods, while novel and inventive, have not met with commercial success. Display market segmentation is shown in figure 2. Note that the market today is dominated by the CRT and will no doubt continue to be for some time to come.

CRTs can display large amounts of data at very low cost. This is in large part due to its inexpensive analog addressing. When a particular display element is to be activated, all that need be done is supply the appropriate currents to the deflection yoke and pulse the grid. This can be done at high resolution and very low cost. Monochrome 500-line resolution CRT modules can be purchased for $100 or less, corresponding to 0.04 cents/pixel. This cost is not approachable by any other technique. In addition CRTs have the capability of presenting information in full color, at high brightness, and with rapid update. The main disadvantages of the CRT are its relative bulk, limited size, and the necessity of supplying high voltage. The cost of CRT displays is also relatively insensitive to the amount of data presented; that is, a 50-line resolution module would cost about the same as a 500-line resolution module.

Applications requiring just a few numerals or lines of text are the areas where non CRT based technologies have had their greatest impact. These displays typically have the asset of being flat and not taking up much room. Flat panel displays are typically addressed by an orthogonal matrix of leads with a display element at each intersection. When a particular x and y lead is energized with $\pm$V, the voltage across the intersection will be 2 V; at least twice that across any other intersection in the matrix. If the display medium

Light Emission
- Plasma Discharge
- Cathode Ray Tubes
- Electroluminescence
- Glowing Filament

Material Transport
- Deformed Membranes
- Rotating Balls
- Electrophoretic
- Colloidal Needle
- Oil Film

Optical Activity
- Liquid Crystal
- Ferroelectrics
- Magneto-optic

Electrochemical
- Electrochromics
- Reversible Electroplating
- Reversible Redox

Figure 1. Physical phenomena researched for display applications.

CRTs (all types)	$4,650
LED	$ 550
LCD	$ 486
Gas discharge	$ 110
VFDs	$ 71
Incandescent	$ 17
Other	$ 5
Total	$5,889

Industrial	$ 330
Color TV consumer	$4,080
Black & white TV consumer	$ 240
Total	$4,650

Source: *Stanford Resources*

Figure 2. a) Estimated worldwide market for electronic displays (millions of dollars, 1982). b) Estimated worldwide market for cathode ray tubes (millions of dollars, 1982).

is chosen with a sufficiently nonlinear response, only that element will be activated. Examples that work well with this technique are plasma and electroluminescence. This simple matrix addressing technique is by far the lowest cost but requires very special properties of the display media.[6] The alternative is to place active nonlinear circuit elements in series with each display pixel by, for example, fabricating the display on a wafer of silicon patterned with the appropriate circuits. An example of this technique is the miniature liquid crystal pocket TVs under active development in Japan.[27]

Light emitting diodes (LEDs) have emerged in the last decade as the most successful technology for numeric indicators. Their strength lies in their reliability, low cost, high brightness and compatibility with low voltage integrated circuits. Major disadvantages of LED displays are their high power consumption and difficulty in fabricating for high information content. Liquid crystal (LC) displays, the next largest market segment, are growing at a rapid rate. A major asset of liquid crystal displays is their extremely low power consumption. In addition, being passive, they have good visibility in high ambient light conditions. The market for non-watch LC displays is growing at 30 percent per year. Matrix-addressed LC displays are available capable of presenting 4 lines of 40 characters per line. I expect matrix-addressed LC displays to dominate the flat panel market by 1990. Plasma or plasma gas-discharge displays have been highly developed over the last few years. For large alphanumeric applications plasma displays represent the main challenge to CRTs. Vacuum fluorescent displays are used extensively as indicators in TVs, stereos, and electronic instrumentation. In recent years high-information-content vacuum-fluorescent displays have become available, although their size will always be limited by vacuum considerations.

CRT Displays

The conventional CRT is a well known device, widely used in a variety of systems ranging from oscilloscopes to the home television receiver. All CRTs consist of: a cathode, or source of electrons; a beam forming structure, consisting of various electrodes; a deflection apparatus to steer the beam; and a screen, consisting of a thin phosphor that is luminescent under electron bombardment. CRTs come in many forms but they are most easily characterized by whether they use electrostatic or magnetic deflection.

Electrostatic deflection is used where very high speeds are required in the deflection, such as in oscilloscope tubes. Scan expansion techniques are used to get the high deflection sensitivity required by high-speed amplifiers. In the past, high deflection sensitivity has been achieved with a domed mesh separating the electron gun from a high voltage (20 kV) accelerating field applied to the tube envelope and phosphor screen (see figure 3). This configuration results in a diverging lens which expands the scan and also results in high brightness due to the high energy imparted to the electron beam. The mesh causes some undesirable results such as reduced beam current due to interception on the mesh wires, halo due to scattered electrons off the mesh, and a reject rate due to contamination.

Recently Tektronix has introduced a new CRT product line based on a meshless scan expansion technique.[12] Meshless scan expansion is accomplished with a lens between the deflection plates and the screen which applies an accelerating field with a large quadrupole component to the beam. The scan in one axis is diverged and in the other axis is strongly converged so that the beam trajectories cross the axis and then expand again.[5] This configuration achieves a net scan expansion in both axes without the undesirable features of a mesh. This relatively new design also achieves a higher deflection sensitivity, higher brightness, and a smaller trace width than the domed mesh design.

To achieve very high beam brightness, other electrostatic tubes use a microchannel plate multiplier that is placed in close proximity to the screen.[11] The microchannel plate consists of a large number of small (25 micron) diameter channels etched in a thin sheet of glass. The walls of the channel are made slightly conductive. When a voltage is applied across the plate input electrons cascade down the channel and are multiplied by secondary emission. Gains of 10^4 are easily achieved.[16] Microchannel plate multipliers are typically used to display very high speed transients or low duty factor events. A major factor limiting their more widespread use is the high cost of the multiplier plate.

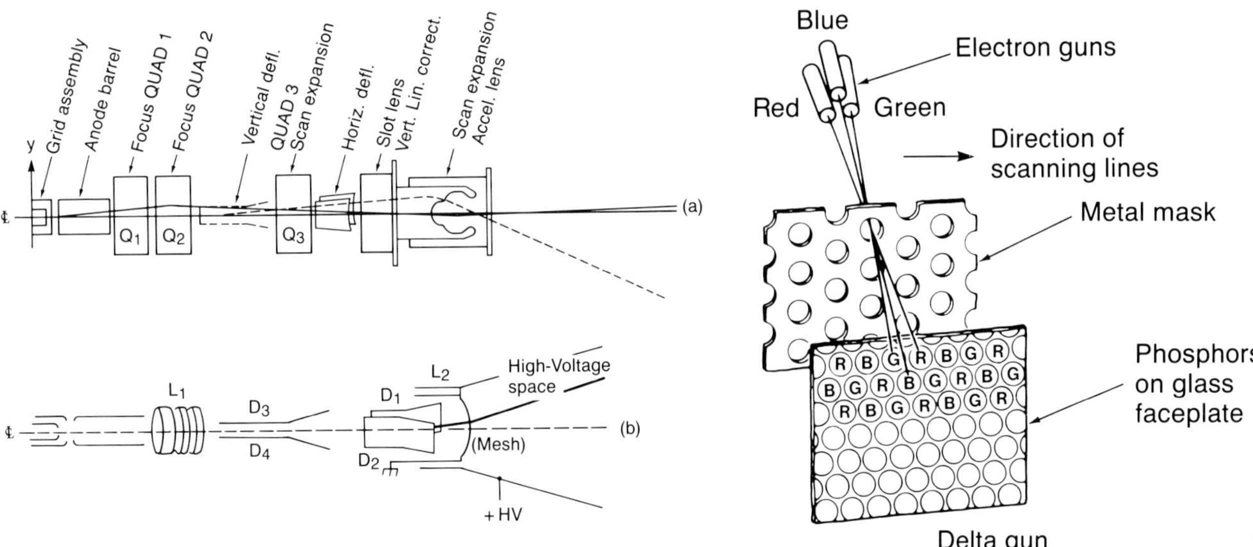

Figure 3. Electrostatic deflection CRTs commonly used in oscilloscopes; a) Tektronix meshless scan expansion CRT b) Mesh-lens post deflection acceleration CRT.

Figure 4. Principle of color selection of the delta gun shadow mask color tube.

Magnetically deflected CRTs form the greater bulk of the entire display market. These CRTs can be characterized by whether they are stroke written or raster scanned and whether they are monochrome or color. The trend in recent years is toward high resolution raster-scan color shadow-mask CRTs.

One type of stroke-written CRT is the bistable storage tube introduced by Tektronix in the early 1970s in a line of graphic computer terminals. With this tube an image is stored on the phosphor using its secondary emission properties and viewed by flooding the screen with a low voltage (200 V) spray of electrons.[14] The bistable storage technology offers resolution up to 4000 × 4000 pixels and gives superior graphics capability at low cost. Disadvantages are low brightness, lack of grey scale, slow update, and monochrome only display. For these reasons direct view storage tubes (DVST), as they are called, are rapidly losing market share. Conventional stroke-written CRTs can be either color or monochrome. Color can be achieved by beam penetration or by use of a shadow mask. The phosphor of a beam penetration tube consists of two layers, each layer luminescent with a different color. With a low energy electron beam (8 kV) only the outer layer of phosphor is excited. When the beam energy is higher (15 kV) both layers are excited and the display has a different color. Stroke-written CRT display systems are typically used to present stick-figure images in computer-aided design work stations, a task they do very efficiently with a minimum of computer power. This conservation of computer power is especially important where dynamic images are to be presented. With the advent of increased computer power this advantage is of less importance and stroke-written systems are losing market share to raster-scanned systems.

High-resolution raster-scanned color shadow-mask tubes are the most rapidly growing display market segment. The principle of color selection is shown in figure 4.[7] The red, green, and blue beams, typically arranged in a delta configuration, come through the shadow mask holes at slightly differing angles and therefore can only illuminate their respective phosphor dots. Each screen is individually fabricated for a particular mask/panel combination with an optical exposure that uses the shadow mask itself as the exposure mask. The process requires tight process control but lends itself well to mass production. In recent years the Japanese have become the dominant supplier by a significant margin. Since three beams are used some means must be provided for convergence so that all three rasters fall on top of one another. This is typically done by dynamic convergence coils on the neck of the CRT and a rather complicated calibration procedure. In recent years the trend in color tube design has been toward higher resolution and to the use of in-line self-converging guns and yokes.[19] Screens with resolution of 0.012 in (0.3 mm) between triads are available from a number of suppliers in sizes up to 19 inch (482.6 millimeter) diagonal. Screens with a resolution of 0.008 in (0.2 mm) are beginning to become available. This dimension is near the resolution limit of the eye at normal viewing distances. The basic advantage of the in-line tubes is self convergence. The combination of in-line guns and a yoke especially designed to give higher-order harmonics in the deflection field allows convergence of the three beams over the entire screen. The need for dynamic convergence circuits and magnetic drivers on the CRT neck, as used on the delta gun tubes, has been eliminated at a substantial cost reduction and ease-of-use improvement. A disadvantage of self-converged, shadow-mask tubes is a somewhat larger spot size compared to non-self-converged tubes. Recently an autoconvergence technique has been introduced based on a photomultiplier pickup of a pattern deposited on the gun side of the shadow mask.[3] This technique eliminates user-controlled convergence adjustments and may breathe new life into the delta gun approach.

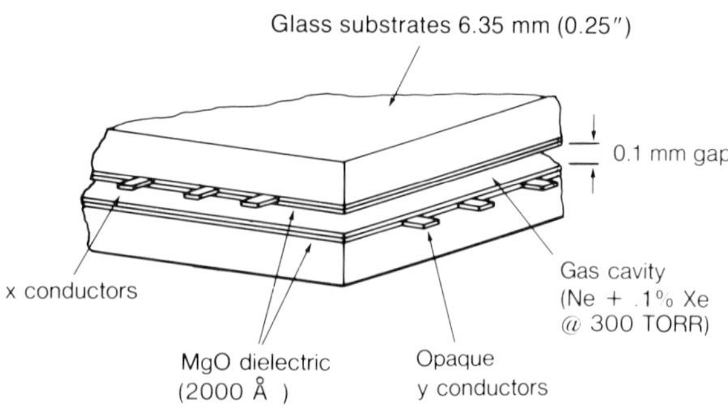

Figure 5. AC plasma panel construction.

Plasma Displays

Gas-discharge displays utilize the light output of a cold-cathode discharge.[22] The process is relatively efficient; especially in neon for which the characteristic orange emission has an output efficiency of .05 Lumens per Watt. A major advantage of gas discharge for display devices is the existence of a sharp threshold. This allows a large number of lines of information to be matrix addressed without crosstalk. An example of a plasma display cell is shown schematically in figure 5. The typical plasma display panel consists of a matrix of gas discharge cells defined by two sets of orthogonal electrodes. These are deposited on two glass substrates which are spaced and filled with a neon-argon gas mixture. Spacing of 0.004 in (0.1 mm) and pressures of 100 torr are typical values. Plasma panels may be operated in two basic modes, AC and DC.

In the AC mode the electrodes on the device are covered with an insulator film.[21] A charge deposited on the insulator during the gas discharge aids the polarity-reversed applied voltage on alternate half-cycles to break down the gas again. Thus, once initiated by a writing pulse, the gas discharge can be maintained with a lower value of applied sustaining voltage and the panel exhibits memory. The use of the insulator eliminates the need for current limiting resistors that would be required for bare electrodes. AC plasma panels typically do not exhibit grey scale since they are essentially bistable memory devices. This memory eliminates the need for display refresh and the associated flicker. In addition, the brightness is independent of display area. The maximum size of AC plasma displays is therefore limited only by manufacturing constraints such as the ability to manufacture large fine electrode grids at high yields. AC plasma technology is the basis for the largest non-projection displays in the industry. Photonics Technology offers a display panel measuring one meter diagonally and has plans for even larger panels in the future. The IBM 3290 terminal, recently announced, uses an AC plasma display. This display has a viewing area of 340×274 mm (13.4×10.8 in) and a resolution of 960×768 lines. This viewing area gives the terminal the ability to display 69 lines at 160 characters/line. Numerous halftone patterns are available to give the appearance of grey scale. The display subassembly itself costs $4,500 including associated electronics.[2]

DC plasma panels were widely used in the past under the trade names "Nixie" and "Self-Scan" manufactured by Burroughs Corp.[17] Burroughs has recently dropped this product line and is concentrating on a modification of the AC panel for future products. DC plasma panels typically do not exhibit memory and must be constantly refreshed. Display brightness is inversely proportional to the number of lines addressed since each line is only on that fraction of the time. There are several companies looking at DC plasma devices especially for TV application. This is because grey scale is relatively easy to achieve through current or duty cycle modulation. Sony Corp. is very active in DC plasma and has reported on a high-resolution panel that includes a trigger electrode in addition to the usual x y matrix. This trigger electrode greatly reduces the voltage and thus the cost of the drive electronics.[1]

Major disadvantages of plasma panels are fabrication and electronics costs and the difficulty in achieving full color output. Nevertheless plasma panel applications will continue to grow where the form factor is of chief importance such as for highly portable instrumentation and military applications.

Liquid Crystals

Many organic molecules, due to their polar nature and elongated shape, exhibit long range orientational structure in their liquid state. This intermediate state between solid and liquid is known as the mesomorphic-or-liquid-crystal state. This state occurs in a relatively narrow temperature range, the material being either solid or liquid outside this range. Liquid crystals exhibit an anisotropic index of refraction and dielectric constant typically described by two constants, one along the long axis and one perpendicular to it. As a result the material exhibits a birefringence effect that can be controlled by external electric fields or heat. This effect is of prime importance in display applications.

Liquid crystals may be classified into three basic types: nematic, smectic, and cholesteric. In nematic liquid crystals the rod-like molecules line up parallel to one another. Smectic liquid crystals exhibit a layered structure with little order within a layer. Cholesteric liquid crystals are similar to nematic with a gradual helical twist normal to the director.

The most commonly used liquid crystal displays use the twisted nematic field effect.[15] The display consists of two glass plates with rows and columns of transparent conductors. Sandwiched between the glass plates is the liquid crystal mixture 10 to 50 microns thick. The alignment of the molecules is arranged parallel to the glass plates by evaporating a thin layer of SiO onto the glass at an oblique angle or by rubbing the surface. On assembly the plates are mounted with the alignment perpendicular to one another. As a result the molecules exhibit a 90 degree twist across the cell (see figure 6). This structure has the effect of giving light, plane polarized in the direction of one rubbing axis, a 90 degree twist as it passes through the cell. When a voltage is applied to the cell the molecules line up perpendicular to the glass surfaces, and plane polarized light passes through unaffected. The addition of polarizers on both sides of a display gives either a white-on-black or a black-on-white display. Since the effect is entirely due to the field, the total power can be a few microwatts per square centimeter. The twist cell has a slow response (50 ms) and a soft threshold which limits the number of lines that can be addressed in a simple matrix fashion. Nevertheless, panels are available with a resolution of 4 lines of 40 characters at low cost; and this is improving month by month.[20]

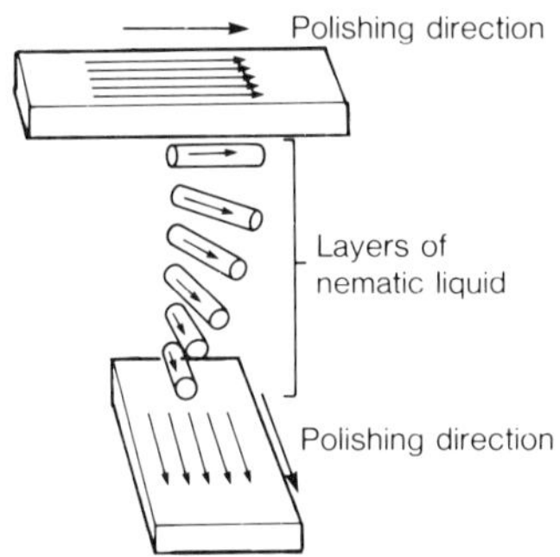

Figure 6. Alignment of molecules in twisted nematic liquid-crystal cell.

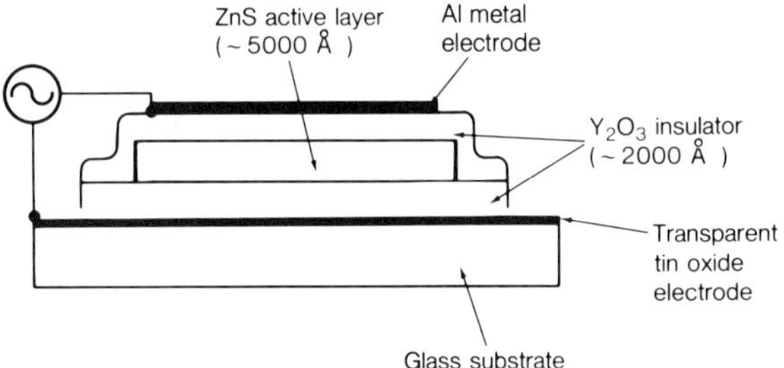

Figure 7. Structure of thin film EL device.

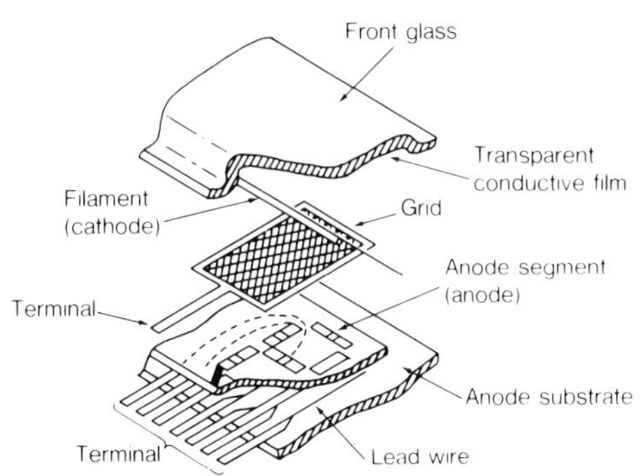

Figure 8. Vacuum fluorescent display.

The dye liquid crystal cell is another approach. In this case a plechroic dye, which has molecules exhibiting anisotropic absorption, is dissolved in the liquid crystal host. Light is attenuated by differing amounts depending on the orientation of the liquid crystal molecules and no polarizer need be used.[25]

One solution to the matrix addressing problem is the active substrate liquid crystal displays. This approach uses an array of transistors, often fabricated on a large single-crystal silicon wafer. One transistor is provided for each pixel allowing faster switching, better grey scale, and a much higher information content. Difficulties in making large defect-free transistor arrays has so far limited the commercialization of this technique, but it is an area of very active research, especially in Japan.[13,18]

Thermally-addressed smectic liquid crystal displays require the presence of both heat and electric field for activation. For these a smectic liquid crystal is used that freezes either clear or cloudy depending on whether or not a voltage is applied during cooling. This effect yields a storage display. The image is normally viewed in projection and is capable of very high resolution. Recently IBM has reported on a 64 million pixel display which uses an array of scanned solid-state lasers to supply the heat.[4]

The advantages of liquid crystal displays, namely low power, low voltage, passivity, and low cost, outweigh the disadvantages. So the use of liquid crystal displays is rapidly increasing.

AC Electroluminescence

AC electroluminescence (ACEL) is a phenomenon where light is more or less uniformly produced in the bulk of a material under the application of a strong electric field. It has been investigated by numerous researchers in various configurations.[9] The most popular structure of the device is shown in figure 7. The light generating mechanism is believed to be due to impact excitation of the manganese ion. The presence of the dielectric layers prevents current runaway and promotes a more uniform current distribution. The advantages of the structure were first demonstrated by Inoguchi et al in 1974.[10] They produced electroluminescent devices and ran them more than 20,000 hours at high-brightness levels up to 1000 fl (3400 candela/meter^2)without degradation. The encapsulating feature of the dielectric films protecting the active layer is no doubt partly responsible for the long life.

An attractive feature of ACEL is the extreme nonlinearity of the brightness voltage curve coupled with high peak brightness. This feature allows a large number of lines to be addressed, without crosstalk, in a simple matrix fashion before the display becomes too dim. Peak brightness of several thousand foot lamberts is readily attainable making displays of several hundred lines practical. A major disadvantage of ACEL devices is the high capacitive reactance (Xc) that must be driven and the relatively low efficiency of light generation.[23] This Xc together with high voltage requirements leads to a high cost for the drive electronics. The extreme ruggedness and excellent viewability of ACEL displays makes them very attractive for certain applications such as for military devices.

Vacuum Fluorescent

A vacuum fluorescent display consists of a matrix-addressed triode structrure in a glass enclosure. A schematic of a vacuum fluorescent display is shown in figure 8. When positive potentials are applied to a grid lead and to a phosphor-coated anode lead the phosphor at the intersection will glow and nowhere else. Thus multiplexing is very easy in vacuum fluorescent devices. The manufacture of large, uniform high-resolution structures in the vacuum envelope is very difficult, however, and limits both resolution and cost. The display is a strong competitor in the medium size display market due to its high brightness and wide viewing angle.[26] Color will be available, at a loss of resolution, by the use of red, green, and blue phosphor strips.

Projection and Light Valve

In the past few years there has been considerable development in projection television displays. Two basic types are being investigated. One type is direct CRT projection using either three tubes or a single very bright shadow mask tube.

Several manufacturers are producing these systems in the $3000 range for use in home entertainment at a screen size of 3 to 6 feet (approximately 1 to 2 meters). Brightness has improved dramatically and resolution today seems fully compatible with standard 500 line NTSC TV.[8] Brightness has been acccomplished with very high voltages (up to 50 kV) and improved light gathering optics. Future developments are likely to be rather less dramatic, however, and the system performance is probably not compatible with displaying 1000-line high definition TV.

Another type of projection system uses a medium to gate the illumination from a very bright light source called a light valve. General Electric is the dominant supplier in this area. In the GE system an electron beam deposits charge on a thin oil film causing the film to ripple. The projector uses an incredibly novel schlieren optical system to transform the ripple pattern into a full color projection image at high brightness and high resolution. The system can project a 1000 lumen image at 800 line resolution but costs $80,000.[24] The device is commonly used to project closed circuit sporting events in auditoriums and has become quite popular. Improvements are underway to increase resolution and reduce system cost. Other light valve systems under development use an electron beam or light-addressed liquid crystal target as the means to form images. It is felt that light valve systems will increase their market dramatically by lowering system cost and with the market need presented by high definition TV.

References

[1] Amano Y, Yoshida K, Shionoya T, Yokono S, (1982) "New DC plasma display panels and their applications," *Displays,* Vol. 3, No. 4, pp. 187-191.

[2] Apperly N, Pleshko P, Pearson K, Pierre E, Sherk T, Hairabedian B, Bradney F, Foster R, (1984) "Design of a plasma flat-panel large-screen display for high-volume manufacture," *Displays,* Vol. 5, No. 1, pp. 21-31.

[3] Denham D, Murch J, Meyer B, Singer M, (1984) "Autoconvergence enhances high-resolution display," *Electronic Imaging,* Vol. 3, No. 1, pp. 30-34.

[4] Dewey A, Anderson S, Cheroff G, Feng J, Handen C, Johnson H, Left J, Lynch R, Marinelli C, Schmiedeskamp R, (1983) "A 64 million PEL liquid-crystal-projection display," *Society for Information Display Technical Digest,* Vol. 14, Los Angeles, CA 90049, pp. 36-37.

[5] Franzen N, (1983) "A quadrupole scan-expansion acceleration lens system for oscilloscopes," *Society for Information Display Technical Digest,* Vol. 14, ibid., pp. 120-121.

[6] Goede W, (1978) "Flat-panel displays: A critique," *IEEE Spectrum,* Vol. 15, pp. 26-32.

[7] Herold E, Morrell A, Law H, Ramberg E, (1974) "Color Television Picture Tubes," *Academic Press,* NY, pp. 42-129.

[8] Hochenbrock R, Rowe W, (1982) "Self-converged three-CRT projection TV system," *Society for Information Display Technical Digest,* Vol. 13, ibid., pp. 108-109.

[9] Howard W, (1981) "Electroluminescent display technologies and their characteristics," *Proc. Society for Information Display,* Vol. 22, pp. 47-56.

[10] Inoguchi T, Takeda M, Kakihara Y, Nakata Y, Yoshida N, (1974) "Stable high-brightness thin-film electroluminescent panels," *Society for Information Display Technical Digest,* Vol. 5, ibid., pp. 84-85.

[11] Janko B, (1979) "A new high-speed CRT," *Proc. Society for Information Display,* Vol. 20, No. 2, pp. 55-56.

[12] Janko B, Franzen N, Sonneborn J, (1983) "CRT architecture with meshless scan expansion," *Society for Information Display Technical Digest,* Vol. 14, ibid., pp. 118-119.

[13] Kasahara K, Sakai K, Okada Y, Yanazisawa T, Matsumoto S, Hori H, (1983) "A 220 × 180 element MOS-LCD for 2 inch picture displays," *Proc. 3rd International Display Research Conference, Japan Display, '83,* Lewis Winner, Coral Gables, Florida, 33134, pp. 408-411.

[14] Kazan B, Knoll M, (1968) "Electronic Image Storage," *Academic Press,* NY, pp. 225-230.

[15] Kemtz A, Von Willisen F, (1976) "Nonemissive Electro-optic Displays," *Plenum Press,* pp. 9-24.

[16] Leskovar B. (1977) "Microchannel plates," *Physics Today,* Vol. 30, pp. 42-48.

[17] Maloney T, (1978) "Self-scan imaging display panel," *IEEE IEDM Technical Digest,* pp. 198-200.

[18] Morozumi S, Oguchi K, Yazawa S, Kodaira T, Ohshima H, Mano T, (1983) "B/W and color LC video display, addressed by poly Si TFTs," *Society for Information Display Technical Digest,* Vol. No. 13, ibid., pp. 156-157.

[19] Morrell A, (1983) "Future development of cathode ray tubes," *Proc. 3rd International Display Research Conference, Japan Display,* ibid., pp. 264-267.

[20] Odawara K, Ishibashi T, Kinuzawa K, Sakurada H, Tanaka H, (1980) "An 80-character alphanumeric liquid crystal display system for computer terminals," *Proc. Society for Information Display,* Vol. 21, pp. 79-83.

[21] Pleshko P, (1979) "AC plasma display technology overview," *Proc. Society for Information Display,* Vol. 20, pp. 127-130.

[22] Sobel A, (1977) "Gas discharge displays: The state of the art," *IEEE Trans. Electron Devices,* Vol. Ed 29, pp. 835-847.

[23] Takahara K, Kawada T, Yamaguchi H, Andoh S, (1983) "A refresh addressing technique for thin-film EL panels," *Proc. 3rd International Display Research Conference, Japan Display '83,* ibid., pp. 578-581.

[24] True T, (1979) "Recent advances in high-brightness and high-resolution color light value projectors," *Society for Information Display Technical Digest,* Vol. No. 10, ibid., pp. 20-21.

[25] Uchida T, (1981) "Bright dichroic guest-host LCDs without a polarizer," *Proc. Society for Information Display,* Vol. 22, pp. 41-46.

[26] Uchiyama M, Masuda M, Kiyozumi K, Nakamura T, (1982) "High resolution vacuum fluorescent display with 256 × 256 dot matrix," *Proc. Society for Information Display.* Vol. 23, No. 3, pp. 163-168.

[27] Williams H, (May '83) "Liquid crystal television – here at last," *Electro Optics,* pp. 55-57.

Electron Optical Systems (pp. 237-244)
SEM Inc., AMF O'Hare (Chicago), IL 60666-0507, U.S.A.

0-931288-34-7/84$1.00+.05

A COMPUTER CONTROLLED SCANNING TRANSMISSION ELECTRON MICROSCOPE EQUIPPED WITH AN ENERGY ANALYZER FOR SPECIAL INVESTIGATIONS ON ELECTRON DIFFRACTION- AND CHANNELING PATTERNS

W. Hylla, H.-J. Kohl, H. Niedrig, D. Wendtland

Optisches Institut der Technischen Universitat Berlin
Sekretariat P 11
Strasse des 17. Juni 135
D-1000 Berlin - 12
W. Germany

Abstract

A scanning electron microscope was equipped with a double tilting stage, driven by stepping motors, to investigate electron channeling patterns (ECPs) and large angle convergent beam patterns (LACBPs) of single crystals. Transmitted electrons may be energy-selected by a magnetic sector-field energy analyzer. The recording of experimental data and the experimental arrangement are controlled by a microprocessor system, including a picture storage unit of 512 x 512 pixels of 16 bit. Recorded patterns can be stored on 1 Megabyte floppies.

A set of useful programs allows one to perform calculations with stored patterns, e.g., contrast enhancement or -inversion, noise reduction, difference or quotient of two patterns etc. The possibility of background subtraction (e.g., in patterns recorded with characteristic energy loss electrons) allows one to get true K-loss convergent beam patterns. Other recording modes allow one to get two CBPs simultaneously recorded with electrons of different energy losses, to measure angle dependences of energy selected electrons, or to take electron energy loss spectra.

A special processor program generates a theoretically calculated CBP or ECP on the TV screen and prints out a list of all band edges up to a chosen limit of Miller indices (hkl). The program requires the coordinates of two known poles and some crystallographic properties of the investigated material. Thus complete indexing of recorded diffraction patterns is easily possible.

The system has been applied, e.g., to investigate localization effects of electron Bloch-waves in graphite.

Key Words: Convergent beam patterns; crystal orientation, digital image recording, electron channeling patterns, electron diffraction, electron energy loss spectroscopy, inelastic scattering, rocking-crystal.

Address for correspondence:
H. Niedrig, Optisches Institut der TU Berlin
Strasse des 17. Juni 135, D-1000 Berlin - 12
W. Germany
Phone No.: 030 - 314 2735

Experimental Set-up

Electron Optics

An electron optical arrangement has been built up, which offers some very special possibilities of investigations on electron scattering processes. The column consists mainly of the probe forming system (tungsten cathode, double condenser, final lens with scan-coils inside), a double-tilting stage for backscattering and transmission, driven by stepping motors, and an energy-selecting double focussing magnetic sector-field analyzer /2/.

Tilting-stage

The specimen is mounted on a stage in a Cardanic suspension, which permits tilting of ± 20° in both axes. Additional stepping motors for sample shift and the possibility of mechanical shift of the whole assembly allow a) to adjust the selected specimen area into the pivot point of the tilting-stage, and b) to adjust the pivot point into the electron optical axis. Two secondary electron detectors of Everhart-Thornley type and a surface barrier semiconductor ring detector are mounted above the specimen. Below the ground plate, a second semiconductor ring detector collects transmitted electrons, scattered into angles greater than ≈ 20 mrad (dark field detector). A magnetic sector-field energy analyzer with a maximum acceptance angle of $2\alpha = 60$ mrad provides the energy distribution of the transmitted electrons, or allows to get energy-selected transmission diffraction patterns /4/.

Microprocessor control

The stepping motors for the tilting movement of the specimen and the magnetic coil current of the energy analyzer are controlled by a microprocessor system (Z 80 based), which also manages data acquisition /3/. Accessing to a 256 kbyte memory, video pictures containing 512 x 512 pixels of 16 bit (65,536 grey levels) can be stored and read out with normal TV frequency. Permanent storage is possible on 1 Mbyte floppies with a transfer time of around 50 seconds. The processor was programmed to perform several calculations using an implemented APU (arithmetic processing unit). Offset subtraction, optimum contrast expansion, noise reduction by averaging, difference- or quotient - patterns and so on can be performed quickly (calculation times up to 3 min.). The set-up in principle is shown in fig. 1.

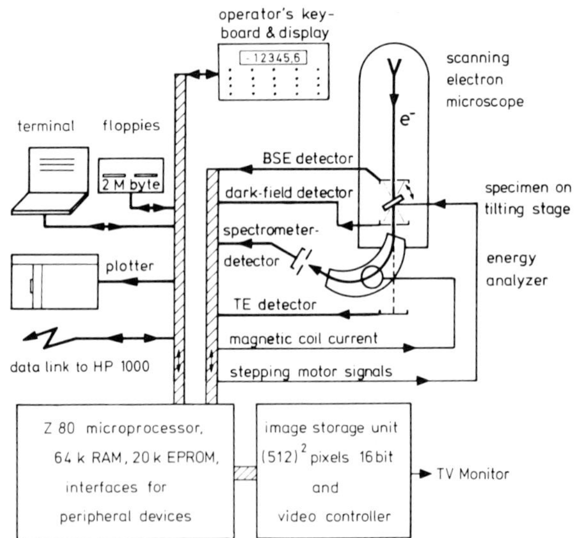

Fig.1: Experimental set-up for computer-controlled recording of energy-selected electron diffraction patterns.

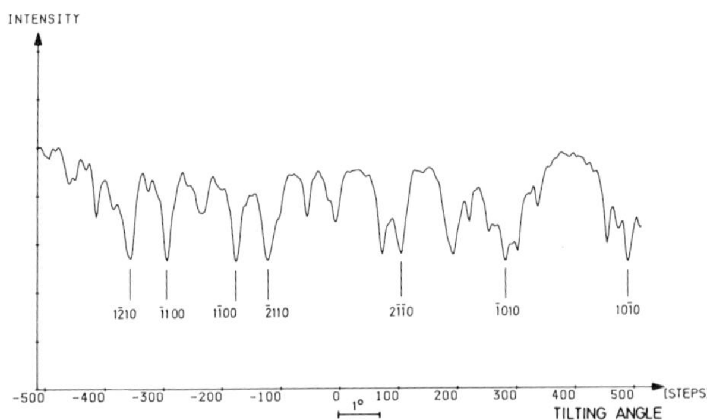

Fig.2: Rocking-curve (linescan) through a graphite large angle convergent beam pattern near the 001-pole. Band edges of low order are marked with their Miller indices.

Energy-selected Convergent-Beam-Patterns (CBPs)

Recording modes and some examples

Generally, three different kinds of recordings can be performed under control of the microprocessor: 1) Rocking-curves (linescans) are obtained by tilting the crystal in one axis (see fig. 2). 2) Diffraction patterns are obtained by rocking the crystal sequently in both axes: Electron Channeling Patterns, using the backscatter detector, Large-Angle-Convergent-Beam Patterns, using the transmission detector (see fig. 3 a,b). 3) Energy-loss spectra of transmitted electrons are obtained by ramping the magnet-coil current of the energy-analyzer (see fig. 4 a,b,c). By suitable programming of the microprocessor, some very special investigations become possible, e.g.: simultaneous recording of two energy-selected convergent-beam-patterns at different energies; creating pictures of energy-selected rocking-curves, lying one upon another, recorded with increasing energy-loss ΔE, thus showing the dependence of different energy losses on the diffraction contrast in a rocking-curve.

Comparison of two energy-selected LACBPs, recorded at different energy-losses

The simultaneous recording of two LACBPs at different energy losses is managed by the processor by switching the energy-analyzer for each increment of tilt between the two preselected energies. In this way, the two recorded patterns become directly comparable as they are not affected (or both in the same way) by long time effects like contamination or intensity drifts etc. That is important if it is intended to do quantitative calculations with the two patterns. The example, given below (fig. 6) points out the differences in two LACBPs, the first taken with electrons of the multiple scattering background, and the other with electrons which have suffered a characteristic

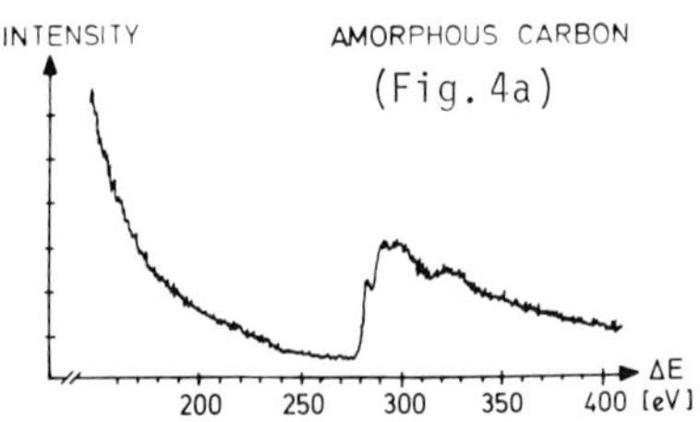

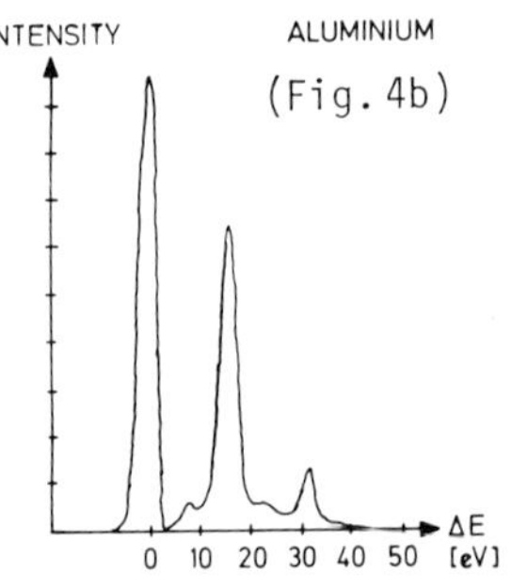

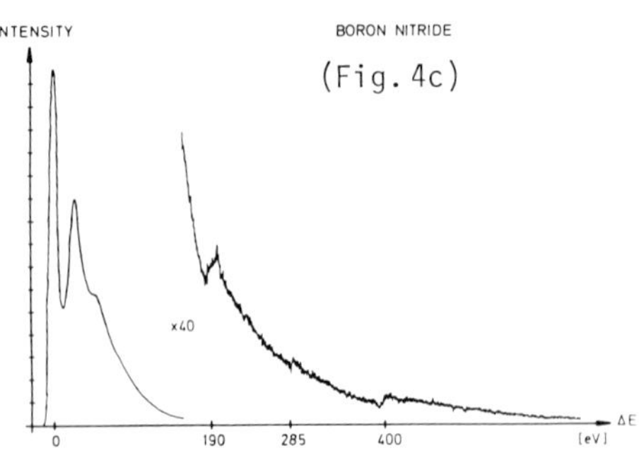

Fig.4: Three examples of electron energy loss spectra:
a) Carbon K-loss of an evaporated thin film (E_0= 20keV)
b) Aluminium plasmon losses of a polycrystalline film (E_0= 20keV)
c) Spectrum of tapered Boron Nitride (E_0= 30keV).

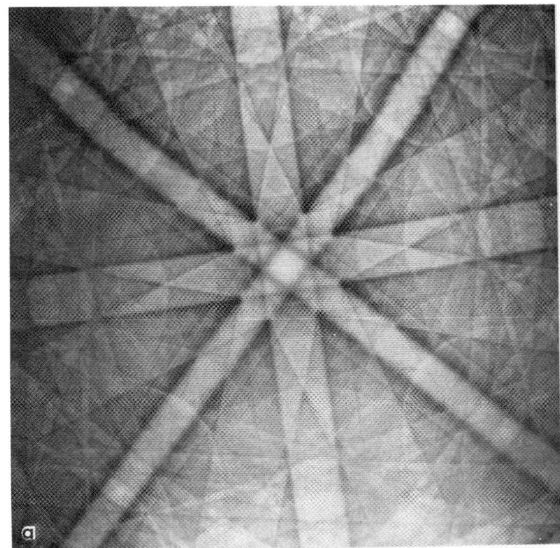

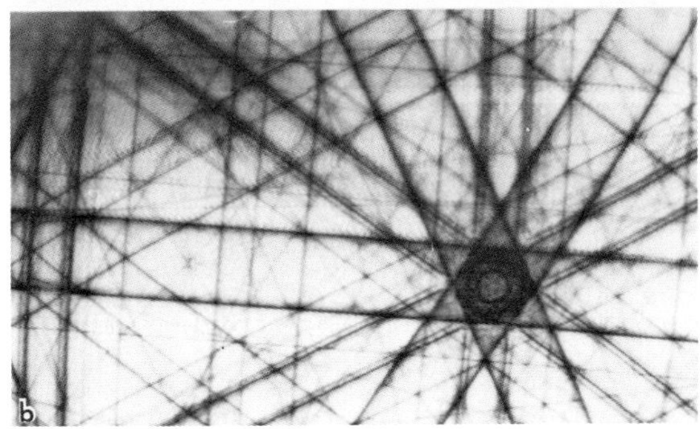

Fig.3: Electron diffraction patterns obtained through the rocking-crystal method:
a) Electron channeling pattern of 100-orientated Germanium (28°x28°, E_0 = 40keV)
b) Large angle convergent beam pattern of graphite (28°x14°, E_0 = 40keV).

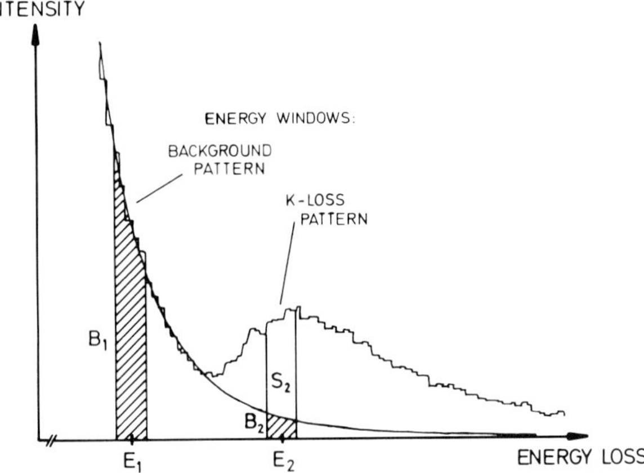

Fig.5: Computer performed backround-fitting and subtraction for recording of true characteristic energy loss patterns. The background pattern at E_1 and the K-loss pattern at E_2 are recorded simultaneously.

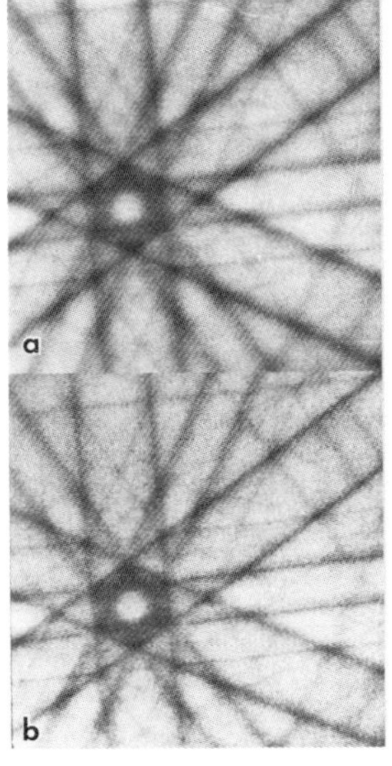

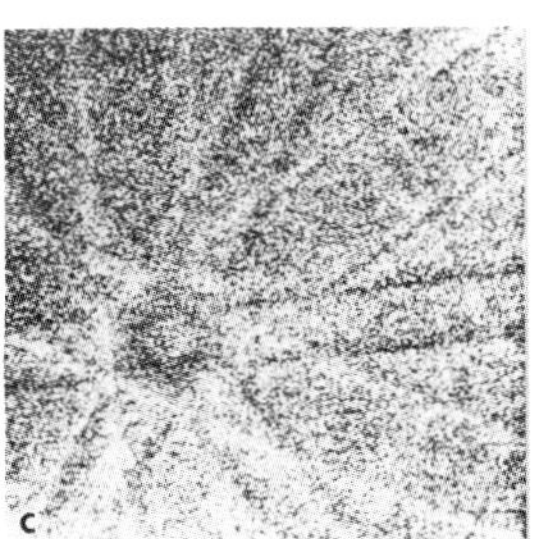

Fig.6: Comparison of energy-selected LACBPs of graphite:
a) Background pattern recorded around 50eV below K-loss energy
b) K-loss pattern (background stripped) recorded simultaneously with a) at 15eV above K-loss energy
c) Difference pattern of the equalized patterns a) and b).

energy-loss. To obtain a true characteristic energy-loss pattern, it is necessary to subtract the background, due to multiple scattering processes. This is done as follows:

- Processor performed background fitting, using the inverse power law $I = A\cdot(E-Eo)^{-r}$ with I: Intensity, E: Energy loss, Eo: Energy offset; A, r: fitting parameters. This formula is slightly modified, compared to that in common use /7/.
- Calculation of the ratio B_1/B_2 from the spectrum (see fig. 5)
- Taking the intensity of a pattern pixel, recorded at E_1 and calculating the corresponding B_2, using B_1/B_2.
- Subtracting the calculated B_2 from the pixel intensity, recorded at E_2.

This procedure is performed for all of the 131,000 pixels of the characteristic loss pattern. The only assumption of this procedure is, that the shape of the multiple scattering background remains always the same, independent of the angle of incidence of the primary electrons. To minimize the error due to long time effects (such as increasing background caused by contamination of hydrocarbons) a new spectrum is recorded and stored at the end of each rocking-curve, i.e. each line of the picture. Fig. 6 shows two LACBPs of a graphite crystal, taken simultaneously in that way, the K-loss pattern background stripped /6/. For direct comparison, both patterns were pushed to the same value of mean intensity and then a difference pattern

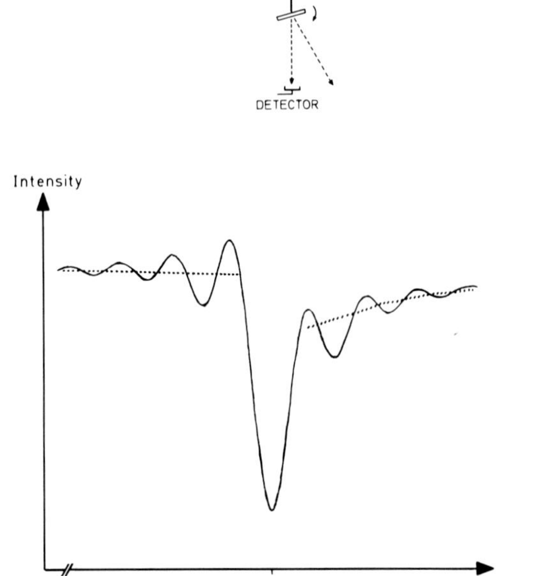

Fig.7: Two-beam approximation of K-loss intensity in a rocking-curve of graphite in the region of 01$\bar{1}$0 Bragg reflection.

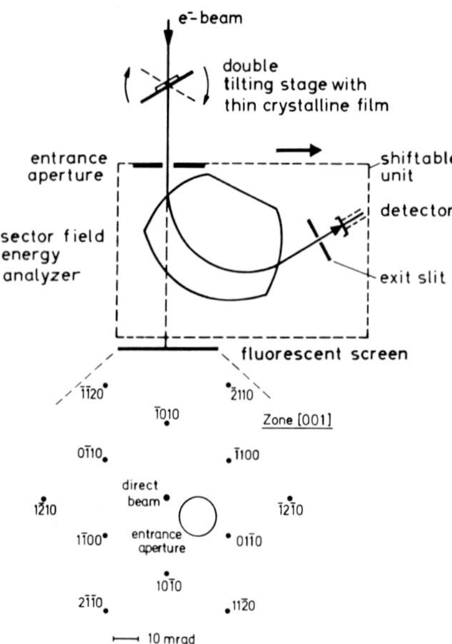

Fig.8: Diagrammatic sketch of the shiftable energy analyzer and enlargement of the graphite diffraction pattern in correct scale. The position of the entrance aperture to record fig.9c is marked.

"K-loss picture minus background picture" was calculated (fig. 6 c). This pattern clearly points out some differences of K-loss and background pattern: a) the dark band edges occur even in the difference pattern; that means a slightly higher diffraction contrast is present in the K-loss pattern. That can be understood, considering that electrons, contributing to the K-loss pattern, suffered only one scattering process, whereas electrons of the diffuse background were scattered several times, causing a loss of information of the elastic intensity distribution (diffraction contrast); b) the dark lines seem to be hemmed in by light seams, pointing out, that the dark lines (band edges) in the background pattern are slightly broader than those in the K-loss pattern. That also may be an effect of multiple scattering of the background electrons: The electron beam is broadened by repeated scattering which occurs as a softening of the Bragg condition.

Visualization of the localization effect in light elements

The experimental set-up of the energy analyzer provides a lot of freedom in adjusting the system in respect to the electron optical axis. Shifting of the whole analyzer assembly allows one to collect electrons which are scattered off the direct beam. The maximum scattering angle which can be accepted by an entrance aperture of 2α = 10 mrad is 30 mrad out of the direct beam. That can be used to emphasize the localization effect of Bloch-waves in light elements such as graphite. Localization effect of Bloch-waves means, that for angles of incidence deviating a small amount from the exact Bragg angle, the Bloch-waves arising under the periodic potential of the crystal are concentrated with their maxima of probability amplitude on or between the atomic sites, respectively. That leads to an asymmetry in intensity of rocking-curves through CBPs in the region of band edges (fig. 7). For the crystal thicknesses, used in our experiments ($t \approx$ 100...300 nm) the superimposed elastic diffraction contrast covers this weak effect, so that an additional suppression of diffraction contrast seems to be appropriate /8/.

This can be done for one selected band edge by shifting the entrance aperture of the energy-analyzer half the way between direct beam and the Bragg reflection corresponding to the band edge to be suppressed. Then, because of the symmetric position between direct beam and Bragg-reflection, the same amounts of scattering reach the shifted entrance aperture, no matter whether the main part of intensity is in the direct beam or in the Bragg-reflection, thus leading to elimination of diffraction contrast for this band edge.

As an example, we chose the (01$\bar{1}$0) band edge of a graphite LACBP to be suppressed. Fig. 8 shows the position and the acceptance angle of the entrance aperture, which can be adjusted by watching the diffraction pattern on a fluorescent screen below the analyzer with magnet coil current switched off. This figure shows the real proportions on the screen at 40 keV primary energy. Fig. 9 demonstrates the resulting changes in the K-loss LACBP. Fig. 9a shows the zero-loss pattern recorded in the direct beam, for indexing and comparison. In figs. 9b, c, the background-stripped K-loss patterns, recorded in the direct beam and with shifted entrance aperture, respectively, are compared. The pattern in fig. 9c

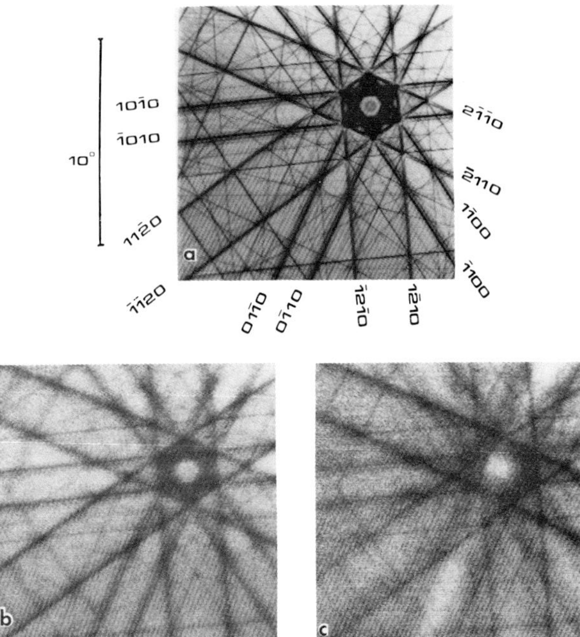

Fig. 9: Visualization of Bloch-wave localization effect:
a) Zero-loss LACBP (graphite, E_0 = 40keV), band edges of lowest indices are given
b) K-loss pattern (background stripped) recorded in the direct beam
c) K-loss pattern (background stripped) recorded with detector aperture shifted half the way between direct beam and 01$\bar{1}$0 reflection.

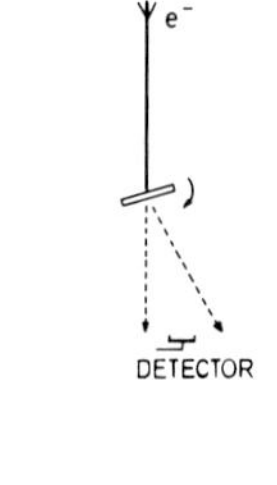

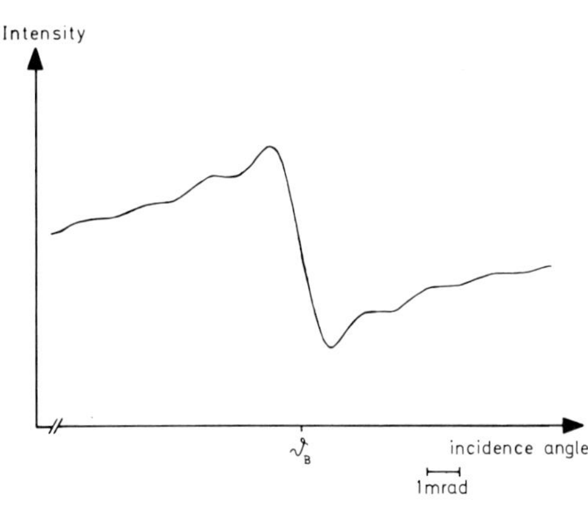

Fig.10: Two-beam approximation of K-loss intensity in a rocking-curve of graphite in the region of 01$\bar{1}$0 Bragg angle, detector shifted between direct beam and Bragg reflection.

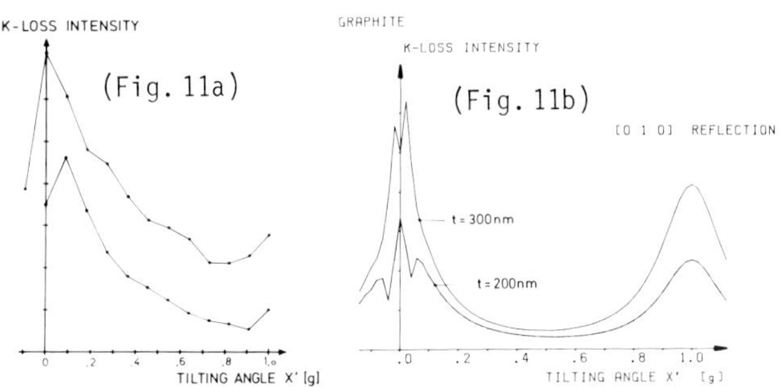

Fig.11: Angle dependence of Carbon K-loss electrons:
a) experimental results obtained for two different film thicknesses
b) two-beam approximation for 200nm and 300nm film thickness.

clearly points out the predicted effects: 1.- the sharp (01$\bar{1}$0) band edge, representing the diffraction contrast, has vanished, 2.- a remarkable asymmetry in the region of the suppressed band edge arose. That indicates the stronger scattering for incidence angles smaller than Bragg angle, corresponding to the excitation of mainly the strongly absorbed type II Bloch-waves with their maxima of probability amplitude just on the atomic sites. The predominant occurrence of the weakly absorbed type I Bloch-waves for incidence angles somewhat larger than Bragg angle, causes a region of decreased intensity outside the vanished band edge in fig. 9c. That experimental result is in good agreement with a two-beam calculation shown in fig. 10, based on Bethe cross-sections for K-shell ionization of carbon /1/, /5/.

This mode of recording may be of further interest for investigations on the crystal structures, the atomic sites or the inelastic scattering cross-sections of di- or poly-atomic crystals /11/.

Measurements of angle dependence of inelastically scattered electrons

The possibility of shifting the energy-analyzer in respect to the electron optical axis allows one to investigate angular distributions of inelastically scattered electrons, e.g., those which suffered a characteristic energy loss through inner shell ionization. Similar measurements have been carried out for aluminium and silicon by Oikawa et al. /10/. We made first measurements on thin graphite crystals, tilted to a Bragg position, so that nearly a two-beam case was realized. The angular dependence of the K-loss signal in the energy range from 295 up to 305 eV is shown in fig. 11a for two different film thicknesses. The site of maximum intensity is very sensitive with respect to the film thickness, because it is an effect of the subsidiary lines. This is also stated in calculated two-beam approximations which are given in fig. 11b. These results differ on principle from those of Oikawa et al. /10/, who found, that Bragg-reflected electrons do not excite inner-shell electrons, contrary to our measurements. For clarifying the problem, further measurements have to be carried out.

Processor program for complete indexing of channeling- and convergent-beam-patterns

The program ORIENT

Because in the majority of experiments we use single crystalline materials, we developed a simple computer program, written in FORTRAN IV, which allows complete indexing of channeling- and convergent-beam-patterns. This program is based on the geometrical conditions for Bragg reflection, and the main input data are: - coordinates of two known poles (e.g. taken from an experimentally obtained pattern), - dimensions of the crystal unit cell, - relative sites of the atoms in the unit cell, - covered angle range of the pattern to be generated, - primary electron energy, - number of band edges and lowest (hkl) may be selected, - all included poles or only poles of a certain type may be marked.

The program starts with the set of lowest allowed (hkl) and checks, if Bragg condition is fullfilled for any two points on the margin of the pattern. In this case, the related band edge (hkl) crosses the pattern, and the program runs in 512 steps through the total range of one tilting angle, calculating for each angle-coordinate the corresponding other one, which satisfies the Bragg condition. All pairs of coordinates found this way, representing the band edge (hkl), are set to a white (or dark) level in the image memory. The equations for calculation of the incidence angle between electron beam and tilted crystal coordinates are matched to the constructional realities of the Cardanic suspension of the double-tilting stage, leading to an according high conformity between experimental and calculated patterns.

As an example, fig. 12 shows an experimentally obtained LACBP of graphite, in comparison with a superposition of the calculated pattern. With the aid of the computer output, (fig. 13) every band edge can easily be identified, looking for its intersections with the margin of the pattern, and finding these tilting-step coordinates out of the given listing.

Orientation determination with the aid of ORIENT

We applied this possibility of complete determination of crystal orientation for realization of special investigations of anisotropic emission of atoms under ion bombardment /9/. For these investigations, it was necessary to cut a slice of single crystalline material out of a crystalline rod /12/ with a surface normal to the [11 3 1] direction. To determine the required tilting angles of the goniometric stage of the cutting device, a large angle ECP simulation with the same orientation as the face of the crystalline rod was calculated, and the angle coordinates of all [11 3 1] poles were printed out (fig. 14a). Then the goniometric stage was tilted by exactly these amounts which lead to an [11 3 1] pole, and a slice was cut out. The ECP we got from this slice is shown in fig. 14b. The [11 3 1] poles are marked, and we found the central one to deviate around 2.9° from the surface normal. These faults depend on the precision in adjusting the crystal on the tilting-stage, that means a) alignment of a marked direction on the crystal surface parallel to one tilting axis b) adjusting the surface normal parallel to the electron optical axis for zero-tilt position. We estimate that these faults can be minimized to less than ± 1° in total.

Input poles:

```
1. U V W =  1  1  2      X, Y =  -266.,   -154.steps
2. U V W =  0  1  2      X, Y =   410.,    172.steps
```

Pattern tilting range:

```
degrees: ( X:   -7.1  to    7.0 ) * ( Y:   -7.1  to    7.0 )
steps:   ( X:  -512.  to  510.  ) * ( Y:  -512.  to  510.  )
```

Listing of drawn band edges:

H	K	I	L	X-Y-Bragg (steps) 1.intersect.pt. with pattern margins		2.intersect.pt.		width (deg)	relative intensity
0	1	-1	0	-259.	510.	-512.	388.	1.62	.340
1	0	-1	0	-187.	510.	510.	30.	1.62	.340
-1	0	1	0	21.	510.	510.	174.	1.62	.340
1	-1	0	0	-348.	-512.	-277.	510.	1.62	.340
-1	1	0	0	-230.	-512.	-159.	510.	1.62	.340
0	-1	1	1	285.	-512.	510.	-405.	1.70	.554
1	1	-2	0	-392.	510.	510.	439.	2.80	.508
1	-2	1	0	-394.	510.	-512.	336.	2.80	.508
-1	2	-1	0	-148.	510.	-512.	-31.	2.80	.508
2	-1	-1	0	209.	-512.	-282.	510.	2.80	.508
-2	1	1	0	433.	-512.	-56.	510.	2.80	.508
0	2	-2	0	-123.	510.	-512.	322.	3.23	.086
2	0	-2	0	-290.	510.	510.	-42.	3.23	.086
-2	0	2	0	125.	510.	510.	245.	3.23	.086
2	-2	0	0	-406.	-512.	-335.	510.	3.23	.086
-2	2	0	0	-172.	-512.	-100.	510.	3.23	.086
0	-2	2	1	-512.	-141.	510.	354.	3.28	.143
0	2	-2	-1	-512.	-407.	510.	88.	3.28	.143
2	-2	0	1	226.	-512.	319.	510.	3.28	.143
-2	2	0	-1	464.	-512.	510.	-35.	3.28	.143
-2	0	2	1	492.	-512.	-512.	155.	3.28	.143
2	0	-2	-1	63.	-512.	-512.	-133.	3.28	.143
0	-2	2	2	141.	-512.	510.	-336.	3.39	.074

Poles:

U	V	W	P	Q	R	S	X, Y (steps)	
0	1	2	-1	2	-1	6	410.	172.
1	1	2	1	1	-2	6	-266.	-154.
1	2	4	0	1	-1	4	72.	9.

Fig.13: Computer output for identification and indexing of poles and band edges.

Summary

An electron optical device has been built up, which allows by means of a Z 80 microprocessor system some special investigations on electron scattering and diffraction processes. Energy-selected Large-Angle-Convergent-Beam-Patterns were taken and compared one to another. A method for visualization of localization contrast in light elements was pointed out. Angular distributions of inelastically scattered electrons were measured. ECPs and LACBPs were used to get a complete determination of crystal orientation. This was done by applying a developed microprocessor program.

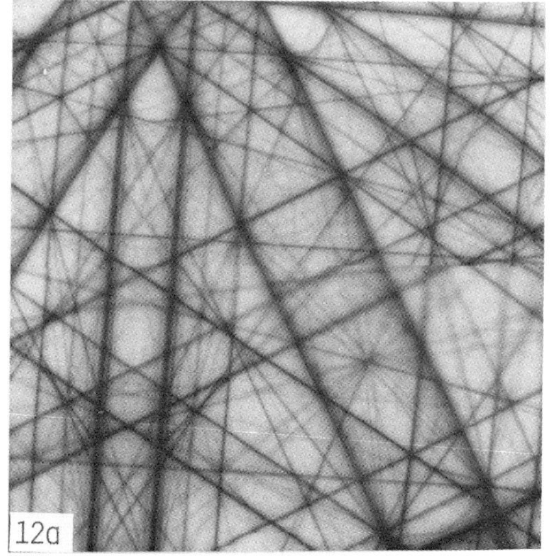

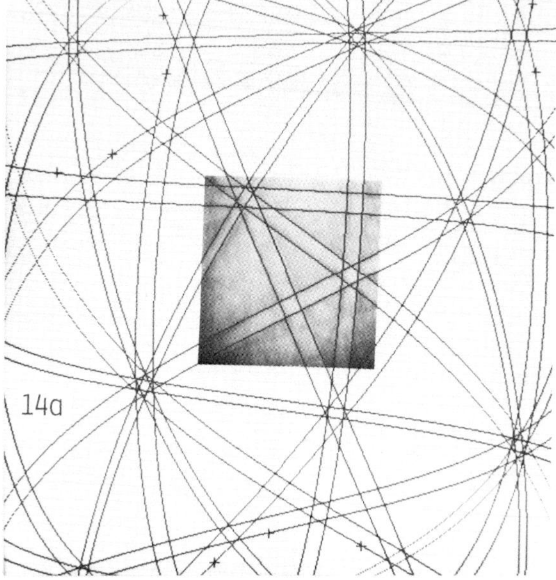

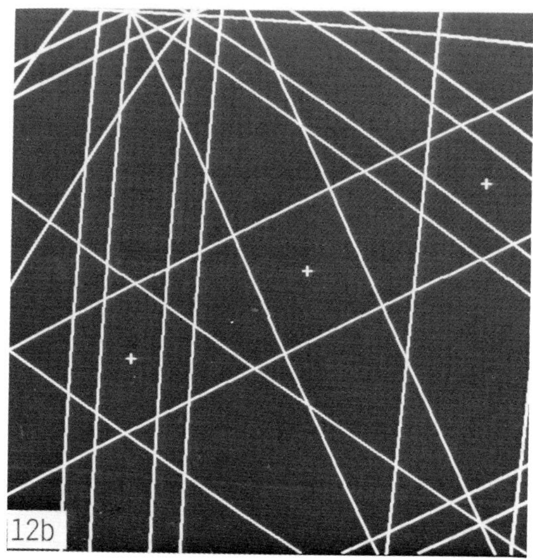

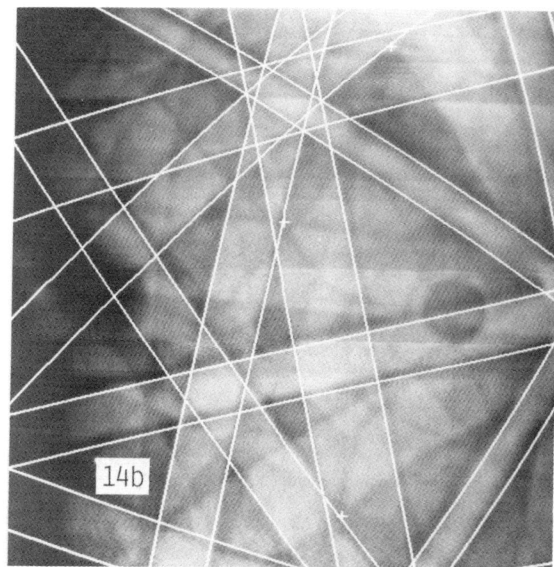

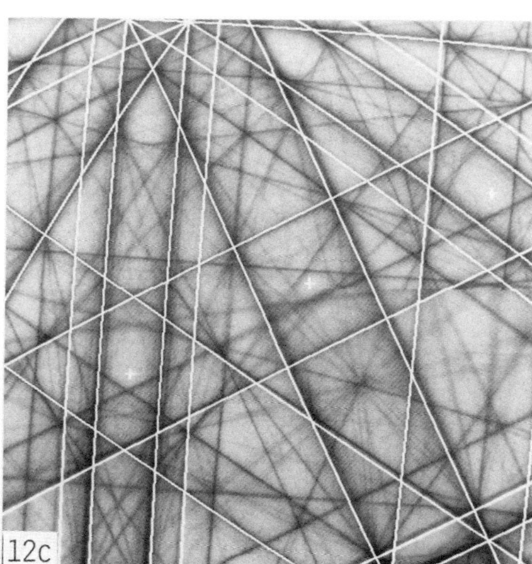

Fig.14: Application of orientation determination:
a) experimentally obtained ECP of a silver rod (central field, 14°x14°), overlaid with a large angle simulation in the same orientation to get the angle coordinates of poles of[11 3 1]- type (marked)
b) ECP of a silver slice cut with a surface normal nearly parallel to the[11 3 1]direction. The central pole (white cross) deviates 2.9° from surface normal.

Fig.12: Comparison of experimentally obtained and computer generated diffraction patterns:
a) LACBP of graphite near 001 pole
b) computer simulation of the 23 lowest indexed band edges
c) overlay of both patterns in the image storage unit.

Acknowledgements

This work has been supported by grants of the "Deutsche Forschungsgemeinschaft".

References

/1/ Bethe H. (1930). Zur Theorie des Durchgangs schneller Korpuskularstrahlen durch Materie. Ann. Phys. 5, 325-400

/2/ Brunner M. (1981). Rocking Crystal Electron Channeling Patterns, Backscattering and Transmission. Scanning Electron Microscopy 1981; I: 385-396

/3/ Brunner M, Burchard D, Hylla W, Kohl H-J, Niedrig H, Wendtland D. (1982). A Digital Storage Unit Combined With an Electron Microscope for Recording Comparable ECPs Obtained With Different Signals. 10th Intern. Congr. Electr. Microsc., Hamburg, Germany (West), JP LePoole et al. (ed.), (Dtsche. Ges. f. Elektr. Mikr., Frankfurt, W. Germany), I: 443-444

/4/ Brunner M, Hylla W, Kohl H-J, Niedrig H, Wendtland D. (1983). Electron Channeling Contrast in Characteristic Energy-Loss Intensity Investigated By the Rocking-Crystal Method. Scanning Electron Microsc. 1983; I: 99-104

/5/ Egerton RF. (1979). K-Shell Ionization Cross-Section for Use in Microanalysis. Ultramicroscopy, 4, 169-179

/6/ Hylla W, Kohl H-J, Niedrig H, Wendtland D. (1983). Direct Observation of Bloch-Wave Localization Effect in Light Elements by Suitable Detector Arrangement. Beitr. elektronenmikroskop. Direktabb. Oberfl. 16, G Pfefferkorn (ed.), Verlag R.A. Remy, Muenster, Germany (West), 67-74

/7/ Joy DC, Maher DM. (1979). 'The Basic Principles of EELS' and 'Elemental Analysis Using Inner Shell Excitations' in: Introduction to Analytical Electron Microscopy, J. Hren et al. (ed.), Plenum Press, N.Y., 223-244 and 259-294

/8/ Kohl H-J. (1983). A Matrix Formulation of the Inelastic Scattering of Electrons in Crystals. Beitr. elektronenmikroskop. Direktabb. Oberfl. 16, G Pfefferkorn (ed.), Verlag R.A. Remy, Muenster, Germany (West), 75-80

/9/ Linders J, Niedrig H, Sternberg M. (1984). Undistorted Measurements of Differential Sputtering Yields Using the Collector Method by Means of Electron Backscattering. Proc. 10th Int. Conf. Atomic Collisions in Solids, Bad Iburg, Germany (West), publ. in: Nucl. Instr. Meth. Phys. Res. (North Holland, Amsterdam), B2: 649-654

/10/ Oikawa T, Hosoi J, Inoue M, Honda T. (1984). Scattering Angle Dependence of Signal/Background Ratio of Inner Shell Electron Excitation Loss in EELS. Ultramicroscopy, 12, 223-230

/11/ Spence JCH, Taftø J. (1982). Atomic Site and Species Determination Using the Channeling Effect in Electron Diffraction. Scanning Electron Microscopy 1982; II: 523-531

/12/ Whitton JL, Carter G. (1980). The Development of Surface Topography by Heavy Ion Sputtering, Perchtoldsdorf, Austria, P Varga et al. (ed.), Inst. f. Allg. Phys., TU Wien, 552-572

Electron Optical Systems (pp. 245-251)
SEM Inc., AMF O'Hare (Chicago), IL 60666-0507, U.S.A.

0-931288-34-7/84$1.00+.05

ENERGY SELECTING ELECTRON MICROSCOPY

F.P. Ottensmeyer

Ontario Cancer Institute
500 Sherbourne Street
Toronto, Ontario M4X 1K9
Canada
Telephone: (416) 924-0671

Abstract

One of the major improvements in transmission electron microscopy over the last years is the addition of the capability of producing images with electrons that have specific narrow energy bands out of the total spectrum of energies they possess after having passed through the specimen. Though the idea is not new, the power of this application is only beginning to be recognized. Most simply, selection of elastically scattered electrons permits increased contrast in high resolution images in bright field, dark field, and diffraction. The use of combined elastic and inelastic signals adds entirely new contrast mechanisms, partially independent of thickness, partly Z-related. Finally, selection of element specific inelastic events permits elemental mapping with spatial resolutions of 0.3 - 0.5 nm and detection sensitivities of about 30 to 50 atoms. Consideration of resolution, sensitivity, image points of analysis and acquisition time leads to a combined improvement of about 10^{13} times over X-ray microanalysis.

Key words: microanalysis, electron energy filter, elastic scatter, sector magnet, imaging filter, transmission electron microscopy, electron diffraction.

Introduction

Major advances in electron microscopy have come about by modifications of the basic design of the instrument which were simple to adapt: the introduction of the stigmator, or the double condenser system; by changes in technique that were easy to implement: sectioning of tissues or staining for contrast; or on the other hand by radical alterations which were so dramatic in effect that the effort in adaptation both by the manufacturer and the user was more than compensated by the result: the introduction of the scanning transmission electron microscope.

One area which has been puzzlingly neglected is the addition of energy selection to the imaging capabilities of the electron microscope. Certainly the problem of chromatic aberration in the image of a specimen has been recognized, with much effort being expended to design round lenses that would not limit resolution due to the natural energy spread of the electron gun. The secondary effect of energy spread due to inelastic scatter of electrons within the specimen being imaged has been less emphasized, although a number of multipole lenses have been designed to correct chromatic aberrations for at least part of the change in energy encountered in the specimen (16). However, the complexity of these systems has to the present prevented their proliferation in non-experimental situations.

An alternative to correction of lenses is the filtering of the energies in the electron beam after it has passed through the specimen. In addition to obviating chromatic aberrations, such an approach has the added benefit of being able to extract more information from the beam on fundamental electron scattering phenomena as well as on the chemistry of the specimen via characteristic interactions.

Energy selecting systems

Probe forming

A number of such energy selecting electron microscopes have been built experimentally. The simplest in principle is the extension of the use of a probe forming system equipped with an electron spectrometer such as an attached sector magnet. Scanning of the probe while selecting a specific energy band in the electron energy spectrum permits energy selected microscopy for bright field elastic images and images with a selected energy loss under axial illumination. Pure elastic dark field images are not possible with present systems, though due to a partial angular separation of elastic and inelastic scatter, a reasonable approximation to elastic dark field can be obtained for thin specimens using the signal from the annular aperture detector in the STEM. High resolution (0.5 nm) elemental maps using probe forming systems are only possible in STEM's equipped with field emission guns. Nevertheless, even for such systems the total beam current in the small probe is still so feeble that mapping of a small area of, say, 128 x 128 picture points still requires about 15 minutes (15). This is in sharp contrast to the use of vastly more efficacious parallel filtering systems below.

Line scanning

An approach to improvement over a point by point image analysis was the passage of an entire image line simultaneously through an electrostatic filter lens (20,25). Information in the image line was spread out in energy at right angles to the line; and a long slit could select the desired energy band. Scanning of the image a line at a time over an entrance slit followed by simultaneous "unscanning" of the energy selected line emerging from the exit slit built up an energy selected micrograph.

Two-dimensional filtration

However, the most efficacious approach is to filter an entire two-dimensional image simultaneously. If elastic bright field or dark field images only were desired, then a threshold filter, a passage of electrons through a grid that presented a voltage barrier which could only be surmounted by electrons that had not lost energy, would be sufficient. Much more versatile is an imaging spectrometer. We have found that a development of a prism-mirror-prism system as an integrated part of a fixed beam electron microscope (8,14,22) not only does not deteriorate the performance of the microscope, but instead provides superior images to the unmodified microscope in all modes of operation, in addition to offering the unique features of energy analysis, microanalysis and elemental mapping that only an imaging energy selecting system can provide. Similar systems have been built (13), while other devices have included a system of crossed electric and magnetic fields (10), as well as two all-magnetic imaging electron energy filters incorporating sector magnets (18,26). The latter eliminate the requirement for an electrostatic mirror operating at the high tension of the electron gun, and have manifold more possibilities of introducing corrective elements. To date however, they have not performed as well in imaging as the mirror-prism system, possibly due to their slightly younger state of development or the difficulties arising from their greater structural complexity.

Our system is shown in principle in Fig. 1. The system is installed as an integral permanent component in the column of a transmission microscope (Siemens Elmiskop 102) above the projector lens. As such it must perform two functions. First, it must act like a spectrometer. This is not too difficult, since any reasonable sector magnet does this superbly. Second and much more crucially, it must act like a lens. Since the system is neither centro-symmetric nor round as are most familiar lenses, it is *a priori* not at all clear that it should do so. Nevertheless, properly aligned, it does. The symmetry it posesses about the axis of the mirror cancels many even of the second order aberrations in the image, while its position in the microscope reduces remaining errors by prior magnification of the image to such a degree that at any final magnification such errors are smaller than the electron impact size in the photographic material that records the image.

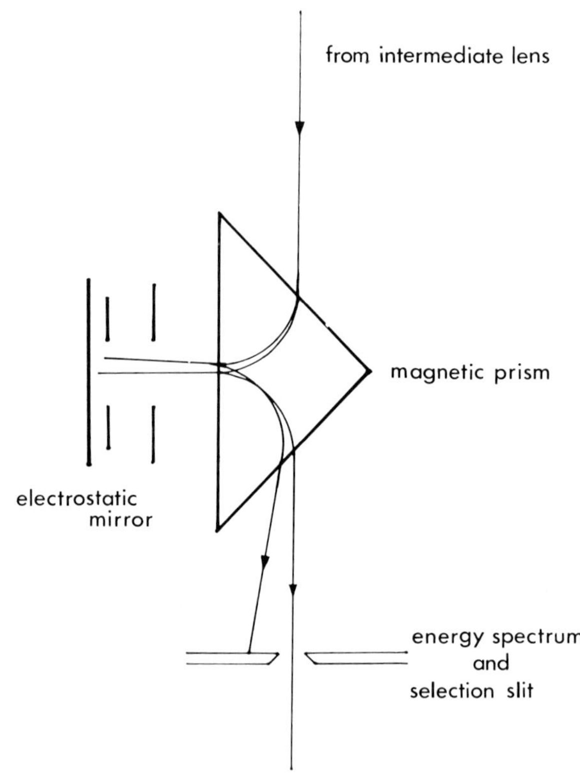

Figure 1. Schematic of imaging energy filter. The electron energy spectrum occurs at the electrostatic mirror and, with twice the dispersion, at the level of the selecting slit. An achromatic image is formed at the cross-over of the beams on the second pass through the prism.

Application

Fig. 2 shows the 0.204 nm lattice lines of gold taken through the filter as an indication that the performance of the original microscope is not deteriorated. On the contrary, the opposite is true, since the normal operating mode of this microscope uses only unscattered or elastically scattered electrons. The effect in every case is a small or large increase in contrast (Fig. 3), depending on the size of the inelastic contribution to the conventional image. Such a conventional image can still be produced by retracting the energy selecting slits (Fig. 1,

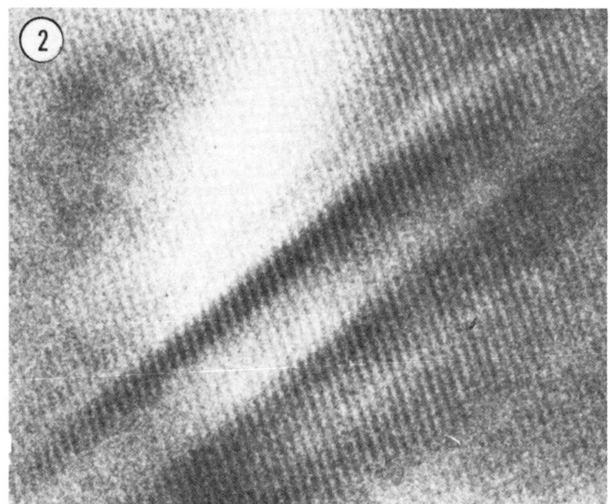

Figure 2. Energy filtered elastic image of a single crystal gold foil showing 0.204 nm lattice lines.

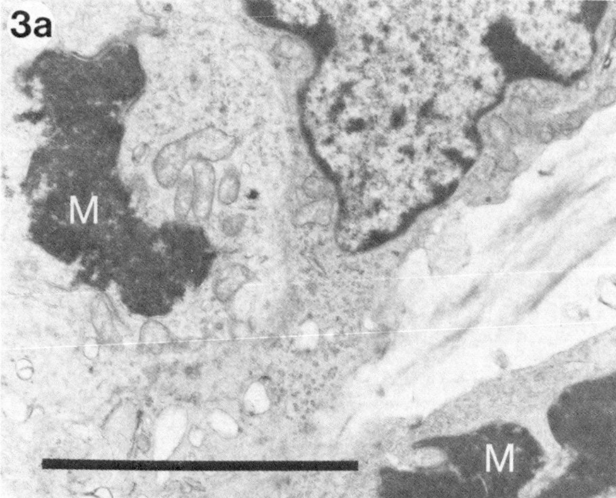

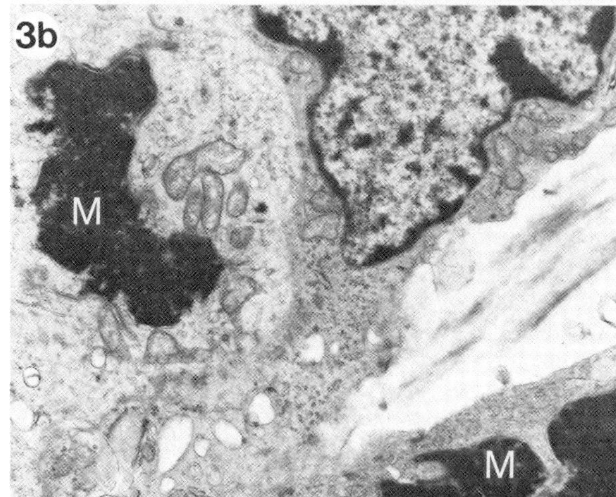

Figure 3. Section through a malignant testicular tumor showing two of many mitotic figures (M) of rapidly dividing cells. The specimen is a standard pathological tissue preparation, glutaraldehyde fixed, post-fixed with osmium tetroxide, stained with uranyl acetate and lead citrate. (a) Bright field electron micrograph using all energies. (b) Elastic bright field image. Both images were printed identically on soft (zero) paper. The increase in contrast in (b) is due to removal of inelastically scattered electrons. Bar = 5 μm.

Fig. 3a). The advantage of energy selection is perhaps more important in low contrast unstained material and thicker specimens such as used in enzyme cytochemistry or X-ray microanalysis. Again in bright field the elastic image provides somewhat greater contrast (Fig. 4a,4b), but for such thin unstained specimens as this section the elastic dark field mode (Fig. 4c) gives the expected huge increase in contrast (24), though now without any deleterious effects of chromatic aberration that occurs in the conventional dark field image.

Of course, being a compleat instrument, the energy selecting electron microscope permits diffraction mode operation like a normal microscope, with the additional choice of total, elastic, total inelastic or energy selected options (Fig. 5). The example also demonstrates the effect of modifying the effective acceptance angle of the filter by means of the pre-spectrometer lenses. Though the spectrometer itself has an acceptance half-angle of about 15 mradians, the half-angle passed by the prism referred to the specimen in these diffraction patterns is easily about 100 mradians. The analysis of fundamental scattering phenomena in electron-matter interactions becomes not only possible with such a device, but easy.

Recently interest has revived in an imaging mode using elastic dark field images divided point by point by the corresponding energy loss signal (6). This was first demonstrated in the STEM (7,9), but is of course also possible in a fixed beam energy selecting microscope (23). Simple theory indicates that the combined signal is proportional to the atomic number of the constituent atoms in the specimen and independent of the thickness (19). However, more careful consideration indicates a partial thickness dependence (12), a strong effect of unsharp masking due to non-local scatter in the inelastic signal, and specific variations due to the presence of ionization edges which tend to reduce the signal of the corresponding atomic element (23). The total effect is interesting, and may be useful; but it must be more thoroughly investigated in order to be understood more fully.

However, the most emphasized characteristic of energy selected microscopy is elemental microanalysis. Certainly, incorporating an electron spectrometer, the microscope provides point or area analysis (Fig. 6). But more important, pairs of energy selected images (ESI, electron spectroscopic images) taken at energies such as the hashed areas in Fig. 6, one above a specific atomic ionization edge, another just below this edge in energy, immediately provide a qualitative indication of the presence and distribution of the specific corresponding atom. This is demonstrated for titanium and copper in the extraction replica of a low alloy steel in Fig. 7, and for phosphorus in the section of a

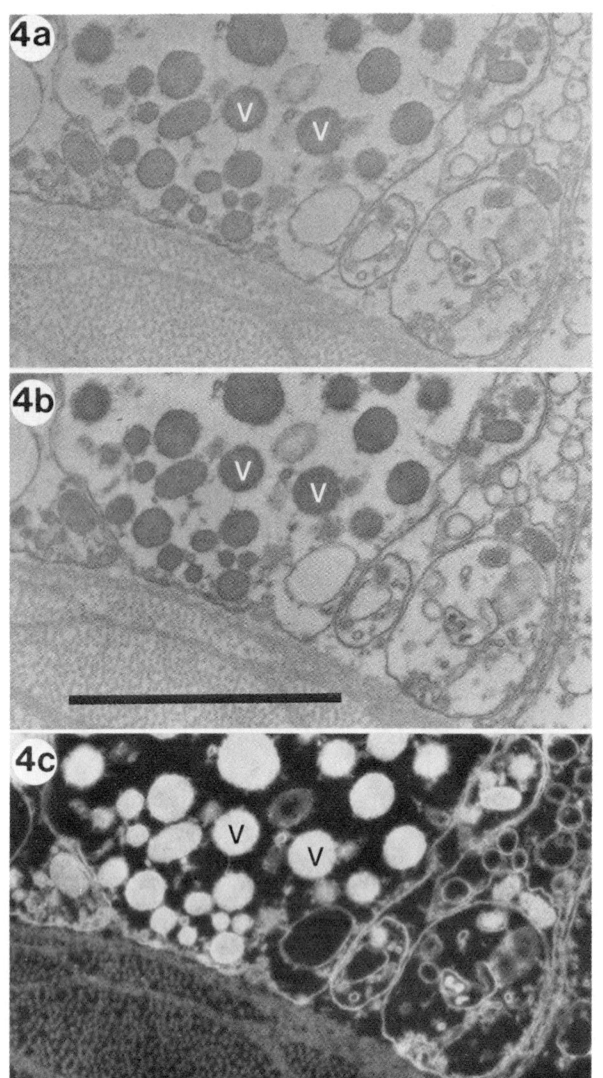

Figure 4. Secretory vesicles (v) in the tegument of the lobster Homarus americanus. The section, dark gray by interference colours, is glutaraldehyde fixed and post-fixed with osmium tetroxide, unstained. (a) Bright field micrograph using all energies. (b) Elastic bright field image. (c) Elastic dark field image. The bright field micrographs were printed identically on soft (#1) paper, the dark field image on still softer (zero) paper. Bar = 1.0 μm.

Figure 5. Diffraction pattern of a thin foil of aluminum, using all energies (total), elastic electrons (el), all inelastic electrons (inel) and various selected energy losses (∓6 eV) as indicated. A band of intensity corresponding to the Bethe ridge (11) moves out from the centre with increasing selected energy losses (arrows).

Figure 6. Electron energy spectrum of an extracellular deposit from an osteoblast (bone-forming cell) in culture (22). The ionization edges occurring in this specimen are identified. The two hashed areas indicate the energy regions with which energy selected images must be taken as a minimum requirement (see text) for elemental mapping of calcium.

Figure 7. Energy selected images of a carbon extraction replica containing precipitates from the surface of a low alloy steel. (a) Elastic bright field image showing precipitates of various sizes and shapes. (b) and (c) A selected area from (a) taken below and above the L-shell ionization edge of titanium. The local increases in brightness in (c) compared to (b) indicate that the larger rectangular precipitates (arrows) contain Ti. (d) and (e) The same area taken with energies below and above the L-shell ionization of copper. The local increases in intensities in (e) compared to (d) indicate the presence of Cu in the small precipitates. The larger Ti-containing precipitates are unchanged in intensity relative to the background, indicating an absence of Cu in these. Micrographs (b) to (e) were printed identically on soft (zero grade) paper. Bar = 1.0 μm.

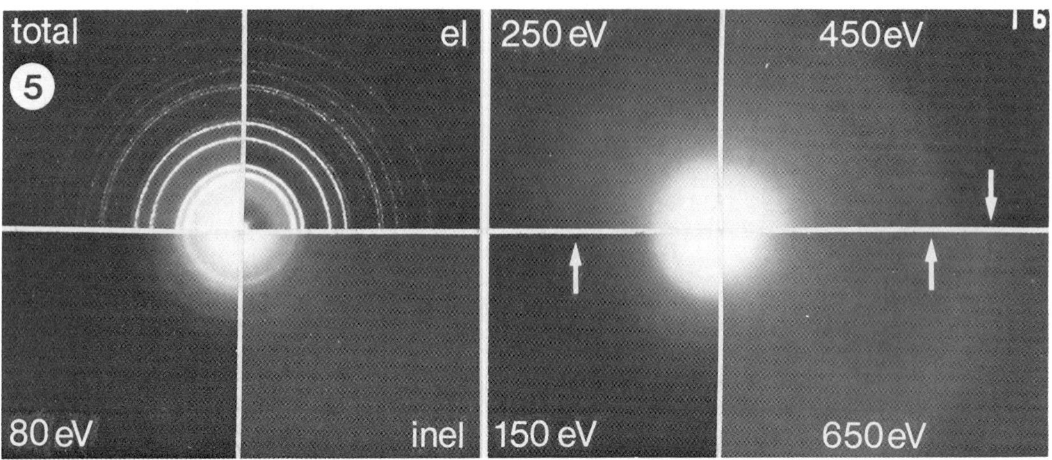

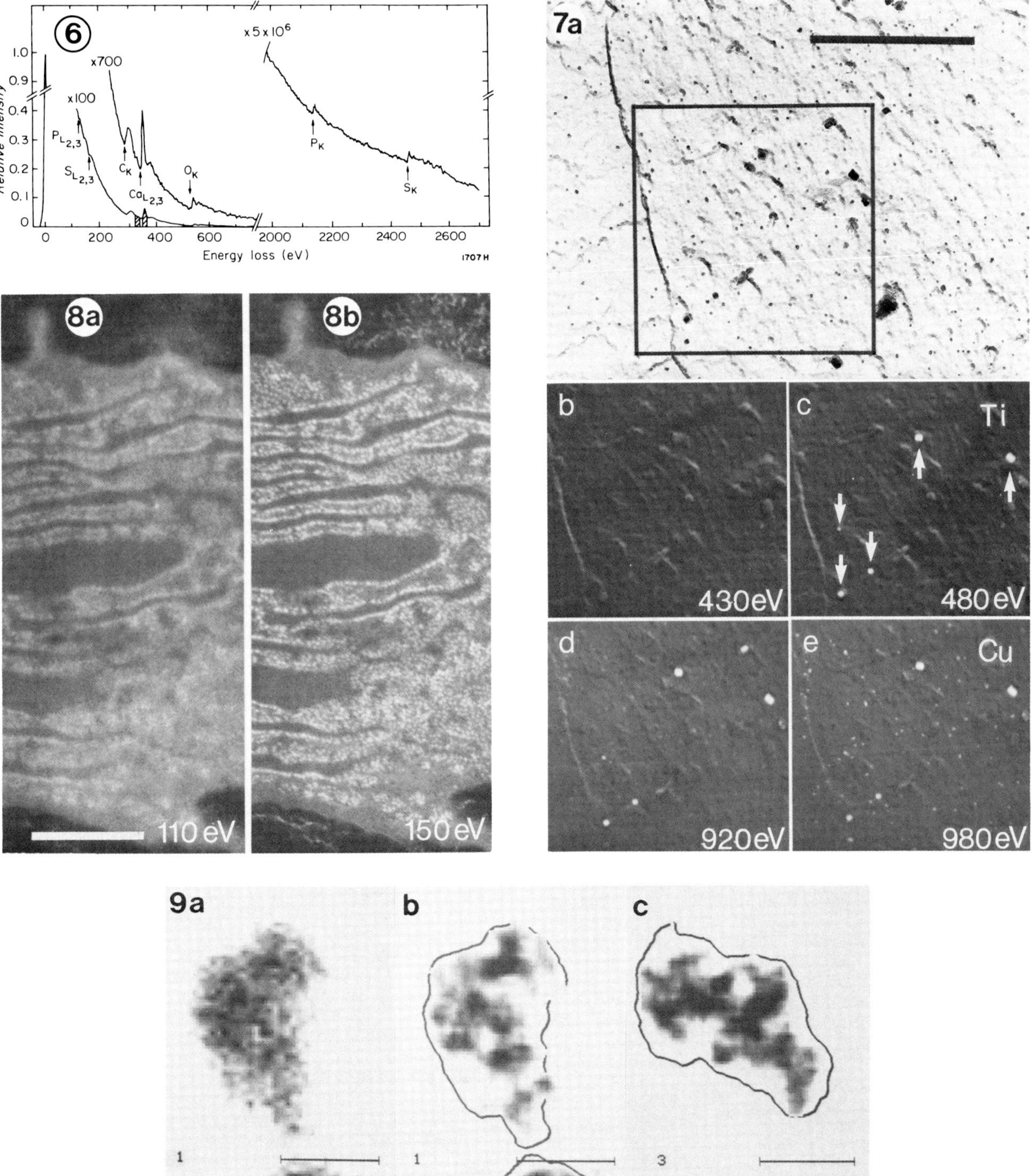

Figure 8. Energy selected micrographs of an unstained section through a chondrocyte in the developing cartilage of the mouse, taken with energies below (a) and above (b) the phosphorus L-shell ionization edge. The increase in local brightness in (b) indicates the presence of P predominantly in the ribosomes of the rough surfaced endoplasmic reticulum. Bar = 0.5 μm.

Figure 9. Phosphorus map within 20 nm long E. coli 30S ribosomal subunits. (a) Energy selected image at 150 eV, just above the P L-shell ionization edge at 132 eV. Phosphorus distribution within the particle in (a) obtained by computer-assisted subtraction of the image in (a) minus its corresponding image taken at 110 eV, below the P L-edge. The contour of the particle in (a) is shown for reference (c). Phosphorus map within another 30S particle at a different orientation. Bar = 8.0 nm.

chondrocyte in Fig. 8. In each case the extraordinary local increase in intensity or contrast relative to the surrounding background for the image above the ionization edge, compared to the image below the edge in energy, signals the presence of the corresponding atomic element. Pure elemental maps can be obtained over entire micrographs by suitable normalization of the two images, alignment and photographic subtraction (22), or by densitometry and computer-assisted image processing of smaller regions selected a posteriori in the image. The latter process was carried out for the phosphorus map within 20 nm long 30S ribosomal subunits in Fig. 9 (17).

Any element from Z = 1 to Z = 92 can be examined by the technique, using low energy signals of the K-shell ionizations for low atomic number elements and L-, M-, N-, and O-shells for higher Z atoms. While we have not tried to be exhaustive, our normal work has so far included the analysis and mapping of B, C, O, N, F, Mg, Al, Si, P, S, Ca, Ti, Fe, Cu, Zn, Ag, W, and U. Applications have ranged from the investigation of Al and W in semi-conductors, P- and Si-containing inclusions in reactor tube cladding, Si deposits in plant fungal infections (21), Ca, S, and P distributions in normal and diseased developing cartilage and bone (2,3), to high resolution maps of the phosphorus distribution within ribosomal particles (6,17) and within nucleosomes, the 10 nm subunits of the fine structure of chromosomes(4,5).

At present the local concentration of the elements examined has been rather high (e.g. 3 M phophorus within the double helix of DNA) even though the absolute amounts detected are only a few tens of atoms. Thus the use of a single pair of images bracketing the specific ionization edge has been sufficient to delineate and quantify a specific element. For a more accurate quantitative approach two or more images displaced in energy below the selected ionization edge could be used to extrapolate the background signal under the edge more precisely. But a compromise here is necessary, since preceding ionization edges and their extended fine structure (Fig. 6) can strongly influence such an extrapolation. Moreover, for a reliable extrapolation even without such complications some averaging over many pixels over and around the point of interest is necessary to overcome noise fluctuations in individual pixels, that are incurred by low electron exposures used to spare sensitive specimens. This practical limitation unfortunately negates to some extent the advantage of a scanning approach in which ideally such an extrapolation could be performed by curve-fitting to a continuous spectrum at every point.

Electron spectroscopic images, which contain as many as 50 million picture points at the resolution of the photographic plate, are taken in a few seconds in a fixed beam transmission electron microscope, compared to a 15 minute exposure for as few as 128 x 128 pixels in even a dedicated STEM with sector magnet spectrometer and field emission gun (15). This major difference is due to the fact that in spite of the high directed brightness of the field emission gun, the total current in the beam at the specimen is of the order of 10^{-9} A compared to 10^{-6} A for the thermionic gun. Operationally the beam in the STEM has to be apertured and focused into a small spot to achieve high spatial resolution in scanning. In contrast the beam of the thermionic gun in a fixed beam microscope does not have to be focused, but indeed has to be spread over the image for most applications we have encountered so far. Effectively, for elemental mapping via an imaging filter the probe size in the fixed beam microscope is not the beam size, but the size of the wave packet of every individual electron in that beam.

Ideally, for elemental mapping in an energy selecting microscope with imaging filter the shortest exposure times could be obtained by matching the minimum possible spot size of the fully condensed beam to the field of view at any magnification. Since for magnifications commonly used in biology this would mean minimum spot sizes as large as several tens of microns, this would be a reversal in thought in electron gun design, which over the last years has striven for small spot sizes.

Conclusion

An energy selecting electron microscope based on an integrated imaging electron spectrometer in a fixed beam microscope offers unhindered normal operational functions with improvement in contrast, coherence and resolution in the image. Pure elastic bright field, dark field and diffraction become standard operation modes, obviating problems due to chromatic aberrations in the image. In addition energy selection in the electron energy loss spectrum provides fundamental data on dispersion of electrons with energy loss, as well as producing high contrast images for thin specimens, that form the basis for elemental mapping with spatial resolutions as good as 0.3 to 0.5 nm and sensitivities of detection in analysis as low as 30 to 50 atoms (1,4). Energy loss spectra over selected areas, recorded in parallel or sequentially, are part of the normal capabilities. Our own microscope (Siemens 102) is not equipped with a probe forming lens, making very small spot analysis via spectra impossible. In more modern instruments this is no longer a limitation. In addition, since the device disturbs neither the specimen area nor the column exit, any adaptations such as goniometers, cooling stages, X-ray microanalysis (as complementary technique of microanalysis for thick specimens), TV cameras etc. are entirely compatible with the design.

Acknowledgements

Thanks are due to H. Hashimoto, W. Rothenburger, A.L. Arsenault, G.C. Weatherly and A.P. Korn for the specimens in Figs. 2, 3, 4 and 8, 7, and 9 respectively. The work was supported by the Ontario Cancer Treatment and Research Foundation, the National Cancer Institute of Canada and grant MT-6337 from the Medical Research Council of Canada.

References

1. Adamson-Sharpe, K.M., Ottensmeyer, F.P. (1981). Spatial resolution and detection sensitivity in microanalysis by electron energy loss selected imaging. J. Microsc. 122:309-314.

2. Arsenault, A.L., Ottensmeyer, F.P. (1983). Quantitative spatial distribution of calcium, phosphorus and sulfur in calcifying epiphysis by high resolution spectroscopic imaging. Proc. Natl. Acad. Sci., U.S.A. 80:1322-1326.

3. Arsenault, A.L., Ottensmeyer, F.P., (1984). Visualization of early intramembranous ossification by electron microscopic and electron spectroscopic imaging. J. Cell. Biol. 98:911-921.

4. Bazett-Jones, D.P., Ottensmeyer, F.P. (1981). Phosphorus distribution in the nucleosome. Science 211:169-170.

5. Bazett-Jones, D.P., Ottensmeyer, F.P. (1982). DNA organization in nucleosomes. Can. J. Biochem. 60:364-370.

6. Boublik, M., Oostergetel, G.T., Frankland, B., Ottensmeyer, F.P. (1984). Topographical mapping of ribosomal RNAs in situ by electron spectroscopic imaging. Proc. 42nd Ann. Meet. EMSA, G.W. Bailey (ed.), San Francisco Press, pp. 690-691.

7. Carlemalm, E., Kellenberger, E. (1982). The reproducible observation of unstained embedded cellular material in thin sections: visualization of an integral membrane protein by a new mode of imaging for STEM. Europ. Molec. Biol. Org. J. 1:63-67.

8. Castaing, R., Henry, L. (1962). Filtrage magnetique des vitesses en microscopie electronique. C.R. Acad. Sci., Paris B255:76-78.

9. Crewe, A.V., Wall, J., Langmore, J. (1970). Visibility of single atoms. Science 168:1338-1340.

10. Curtis, G.H., Silcox, J. (1971). A Wien filter for use as an analyser with an electron microscope. Rev. Sci. Instrum. 42:630-637.

11. Egerton, R.F. (1979). K-shell ionization cross-sections for use in microanalysis. Ultramicroscopy 4:169-179.

12. Egerton, R.F. (1982). Thickness dependence of the STEM ratio image. Ultramicroscopy 10:297-299.

13. Egerton, R.F., Philip, J.G., Turner, P.S., Whelan, M.J. (1975). Modification of a transmission microscope to give energy loss spectra and energy selected images and diffraction patterns. J. Phys. E8:1033-1037.

14. Henkelman, R.M., Ottensmeyer, F.P. (1974). An energy filter for biological electron microscopy. J. Microscopy 102:79-94.

15. Jeanguillaume, C., Tence, N., Trebbia, P, Colliex, C. (1983). Electron energy loss chemical mapping of low Z elements in biological sections. Scanning Electron Microsc. 1983; II:745-756.

16. Koops, H. (1978). Aberration correction in electron microscopy. In: Electron Microscopy 1978. Vol. III State of the Art. Proc. 9th Intnl. Congr. Electron Micr., J.M. Sturgess (ed.), Microsc. Soc. Canada (publ.), Toronto; pp. 185-196.

17. Korn, A.P., Spitnik-Elson, P., Elson, D., Ottensmeyer, F.P. (1983). Specific visualization of ribosomal RNA in the intact ribosome by electron spectroscopic imaging. Eur. J. Biochem. 31:334-340.

18. Krahl, D., Hermann, K.-H., Kunath, W. (1978). Electron optical experiments with a magnetic imaging filter. In: Electron microscopy 1978 (Proc. 9th Intnl. Cong., ed. J.M. Sturgess; Microscopical Soc. Canada, Toronto). Vol. 1, pp. 42-43.

19. Lenz, F. (1952). Zur Streuung mittelschneller Elektronen in kleinste Winkel. Z. Naturforsch. 9a:185-204.

20. Moellenstedt, G. (1949). The electrostatic lens as a velocity analyser of high resolving power. Optik 5:499-517.

21. Ottensmeyer, F.P. (1982). Scattered electrons in microscopy and microanalysis. Science 215:461-466.

22. Ottensmeyer, F.P., Andrew, J.W. (1980). High resolution microanalysis of biological specimens by electron energy loss spectroscopy and electron spectroscopic imaging. J. Ultrastruct. Res. 72:336-348.

23. Ottensmeyer, F.P., Arsenault, A.L. (1983). Electron spectroscopic imaging and Z-contrast in tissue sections. Scanning Electron Microsc. 1983;IV:1867-1875.

24. Ottensmeyer, F.P., Pear, M. (1975). Contrast in unstained sections: a comparison of bright and dark field electron microscopy. J. Ultrastruct. Res. 51:253-260.

25. Watanabe, H., Uyeda, R. (1962). Energy selecting electron microscope. J. Phys. Soc. Japan 17:569.

26. Zanchi, G., Sevely, J., Jouffrey, B. (1977). An energy filter for high voltage electron microscopy. J. Microsc. Spectrosc. Electron. 2:95-104.

Electron Optical Systems (pp. 253-272)
SEM Inc., AMF O'Hare (Chicago), IL 60666-0507, U.S.A.

0-931288-34-7/84$1.00+.05

LOW VOLTAGE SCANNING ELECTRON MICROSCOPY

James B. Pawley

Zoology Department and
High Voltage Electron Lab
University of Wisconsin
1675 Observatory Drive
Madison, WI 53706
Phone No.: (608) 263-3147

Abstract

The scanning electron microscope (SEM) usually operates with a beam voltage, V_o, in the range of 10-30 kV, even though many early workers suggested the use of lower voltages to increase topographic contrast and to reduce specimen charging and beam damage. The chief reason for this contradiction is low instrumental performance when V_o = 1-3 kV. The problems include low source brightness, greater defocussing due to chromatic aberration, greater sensitivity to internal and external stray fields and difficulty in collecting the secondary electron signal without defocussing the probe. Recently considerable efforts have been made to overcome these problems because the semiconductor industry, which is now the major user of the SEM, has found that low V_o is necessary to reduce beam damage. The resulting equipment has greatly improved performance at low kV and substantially removes the practical deterrents to operation in this mode on other types of samples. This paper reviews the advantages of low voltage operation for topographic imaging, recent progress in instrumentation and describes a prototype instrument designed and built for optimum performance at 1 kV. Other limitations to high resolution topographic imaging such as surface contamination, the de-localized nature of the inelastic scattering event and radiation damage are also discussed.

Key words: Low Voltage Scanning Electron Microscopy, Secondary electron contrast, High resolution, Topographic imaging

Introduction

As the beam voltage, V_o is reduced in the range 30-1 kV, three physical parameters relevant to specimen damage, surface charging, and topographic contrast also change. The secondary electron coefficient, δ, increases to >1 while the electron range (R) and the energy deposited per electron (eV_o) both decrease. When δ is large there is less surface charging on insulating samples and there is more signal per beam electron, while a smaller R means the beam/specimen interaction is more localized and topographic contrast is higher. This was recognized early by Thornley (1960), but widespread use of the SEM in the 1-3 kV range was delayed by technical limitations that can be grouped in four categories: 1) low source brightness; 2) increased effect of chromatic aberration; 3) increased sensitivity to stray fields; 4) defocusing of the probe by the secondary electron collection field. With a few exceptions (Welter and Coates, 1974) these disincentives prevented significant efforts to improve low voltage SEM (LVSEM) performance. More recently, however, SEM studies of semiconductors were found to be limited by the damage caused by the beam and the most effective way to limit this damage was to use V_o in the range 0.5-1.5 kV (Keery et al. 1976, Miyoshi et al. 1982). The fact that the semiconductor industry, which presently represents over 80% of the SEM market, urgently needed to monitor production procedures and final performance without damaging the specimen provided a new impetus for the development of equipment optimized for low voltage operation. (Tamura et al. 1980; Todokoro et al. 1980, 1983, 1984; Buchanan, 1982, 1983; Buchanan and Menzel, 1984; Pomposo and Coates, 1983; Pawley and Wall, 1982; Boyes 1984; Pawley, 1984a,b, other papers in this volume).

It is the purpose of this paper to draw attention to these developments in the belief that the low voltage capability of this new equipment will find widespread application outside the field of semiconductor research. The advantages of low-voltage operation will be discussed particularly in regard to the possibility that it may eventually provide the ultimate in high resolution topographic images of biological samples (Pawley, 1984a,b). This will be followed by a description of both the instrumentation problems associated with operating the SEM at high resolution with V_o ≃ 1 kV and the strategies that have been developed or proposed to overcome these problems. A prototype instrument designed to produce a 1-2 nm beam at 1 kV will be described along with a discussion of preliminary results and problems encountered. Finally there is a brief discussion of the limitations posed to the ultimate topographic resolution of an ideal instrument by radiation damage, sample-derived surface contamination and the delocalized nature of the in elastic scattering event.

The Advantages of LVSEM

Most of the advantages of using an SEM with Vo ≃ 1 kV derive directly from the fact that electrons impinging on the surface of a solid with less energy, penetrate into it a shorter distance and also have a higher cross-section for producing secondaries near the surface where they have a higher chance of escaping (Kotera et al. 1981). As a result, δ approaches unity, charging artifacts on insulating surface become less pronounced and the signal/beam-electron is increased. Also less energy (eV_o) is deposited in the sample and on insulating samples, charge is not injected and trapped so far beneath the surface.

All of these features are important for the study of uncoated resist patterns or passivation layers on semiconductor devices or for viewing voltage contrast effects. Relative freedom from charging artifacts is an obvious advantage, but it is even more important to avoid high surface potentials that might cause breakdown in the device and to reduce beam penetration because charges trapped in insulating regions can distort the energy band structure of the device, degrading and possibly destroying it. At 1 kV, charge injection is restricted to the outer 0.02 μm or so, rather than 130 times that depth at 30 kV--a crucial difference in devices only a few micrometers deep. Finally, both the silicon and the resist layer on its surface are composed of materials having relatively low atomic number (Z). At high beam voltages, very little surface detail can be seen on uncoated low Z samples, but at 1 kV the energy is deposited nearer to the beam entry point and so contrast produced by topographical variations is proportionately larger. Aspects of the study of semiconductors in the SEM that emphasize the utility of low voltage operation are discussed by Pfeiffer (1982), Todokoro et al. (1983,1984), Tamura et al. (1980), Buchanan and Menzel, 1984 and Brandis et al (1984) and low voltage electron lithography is described by Yau et al. (1981), Varnell (1981) and Polasko et al. (1983) Newman et al (1984) by contributions from Pfeiffer, Russell, Orloff and Murray in this volume. Although the subjects covered by these authors are in large part responsible for recent instrumental improvements in the LVSEM, they will not be discussed specifically further here. Contrast and charging will now be considered in more detail.

Topographic Imaging in the SEM

At low magnification, the secondary electron image from the SEM is easily interpreted by the brain to yield a fairly accurate understanding of the shape of the surface of the specimen. It does so because there is a rough equivalence between the secant laws relating the apparent brightness of a diffusely-illuminated matte surface and its angle with the line of sight on the one hand and the variation of δ with incidence angle on the other (Everhart et al., 1959). Unfortunately, this encoding relation breaks down as the magnification increases and the beam interaction volume on the sample becomes appreciably larger than the corresponding pixel size in the image (Fig. 1). The effective radius of the interaction volume, r is about 40% of the electron range, $R = kV_o^{3/2}$ (Joy, 1984a,b). Therefore, as V_o is reduced from 20 kV to 1 kV, the radius of the beam interaction volume is reduced by 90:1 and topographic contrast increases. The possible effects of this increased contrast on image formation at high spatial resolution is only now beginning to be assessed (Boyes, 1984, Pawley, 1984a,b, Joy, 1984a,b).

Clearly, resolution in the secondary electron mode depends entirely on the size of the area sampled by the beam. This in turn depends on both the size of the probe and the scattering properties of the specimen.

Topographic Resolution/Contrast in the SEM

The detected secondary electron signal from a given pixel on the sample is a function not only of δ and the surface angle, but also of the average, Z, of the volume of beam penetration (Everhart et al., 1959; Seiler, 1976; Ball & McCartney, 1981), the crystallographic orientation (LeGressus et al., 1983), the surface potential (Oatley & Everhart, 1957; Oatley, 1969; Banbury & Nixon, 1970; Pawley, 1972; Kursheed and Dinnis, 1983), the presence of nearby surface features which may affect the collection efficiency (Everhart et al., 1959; Pawley, 1972), the presence of second or third surfaces of the sample within the penetration volume from which additional electrons may be produced and collected (Wells, 1978), the efficiency with which high energy backscattered electrons are converted into collectable secondaries by collisions within the sample chamber (Oatley et al., 1965; Reimer & Volbert, 1979; Peters, 1982) and, finally, various arcane variables such as the presence of subsurface charge that may effect δ on uncoated insulators (Shaffner & Hearle, 1976).

The complexity of the interaction of these variables as they affect the high magnification image of a typical sample is considerable and is perhaps the reason contrast and resolution in the secondary electron mode have been the object of so much research and interest, for example: Everhart et al., 1959; Oatley et al., 1965; Pease & Nixon, 1965; Clarke, 1970; Oatley, 1972; Catto & Smith, 1973; Wells, 1974a,b; Haggis & Bond, 1979; Peters, 1979, 1982; Reimer, 1978; Joy, 1984a,b).

The three main theoretical assumptions that have guided inquiries into topographic resolution in the SEM are: 1) That secondary electrons (0-50 eV) are produced by inelastic collisions between electrons from the beam and those in the sample but that only collisions within a few nm of the surface have any chance of producing secondaries that can escape and be collected. 2) That the local surface angle modulates the number of secondaries so produced in a way which produces an image that is easily interpreted by the brain as topography when the secondary electron signal is presented as an intensity-modulated image. 3) That while collectable secondary electrons are produced by beam electrons striking the sample (Type 1) or the objective aperture (Type 4) and also by backscattered electrons emerging from the sample (Type 2) or striking the specimen chamber (Type 3), only the Type 1 signal carries high resolution topographic information. (Several other minor electron currents are described by Oatley (1983.).

The distribution of secondary electrons leaving the sample as a function of the distance from the point of impact is assumed to have a small peak within a few nm of the beam axis and a long 'tail' extending many micrometers in all directions and corresponding to the probability of a secondary electron being produced by a re-emergent backscattered electron (Joy, 1984a,b). For fairly high V_o, such a distribution has been directly observed by using the surface of a tilted SEM sample as the source of an emission microscope (Hasselbach & Rieke, 1982; Hasselbach et al., 1983). In visualizing the relative dimensions of these two parts of the distribution, it is important to keep in mind the large magnitude of the difference between the range of a 20 kV backscattered electron and a 4-50 eV secondary electron. On metal-coated, dried biological material (density 0.2g/cm^3, Z = 7) the former may be 100 μm and the latter 0.002 μm (Joy, 1984a). To obtain high resolution topographical information it is necessary to somehow separate the relatively small peak signal (Type 1) from the much larger slowly varying signal produced by the tail (Types 2 and 3).

Initially, it was thought that this separation could be accomplished simply by raising the beam current and treating the tail signal as a DC noise signal that could be electronically removed by analog subtraction. This approach was not very successful, probably because of the difficulty of increasing the beam current in a small spot sufficiently and because, even though the average value of the DC offset could be removed, the noise associated with statistical variations in the number of electrons that this signal represented could not be removed and soon this noise swamped variations in the Type 1 signal (Wells, 1974a and b).

To reduce the Type 3 and, to a smaller extent, the Type 2 signal, (Peters, 1982, Peters et al, 1983) has recommended placing backscattered electron absorbers below the polepiece and coating the surface of biological samples with very thin layers of low Z metals. His results, using a field emission SEM at 30kV show a clear improvement over normal operation but the approach does not tackle the problem of removing the Type 2 signal very directly. On the other hand, Crewe and Lin (1976) recommended detecting the backscattered signal independently, using a semiconductor detector attached to the polepiece, and then subtracting some fraction of this from the signal derived from the normal scintillator-photomultiplier detector. The logic is that the Type 2 and Type 3 signals should be proportional to the backscattered detector output and subtracting this from the normal detector output should leave only the Type 1 signal. As the backscatter detector used in this work covered about π steradians, this is a reasonable analysis but the correspondence is not perfect because the detector has a large hole in the middle to allow the beam to pass through and the energy and angle of a backscattered electron may effect its chance of producing a collectable secondary in a way not proportional to the signal it produces in the semiconductor detector. Nonetheless, these authors also show a clear improvement (pp. 236) and the technique has been used by others (Volbert 1982a&b).

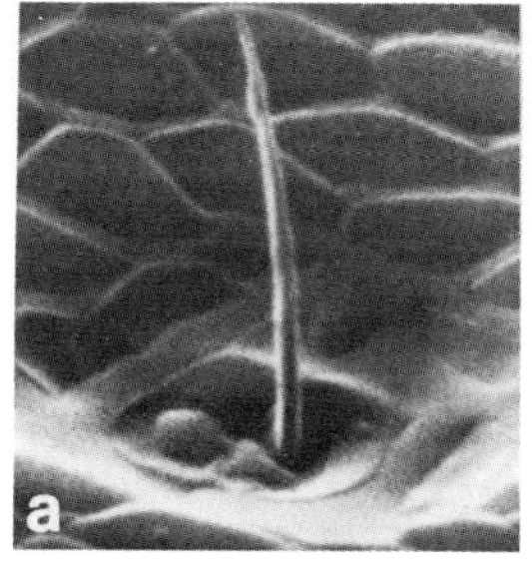

2 kV

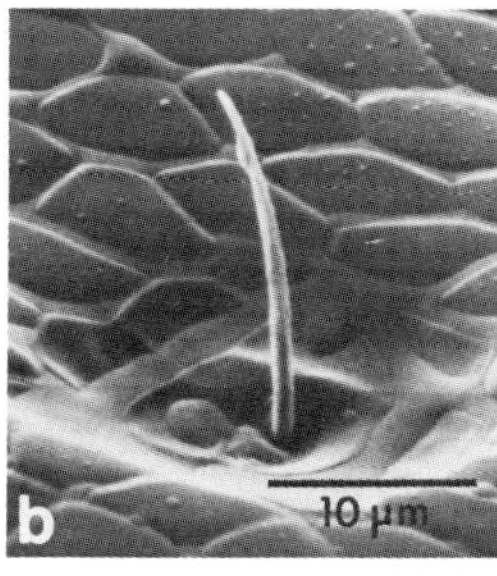

5 kV

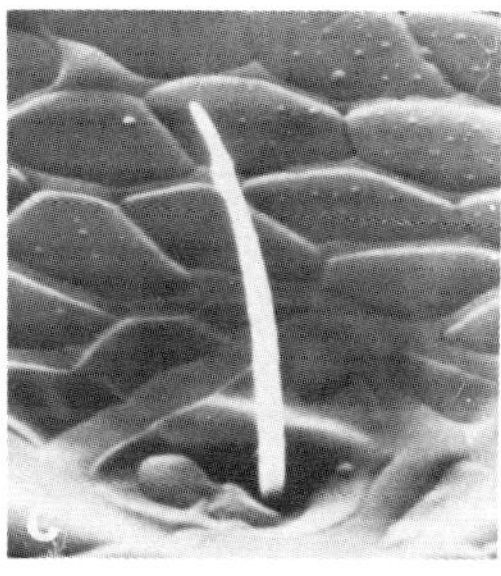

10 kV

Figure 1: Three micrographs made with a 'conventional' SEM and showing a hair on a flour beetle which has been sputter-coated with gold and imaged with V_o = 2, 5 and 10 kV. In a, details on the surface of the hair can be seen and the near side is approximately the same shade of grey as comparably oriented surfaces on the bulk of the specimen. In c, the hair appears much brighter than these adjacent areas because considerable signal is generated as the beam emerges from the far side of the hair and the coding of the image is no longer strictly topographic. Figure 1b shows an intermediate condition.

Finally, there is a large group of investigators who, untroubled by theoretical misgivings, have made images of a variety of samples that appear to demonstrate topographic resolution far in excess of that which would be possible if the Type 1 signal is indeed likely to be swamped by Types 2 and 3. Some of these studies have used scanning attachments on the transmission electron microscope operated at (TEM/SEM) 20-80kV (Koike et al., 1971, 1973; Arro et al., 1981; Haggis & Bond, 1979; Haggis, 1982; Haggis et al., 1983). In the TEM/SEM, the sample is immersed in the lens field and the secondary electron signal consists of those electrons that spiral up the field lines and out through the upper pole piece. This process may preferentially exclude the Type 3 signal and definitely provides a distinctly different image of the sample than does a conventional detector (Buchanan, 1982, 1983). Other workers have utilized field emission SEMs (Lin & Lamvik, 1975; Watabe et al., 1978; Sawada, 1981; Peters et al., 1983) which in some cases were modified to permit secondary electrons to be collected from a sample located in the lens field (Tanaka, 1980, 1981).

A third approach involves looking at what might be considered the inverse of the Type I signal, namely the signal derived from backscattered electrons that have only lost a small amount of energy in a glancing collision with a steeply tilted sample. This low-loss backscattered signal can be detected in a normal SEM with an appropriately placed backscattered electron detector (Wells, 1979) or from a sample immersed in the field of a short focal-length condenser-objective lens which also serves as an energy filter (Wells et al., 1973; Joy & Maher, 1976; Kokubo et al., 1975). The latter method is capable of producing very high resolution images of metal-coated samples because electrons that have lost only 200-400eV have only participated in interactions near the sample surface (Broers et al., 1975).

<u>Topographic or Z contrast?</u>

It is not clear that any of these signals is really a topographic signal rather than a Z signal that chiefly responds to variations in the granularity or the effective thickness of the metal coating on such samples and is further modified by differences in the signal collection efficiency from point to point on the sample (Fig 2). In fact Wells points out that a layer of carbon contamination, artificially produced to cover the surface of such a sample, is barely visible using the low loss mode (Wells, 1979, pp. 213). A similar lack of fine detail on flat surfaces can be seen in the images of carbon black particles shown in the TEM/SEM by Koike et al. (1973). Though these images appear to be topographic, they are not topographic in the same sense as is the case at low magnification. They may indeed provide useful information about the sample, but it is important to realize that they are in fact analogous to TEM images of shadowed replicas and should likewise be viewed with caution. They may, for instance, reveal more about the nucleation of metal particles than they do about sample topography and they discriminate against small features on flat surfaces and in favor of similar features suspended over the cavities.

Our acceptance of Z-contrast images as 'topographic' can be traced to the need of manufacturers to demonstrate real improvements in instrument performance. In the early 1970's resolution in images made using the signal from the secondary electron detector fell below about 20 nm and the criterion ceased to be the smallest discernible <u>surface</u> object and became instead the smallest discernible object. Subsequently, the probe diameter was greatly reduced and the test objects were chosen to demonstrate this improvement, rather than to demonstrate that smaller surface features could be resolved (Ballard, 1972). As a result the 'resolution' in the secondary electron mode is now often quoted to be 1.5-3 nm while the best results show images of biological objects such as intermediate filaments and ribosomes in the size range of 10-25 nm (Tanaka, 1981; Haggis, 1982; Haggis et. al., 1983; Peters, 1982; Peters & Green, 1983).

Many popular test objects can be modelled as a series of heavy metal particles covering the surface of a low Z substrate such as a carbon film or a dried biological sample. In this case, the main contrast is Z contrast, either between high Z metal grains and low Z inter-grain spaces, or, in the case of uniform coating on a bumpy surface, variations in the effective thickness of this coating as the beam traverses the coating at different angles. On a highly convoluted surface, this signal may also reflect variations in coating thickness and the large variations that exist in the efficiency with which electrons emerging from a given area are collected. These effects are diagrammed in Fig. 2 which shows a hypothetical coated surface and a corresponding image shaded solely in response to changes in effective coating thickness.

The problem of low contrast can also be approached by reducing the beam voltage as suggested by many early authors (Thornley, 1960; Kosuge et al., 1970; Boyde, 1971; Catto & Smith, 1973; Welter & Coates, 1974; Wells, 1974b pp. 127; Dilly, 1980). At $V_o < 10$ kV there is a sharp increase in Type 1 signal (Joy, 1984a). In addition R at 1 kV is only about 2% of that at 20 kV and so the area of sample from which Type 2 secondaries are produced is 2500 times smaller. As the backscatter coefficient diminishes only slightly with voltage (Niedrig, 1978; Reimer, 1979, Fig. 8; Kotera et al., 1981), the number of Type 2 electrons may still be significant but they will emerge from a smaller area. There have been few attempts to produce high resolution SEM images using beam energies near 1 kV because of the electron-optical constraints mentioned in the Introduction and discussed in the next section, so it is still not certain that, on the finest scale, LVSEM has a clear advantage. On the other hand, results at somewhat lower resolution show a clear increase in the contrast of small details (Fig. 3) and so there is reason to expect that the same will hold if LVSEMs with smaller probes can be made.

An analysis of the effect of V_o on high resolution topographic contrast at present depends on estimates of how V_o affects the modulation of Type 1 electrons by the local surface angle and how it

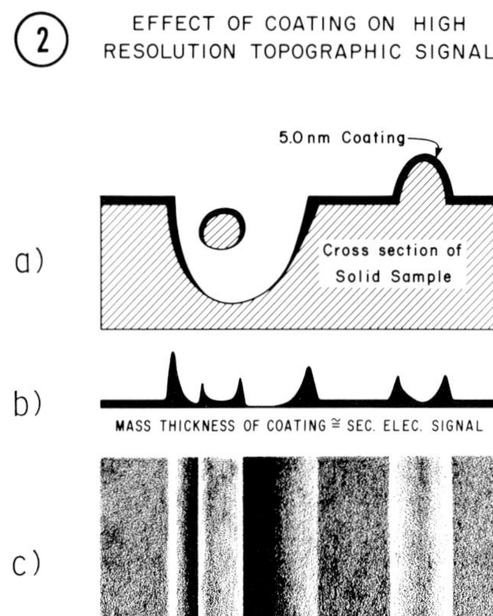

Figure 2: Coating thickness contrast: a) cross-section through a rough surface that has been metal coated is shown schematically, b) the effective thickness of this coating as a function of position, c) what appears to be a topographic image of an extended specimen having the cross-section shown in a. It results from coding image lightness as being proportional to coating thickness rather than surface angle. (The image may be more easy to interpret if viewed from a distance.)

Figure 3: Critical point dried blood cells on a grid covered with a thin film, coated with carbon and imaged at 1 kV (a,c) and 20 kV (b,d) in a FE SEM. The upper pair clearly shows the loss in contrast of small surface detail in the background film (arrows) and on the surface of a red blood cell. In the lower pair the cells are located over a grid bar which produces a considerable background signal at the higher voltage. Again small surface details are more visible at the lower voltage (arrows) as are details on the ruffles on the surface of this platelet. (Sample kindly provided by Dr. E. de Harven).

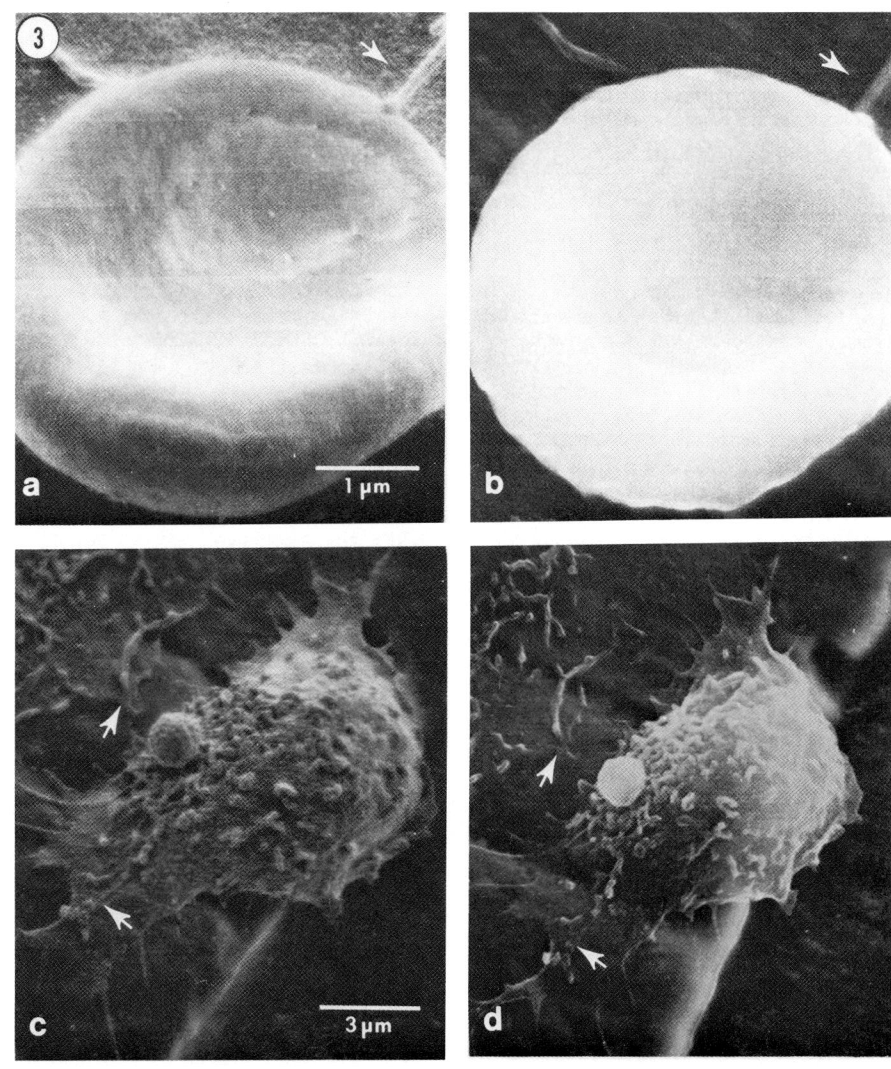

effects the relative strengths of the four types of detected signal. Because there is a lack both of suitable test objects and of probe forming columns with the necessary capabilities, such analyses typically involve the use of Monte Carlo techniques to simulate the scattering interactions of a beam as it passes over a homogeneous surface having some analytically simple topographic feature such as a cube (George and Robinson, 1975) or a gaussian asperity (Catto and Smith, 1973). To increase computational speed, Monte Carlo methods greatly simplify the interactions between the beam and the sample. They usually consider only the uncontaminated surface of a homogeneous (non-crystalline) metallic object and assume that both elastic and inelastic collisions are highly localized. They usually ignore a host of other interactions, such as those producing X-rays, Auger electrons and other characteristic interactions, even though these interactions may produce other collectable secondaries from remote locations. Finally, as the rate of energy transfer increases strongly when the electron slows down near the end of its track, the last 200-500 eV is usually modelled as being deposited in a single point. When these simplifying assumptions are used to estimate contrast on the size scale of nm with V_o = 1 kV, important errors are introduced. For instance, inelastic collisions are known not to be highly localized (Isaacson and Langmore, 1974) and 200-500 eV represents too large a fraction of eV_o to be approximated by a single event.

In spite of these limitations, some interesting trends are evident in the most complete of these early studies, that by Catto & Smith (1973). These authors note an increase in the size of the smallest discernible feature as V_o is increased in the range 10 kV-30 kV. Unfortunately, they do not continue their analysis to lower V_o for reasons that are not clear. More recent simulations (Murata et al, 1981, Joy, 1984a,b) include the effects of the production of fast secondaries (Joy et. al., 1982) which may have energies up to $eV_o/2$ and can therefore excite additional low energy secondaries along their tracks. Though not produced in great numbers, they are important because they often travel almost perpendicular to the primary beam and therefore produce secondaries at some distance from the probe. Figure 3 from the 1984b paper by Joy shows secondary emission vs distance from the axis for 2, 5 and 30 kV, and is reproduced here as Fig. 4a. The intensity in electrons is normalized to the number produced at a given voltage on axis. While all three curves drop to ~25% by 2 nm, they have distinctly different shapes beyond this distance. In particular, the intensity of the Type 2 signal (i.e. that emerging more than 2 nm from the axis) appears to be substantially higher at the lower voltages, implying a reduction in topographic contrast at lower V_o. This is not actually the case because 1) The graph is 100 times too small on both axes to show the full curve for 30 kV while it does show the 2 kV curve out almost to the edge of the interaction radius, r. In figure 4b the same data has been extrapolated to the larger size range. Although these curves only represent an estimate of the probable shape and should not be taken as hard data, they do emphasize the fact that the central peak in the distribution does sit on a large iceberg of poorly localized emission. If the curves were plotted for dried biological tissue (density = 0.2 g/cm^2) rather than for solid carbon, the distance scale would be expanded by a further factor of 10. Each distribution has been normalized at its maximum, and therefore no allowance has been made for the increase in total δ at the lower voltages.

If the curves in 4b were more accurate, we would integrate the signal under the curve in two regions: the local region up to 2μm and the non-local region beyond this distance. It seems likely that the ratio of local to non-local signal might be 5-6 times higher at 2 kV versus 30 kV.

This difference represents a potential increase in detectable topographic contrast and it has some interesting consequences. As V_o is reduced from 20 - 1 kV, δ (normal incidence) increases from 0.1 to ~ 1 and therefore generally recognized that, the beam current (I_b) required to produce an image of a given quality is correspondingly reduced by a factor of 10. However, the effect of increased contrast on the required current is usually not specifically considered.

Wells (1974b eq. 2.21b) calculated the relation between required current I_b and signal contrast (S_a) for low contrast objects as:

$$I_b = K \frac{\delta_b}{\delta_a^2} \quad (1)$$

where K depends on the raster size, recording time and number of statistically defined grey levels in the image, ($K \simeq 10^{-12}$ for 500x500, 100 sec, 10 levels), and δ_b is the secondary electron coefficient of the D.C. background signal (the 'black' signal) and δ_a is the effective contrast of the normalized signal, (i.e. the peak signal minus δ_b). This equation includes the assumption that $\delta_b > \delta_a$ and this is true for small surface features seen at high magnification. Equation 1 is important to an analysis of the LVSEM because it shows that the required current depends directly on the DC offset of the signal and inversely with the square of the contrast. At low Vo both of these quantities change so as to reduce the beam current required to produce an image of a given quality. This is important because, as we shall see below, gun brightness is significantly reduced at low beam voltage.

Specimen Charging in the LVSEM

Many objects of microscopic interest are electrical insulators. When the surface of such a sample is scanned by a 1-30 kV electron beam, collisions in the layer immediately adjacent to the surface cause it to become somewhat depleted in electrons, while the next layer immediately below becomes negatively charged because beam electrons are trapped as they reach the end of their range. The deposition of a net charge in the sample depends on δ, which in turn depends on the type of material, and the local surface angle. (at glancing incidence δ can be as much as 4 at 20 kV. Pawley, 1984b). Around 10-30 kV, δ for most samples is less than unity and the sample accumulates a negative charge which may degrade the image by defocusing the beam or by distorting the collection field so as to produce the anomalous changes in apparent brightness familiar as the most common form of charging artifacts (Pawley, 1972). As higher surface potentials are reached (> 10's of volts), other, more extreme phenomena are recorded as described by Shaffner & Hearle (1976). Other variables that exacerbate the charging problem are high specimen resistivity (so-called insulators vary in resistivity over a range of 15 orders of magnitude), low specimen dielectric constant and slow scan speeds (Welter & McKee, 1972).

The situation is somewhat different in the LVSEM because on a variety of samples δ at normal incidence becomes greater than unity in the range of approximately .5 to 3 kV and so the sample charges positively. This is a far more stable situation because low energy electrons are constantly being evolved from the surface and so even a small positive charge imbalance can attract an appropriate neutralizing charge without the necessity of developing a surface potential higher than a few hundred millivolts. Because this self-regulating process is so efficient, it is often claimed that charging artifacts do not exist when $\delta \geq 1$ and the rapidity with which TV-rate images of such samples stabilize is offered as proof of this contention (Welter & McKee, 1972). However, this analysis is only strictly true for the trivial case of a flat, featureless sample with V_0 adjusted for $\delta = 1$. More topographically interesting samples show contrast and hence, $\delta = 1$ cannot be satisfied everywhere. In practice, areas where the beam incidence approaches normal may become slightly negative while areas of glancing incidence, or where the beam penetrates porous surface features will tend to become positive. Lateral electrostatic fields will exist between neighboring charged areas and vertical fields will exist between the electron-depleted surface layer and the trapped charge below. A small amount of current flows between these areas using free subsurface electrons, ionized by the beam, as charged carriers (Bresse, 1982). Clearly the situation is far more complicated than is implied by the simple statement that there are no charging artifacts whenever V_0 is set so that $\delta \geq 1$. The stability of an image scanned at TV rate is only evidence that a particular charge distribution is stable, not that there is an absence of charging.

The details of this process are of interest here because they involve the establishment of surface and subsurface potentials which, though small when compared to those found with higher V_0, may still be capable of defocusing or deflecting a beam of only 1 kV and thereby degrading the image. Surface potentials of the same size as ΔV (0.2V for FE guns) will defocus the beam and even smaller potentials could deflect it a few nm, perhaps in an erratic manner. In fact some investigators claim less trouble with charging at high voltage in the TEM/SEM than at lower voltages in the SEM (Haggis, 1982). This can be attributed to increased beam-induced conductivity at higher V_0 and to the partial immunity of the TEM/SEM detector to voltage contrast.

At present, the magnitude of these effects has not been investigated in the range of resolution and V_0 discussed here. It can be expected that when $V_0 \cong 1$kV, charging effects on insulating samples scanned at TV rates should be insufficient to produce large variations in signal collection efficiency, but not that they will be totally absent. Very thin (1-1.5nm) layers of coating material such as those used by Peters (1979, 1982) may be necessary. Fortunately, the procedures for applying these coatings have greatly improved in the past few years particularly with the introduction of ion-gun based sputter sources (Adachi et al., 1983; Evans & Franks, 1981; Kemmenoe & Bullock, 1983). As a result problems with decoration artifacts should be less common and the pseudo-topographical contrast caused by the coating and discussed above, should be minimized by the use of very thin coatings.

Technical Limitations and Possible Solutions in LVSEM

The technical difficulties that must be overcome in order to produce high resolution information from an LVSEM are similar to those noted for the low voltage TEM by Wilska (1964, 1965).

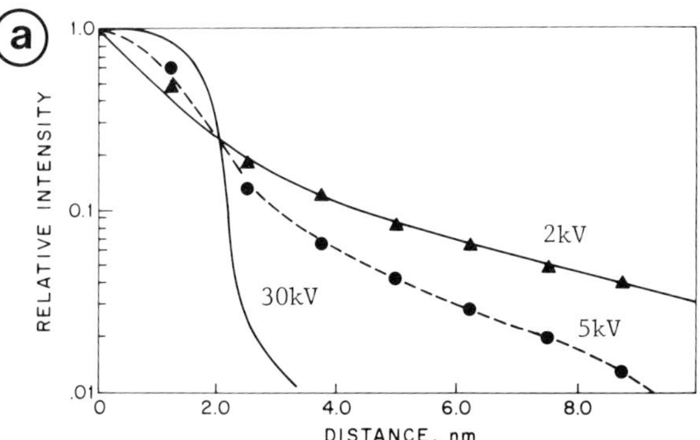

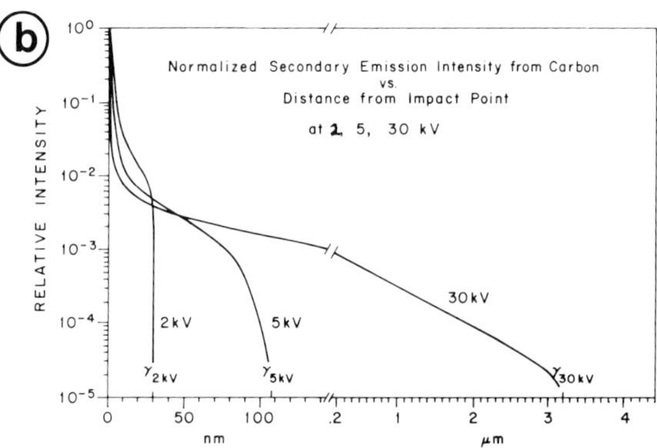

Figure 4: a) Normalized intensity of secondary electron production from carbon as a function of distance from the point of impact at 2, 5 and 30 kV (from Joy, 1984a). b) the same data as in a), extrapolated to the total emission range at each voltage and plotted on different scales.

They were first listed for the SEM by Oatley et al. (1965, p.215-217). They can be lumped into four areas: 1) low source brightness, 2) increased effect of chromatic aberration, 3) increased susceptibility to stray fields, 4) interactions between the beam and the signal collection field. Though these problems and their solutions sometimes interact, they will be treated separately below.

Brightness of Thermionic Sources at Low Voltage

It was early recognized that source brightness was the practical limit on the performance of an SEM with a heated tungsten source (Broers, 1974, 1982). In principle, the effect of spherical or chromatic aberration on spot size can be minimized by reducing the acceptance angle of the final lens, α, until the lens becomes diffraction limited. However, in instruments with conventional tungsten sources, the image becomes too noisy for convenient use long before α is reduced to the diffraction limit.

The brightness (β) of a thermionic electron source is limited by the Langmuir Equation (Langmuir, 1937) for small α.

$$\beta = j_o \frac{(11,600)}{\pi T} V_o \text{ amps/cm}^2 \text{ ster} \qquad (2)$$

Where j_o = current density at the source surface in amp/cm^2, T = source temperature in °K, and V_o = beam voltage in volts.

From equation 2 it follows that low V_o operation will produce reduced brightness. In practice the brightness actually obtained is even lower than we might expect from (2) because this equation is only valid in the absence of space charge near the cathode surface. Such space charge shields the cathode from the accelerating field and further reduces β. Though a given gun geometry may be virtually free of space charge effects near its highest operating voltage (20-30 kV) (Broers, 1974; Oatley, 1975), the fields present at the filament surface are proportionately less at 1 kV and the gun brightness will be limited by space charge unless the geometry is changed.

Practical measures to improve thermionic gun brightness at low kV therefore include changing gun geometry and the use of LaB_6 cathodes. The latter have a brightness about 6 to 10 times that of tungsten for comparable lifetime, a tip which is more pointed and which reduces the effect of space charge and operates at a lower temperature than normal tungsten (T = 1800 °K vs 2700 °K). Changes in gun geometry may involve simply reducing the gap between the Wehnelt and the anode by using an anode spacer or a mechanism to actually move the anode towards the cathode or it may involve adding additional anodes to the gun. A description of this 'double-anode' approach was recently given by Yamazaki et al. (1984). This group installed extraction anodes of various shapes and spacings between the Wehnelt and the normal anode. At low V_o this electrode is run a few kV above ground to produce a higher constant voltage between it and the cathode (V_1) and thereby reduce the effects of space charge.

Careful measurements verified that, at 1 and 2 kV, this produced 10 times the brightness of a normal 30 kV gun with tungsten and eight times the brightness with LaB_6. They reached the theoretical brightness specified by equation 1 at both these voltages using their best geometry, which is diagrammed in Fig. 5. Although it is not specifically pointed out in their paper, the acceleration/deceleration electrode system acts as a weak positive lens and this is probably why these workers found that β went through a maximum at V_1 = 1.5 kV. The effect of this lens on the imaging system was unclear but in images of a test sample, a distinct improvement was associated with the use of the double-anode system.

Brightness of Field-Emission Sources at Low Voltage

Field-emission (FE) sources have long been identified with high brightness (Crewe et al., 1968, 1970; 1971; Crewe, 1973; Hainfeld, 1977 has a good introductory review), but few commercial instruments have capitalized on this feature because of the stringent requirement for vacuum in the range of 10^{-8} pa(10^{-10} torr) around the emitter tip and because the current produced in a fine beam is subject to some temporal instability which tends to produce streaky images (Tuggle & Watson, 1984; Orloff this volume). Most FE sources utilize a double anode design. The V_1 is normally 3 to 7 kV and is used to adjust beam current, which is otherwise only dependent on the work function, ϕ, and the tip radius, r_o. A second supply between ground and the cathode adjusts the beam voltage (V_o) to the desired level and, in the case of the LVSEM, this means reducing it and thereby again producing an electrostatic lens.

In principle, the geometric parameters can be adjusted so that the tip emits efficiently with V_1 =1 kV. This would make the second anode unnecessary and avoid the consequent lens action. However, in practice, tips with sufficiently small r_o usually prove unstable and subject to catastrophic failure while a suitable choice of the spacing and shape of the two anodes can reduce the lens effect to a low level.

In the range of voltages discussed here, the brightness of the source depends only on V_1 and not directly on V_o except to the extent that the lens effect degrades the source image (Hainfield, 1977).

$$\beta = \frac{a\ V_1^3 \exp}{\phi r_o^2} \frac{(-b r_o \phi^{3/2})}{V_1} \qquad (3)$$

Where a and b are constants (a = 8.7 x 10^{-8}, b = 2.1 x 10^8). Several FE guns have been designed to incorporate a magnetic lens which operates on the beam as it leaves the tip (for example Kuo & Siegel, 1976; Ichinokawa et al., 1982). These lenses have superior electron optical characteristics to the electrostatic lenses and are said to produce improved performance especially when operating at high current and low beam voltage. However, because of their high current these guns are more susceptible to the lateral electron-electron interactions discussed below and so it is not clear they would be suitable to high resolution LVSEM.

Measurements with V_1 = 3kV, V_2 = 1kV have yielded values of β = 3-70 x 10^6 amp/cm^2, ster or about 1000 times that measured by Yamazaki et al. (1984) on the LaB_6 double-anode gun.

Although there are reports that electron-electron interactions within the beam can degrade the expected performance of FE guns both in terms of reducing the brightness and increasing the effective energy spread (Bauer & Speidel, 1981; Van Der Mast, 1983) these effects seem to be most serious on heated FE sources, sometimes referred to as Schottky or TF guns and hence the total beam current, unless the emitting area, can be restricted (Orloff, this volume). This effect is less serious on room temperature FE guns operated at moderate tip currents of about 10μA. Other workers have observed no such effects as long as high current density crossovers are avoided (Crewe et al., 1971). Clearly the electron-electron interactions near the cathode surface are reduced by the fact that the FE cathode has a tip radius 20-50 times smaller than LaB_6. This subject is considered further in the next section.

Noise in FE Sources

If the current present in the final beam of a high resolution FE SEM is traced back to the tip, it is found to arise from an area of only a few nm^2. The adsorption and desorption of individual molecules from this small surface can therefore produce a significant variation in its average work function while the etching produced by the collision of a single ion can change the microtopography and hence the local surface field. As a result, the current in the final probe is found to have a noise component unrelated to shot noise of between 3 and 10% (Hainfeld, 1977).

This noise drops in magnitude with increasing frequency and therefore is most troublesome at low frequencies. Efforts to stabilize the beam current by measuring the current striking an aperture and using this as a feedback signal to readjust V_1 (Nomura et al., 1973) are quite effective but not wholly successful because, due to the localized nature of the disturbance at the tip, the current striking the aperture is not necessarily a good measure of the current passing through it. Also, the changes in V_1 necessary to stabilize the current change the optical properties of the electrostatic lens. More recent systems avoid this optical effect by applying the signal from the aperture to an analog multiplier which directly normalizes the video signal for changes in beam current (Saito et al., 1982).

Another approach involves rapid, multiple scanning of the sample with the idea that low frequency variations will average out (Welter & McKee, 1972). The TV scan rate has other advantages with respect to ease of operation, stabilization of charging artifacts and quasi-immunity to stray field but it requires very high detector bandwidth (40 MHz for a 1000 x 1000 raster, 1/30 sec.) and careful scan coil design to avoid image distortion.

The matter of source noise can be crucial to the final performance of a high resolution LVSEM. There will be little net gain in contrast by going to lower V_0 if the improved contrast at the sample is swamped by false contrast due to source instability. It is possible that this limitation lead to further consideration of the Zr/W, TF source developed by Orloff et al and described elsewhere in this volume.

Electron Optics for the LVSEM

Figure 6 shows effect of diffraction and spherical and chromatic aberratiion on attainable probe diameter for a 'conventional' SEM operating at 20 kV and an SEM operating at 1 kV using a lens having aberration constraints typical of a high quality TEM. In both cases the dominant lens defect is chromatic aberration. The diameter of the disk of confusion due to this defect, d_c, is:

$$d_c = C_c \alpha \frac{\Delta V}{V_o} \quad (4)$$

where C_c is the chromatic aberration coefficient and ΔV is the energy spread of the beam. Clearly, $\Delta V/V_o$ increases rapidly at low V_o, hence the problem. It can be attacked by lowering C_c, α , or ΔV. Lenses can be designed to reduce C_c by shortening their focal length. While it is relatively easy, in terms of the total magnetic flux required, to construct a lens of short focal length at this low energy, the sample is soon immersed in the lens field so it may become more difficult to collect the secondary electron

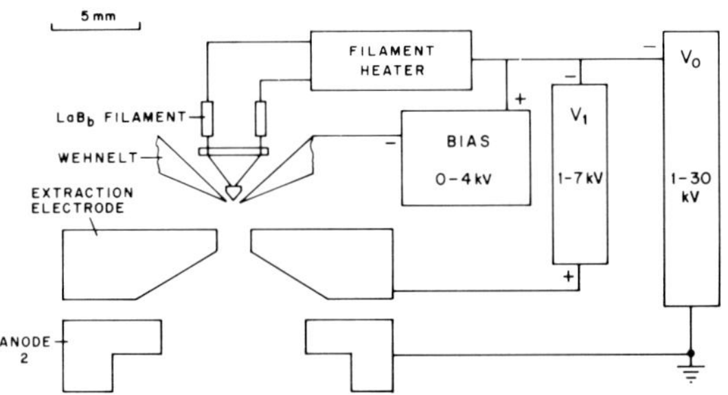

Figure 5: Double anode system for improved brightness from thermionic cathodes at low kV (after Yamazaki et al., 1984).

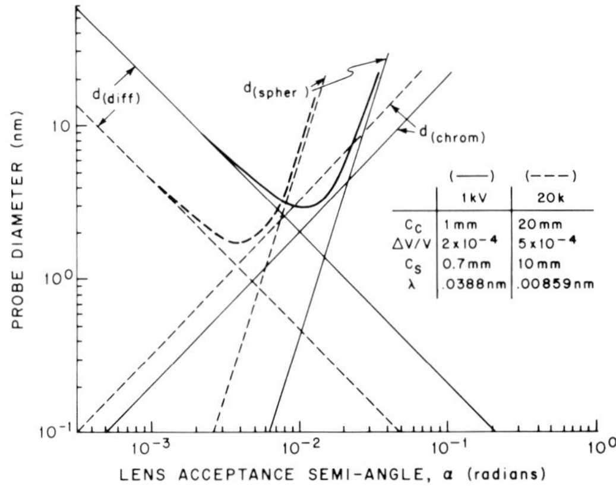

Figure 6: Electron optical limits to resolution in the conventional SEM at 20 kV (dotted) and in a theoretical SEM using a low aberration lens similar to that used in a modern TEM but operating at 1 kV. Diffraction and spherical and chromatic aberration are the only limits considered. The smaller spot size minima are found to be both fairly similar and smaller than the size of actual biological objects that can be imaged at present in the SEM.

signal. At the ultimate, it would probably be very difficult to design a practical system where C_c was much less than 0.2 mm, (Pawley & Wall, 1982; Barth & LePoole, 1976). This compares with the 5 to 10 mm found on most commercial instruments and the 1-3 mm found on some sample-in-lens SEMs (Koike et al., 1971; Buchanan, 1982).

The semi-angle, α, can also be manipulated, but because of the relatively long wavelength,λ, of 1 kV electrons (37 pm), the diffraction limit, d_d, is soon reached.

$$d_d = \frac{0.6\lambda}{\alpha} \text{ or } \frac{0.02 \text{ nm}}{\alpha} \text{ at 1 kV} \qquad (5)$$

Finally, there is some control over ΔV. The energy spread of the beam has many sources: the intrinsic energy spread of electrons leaving the cathode, power supply instabilities and energy broadening caused by lateral electron-electron interactions in high-current crossovers known generally as the Boersch effect (1954).

The expected energy spread at the cathode surface for thermal emitters is kT and this again emphasizes the advantage of LaB_6 vs tungsten because of its lower operating temperature (kT = 0.13 eV vs 0.2 eV). Intrinsic energy spread in FE sources depends on the shape, the tip material and the crystallographic orientation of the tip but it is usually quoted as about 0.2 eV for tungsten (Crewe et al., 1971; Hainfeld, 1977).

Lateral interactions between electrons are more noticeable when high current beams must form crossovers and this is often the case in thermal sources where large beam currents are often a byproduct of efforts to increase β by reducing the effect of space charge (Oatley, 1975). The problem is compounded by the fact that this large current is usually focussed into a small gun crossover by the effect of the Wehnelt. Under these circumstances, a considerable improvement can be gained by employing pointed cathodes as these permit a high extraction field over only a small emitting area and therefore a lower total current (Wiesner, 1973; Wiesner & Everhart, 1973; Ohshita et al., 1978). Measured values of ΔV from thermal sources usually average about 2 to 4 eV but the measurements are usually made at voltages much higher than 1 kV, where the Boersch effect is likely to be less strong because the electrons move faster and therefore have less opportunity to interact (Pfeiffer, 1972, 1982).

The FE and TE guns have clear advantages in this regard. They not only have low intrinsic energy spread but they can operate well at low total beam currents and because of their small virtual source size they can, in principle, operate with no crossover before the sample. Taken together with the higher brightness of FE and TF at low kV, the reduced energy spread provides a convincing rationale that any serious effort to produce optimum performance in the LVSEM will necessarily require either a FE source and a method for compensating for its temporal instability or a TF source.

Stray Fields

A 1 kV electron takes 5 times as long to travel down a given column as does a 25 kV electron. For this reason, in simple terms, it is 5 times as susceptible to transverse stray electrostatic or magnetic fields. This effect is sometimes exaggerated on large, older instruments because they are often run with less demagnification in the intermediate lenses to compensate for low source brightness and as a result, stray fields acting on the upper column, which usually have no visible effect, begin to be noticeable. Field emission systems are similarly susceptible because they normally operate with little or no demagnification.

Both AC and DC electromagnetic fields can degrade the performance of electron optical instruments. DC magnetic fields are produced by ion pumps, lenses, the earth and any stray ferromagnetic materials that may have been built into the apparatus by mistake. Usually their only effect is to cause misalignment between the mechanical and the electrical axis of the instrument but they can also be responsible for saturating high permeability shielding materials, thereby rendering them ineffective for shielding AC fields. At low voltage, stray electrostatic fields can also be very troublesome. Any insulator which can be encountered by the beam will develop a surface charge and this charge will in turn produce a field that deflects the beam. To avoid this, the column must be designed so that all insulators are shielded from the beam and great care must be taken to exclude even the most minute particles of dust or lint from the apparatus. Even the choice of materials is important because many metals commonly used for high vacuum applications such as stainless steel, Mo, Ti and Al are normally covered with a layer of nonconductive oxide. The effects of surface charge accumulation on many of the metals used in electron microscopy is well described by Anger et al. (1983). Particular attention should be given to the fabrication of beam tubes and apertures and these should probably be made of acid-cleaned Pt.

Although there are instances of electrostatic pickup from nearby radio stations producing noise in the scanning circuits, most AC fields of interest to the microscopist are magnetic and are linked to the mains frequency. Their effect can be reduced by synchronizing the scan frequency to the mains which has the effect that the stray field becomes an image distortion rather than a blurring function. Even so, stray AC magnetic fields remain one of the most persistent practical problems associated with operating the SEM at low voltage, especially when small-area, rapid-scan rates are used for focusing and astigmatism correction. To avoid them, great care must be taken in selecting the installation site, in making the column as short as practical and shielding it with several layers of high permeability magnetic materials, especially the gun region and the sample chamber. Several small commercial FE SEMs have been entirely enclosed in a box of shielding material. In addition, it is necessary to ensure that no stray fields are introduced to the column by currents flowing in ground loops through the equipment or by ripple on supplies feeding the scanning, stigmator or alignment coils or the field used to collect the signal electrons. These stray currents may be insignificant when the instrument operates at high voltage and only become noticeable when the magnitude of

the current in the deflection coils is reduced to operate at low voltage.

The ability to detect and eliminate stray fields is crucial to successful operation of the LVSEM, but unlike the design of the microscope column, it is at least in part susceptible to improvement by the efforts of a well-informed operator. The techniques by which this may be done are discussed in more detail elsewhere (Pawley, 1984c).

Problems of Signal Collection in the LVSEM

In most SEMs, secondary electrons produced by collisions between the beam electrons and the sample are collected by a field imposed by a grid at about +300 volts, which attracts electrons to the entrance of the scintillator/photomultiplier signal amplifier (Everhart & Thornley, 1960). This works well when the sample is 5 to 10 mm below the objective lens pole-piece, but as the working distance is reduced in an attempt to diminish C_c, the same horizontal field at the sample surface becomes less efficient at collecting signal electrons. On the other hand, this field becomes relatively large compared with the beam energy and it therefore can produce some distortion of the beam.

There have been four strategies to overcome this problem. The simplest is to use an objective lens with a sharply conical lower pole-piece which permits the collection field to penetrate to the electron optical axis more easily (Nakagawa et al., 1982; Pomposo and Coates, 1982). This approach also permits observation of large highly-tilted flat samples but it has the disadvantage that conical lenses often have reduced electron optical properties due chiefly to flux leakage in the region where the conical pole-pieces taper together. The second method is to use the signal collection system employed in the TEM/SEM where the sample is immersed in the lens field and the low energy electrons spiral up the field lines through the hole in the upper pole-piece where they are then collected by a small transverse electric field (Koike et al., 1971, 1973; Buchanan, 1982; Tamura et. al., 1980). This system has many advantages: 1) It will work with very short focal length lenses. 2) The transverse field occurs in an area where it can be carefully controlled and is not subject to inhomogeneities produced by irregularities in specimen topography; a consideration that becomes more important on samples which are not flat semiconductors. 3) It seems to selectively exclude at least some of the low resolution Type 2 and 3 signals produced by backscattered electrons (Buchanan, 1983). 4) It seems to be somewhat immune to the variations in collection efficiency caused by differences in specimen surface potential that are responsible for most simple charging artifacts. There are also some disadvantages: magnetic samples cannot be viewed and because there is no directional collection field at the sample, the 'shadowing' familiar from normal SEM micrographs is absent.

The method described by Volbert and Reimer (1980) involves the use of a pair of scintillator/photomultiplier detectors, one on either side of a sample with the result that there is no field on the axis. We have used this approach with a sample immersed in the lens field (Pawley & Wall, 1982). An axial metal tube protects the beam from the effects of the collection field until just before it reaches the sample. This detector will be described further in the next section.

A final and very promising possibility is described by Russell elsewhere in this volume and grows out of our early design by Venables and Harland, (1973.). It involves the use of a microchannel plate amplifier mounted above the sample and having a hole in the middle to let the beam pass through. The front surface can be biassed slightly positively or negatively without degrading the beam as the resulting field is cylindrically symmetric. Positive bias permits detection of secondaries with high efficiency and the results described by Russell show great promise for LVSEM.

None of these schemes represents an ideal solution in that all have the potential to degrade the beam before it reaches the sample. Their efficiency, in terms of fraction of emitted secondary electrons actually collected, has not been reported, but, of course, electrons lost at this stage can only be replaced by higher beam current and a larger spot so this is an important parameter.

An LVSEM Test Bench

In 1977 we reported on a freeze-fracture chamber directly attached to an SEM with an LaB_6 source and designed so that the coated fracture surface could be directly viewed using a cold stage (Pawley & Norton, 1978). Though we had hoped that such a system would provide an image similar to that obtained from freeze-fracture TEM, we found that the resolution/contrast at 20-30 kV was insufficient to resolve even the 10-12 nm intermembrane particles normally found on fractured membrane surfaces (Pawley et al., 1980). We then tried to image an actual shadowed freeze-fracture replica suspended over a Faraday cage at room temperature (Pawley et al., 1978). Such a sample should permit very high resolution SEM imaging because, as the sample is very thin, the Type 2 and 3 signals are almost absent. However, images of the replicas showed no trace of the particles. Indeed the signal from the replica was very small altogether, about 5% of that on a solid metal surface and it was to this low signal level that we attributed our failure. The only possible solution seemed to be to go to lower beam voltage and as there was no high resolution LVSEM equipment commercially available at that time, we began a modest program to develop a prototype instrument in 1980. This instrument was designed to overcome some of the problems discussed above and to produce a 1-2 nm probe at 1 kV in order to determine whether or not images made with it were superior to those made at higher voltage (Pawley & Wall, 1982; Pawley & Winters, 1983).

Design

A diagram of the present version of this instrument, and photos of the entire assembly and the column itself are shown in Fig. 7. The electron source is a cold FE cathode, using single crystal tungsten in the (1,1,1) orientation

(F.E.I. Inc., Hillsboro, Oregon) and the gun is pumped from above with a 220 l/sec. ion pump. It has been designed to be rigid, compact and well-shielded from internal and external magnetic fields. The gun block has side ports for a window and to feed through both high voltage and heater currents. It is machined from a single block of stainless steel to avoid the possibility that the welds might become ferromagnetic and the cylindrical part of the first anode is made of mu-metal. The anode itself is made of a thin M_o foil and can be heated by radiation and electron bombardment from a tungsten filament located below it to speed out-gassing. The gun isolation valve and the movable, three-position, aperture are bellows-sealed into the lower part of the gun block. Below the aperture the beam enters a platinum vacuum liner tube outside of which are situated stigmator, alignment, and double-deflection scanning coils and also a small condenser lens. The beam tube is brazed to the specimen chamber which has side-ports for two scintillator/photomultiplier secondary electron detectors, each employing a single crystal Yttrium Aluminium Garnet (Ce^{+++}) hemispherical scintillator (Pawley, 1974; Autrata et al, 1978, 1983), a 20 l/s, water-cooled ion pump and the controls for a specimen stage holding two 3 mm grids. A viton-sealed port on the bottom permits specimen exchange. The objective lens is of the pancake type as described by Mulvey (1982). It is excited by a 225 turn tape winding and cooled by laminar water flow past the bottom of the lower lens pole plate. The pole tip radius is 1 mm and the calculated lens parameters are: f = 0.54mm, C_c = 0.22 mm. Mechanical alignment of the tip is performed by adjusting set screws in the spider which holds the filament assembly while looking up the axis at the tip. After tip alignment, all of the gun components are clamped rigidly together by the upper threaded ring. All other components are prealigned and clamped by bolts to the gun block except for the objective lens which can be translated $\pm$ 1mm, X and Y.

The electronics are a modified version of those supplied for an AMRAY 1200 microscope (AMRAY, Bedford, Mass.). Separate controls are provided for the acceleration and PMT voltage of each electron detector. The entire column is surrounded by a 400mm dia. alloy cylinder which is lined with 1mm thick mu-metal and to which is attached part of the isolation valve mechanism. The cylinder and the column hang from the gun ion pump which is supported by a large steel plate. This is, in turn, suspended from a steel frame using a set of pulleys and a total of 22 thicknesses of elastic cordage which provides vibration isolation (vertical resonant frequency $\cong$ 2 Hz).

Operation

Initially, very sharp cathodes were used to permit operation with $V_1 = V_0 = 1$ kV. However, after a few hours of operation these tips would fail and so they were replaced by tips operating at $V_1 = 3$ kV for a tip current of 20μA.

The apertures used are carefully cleaned and then coated with evaporated gold immediately before use. A 1000 μm aperture is used for the coarse set-up and 100 μm is used when operating the condenser lens only. A picture of a 1000 mesh Cu grid made using the 100 μm aperture and the condenser lens only is shown in figure 7d and shows a resolution of about 50 μm. This is a reasonable result considering the long working distance.

Though the lower lens produces approximately the expected magnetic field (2.3 k gauss, on axis 75 μm above tip at 5A), we have not yet operated it successfully as part of the microscope. To begin with, the level of the signal is very low when the lens is excited and this is true even when the collection field is increased by applying 12 kV to only one scintillator so that there is no null point on the axis. Secondly, the beam enters the fringe field well before it reaches the sample (7 gauss at the distance of the FE cathode) and this pre-field is so strong that the optical properties of the lens seem not to be as calculated. Similar problems were encountered by Hill & Smith (1982) when they used a similar lens in a conventional SEM.

Unfortunately, no attempt to achieve high resolution images can be made with this instrument until this problem is overcome. Therefore we plan to modify the lens by adding an upper pole piece, thereby making it more similar to the lens used in the TEM/SEM (Fig. 8). The calculated aberration constants of this lens are somewhat larger than those of the present lens but we still hope to obtain a beam of 2.5nm at 1 kV.

Other Limitations to High Resolution Performance

Spatial resolution in the topographic image from the SEM is so central to one's assessment of the instrument's capabilities it has been much studied and discussed (Oatley et al., 1965; Pease & Nixon, 1965; Wells, 1974a,b; Broers, 1982; Catto & Smith, 1973; Watabe et al., 1978; Peters, 1979, 1982). As has been mentioned above, many of these analyses, when evaluating the final images, have mistaken the topographically modified Z-contrast produced by the coating material for true topographic contrast. Catto and Smith (1973) avoid this but their theoretical analysis deals strictly with the information theory aspects of the beam/specimen interaction given certain ideal conditions. Their analysis uses basic electron scattering theory to calculate the signal-to-noise ratio of the signal from small Gaussian-profile asperities on a solid gold sample versus the radius of and distance between these asperities. The analysis is performed at 10, 20, and 30 kV and assuming a probe diameter of 0, 0.5nm, and 5nm. Not surprisingly their results show the best performance at the lowest voltage where a 1.0nm asperity should just be visible using a 0.5nm beam.

Comforting though it is to know that such resolution is not impossible from the point of view of scattering and information theory, it should be kept in mind that there are several practical limitations to actually obtaining this performance on real samples besides probe diameter and current. Specifically these include 1) surface contamination, 2) radiation damage, 3) the delocalized nature of inelastic scattering and 4) beam tailing.

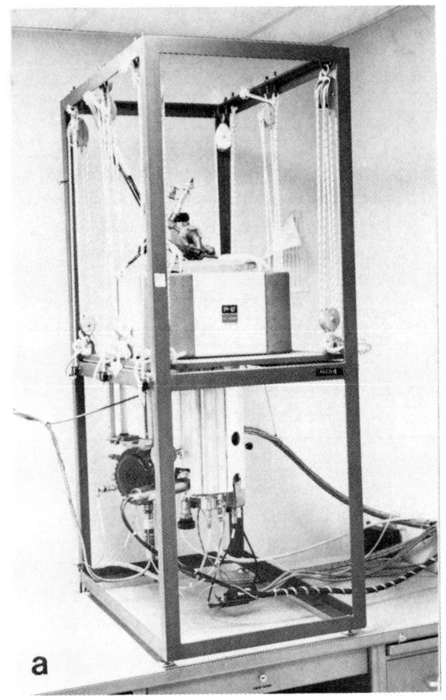

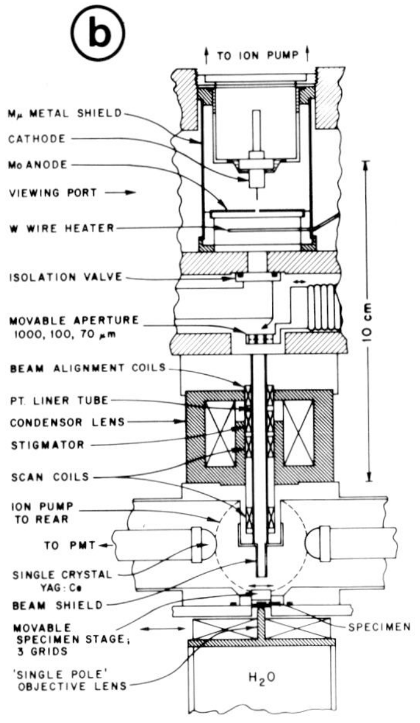

Figure 7: Low voltage SEM test bench: a) The entire column assembly showing the two ion pumps, the outer magnetic shield and the vibration isolation system. b) A diagram of the major components of the E.O. column. c) A photo of the E.O. column with the outer shield removed. (1) Bellows for isolation value. (2) Aperture motion control from side. (3) Stage motion control, similar to (2). d) An early micrograph of a 1000 mesh grid made at 1 kV using only the condenser lens.

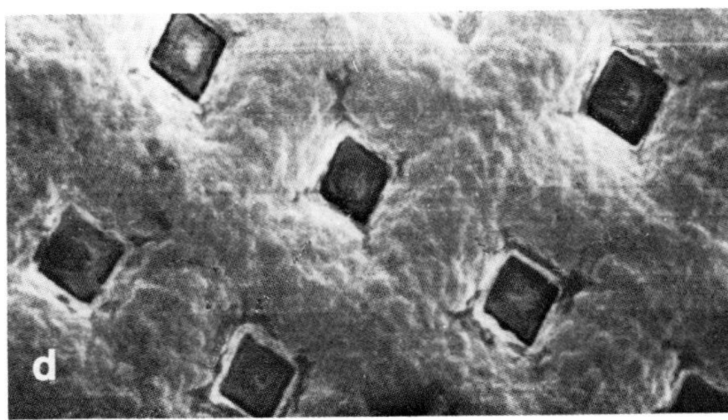

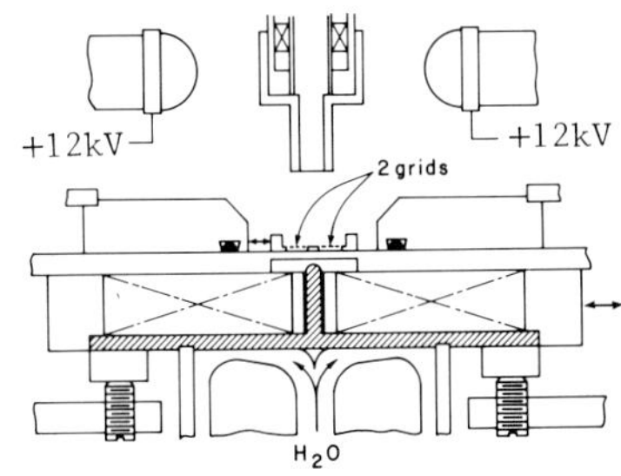

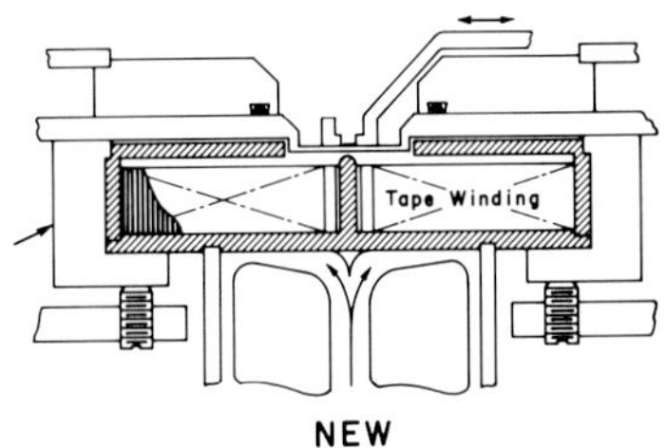

Figure 8: Present and proposed magnetic circuits for the objective lens of the LVSEM test bench.

Contamination

Layers of organic contamination accumulate on surfaces subjected to electron beam bombardment and the problem is more severe when small, high-current probes are used (Fourie, 1981). The presence of such a film is much more noticeable in the secondary electron image if a low V_o is used. Figure 9a and b show two micrographs of the same area of Type 2 cell on a lung alveolar wall. The contaminated area can be distinguished in 9a, made with V_o = 1 kV though it is not evident with V_o = 10 kV in 9b.

It has been assumed that any effort to produce the ultimate in topographic resolution will entail a FE source and therefore an oil-free, generally bakeable, vacuum system where these problems would be less serious. In such an instrument the sample itself becomes the major source of contamination. Even using the cleanest possible grids and support films, a layer of contamination rapidly builds up on biological samples unless they and their surroundings are cooled sufficiently (about -60°C) to arrest the process of surface diffusion (Wall et at., 1977; Voreades & Wall, 1979). Therefore any effort to obtain high resolution surface images from organic materials, rather than from a metal such as gold, will probably require a cooled sample and lens assembly.

Radiation Damage

The kilovolt electrons impinging on an SEM sample undergo inelastic collisions that may result in the transfer of more than a few electron volts of energy. As such, they constitute an intense source of ionizing radiation. The damage caused by this interaction to covalently bonded samples viewed in the TEM has been widely studied (Glaeser 1971, 1975, Cosslett 1978) and found to seriously limit structural information retrieval below 2nm. The situation is even more serious in the SEM because the entire beam energy is absorbed in the sample. Even with V_o = 1kV and a 10^{-11}A beam the power of the beam is $I\ V_o = 10^{-8}$ watts. If we assume that one half of this energy is absorbed in the upper 10nm of a sample with density 1 scanned with a raster 1000nm on a side, the dose rate, D_r, is

$$D_r = \frac{10^{-8}\ \mathrm{j/s}}{2((10^{-4})^2 \times 10^{-6})\mathrm{g}} \ \frac{\mathrm{g}}{10^{-5}\mathrm{j}} = 5\mathrm{x}10^{10}\ \mathrm{Rads/sec.} \qquad (6)$$

where 1 Rad = 10^{-5}j deposited/gram of irradiated sample. This is a very high dose rate and, assuming a 100s scan, it is more than 10^3 times that common in the low-dose TEM studies designed to preserve structures below 2nm. It is reasonable to assume that biological samples exposed to this flux of ionizing radiation will be rapidly reduced to a highly traumatized carbon skeleton of the original structure. The image obtained will be an image of ashes and the relation it bears to the original structure will be unknown. Certainly any structure of less than 2-3nm should be initially treated very skeptically. On the other hand, the acceptable level of damage depends on the end-point. The 100 electrons/nm^2 thought acceptable for low dose TEM is far more than that required to inactivate all enzymes, while gross molecular shape is sometimes preserved at much higher doses (Ottensmeyer et al., 1978). As we are not seeking atomic resolution, it is not unreasonable to expect that the number and position of specific features in the original sample may be retained as lumps on the surface of the ash and, of course, the situation is less severe if the sample is really a thin metal coat covering the organic material of interest.

De-Localization of the Inelastic Scattering Event

The secondary electrons that provide the SEM signal are produced by inelastic collisions with electrons in the sample. This process is not highly localized in that it can occur when the probe electron passes some nm away from the electron being excited (Isaacson & Langmore, 1974). Barth & LePoole (1976) point out that, as the delocalization is proportional to V_o, better results are to be expected at low voltage. In particular, they predict that 0.6nm localization should be possible at 1 kV under somewhat optimistic experimental circumstances. (C_c, C_s = 0.07mm, $\Delta V \leq 0.2$V, $\alpha = 4\mathrm{x}10^{-2}$rad.)

Another consideration seldom discussed in terms of topographic imaging involves the effects of low-energy X-rays. The spectrum of Bremsstrahlung X-ray production increases exponentially at low energy (1000-10 eV) but these interactions are usually ignored because the X-rays so produced are absorbed so strongly by the sample that very few of them leave it. Their generation is of interest here because their absorption in the sample can result in the production of a secondary electron a few nm away from their generation site. We are unaware of quantitative data at these voltages and on this size scale that would permit an accurate estimate of the size of this effect but it could be an important factor.

As these effects act independently from all the electron optical blurring functions, they should reduce the actual point-to-point resolution by at least an additional 0.5 nm from that theoretically calculated from electron optical and electron scattering considerations.

Beam Tailing

When speaking of the beam diameter of a focused spot, it is customary to assume that the current density resembles a Gaussian distribution or an Airey disk and to refer to its diameter at half maximum or the distance over which the intensity drops from 80% to 10% of the peak value. As has been pointed out by Cliff & Kenway (1982), beams in probe forming instruments are often non-gaussian for various reasons, and in particular, they often have a small central peak surrounded by a much wider "shadow" of lower but significant intensity. This shadow greatly complicates the criteria for visibility as measured by Catto and Smith (1973). When discussing small probes it is essential to keep in mind the fraction of the total current actually in the central spot.

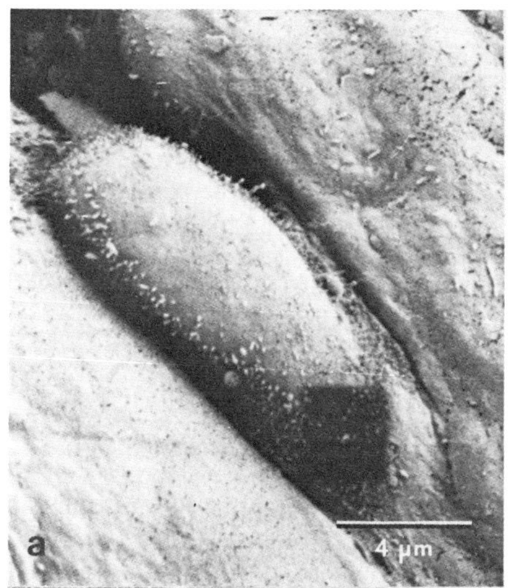

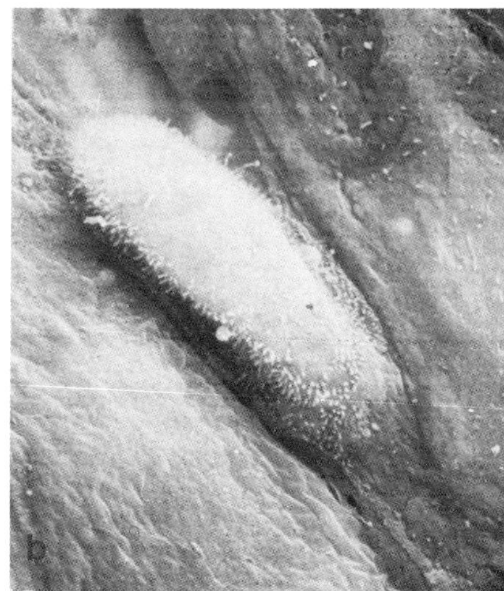

1 kV

10 kV

Figure 9: The contamination raster laid down during a previous high magnification examination and visible in a) taken with V_o = 1 kV is not evident in b) where V_o = 10 kV. (Sample kindly provided by Dr. J. Bastacky.)

The Prospects for Topographic Imaging in the LVSEM

In light of all these problems, what performance is it reasonable to expect under the best possible conditions and what are these conditions? As discussed above, the ideal microscope should probably employ a low-current cold-field-emission source and a lens with short focal length which also permits collection of the secondary electron signal. Beyond that, there are many theoretical advantages of operating it at liquid helium temperature. Not only is surface contamination negligible but superconducting materials are also perfect shields against stray electromagnetic fields (Dietrich et al., 1977). Furthermore, the low frequency noise and energy spread of the gun would be somewhat less. Though the primary ionizing event that produces radiation damage would not be eliminated, and chemical reactions would still occur following the accumulation of sufficient beam-produced free radicals, it is still probable that many low molecular weight species produced by the interaction might remain frozen nearly in place at these temperatures. This would not preserve molecular integrity but to some extent, the lumps would not move.

One disadvantage, aside from considerable complexity, might be increased charging artifacts. Even semiconductors become insulators at these temperatures and it might be necessary to lightly coat all samples. Furthermore, trace amounts of residual gas could create unwanted insulators if they became frozen onto sensitive surfaces.

With such an instrument it would seem that uncoated cubic surface features of low Z material as small as 3nm on a side might be detectable on a flat solid surface as long as they were not destroyed by radiation damage. Two such objects could be distinctly imaged if separated by about 5nm center to center. Information from lightly coated (1-1.5nm) samples might be somewhat better, assuming the coating material was chosen for low secondary electron mean-free-path (Everhart, 1970) and that we are now referring to the size of the surface features after coating. Though this is the same size range that is covered by the best replica techniques, it is important to remember the benefits of directly imaging the actual sample. Oatley (1982) points out that in the early 1960's commercial introduction of the SEM was delayed by the logic that, as replica techniques had higher resolution, there would be no market for the SEM. This analysis failed to take into account the extent to which specimen preparation is simplified and the areas of possible application increased by avoiding the necessity of having to produce a replica. Because of a willingness to accept the SEM's lower resolution in order to be able to examine the surface of a larger and more convoluted sample, the instrument came into common use and it has now been improved to the point where it may no longer even have to defer to the replica techniques in terms of resolution. When this happens it will be a considerable achievement.

Apart from ultimate performance at the limit of topographic imaging, important improvements in the low voltage performance of most current instruments should lead to their increased use in the 2-5 kV range by a wide variety of users (Fig. 1b). The next practical step might be the use of a TEM lens with properties similar to those diagrammed in Fig. 6 in conjunction with an FE source to produce the first really high resolution LVSEM results. The important test should be whether or not going to a higher voltage produces more information about the topography of a sample or merely sharper pictures of the metal grains of the coating material.

Acknowledgements

Thanks are due to the AMRAY Corporation (Bedford, Mass.) and to the Wisconsin Alumni Research Foundation for supplying the electronics and vacuum pumps (respectively) used in building the LVSEM test bench and to the National Institutes of Health for support through Grant RR-00570-14 to the Madison HVEM Biotechnology Resource. Personal thanks are also due to Dr. J. Wall (Brookhaven National Laboratory) for his generous assistance and encouragement in the design of the test bench and to Mr. M. Winters, without whom it would not have been completed. Dr. Autrata (Institute of Scientific Instruments, Brno) kindly provided the single crystal YAG scintillators and Dr. E. de Harven (University of Toronto) and Dr. J. Bastacky (Berkeley, CA) allowed me to use their samples for Figures 3 and 9 respectively. Figures 3 and 9 were taken by Mr. Naito (Hitachi, Mt. View, CA) on an S-800 FE SEM while Figure 1 was made by Mr. R. Kreiger (International Scientific Instruments, Mt. View, CA) on an SS-40 conventional SEM. Finally, thanks are due to Mr. W. McInvaille and Ms. G. Krewson for their patience with the manuscript and to Dr. D. Joy who read the manuscript and allowed me to use Fig. 4a from a preprint of his recent paper.

References

Adachi K, Hojou L, Katoh M, Kanaya K. (1983) High resolution shadowing for electron microscopy by sputter deposition. Ultramicros. 12, 17-29.

Anger K, Lischke B, Sturm M. (1983) Material surfaces for electron-optical equipment. Scanning 5, 39-44.

Arro E, Collins VP, Brunk UT. (1981) High resolution SEM of cultured cells: preparatory procedures. Scanning Electron Microsc. 1981; II: 159-168.

Autrata R, Schauert P, Kvapil JS, Kvapil J. (1978) A single crystal of YAG--new fast scintillator in SEM. J. Phys. E: Sci. Instrum., II, 707-708.

Autrata R, Schauer P, Kvapil J, Kvapil JR. (1983). Single-crystal aluminates - A new generation of scintillators for scanning electron microscopes and transparent screens in electron optical devices. Scanning Electron Micros. 1983; II: 489-500.

Ball MD, McCartney DG. (1981) The measurement of atomic number and composition in a SEM using backscattered detectors. J. Micros. 124(1), 57-68.

Ballard DB. (1972) Comparison and evaluation of specimens for resolution standards. Scanning Electron Micros. 1972: 121-128.

Banbury JR, Nixon WC. (1970) Voltage measurement in the scanning electron microscope. Scanning Electron Micros. 1970: 473-480.

Barth JE, Poole JB. (1976) Low voltage electron microscopy - how low? Ultramicros. 1, 387-388.

Bauer B, Speidel R. (1981) Influence of energy spread of field-emitted electrons on resolution in the scanning transmission electron microscope (STEM). Ultramicros. 6, 281-286.

Boersch H. (1954) Experimentele Bestimmung der Energieverteilung in thermisch ausgeloesten Elektronen Strahlen. Z. Phys. 139, 139.

Boyde A. (1971) A review of problems of interpretation of the SEM image with special regard to methods of specimen preparation. Scanning Electron Micros. 1971: 1-8.

Boyes ED. (1984) High Resolution at Low Voltage: The SEM Philosopher's Stone? EMSA (San Francisco Press, SF) 42, 446-450.

Brandis EK, De Stafeno J, Flitch R, Landenberger R. (1984) Low voltage SEM, Auger, and XPS of Surface Contaminants EMSA (San Francisco Press, S.F.) 42, 458-459.

Bresse JF. (1982) Quantitative investigations in semiconductor devices by electron beam induced current mode: A review. Scanning Electron Micros. 1982; IV: 1487-1500.

Broers AN, Panessa BJ, Gennaro, JF. (1975) High-resolution scanning electron microscopy of bacteriophage 3C and T4. Science 189, 637-639.

Broers AN. (1974) Recent advances in scanning electron microscopy with lanthanum hexaboride cathodes. Scanning Electron Micros. 1974: 9-18.

Broers AN. (1982) Resolution in surface scanning electron microscopy of bulk materials. Ultramicros. 8, 137-144.

Buchanan R. (1982) New SEM lens give sharpest micrographs yet. Industrial Research and Development/Aug, 92-95.

Buchanan R. (1983) The scanning electron microscope for semiconductor application. Microelectronic Manufacturing and Testing/Feb., 22-24.

Buchanan R, Menzel E. (1984) Some recent development in Low Voltage E beam testing of IC's. EMSA (San Francisco Press, S.F.) 42, 460-464.

Catto CJD, Smith KCA (1973) Resolution limits in the surface scanning electron microscope. J. of Micros. 98, 417-435.

Clarke DR. (1970) Review: Image contrast in the scanning electron microscope. J. of Material Sci. 5, 689-708.

Cliff G, Kenway PB. (1982) The effect of spherical aberration in probe-forming lenses on probe size, image resolution, and X-ray resolution in scanning transmission electron microscope. Microbeam Analysis, 107-110.

Cosslett VE. (1978) Radiation damage in the high resolution electron microscopy of biological materials: a review. J. Micros. 113(2), 113-129.

Crewe AV. (1973) Production of electron probes using a field emission source. In: Progress in Optics XI (Ed. by E. Wolf), pp. 225-246. (Elsevier Publishers) North-Holland.

Crewe AV, Eggenberger DN, Wall J, Welter LM. (1968) Electron gun using field emission sources. Rev. Sci. Inst. 39, 576-583.

Crewe AV, Isaacson M, Johnson D. (1971) A high resolution electron spectrometer for use in transmission scanning electron microscopy. Rev. Sci. Inst. 42(4), 411-420.

Crewe AV, Lin PSD. (1976) The use of backscattered electrons for image purposes in a scanning electron microscope. Ultramicros. 1, 231-238.

Crewe AV, Wall J, Langmore J. (1970) Visibility of a single atom. Science 168, 1338-1340.

Dietrich I, Rox F, Knapek E, Lefrank G, Nachtrieb K, Weyl R, Zerbst H. (1977) Improvements in electron microscopy by application of superconductivity. Ultramicros 2, 241-249.

Dilly PN. (1980) Enhanced contrast of cilia using low accelerating voltage as an aid to low power survey and counting. Scanning 3, 283-284.

Evans AC, Franks J. (1981) Specimen coating for high resolution scanning electron microscopy. Scanning 4, 169-174.

Everhart TE. (1970) Contrast and resolution in the scanning electron microscope. Third Annual Cambridge Stereoscan Colloquium, (available from Cambridge Instr. Co., Monsey, NY), 1-8.

Everhart TE, Thornley, RFM. (1960) Wide-band detector for micro-microampere low-energy electron current. J. Sci. Intrum. 37, 246-248.

Everhart TE, Wells OC, Oatley CW. (1959) Factors affecting contrast and resolution in the scanning electron microscope. J. Electron. Control 7, 97-111

Fourie JT. (1981) Electric effects in contamination and electron beam etching. Scanning Electron Micros. 1981; I: 127-134.

George, EP, Robinson, VNE. (1975) Topographic intensity profiles in the scanning electron microscope--cubes. J. Micros. 105(3), 289-297.

Glaeser, RM. (1971) LImitations to significant information in biological electron microscopy as a result of radiation damage. J. Ultrastruct. Res. 36, 466.

Glaeser, RM. (1975) Radiation damage and biological electron microscopy. In: Physical Aspects of Electron Microscopy and Microbeam Analysis (Ed: B. M. Siegel and D. R. Beaman), Wiley, New York, pp. 205-230.

Haggis, GH. (1982) Contribution of scanning electron microscopy to viewing internal cell structure. Scanning Electron Micros. 1982; II: 751-763.

Haggis, GH, Bond, EF. (1979) Three-dimensional view of the chromatin in freeze-fractured chicken erythrocyte nuclei. Journal of Microscopy 115(3), 225-234.

Haggis, GH, Schweitzer, I, Hall, R, Bladon T. (1983) Freeze fracture through the cytoskeleton, nucleus and nuclear matrix of lymphocytes studied by scanning electron microscopy. J. of Micro. 132(2), 185-194.

Hainfield, JF. (1977) Understanding and using field emission sources. Scanning Electron Micros. 1977; I: 591-604.

Hasselbach F, Rieke U. (1982) Spatial distribution of secondaries released by backscattered electrons in silicon and gold for 20-70 keV primary energy. Proc. Europ. Reg. Conf. Electron Microscopy I, (Deutsche Gesellschaft fur Elektronenmikroskopie e.V.), 253-254.

Hasselbach, F, Rieke, U, Straub, M. (1983) An imaging secondary electron detector for the scanning electron microscope. Scanning Electron Micros. 1983; II: 467-478.

Hill, R Smith, KCA. (1982) The single-pole lens as a scanning electron microscope objective. Scanning Electron Micros. 1982; II: 465-471.

Ichinokawa T, Sekine M, Gur ZS, Ishikawa, A. (1982) Electron optical properties of low energy field emission gun in the energy range from 100 to 2000 eV. Proc. Europ. Reg. Conf. Electron Microscopy I, (Deutsche Gesellschaft fur Elektronenmikroskopie e.V.), 351-352.

Isaacson M, Langmore, JP. (1974) Determination of non-localization of the inelastic scattering of electrons by electron microscopy. Optik 41(1), 92-96.

Joy DC, Maher DM. (1976) Low-loss images in a STEM/TEM microscope. EMSA 34, (Claitor's Publ. Div., Baton Rouge, LA), 496-497.

Joy DC, Newbury DE, Myklebust RL. (1982) The role of fast secondary electrons in degrading spatial resolution in the analytical electron microscope. J. Micros. 128.

Joy DC. (1984a) Beam interactions, contrast and resolution in the SEM. Microelectronic Engineering (in press).

Joy, DC. (1984b) Resolution in Low Voltage SEM, EMSA 42, (San Francisco Press, SF), 444-445.

Keery WJ, Leedy KO, Galloway KF. (1976) Electron beam effects on microelectronic devices. Scanning Electron Micros. 1976; I: 507-514.

Kemmenoe BH, Bullock GR. (1983) Structure analysis of sputter-coated and ion-beam sputter-coated films: a comparative study. J. of Micro. 132(2), 153-163.

Koike H, Namae T, Watabe T, Milkajiri A. (1973) An approach to microanalysis with the electron microscope. JEOL News 10e(4), 2-8.

Koike H, Ueno K, Suzuki M, Matsuo T, Aita S, Shibatomi K. (1971) High resolution scanning device for the JEM-100B electron microscope. JEOL News 9e(3), 20-21.

Kokubo Y, Ueno K, Iwatsuki M, Koike H. (1975) An application of strongly excited objective 25 lens to low loss image. EMSA (Claitor's Publishing Div., Baton Rouge, LA), 33, 138-139.

Kosuge T, Hashimoto H, Sato M, Komoto S. (1970) Quality of the secondary electron image at low accelerating voltage. In: Microscopie electronique, vol. I (Ed. by P. Favard), pp. 201-202. Societe Francaise de Microscopie Electronique, Paris.

Kotera M, Murata K, Nagami K. (1981) Monte Carlo simulation of a 1-10-keV electron scattering in a gold target. J. Appl. Phys. 52(2), 997-1003.

Kuo HPK, Siegel BM. (1976) A field emission probe forming system with a magnetic pre-accelerator lens EMSA 34 (Claitor's Publ. Div., Baton Rouge, LA), 522-523.

Kursheed A, Dinnis AR. (1983) Computation of trajectories in voltage contrast detectors. Scanning 5, 25-31.

Langmuir DB. (1937) Theoretical limitations of cathode-ray tubes. Proc. I.R.E. 25(8), 977-991.

Le Gressus C, Duraud JP, Massignon D, Deacon OL. (1983) Electron channelling effect on secondary electron image contrast. Scanning Electron Micros. 1983; II: 537-542.

Lin PSD, Lamvik MK. (1975) High resolution SEM at the subcellular level. J. Micros. 103, 249-257.

Miyoshi M, Ishikawa M, Okumura K. (1982) Effects of electron beam testing on the short channel metal oxide semiconductor characteristics. Scanning Electron Micros. 1982; IV: 1507-1514.

Mulvey T. (1982) Unconventional lens design. In: Magnetic Electron lens properties (Ed. P. Hawkes). Springer-Verlag, Berlin, 359-412.

Murata K, Kyser DF, Ting CH. (1981) Monte Carlo simulation of fast secondary electron production in electron beam resists. J. Appl. Phys. 52, 4396-4405.

Nakagawa S, Shibuki Y, Sahara K, Norioka S. (1982) High performance analytical scanning electron microscope provided with C-F mini lens and zoom condenser lens system. Proc. Europ. Reg. Conf. Electron Microscopy I, (Deutsche Gesellschaft fur Elektronenmikroskopie e.V.), 389-390.

Newman TH, Pease RFW, Polasko KJ, Yao YW. (1984). Lithography with low energy electrons. EMSA (San Francsico Press, SF) 42, 468-469.

Niedrig H. (1978) Physical background of electron backscattering. Scanning 1(1), 17-34.

Nomura S, Komoda T, Kameryo T, Nakaizumi V. (1973) Stable field emission gun with an electronic feedback system. Scanning Electron Micros. 1973: 65-72.

Oatley CW (1969) Isolation of potential contrast in the scanning electron microscope. J. Phys. E. 2(2), 742-744.

Oatley CW. (1972) The Scanning Electron Microscope, part I, The Instrument. The University Press, Cambridge, 1-194.

Oatley CW. (1975) The tungsten filament gun in the scanning electron microscope. J. Phys. E.: Scientific instruments 8, 1-5.

Oatley CW. (1982) The early history of the scanning electron microscope. J. Appl. Phys. 53(2), R1-R13.

Oatley, CW (1983) Electron currents in the specimen chamber of a scanning microscope. J. Phys. E. 16(4), 308-312.

Oatley, CW, Everhart TE. (1957) The examination of p-n junctions with the scanning electron microscope. J. Electronics II(6), 568-570.

Oatley CW, Nixon WC, Pease RFW. (1965) Scanning electron microscopy. In: Advan. in Electronics & Electron Physics, pp. 181-247. Academic Press, New York.

Ohshita A, Shimoyana H, Maruse S. (1978) Brightness in the hot cathode electron at high emission densities. J. Electron. Micros. 27(4), 253-257.

Ottensmeyer FP, Bazett-Jones DP, Rust HP, Weiss K, Zemlin F, Engel A. (1978) Radiation exposure and recognition of electron microscopic images of protamine at high resolution. Ultramicros. 3, 191-202.

Pawley JB. (1972) Charging artifacts in the scanning electron microscope. Scanning Electron Micros. 1972: 153-160.

Pawley JB (1974). Performance of SEM scintillator materials. Scanning Electron Micros. 1974: 27-34.

Pawley JB. (1984a) Low Voltage SEM, J. Microscopy 13, 387-410.

Pawley JB (1984b) SEM at Low Beam Voltage, EMSA (San Francisco Press, SF) 42, 440-443.

Pawley JB (1984c) Strategy for locating and eliminating sources of main frequency magnetic stray field, Scanning (in press).

Pawley JB, Hook G, Hayes TL, Lai C. (1980) Direct scanning electron microscopy of frozen-hydrated yeast. Scanning 3(3), 219-226.

Pawley JB, Hayes TL, Hook G. (1978) Preliminary studies of coated complementary freeze-fractured yeast membranes viewed directly in the SEM. Scanning Electron Micros. 1978; II: 683-690.

Pawley JB, Norton JT (1978) A chamber attached to the SEM for fracturing and coating biological samples. J. Micros. 112(1), 169-182.

Pawley JB, Wall J. (1982) A low voltage SEM optimized for high resolution topographical imaging. Proc. Europ. Reg. Conf. Electron Microscopy 1 (Deutsche Gesellschaft fur Elektronenmikroskope e.V.), 383-384.

Pawley JB, Winters MP. (1983) Low voltage SEM. 41st Ann. Proc. EMSA (San Francisco Press, SF) 41, 448-492.

Pease RFW, Nixon WC. (1965) High Resolution SEM. J. Sci. Instrum. 42, 31-35.

Peters K-R. (1979) Scanning electron microscopy at macromolecular resolution in low energy mode on biological specimens coated with ultra thin metal films. Scanning Electron Micros. 1979; II: 133-148.

Peters K-R. (1982) Conditions required for high quality high magnification images in secondary electron-I scanning electron microscopy. Scanning Electron Micros. 1982; IV: 1359-1372.

Peters K-R, Green SA (1983) Macromolecular structures of biological specimens are not obscured by controlled osmium impregnation. EMSA 1983 (San Franscisco Press, SF), 606-607.

Peters K-R, Palade GE, Schneider BG, Papermaster DS. (1983) Fine structure of periciliary ridge complex of frog retinal rod cells revealed by ultrahigh resolution scanning electron microscopy. J. Cell Biol. 96, 265-276.

Pfeiffer HC. (1972) Basic limitations of probe forming systems due to electron-electron interactions. Scanning Electron Micros. 1972: 113-120.

Pfeiffer HC. (1982) Probe forming systems. Proc. Europ. Reg. Conf. Electron Microscopy I (Deutsche Gesellschaft fur Elektronenmikroskope e.V.), 435-442.

Polasko KJ, Yau YW, Pease RFW. (1983) Low energy electron beam lithography. Optical Engineering 22(2), 195-198.

Pomposo TF, Coates VJ. (1983) Computerized electron-beam line-width measuring and inspection: a new tool. Silicon Processing, ASTM STP 804; DC Gupta (ed.), American Soc. for Testing & Materials, 501-508.

Reimer L. (1978) Scanning electron microscopy-present state and trends. Scannning 1(1), 3-16.

Reimer L. (1979) Electron - specimen interactions. Scanning Electron Micros. 1979; II: 111-124.

Reimer L, Volbert B. (1979) Detector system for backscattered electrons by conversion to secondary electrons. Scanning 2, 238-248.

Saito S, Nakaizumi Y, Mori H, Nagatani T. (1982) A field emission SEM controlled by microprocessor. EM I, (Deutsch Gesellschaft fur Elektronenmikroskopie e.V.), 379-380.

Sawada H. (1981) Three dimensional observation on muscular tissues. Scanning Electron Micros. 1981; IV: 7-15.

Seiler H. (1976) Determination of the "information depth" in the SEM. Scanning Electron Micros. 1976; I: 9-16.

Shaffner TJ, Hearle JWS. (1976) Recent advances in understanding specimen charging. Scanning Electron Micros. 1976; I: 61-70.

Tamura N, Saito H, Ohyama J, Aihara R, Kabaya A. (1980) Field emission SEM using strongly excited objective lens. EMSA 38, (Claitor's Pub. Div., Baton Rouge, LA), 68-69.

Tanaka K. (1980) Scanning electron microscopy of intracellular structures. In: International Review of Cytology, Vol. 68, pp. 97-115. Academic Press, New York.

Tanaka K. (1981) Demonstration of intracellular structure by high resolution scanning electron microscopy. Scanning Electron Micros. 1981; II: 1-8.

Thornley RFM. (1960) Recent developments in scanning electron microscopy. In: Proc. Eur. Reg. Conf. Elect. Micros. (Deutsche Gesellschaft fur Elektronenmikroskopie e.V.) (Ed. by A. L. Houwink and B. J. Spit), pp. 173-176.

Todokoro H, Fukuhara S, Sakitani Y. (1980) Low acceleration SEM. EMSA 38, (Claitor's Publ. Div., Baton Rouge, LA), 70-71.

Todokoro H, Fukuhara S, Komoda T. (1983) Stroboscopic scanning electron microscope with 1 keV electrons. Scanning Electron Micros. 1983; II: 561-568.

Todokoro H, Yoneda S, Yamaguchi K. (1984) Stroboscopic testing of LSI's with LVSEM, EMSA 42 (San Francisco Press, SF), 464-468.

Tuggle DW, Watson SG. (1984) A low voltage field emission column with a Schottky emitter, EMSA, 42 (San Francisco Press, SF), 454-457.

Van Der Mast KD. (1983) Field emission, development and possibilities. J. of Micros. 130(3), 309-324.

Varnell GL. (1981) Micron and submicron lithography for VLSI device fabrication. Scanning Electron Micros. 1981; I: 343-350.

Venables JA, Harland CJ. (1973) Electron backscattering patterns - A new technique for obtaining crystallographic information in the SEM, Phil. Mag., 27, 1193-1200.

Volbert B. (1982a) Signal mixing techniques in scanning electron microscopy. Scanning Electron Micros. 1982; III: 897-905.

Volbert B. (1982b) True surface topography: the need for a signal mixing. Proc. Europ. Reg. Conf. Electron Microscopy I (Deutsche Gesellschaft fur Elektronenmikroskopie e.V.), 233-239.

Volbert B, Reimer L. (1980) Advantages of two opposite Everhart-Thornley detectors in SEM. Scanning Electron Micros. 1980; IV: 1-10.

Voreades D, Wall JS. (1979) Contamination and reverse contamination at low temperature. EMSA 37 (Claitor's Pub. Div., Baton Rouge, LA), 358-359.

Wall J, Bittner J, Hainfeld J. (1977) Contamination at low temperature. EMSA 35 (Claitor's Pub. Div., Baton Rouge, LA), 558-559.

Watabe T, Hoshino T, Harada Y. (1978) The visibility of individual ferritin particles in a scanning electron microscope with a field emission gun. Ultramicroscopy 3, 19-27.

Wells OC. (1974a) Resolution of the topographic image in the SEM. Scanning Electron Micros. 1974: 1-8.

Wells OC. (1974b) Scanning Electron Microscopy. McGraw Hill, New York.

Wells OC (1978) Penetration effect at sharp edges in the scanning electron microscope. Scanning 2, 199-216.

Wells, OC (1979) Effects of collector take-off angles and energy filtering on the BSE image in the SEM. Scanning 2, 199-216.

Wells OC, Broers AN, Bremer CG. (1973) Method for examining solid specimens with improved resolution in the scanning electron microscope (SEM). Appl. Phys. Lett. 23(6), 353-355.

Welter LM, Coates VJ. (1974) High resolution scanning electron microscopy at low accelerating voltages. Scanning Electron Micros. 1974: 59-66.

Welter LM, McKee AN. (1972) Observations on uncoated, nonconducting or thermally sensitive specimens using a fast scanning field emission source SEM. Scanning Electron Micros. 1972: 161-168.

Wiesner JC. (1973) Characteristics and applications of pointed cathodes in SEM's. Scanning Electron Micros. 1973: 33-40.

Wiesner JC, Everhart TE. (1973) Point-cathode electron sources - electron optics of the initial diode region. J. Appl. Phys. 44(5), 2140-2148. and 45(6), 2797-2798.

Wilska AP. (1964) Low-voltage electron microscopy: A 6-kV instrument. J. Royal Micros. Soc. 83(1&2), 207-211.

Wilska AP. (1965) Expectations and limitations of low voltage electromiscroscopy. Laboratory Investigation 14(6), 825-828.

Yamazaki S, Kawawoto H, Saburi K, Naktasuka H, Buchanan R. (1984) Improvement in SEM gun brightness at low kV using an intermediate extraction electrode. Scanning Electron Micros. 1984; I: 23-28.

Yau YW, Pease RFW, Iranmanesh AA, Polasko KJ. (1981) Generation and applications of finely focused beams of low-energy electrons. J. Vac. Sci. Technol. 19(4), 1048-1052.

SUBJECT INDEX

AUTHOR INDEX

SELECTED PAPERS ON ELECTRON OPTICAL SYSTEMS AND RELATED TOPICS IN SEM/1983 AND 1984

Papers on various topics related to the subjects covered in this volume have been published in SEM volumes from time to time. The total number of such papers published during the years 1978 to 1984 exceeds 150. In the list below, abridged titles, names of first authors, and page numbers with part and year are given for papers from the 1983 and 1984 volumes of "Scanning Electron Microscopy."

The Early Days of Electron Microscopy; G.E. Pfefferkorn (1/I/1984

Effects of Secondary Electron Detector Position on SEM Image; H. Kawamoto (15/I/1984

Improvement in SEM Gun Brightness at Low kV...; S. Yamazaki (23/I/1984

Computer Archiving and Image Enhancement of Diagnostic Electron Micrographs...; T. Okagaki (513/II/1984

Automated Fiber Counting in the SEM; W.R. Stott (583/II/1984

Quantitative Contrast Evaluation for Different STEM Imaging Modes; R. Reichelt (1011/III/1984

REVIEW: Fundamentals of Liquid Metal Ion Sources ...; G.L.R. Mair (1531/IV/1984

TUTORIAL: A Comparison...Electron Guns...Electron-Beam Inspection; J. Orloff (1585/IV/1984

REVIEW:...Second Order Corrected Spectrometers for Use with the STEM; M. Scheinfein (1681/IV/1984

TUTORIAL: Using a Microcomputer...in the Teaching of...SEM and TEM; J.R. Michael (1723/IV/1984

High Resolution SIMS Using Liquid Metal Field-Ionization Sources; A.R. Bayly (23/I/1983

REVIEW: Secondary Electron Analyzers for Voltage Measurements; E. Menzel (65/I/1983

An Imaging Secondary Electron Detector for...SEM; F. Hasselbach (467/II/1983

A Model for Microcomputer Control of the SEM Used in Basic Research; R.C. Farrow (479/II/1983

REVIEW: Single-Crystal Aluminates...Scintillators for SEM & Transparent Screens; R. Autrata (489/II/1983

Stroboscopic SEM with 1 keV Electrons; H. Todokoro (561/II/1983

TUTORIAL: Scanning Tunneling Microscopy, an Atomic Probe; G. Binnig (1079/III/1983

REVIEW: Interfacing and Computer Control for Surface Science Microscopy; R. Browning (1655/IV/1983

REVIEW: New Methods for Cathodoluminescence in the SEM; A. Boyde (1803/IV/1983